The Search for Life in the Universe

Second Edition

Donald Goldsmith
Interstellar Media
Berkeley, California

Tobias Owen
Institute for Astronomy
University of Hawaii
Honolulu, Hawaii

ADDISON-WESLEY PUBLISHING COMPANY
Reading, Massachusetts . Menlo Park, California . New York
Don Mills, Ontario . Wokingham, England . Amsterdam . Bonn
Sydney . Singapore . Tokyo . Madrid . San Juan . Milan . Paris

To Rachel, David, Jonathan, and Kirill

Library of Congress Cataloging-in-Publication Data

Goldsmith, Donald.
The search for life in the universe / Donald Goldsmith & Tobias Owen.— 2nd ed.
p. cm.
Includes bibliographical references and index.
ISBN 0-201-56949-3
1. Life on other planets. 2. Astronomy. I. Owen, Tobias C. II. Title.
QB54.G58 1992
574.99--dc20 91-26177
CIP

Credits (see page 521)

1 2 3 4 5 6 7 8 9 10 - HA - 95949392

Contents

	Foreword	**ix**
	Preface	**xi**
PART ONE	**Why Do We Search?**	**1**
Chapter 1	**The Search from the Human Perspective**	**3**
	The Quest for Extraterrestrial Life	3
	The Quest for Life's Origins	4
	The Importance of Mars	6
	The Scientific View of the Universe	10
	Cosmic Loneliness	14
PART TWO	**The Universe at Large**	**21**
Chapter 2	**Space, Time, and the History of the Universe**	**23**
	The Distances to Astronomical Objects	24
	The Spectra of Stars	30
	The Doppler Shift and the Expanding Universe	35
	The Big Bang	40
	The Cosmic Background of Photons	41
	The Inflationary Universe	43
	The Dark Matter	44
	Is the Universe Finite or Infinite?	46
	Will the Universe Expand Forever?	48
Chapter 3	**Galaxies**	**55**
	Spiral Galaxies	56
	The Structure of Spiral Galaxies	59
	Dark Matter in Spiral Galaxies	63
	Elliptical Galaxies	65
	Irregular Galaxies	67
	The Formation of Galaxies	68

Star Clusters 68
Radio Galaxies 71
Quasars 72
Accretion Disks and Supermassive Black Holes 75
Could Quasars Be Intergalactic Beacons? 77

Chapter 4 Interstellar Gas and Dust 81

Probing the Interstellar Medium 82
Radio Waves from Interstellar Hydrogen Atoms 86
Interstellar Dust Grains 87
Interstellar Molecules 88
Molecular Clouds 90
The Different Types of Molecules in Molecular Clouds 93
Did Life Begin in Interstellar Clouds? 96
The Birth of Stars 98
How Many Stars Form with Planets Around Them? 98

Chapter 5 Energy Liberation in Stars 103

Stellar Lifetimes 104
How Stars Liberate Energy 106
The Proton-Proton Cycle 107
The Importance of Temperature Inside Stars 110
The Struggle Between Gravity and Pressure 111
The Influence of Mass upon Stellar Lifetimes 112
The Types of Stars 115
The Temperature-Luminosity Diagram 118
Red Giants and White Dwarfs 119

Chapter 6 How Stars End Their Lives 125

Nuclear Fuel Consumption in Stars 125
The Evolution of Stars 126
The Exclusion Principle in White Dwarf Stars 130
Supernova Explosions 132
The Second Type of Supernova 135
The Production of Heavy Elements in Supernovae 135
Cosmic Rays 141
Pulsars: Cosmic Lighthouses? 142

PART THREE	**Life**	**153**
Chapter 7	**The Nature of Life on Earth**	**155**
	What is Life?	155
	Biologically Important Compounds	159
	The Capacity to Reproduce	163
	Further Functions of DNA	164
	Evolution and the Arrow of Time	169
	Energy	171
	The Unity of Life	173
Chapter 8	**The Origin of Life**	**179**
	How the Earth Got Its Atmosphere	181
	The Evolution of the Atmosphere	183
	The Effects of Life on the Atmosphere	184
	Early Ideas about the Origin of Life	188
	The Chemical Evolution Model for the Origin of Life	188
	An Experimental Test of the Primordial-Soup Model	191
	Did Life Really Originate in This Manner?	193
	The External Alternative	195
	Polymerization	196
	Beyond Polymers	199
Chapter 9	**From Molecules to Minds**	**205**
	Prokaryotes	205
	Eukaryotes	208
	The Great Leap Forward	211
	Life on Other Planets	215
	Evolution and the Development of Intelligence	217
	Is Intelligence Inevitable?	219
	Future Evolution on Earth	221
	The Web of Life	222
	Gaia	224
Chapter 10	**How Strange Can Life Be?**	**229**
	The Chemistry of Alien Life	229
	The Superiority of Carbon	231
	Solvents	234
	Nonchemical Life	239
	Black Clouds	240
	Life on Neutron Stars	242
	Gravitational Life	244
	The Advantages of Being Average	245

PART FOUR **The Search for Life in the Solar System** **251**

Chapter 11 **The Origin and Early History of the Solar System** **253**

The Formation of the Solar System 254
Comets 259
Asteroids, Meteoroids, and Meteorites 266
Bombardment of the Inner Planets: A Threat to Life? 267
Meteorites 270
Amino Acids in Meteorites 271
Mercury and the Moon 273
The Early History of the Earth and the Moon 274
Human Exploration of the Moon 276

Chapter 12 **Venus** **283**

The Temperature of Venus 283
The Atmosphere of Venus 285
The Greenhouse Effect 286
Why is Venus So Different from Earth? 289
Life on Venus? 293
Exploration by Spacecraft 295

Chapter 13 **Mars** **301**

Modern Observations of Mars 303
Results from Space Probes 304
The Viking Project 307
The New Mars 315
Phobos and Deimos 319

Chpater 14 **Is There Life on Mars?** **325**

How to Find Martian Microorganisms 327
The Viking Results: Atmospheric Analysis 329
The Viking Results: Soil Analysis 329
The Viking Biology Experiments 331
Results of the Viking Biology Experiments 333
Did the Vikings Land in the Wrong Places? 337
An Ancient Eden? Goals for Future Exploration 339
What Went Wrong on Mars? 341
Epilogue: What About That Face on Mars? 343

Chapter 15 **The Giant Planets and Their Satellites** **349**

Spacecraft to the Outer Solar System 350
The Composition of the Giant Planets 354
Chemistry on the Giant Planets 359

	Could Life Exist on the Giant Planets?	361
	Rings and Satellites	362
	Titan	366
	Iapetus: An Intelligence Test for Earthlings?	370
	Triton: Chemistry at Low Temperatures	371
	Cosmic Messengers	374
PART FIVE	**The Search for Extraterrestrial Intelligence**	**379**
Chapter 16	**Is Earth Unique?**	**381**
	Distinguishing Characteristics of the Earth	382
	The Planetary Systems of Other Stars	384
	Searching for Other Planetary Systems	386
	The Likeliest Stars	391
Chapter 17	**The Development of Extraterrestrial Civilizations**	**403**
	How Many Civilizations Exist?	404
	The Search for the Perfect Restaurant	405
	The Number of Civilizations: The Drake Equation	408
	How Long Do Civilizations Last?	411
	How Eager Are Civilizations for Contact?	414
	How Does Communication Proceed?	416
	Further Advances of Earthlike Civilizations	421
	Colonizing the Galaxy	424
Chapter 18	**How Can We Communicate?**	**433**
	Interstellar Spaceships	436
	Time Dilation	441
	The Difficulties of High-Velocity Spaceflight	443
	Automated Message Probes	446
Chapter 19	**Interstellar Radio and Television Messages**	**453**
	Where Should We Look?	454
	What Frequencies Should We Search?	456
	Frequency Bandpass and Total Range	461
	How Can We Recognize Another Civilization?	464
	The Present State of Radio Searches for Other Civilizations	469
	The Targeted Search	472
	The Sky Survey	473
	What Messages Could We Send or Receive?	474

Chapter 20 **Extraterrestrial Visitors to Earth?** **483**
What Evidence Do We Seek? 483
Recurrent Themes in UFO Sightings 485
The Lubbock Lights 490
Venus in Georgia 490
Landing in Socorro 492
Difficulties in Verifying the Spacecraft Hypothesis 493
Classification of UFO Reports 493
Arguments for the Spacecraft Hypothesis 495
A Cover-up? 496
Are UFOs a Modern Myth? 497
Von Däniken: Charlatan of the Gods? 498

Chapter 21 **Where is Everybody?** **505**
We May Be Alone, or Nearly So 506
Civilizations May Have Little Interest in Communication 510
We Are Still a Primitive Civilization 513
Epilogue: The Search Continues 517

Credits 521

Index 523

Foreword

The search for life in the universe is one of the most fascinating of subjects, be it in the eyes of professional scientists or of lay-people of all ages. Almost every knowledgeable person alive would be thrilled to learn of a contact with extraterrestrial life, especially intelligent life. To many, this is the greatest adventure left to humanity. More than an adventure, such a discovery would lead to a wealth of new information in science, technology, and, most likely, social matters. There could hardly be a more important milestone in history. In fact, we would not only better understand our past history, but we would get a glimpse of the history of times yet to come.

This fascination has turned out to be a magnet that has created in students an interest in science they might otherwise not have. For science education, this is wonderful, because the search for life elsewhere calls for an understanding of a remarkable variety of scientific and technical areas. These areas of study include basic chemistry, the definition of life, the chemistry of life, the origin of planetary systems, the nature of planets and their atmospheres, the evolution of climate, biological evolution, the technology of space travel, the workings of radio telescopes, and the nature of language. It is a heady and provocative list. Obviously then, an interest in extraterrestrial life can motivate students to become broadly educated in science, benefiting not only themselves but the world at large. This has been recognized at many schools, both at the secondary and college level. There are now large numbers of general education science courses, particularly in colleges, that are based on the theme of the search for extraterrestrial life. The enrollments for these courses are large, indicating that this theme attracts students successfully. The need for much broader science education, leading to an informed citizenry, has been served well by this development. Each year sees an emergence of more courses in this area.

As with any sophisticated course, an appropriately sophisticated textbook is vital to the learning process. The first edition of this book has filled this role admirably since 1980. However, the relevant areas of science progress at such a rapid pace that it is difficult for a textbook of this type to have a long and useful life. This is one reason why those of us who teach in this field are delighted to see this newly updated edition.

There is, however, a more important reason for creating an appropriate text. Many people, including many scientists, think that there is so little we know about extraterrestrial life that anyone can become an expert overnight. This results all too often in the publication of books on this subject with major defects; typically, these books are unbalanced in content and approach. More seriously, they are often contaminated with unjustified speculation, or worse, pseudo-science or non-science. These inadequacies are not surprising considering that the subject provides fertile ground from which springs much of science fiction, speculations about the future, and unfortunately, science quackery.

The ideal textbook for the study of extraterrestrial life would carefully present all scientific arguments. Assumptions and weak data would be conspicuously identified. Observational evidence would be evaluated in a fair, objective, and knowledgeable manner, and conclusions would be clearly stated, with all the justified caveats presented. It takes experts in the field to meet these demands, and Goldsmith and Owen, having each lived with this subject for so long, are justly qualified to provide the required knowledgeable yet cautious authorship.

This book meets these demands; the authors have carried out their task well.

This book will go far in building understanding and support for the search for life in the universe. It will also enrich the reader's understanding of the uniquely exciting history and nature of intelligent life on earth. Certainly this text will increase the wonder and delight which all of us find in the search for life in the universe.

Frank D. Drake
Professor of Astronomy and Astrophysics
University of California, Santa Cruz

Preface

In this book, we consider several large questions: What is the nature of the universe in which we live? How did life on Earth begin, and how did it evolve to its present state? Does life exist elsewhere in the solar system? In the Milky Way galaxy? If it does, how can we make contact with other forms of intelligent life? Or have they already attempted to contact us? Could we be the only forms of matter that have ever asked such questions?

In seeking answers to these questions, this book offers a survey for the educated layperson, and for nonscience college students, of the fundamentals of astronomy, plus topics from biology, geology, and planetary science that are essential to understand how life on Earth fits into the cosmic scheme of things. As a result, this book can be used as a text in a one-quarter or one-semester course on astronomy, physical science, general science, or (best of all) the search for life in the universe.

Since we have yet to find life beyond the Earth, the search for life in the universe has often been accused of being a "subject without subject matter." In a narrow sense, this accusation carries weight, but in a wider view, the search for life allows students and instructors to combine a wide variety of topics into a coherent whole. Precisely because the search for life is still in its earliest stages, those involved in it must attempt to interpret ideas, observations, and experiments from the frontiers of several areas of science. Though many of the basic astronomical, geological, and biological facts presented here are well established, we shall often be forced to deal with wide-ranging, though scientific, speculation. We hope to convey the tension between rigorous proof and imaginative speculation, which provides one of the most enjoyable aspects of scientific research, and we hope that our readers will experience some of this pleasure as they consider the problems we address in this book. Certainly few of us remain unmoved by the prospect of establishing communication with another race of beings on a far-distant world.

In assessing the possibilities for such communication, we simultaneously perform a different feat: We come to know ourselves, who we are, from whence we came, what our future may be as potential members of a galactic community. As self-centered infants become socially-oriented adults, we human beings have evolved from an ancient belief in the Earth as the center of the cosmos to reach an understanding of our comparatively insignificant place in the universe. We now face the question of whether human intelligence and self-consciousness are unique (or nearly so), or whether the events that produced intelligent beings from chemical interactions on Earth may have occurred time and again, over and over throughout the universe.

Despite our modern knowledge that we orbit a representative star in an average galaxy, the human sense of our uniqueness dies slowly. The lingering belief that we must be special appears in many of the UFO reports that tell of far more advanced creatures who nevertheless have a deep and abiding interest in planet Earth. Scientific progress proceeds by attempting to recognize and to contain our innate prejudices, replacing emotional bias with experiments that can test hypotheses. In the search for life, our generalizations must remain hypotheses, so long as we have only a single example of life to examine. We must stretch our knowledge and our theories as far as we can, attempting to see which possibilities seem more likely than others. The ultimate experiment is a determined effort to contact extraterrestrial intelligent beings, using the best means currently available. If such contact should occur, rendering this book obsolete, no one would be more delighted than its authors.

A fine source of information and resource materials for teaching astronomy is the Astronomical Society of the Pacific, at 390 Ashton Ave., San Francisco, CA 94112. This society is a worldwide organization of amateur and professional astronomers; it publishes a bimonthly magazine and offers lists of selected readings on various astronomical topics, as well as a yearly catalog of slides, videotapes, and other astronomically interesting items.

In writing the second edition of this book, we have received kind assistance from a host of scientists. We are grateful to Jon Arons, John Billingham, Ken Brecher, Chip Cohen, Marc Davis, Ben Finney, Andy Fraknoi, Margaret Geller, Paul Goldsmith, Sam Gulkis, Eric Jones, Mike Klein, Ed Krupp, Lynn Margulis, Larry Marschall, Chris McKay, David Morrison, Allison Palmer, Michael Papagiannis, J. William Schopf, Seth Shostak, Frank Shu, Michael Soule, John Stolz, Woody Sullivan, Richard Wainscoat, Ben Zuckerman, and especially to Timothy Barker, Barbara Bowman, Frank Drake, Al Glassgold, Jill Tarter, and Joe Veverka for their comments on different parts of the manuscript and their help with the text and illustrations. Jon Lomberg, who has prepared the new line drawings for this edition, has proven a fount of ideas and assistance; his typesetter, Vicki Scott, has been a font of help. Our editor at Addison-Wesley, Stuart Johnson, has dealt with us generously and forthrightly. We thank Mildred O'Dowd and Diane Tokumura for their skill and labor in helping to prepare the manuscript. Any mistakes are, of course, our own, and like Shakespeare's Cassius, we remain conscious that our faults lie "not in our stars but in ourselves."

Donald Goldsmith
Tobias Owen

PART ONE

Why Do We Search?

To see a World in a Grain of Sand
And a Heaven in a Wild Flower,
Hold Infinity in the palm of your hand
And Eternity in an hour.

—WILLIAM BLAKE

The history of human awareness of the universe has brought a steady increase in our desire to find the roots of our existence and to understand how humans fit into the cosmos. We now stand at the threshold of determining how life arose on this planet, and of applying what we know about life on Earth to our quest for life on planets that may orbit other stars. But we ought to pause to ask some key questions: Why do we search? How has the search for our origins, and for evidence of our cosmic kin, proceeded in the past? And what does the search for extraterrestrial life tell us about our attitude toward the universe around us?

In 1976, two Viking spacecraft landed on Mars to conduct the first direct exploration of another planet. These miniaturized laboratories, operated by remote control from a distance of more than a hundred million kilometers, sent a flood of data to Earth for analysis and speculation. This photograph shows frost deposits in the shadows of rocks on the Martian surface.

1

The Search from the Human Perspective

The Quest for Extraterrestrial Life

Throughout history, different cultures have speculated about the variety of intelligent beings that might exist in the heavens. Many stories and pictures describe visitors from the skies, testimony to the human longing to connect with the cosmos. Within the last five centuries, prevailing opinions about extraterrestrial life within what we call Western civilization have varied widely, embracing both the intuitive belief that the Earth is the only inhabited planet, firmly situated at the center of the universe, as well as the notion that the cosmos must contain many planets much like our own.

During the seventeenth century, as science gained a wide appeal as an interesting way to enjoy nature, the idea that all the planets in our solar system are inhabited received widespread acceptance. The Dutch scientist Christiaan Huygens, famous for his achievements in the field of optics, wrote an entire book on the subject of life on other worlds. Huygens speculated on the characteristics that the inhabitants of the different planets must have in order to survive comfortably under the extremes of gravity and atmospheric composition that he imagined. Half a century later, the great French satirist Voltaire imagined a giant inhabitant of the planet Saturn visiting Earth and eating mountains for breakfast. During the next two centuries, belief in the possibility of life outside the Earth grew stronger and weaker, as new discoveries concerning the nature of life on Earth and the conditions on our planetary neighbors emerged from new advances in scientific inquiry.

The second half of the twentieth century has opened new perspectives on the possibility of life elsewhere in the cosmos. Astronomers have found that we live within a grouping of billions of stars, gas, and dust called the

Milky Way galaxy (Color Plates 2 and 3) and that our home galaxy is just one among billions (or many more!) of similar galaxies that sprinkle the visible universe (Fig. 1.1). Although the enormous number of stars in all these galaxies suggests many possibilities for other forms of life, astronomers have come to see that the immense distances that measure the universe imply that we on Earth may well be isolated from personal visits, and that the best way to find our neighbors may be to detect their radio messages.

The Quest for Life's Origins

The search for extraterrestrial life has another, equally important strand. If we could understand the *origin* of life on Earth, we could hope to estimate how common or rare life may be among the starry systems that surround us. Knowledge of how life began would allow us to compare the conditions on Earth that gave birth to life about four billion years ago with those that we hypothesize to exist on other planets, within interstellar

Figure 1.1 Each of the hundreds of galaxies visible in this cluster of galaxies, half a billion light years from Earth, contains on the order of 100 billion stars.

clouds, and in other locations where we can imagine that life may have arisen. But even though biologists have discovered tremendous, unifying principles about life on Earth—from the way that living cells replicate to the evolutionary forces that have brought forth the stunning diversity of life—they have yet to solve the mystery of life's origin. The quest to answer the riddle of life—how was living matter first formed from inanimate material?—has fascinated scientists for centuries.

Nearly 12 decades ago, in May 1876, Her Majesty's Ship *Challenger* returned to port at Sheerness on the Thames after more than three years away from England. Unlike earlier voyages of exploration, which sought chiefly after wealth or alien populations to exploit, the *Challenger* aimed to find the origins of life. During the ship's voyage around the world, the scientists aboard the vessel systematically dragged the ocean bottoms for the first time in human history. Day after day, the ship's crew brought up samples of water and mud from the abyssal deeps, the bottom layers of the oceans of the world, containing marvelous new sea creatures previously unknown to humanity. The scientists on the expedition hoped that they would be the first to find "living fossils," early forms of life on Earth, still thriving in the ocean depths where conditions have barely changed since the time that life began.

But the *Challenger*'s scientists had expectations that went beyond the quest for living fossils. Twenty years earlier, when the first transatlantic telegraph cable had been laid, the crew of the ship laying the cable had discovered a gelatinous ooze from the bottom of the ocean. According to leading scientists, this ooze probably represented the primitive protoplasm, or *Urschleim* (original slime), from which all life had arisen. A careful study of this ooze would surely unlock the secret of how life began on Earth.

But alas! The *Challenger* expedition found no living fossils, and the mysterious *Urschleim* turned out to be totally inanimate. Although the ooze changed its chemistry in ways reminiscent of the processes that living organisms undergo, these changes could be reproduced nicely by adding a strong solution of alcohol to ordinary sea water. In short, chemistry and not biology ruled the ocean bottoms.

Today, 116 years after the *Challenger* expedition, we know far more about the Earth and its oceans. One key discovery has been what we have *not* found: The history of the origin of life on this planet has vanished. No geological records exist of the Earth's first 500 million years, because erosion and the motions of the giant plates that form the Earth's crust have erased them forever. As the plates have moved, slowly but inexorably, they have dragged eroded material that once formed the Earth's surface down below the present crust of our planet (Fig. 1.2). The motion of the crustal plates, called *plate tectonics,* has removed what would have been the best way to discover the earliest terrestrial history.

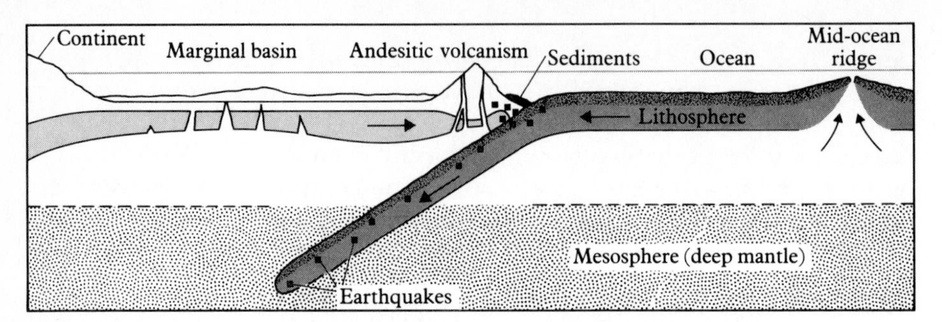

Figure 1.2 The motions of the Earth's crustal plates cause the plates eventually to ride over one another, hiding the geological record from billions of years ago.

What is left to examine in the search for life's origins? We do find an amazing variety of living organisms on Earth today, and we can study a fossil record that extends over more than 3 billion years, although the oldest, simplest organisms have left almost no fossil traces. But we cannot reach back through this record to examine the point at which life differentiated itself from inanimate matter.

Despite our lack of definite information about the origin of life on Earth, no doubt exists concerning the interest of human beings in this subject. Every culture has its own creation myths, and even our own "sophisticated" civilization cares deeply about its origins. How then shall we proceed?

We can imagine two extreme approaches to providing the answers to these questions. One method constructs *theories* of the processes that began life and then attempts to duplicate these processes through laboratory experiments. The second method seeks to find *examples* in nature, in the hope that studying these examples will reveal the essential clues to this remarkable transformation of matter.

The first method of attack, which has been modestly successful, is presented in Chapter 8. We have already seen that the second method cannot be pursued very far on Earth, for the early record of the planet's history no longer exists. But what about the other planets? The space age has enormously broadened our horizons. We can now investigate other worlds, and can even perform experiments within their atmospheres and on their surfaces. These experiments are beginning to answer some of the questions that have fascinated human imaginations since intelligence first developed on the Earth.

The Importance of Mars

For centuries, speculation about life beyond the Earth has had a central theme: The planet Mars has seemed an especially attractive extraterrestrial abode. The reasons for this earthly fixation upon Mars are many. The planet's reddish color, together with its great variations in brightness and its peculiar apparent motions in the sky, won Mars special attention and fear from the dawn of history. These visible characteristics led ancient observers to name this planet after the god of war in the Greco-Roman pantheon. During the late seventeenth century, the first telescopic observations of Mars, made by Christiaan Huygens, showed that the planet's surface has permanent markings, which in turn revealed the planet's rotation (Fig. 1.3). Such revelations have showed us that what we are observing is the planet's firm surface, rather than the changing cloud bands of Jupiter or the featureless cloud cover of Venus. Observations of Mars made during the subsequent two hundred years revealed the polar caps, the Martian clouds, and seasonal changes in the dimensions of the caps and in the contrast of the light and dark markings on the planet's surface.

During the final decades of the nineteenth century, a new dimension appeared in the Mars puzzle. Several European astronomers, among them Giovanni Schiaparelli of Milan, discovered faint, straight markings on the

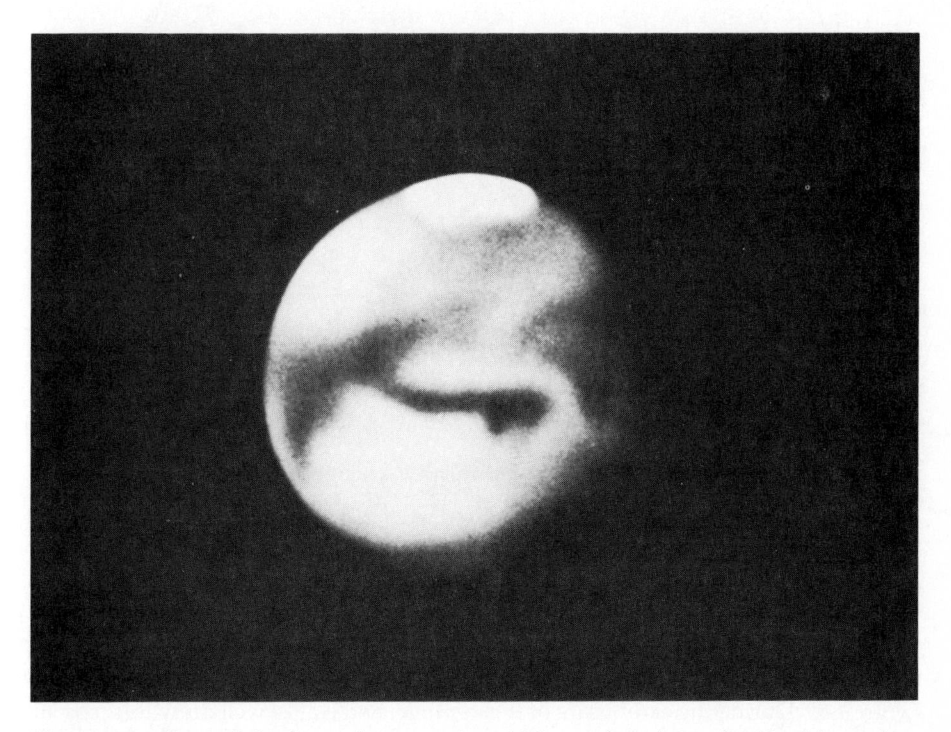

Figure 1.3 One of the best photographs of Mars taken from the Earth's surface shows a polar cap plus vague dark markings against a background of lighter material.

surface of Mars, the famous Martian "canals."[1] Today we know that these canals do not exist. Instead, they are (and were) an optical illusion that makes the eye see lines when only dots are present. But for half a century, the Martian "canals" provoked a series of bitter controversies.

The high-water mark of the belief in the canals of Mars was reached by Percival Lowell, who built his own observatory near Flagstaff, Arizona, where he carefully studied Mars with a 24-inch telescope and drew maps of a fine network of canals that completely covered the planet except for the polar frost caps (Fig. 1.4). Lowell found that this canal network underwent

[1] Schiaparelli used the Italian word canali, which means "channels" as well as "canals."

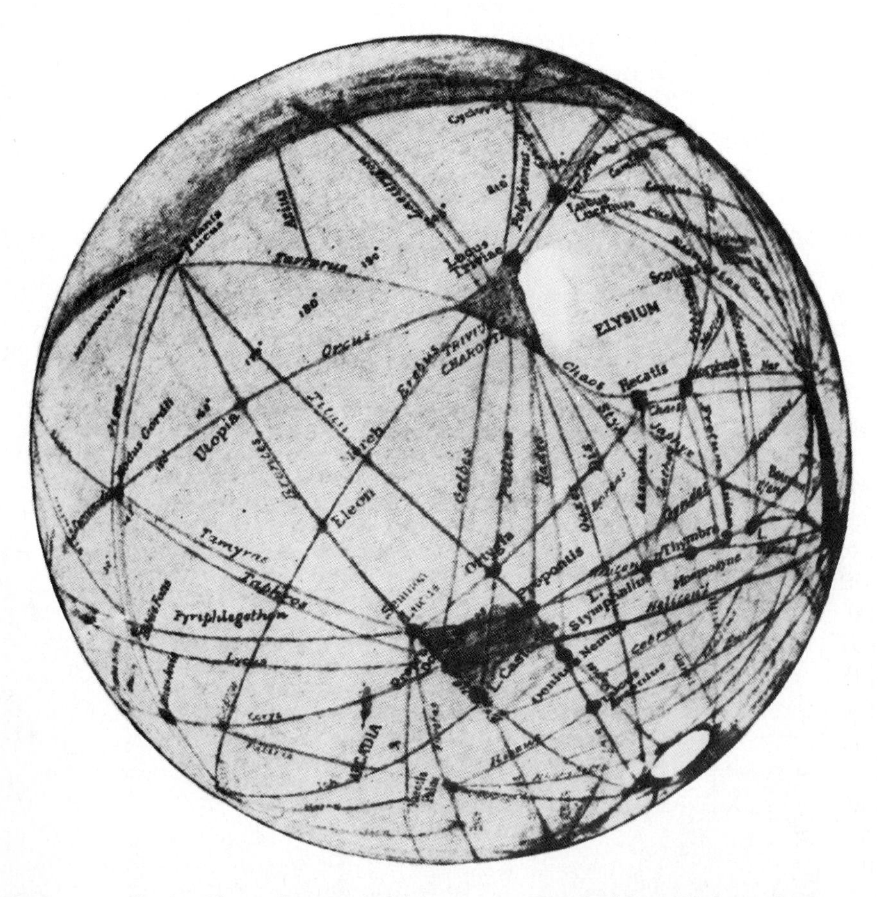

Figure 1.4 During the early years of this century, Percival Lowell drew this map of Mars, based on his telescopic observations from Arizona. The map included dozens of "canals," which Lowell thought must carry water from the polar caps to the planet's equatorial regions.

seasonal changes that were synchronized with changes in the contrast of the large, dark areas on Mars: The canals would always grow darker as the Martian summer began. Lowell concluded that intelligent, technologically capable Martians had constructed the canals to bring water from the melting polar caps to irrigate their fields (the dark areas), which otherwise would remain incapable of producing crops. The seasonal increase in contrast between the light and dark areas, Lowell said, was the straightforward result of plant growth in the dark areas during the summer season.

Lowell's arguments made up in passion what they lacked in proof. During the first few decades of this century, astronomers developed the techniques of modern astrophysics, and their applications to studies of Mars struck hard at Lowell's hypotheses. Measurements revealed that most of the planet's surface is colder than the freezing point of water, and a full 100° C below zero during the Martian night. Astronomers found no water vapor or oxygen in the Martian atmosphere, and the conclusion grew stronger that this atmosphere must be extremely thin.

Despite the scientific evidence that Mars should be hostile to life, and despite the negative results from a search for radio signals from Mars that was made in 1924, the idea that an advanced civilization might exist on the planet became so deeply entrenched in the public mind that on 30 October 1938, a serious panic erupted when Orson Welles presented a radio dramatization of H. G. Wells's novel *War of the Worlds.* Hundreds of thousands of listeners temporarily abandoned the safety of their homes and rushed outdoors to meet, or to flee from, the wave of Martian invaders who, Welles reported, were overrunning New Jersey (Fig. 1.5).

Of course, the explanation that the "invasion" was merely a radio show soon satisfied everyone, but the "War of the Worlds" remains an outstanding example of the depth of human interest in aliens and the fear of their intentions. If the performance were to be repeated today (of course, we would now require facsimiles of Martians for television), we can be sure that millions of viewers would be easily taken in, not simply because we are so easily deceived, but because we retain a deep-seated urge to believe in the existence of other civilizations. The Welles broadcast also reminds us of our basic ambivalence toward extraterrestrial visitors. We expect them to be far wiser, stronger, and more sophisticated than we are, but we cannot decide whether they will be gentle and loving or belligerent and tyrannical. The one thing that we inevitably expect is that they will find us fascinating. We have always feared and loved our gods, and it should not be surprising that similar emotions should arise when we speculate about advanced forms of extraterrestrial life.

Fifty years after Lowell, and 25 years after the Welles broadcast, humanity sent the first terrestrial spacecraft to Mars. The story of *Mariner 4,* which took the first close-up pictures of the red planet, of the succeeding *Mariner*

~~N~~ew York T

Copyright, 1938, by The New York Times Company.

NEW YORK, MONDAY, OCTOBER 31, 1938.

Radio Listeners in Panic, Taking War Drama as Fact

Many Flee Homes to Escape 'Gas Raid From Mars'—Phone Calls Swamp Police at Broadcast of Wells Fantasy

A wave of mass hysteria seized thousands of radio listeners throughout the nation between 8:15 and 9:30 o'clock last night when a broadcast of a dramatization of H. G. Wells's fantasy, "The War of the Worlds," led thousands to believe that an interplanetary con- and radio stations here and in other cities of the United States and Canada seeking advice on protective measures against the raids.

The program was produced by Mr. Welles and the Mercury Theatre on the Air over station WABC and the Columbia Broadcasting Sys-

Figure 1.5 When Orson Welles dramatized H. G. Wells's *War of the Worlds* in a 1938 radio broadcast, panicked listeners believed that invaders from Mars had landed in New Jersey. The next day's *New York Times* reported the incident on the front page.

missions, and of the *Viking* landings on Mars, appears in Chapters 13 and 14. With these and other spacecraft, a few dozen years have brought us from dim and distant views of the planets into the era of direct planetary exploration—with a continuing focus on the search for life on Mars.

The Scientific View of the Universe

This book describes the universe from the moment of its birth to the present day, with an emphasis on the clues we can find to the twin mysteries of life's origin on Earth and its distribution in the universe.

Because we have no definite answers to these mysteries, we must attempt to draw conclusions from life, its fossil record on Earth, and the observations we can make of conditions on other planets. We must then speculate, as best we can, in order to judge the probability of finding life elsewhere in space. We shall see that a great difference exists between unbounded speculation and speculation that draws on what we know about the universe. Scientific speculation—speculation that is directed by knowledge and bounded by physical laws rather than by fantasy—does not have so widespread an appeal as the more traditional forms of human conjecture, completely unfettered by science. In contemplating extraterrestrial life, science may seem mainly a drag on our imaginations, a weight that prevents our fancy from soaring free. Many people find the universe beyond the Earth so strange that they react either by never thinking about it, or by believing that anything goes: that no sort of life should be more improbable than another, or that laws of nature as yet unknown to us appear in everything from extrasensory perception to the Bermuda Triangle.

In contrast to these views, and with a track record of success, scientists hold the view that we must proceed carefully from what we understand—through multiple observations and experiments—toward our speculation about things that we do not understand. How does this differ from Percival Lowell's writings about life amidst the Martian canals? The key to the scientific method is its reliance upon testing and verification rather than upon assertion alone. If scientist A claims a detection of gravity waves, or cold fusion, or high-temperature superconductivity, then scientist B must be able to reproduce this result independently before it will be generally accepted. In fact, scientist A will usually explain how to repeat the experiment, or will suggest new experiments to test the theory. It doesn't matter how famous scientists are, how successful in life, or how impressive their previous work has been. Either the experiment works or it doesn't; either the observations are accurate or they aren't; either the theory makes a correct prediction or it doesn't. These canons of science have on occasion been violated by fraudulent or overly credulous researchers, but eventually the truth wins out, precisely because scientists place such stress on being able to repeat experiments and achieve the same result before they will accept that result as true.

When we look at the world, we rely on the opinions of wise people, but only up to a point; to be completely certain, we must be able to test what they say. This is the challenge we shall try to meet in the following chapters: how to convert centuries of speculation and authoritative statements concerning the existence of extraterrestrial life into an experimental science.

On occasion, some new and startling fact seems to contradict the theories that summarize what we know about the universe. These are intensely exciting moments. In such cases, scientists remain reluctant to change their theories *until* they have become convinced that they can exclude any other explanation that would not require such a change. Using this approach, most scientists will not, for example, regard UFO reports as evidence of extraterrestrial spacecraft until they have eliminated human error, psychological reactions, natural phenomena, or fraud as the causes of the UFO reports. On the other hand, a well-documented UFO sighting that was clearly evidence for a visit to Earth by extraterrestrials would be welcomed with keen interest by the scientific community (Fig. 1.6).

Occasionally this skeptical attitude produces an amusing anomaly. This was true, for example, when the French Academy of Sciences went on record during the late eighteenth century as declaring that the notion of stones falling from the sky (what we now call meteorites) was complete nonsense. Yet a contemporary physicist, Ernst Chladni, was already assembling a case for the extraterrestrial origin of these stones. His theory—that stones do indeed on occasion fall from the sky—received spectacular confirmation through a widely observed shower of meteorites near L'Aigle, France, in 1803. Chladni had been trained as a lawyer before studying science and was therefore more accustomed to dealing with eyewitness testimony than most scientists then or now. As we shall see in Chapter 20, an approach similar to Chladni's is required to address the contemporary conundrum posed by popular reports of UFOs (Fig. 1.7).

Figure 1.6 Hollywood's depiction of alien visitors typically emphasizes the shock that they induce in ordinary people rather than the potential for mutually beneficial exchanges.

Figure 1.7 During the late 1980s and early 1990s, mysterious "crop marks" appeared in sourthern England (and eventually in other countries). To explain them, popular theories suggested extraterrestrial visitors, but scientists developed a theory based on rare types of atmospheric vortices. In fact, pranksters eventually revealed that they had dragged boards through wheat fields to produce the patterns.

The scientific view of the universe does not, of course, provide the only method to experience reality. There are many ways to tell the same story, and only a small fraction of humans attempt to maintain a scientific outlook continuously. To do so often violates human intuition, with which we maintain a system of beliefs that formed long before the scientific outlook emerged. What makes science look good as a worldview is that *science works:* The model of the physical universe that scientists use can successfully explain observations and make accurate predictions of future events. Furthermore, the scientific model allows changes to occur in the framework of our understanding as new discoveries are made. The changes may provoke great debate among scientists, who nevertheless agree upon the *principles* through which they must alter their framework of knowledge.

Because the basic subject matter of this book involves many areas of knowledge where profound changes are now occurring, the reader will

Figure 1.8 Our understanding of physics owes its greatest debts to Galileo Galilei (left), Isaac Newton (center), and Albert Einstein (right). Newton built on the framework of Galileo's ideas and experiments; Einstein refined and enlarged Newton's laws of motion, showing how they apply at speeds close to the speed of light.

continually encounter passages in which we quote more than one informed opinion. This presents the most exciting part of science: the quest to pass beyond the knowledge we have in order to attain new knowledge, knowledge that will change our perspectives about the universe and about ourselves (Fig. 1.8).

Cosmic Loneliness

For millennia, humans have generally held that an ever-watching cosmic force guides all events on Earth. During the past few hundred years, some of the human population has lost this view, believing instead that individuals have responsibility for what they do, even if a deity exists. This change in viewpoint, which scientists in particular have often led, has tended to emphasize the importance of humanity, since a greater sense of our own power over our lives has naturally produced a greater awareness of our ability to affect the world. We are now entering a new phase of this awareness, in which we are beginning to recognize our responsibilities for taking care of the planet we inhabit.

This new perspective has often been accompanied by a vague sense of loneliness, the feeling that we are terribly isolated in a huge, uncaring universe. This view ignores the fact that we are part of the great cosmic web that holds the myriads of stars in our galaxy, and the countless galaxies that journey through the vast dark spaces beyond. Similarly, the unease of "cosmic loneliness" has led many people to reject the idea that our own planet was once entirely devoid of human beings, and that an immensely complex

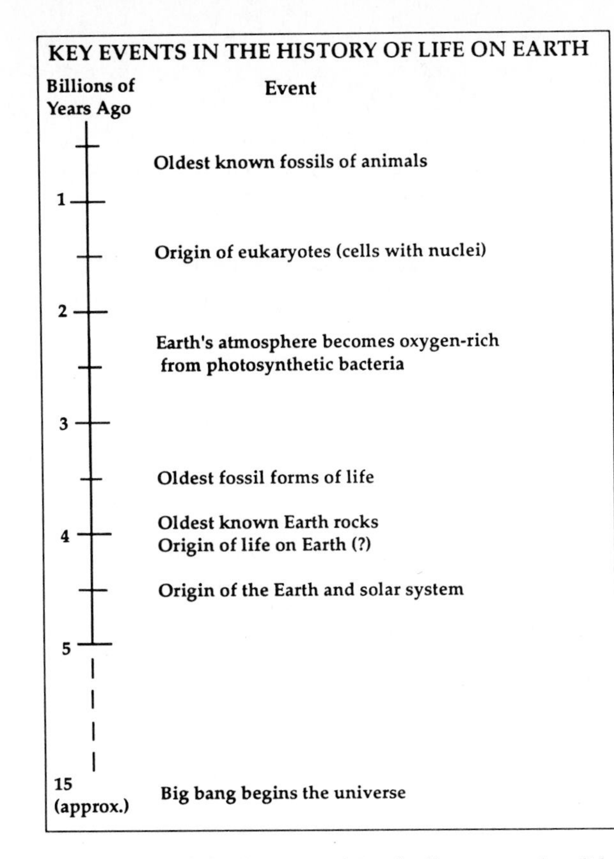

Figure 1.9 A "time line" of the history of Earth illustrates that life has existed for most of the Earth's 4.5 billion years, but complex forms of life appeared only within the most recent one-sixth of that time.

and slow process brought us to our present state. In fact the prehuman period includes more than 99.9 percent of the Earth's history (Fig. 1.9), and therefore has the most to tell us about life on Earth.

Modern theories of how life achieved its present condition date back to 1859. In that year, Charles Darwin proposed a theory of the evolution of species, seeking to explain how a planet without human beings came to acquire them through the transformation of species types. The immediate and widespread opposition to Darwin's theory among the general population demonstrated how much most people cherished the idea of an Earth that has always had humans living upon it, that was in fact *created for that very purpose*. Despite adverse public reaction, Darwin's theory received detailed confirmation as scientists uncovered the fossil history of life on Earth, extending now 3.5 billion years into the past.

Today, die-hard "creationists" continue to oppose the theory of evolution, claiming that sacred texts show the theory must be wrong, or that the loose ends of the theory show its fundamental incorrectness. This desire to reject the model of evolution testifies to our reluctance to abandon the concept of humanity as centrally placed in the structure and history of the universe. The great Russian rocket pioneer Constantin Tsiolkovsky once said that "the Earth is the cradle of humanity—but one cannot live in the cradle forever." This bold thought remains an excellent reminder of the fact that our human attitudes tend to keep us from recognizing the truth about our role in the universe: We *are* part of the cosmos, but a fragile and extremely young part.

Darwin's model for the evolution of life on Earth has been verified in a host of ways. But the existence—let alone the evolution—of life *beyond* the Earth remains completely speculative. Some consider speculations about extraterrestrial life complete foolishness: How can we study what we know nothing about? The answer to this question lies in the fact that we do know a great deal about the general principles of life in the universe, if life on Earth provides a reasonably typical example. In simple terms, *if evolution made us the way we are, then roughly similar conditions should have led to the development of roughly similar forms of life elsewhere.* Our study of the evolution of the universe therefore extends our perspective from life on Earth now, and the fossil record of bygone life, to the still more varied possibilities of life elsewhere, perhaps flourishing now, perhaps in existence only in the past, perhaps only in the future.

To study how life might have evolved far from Earth, we must first study how the universe itself has evolved. Then we can look for the likeliest sites for the origin and development of life, and can see what prelife conditions arise spontaneously in astronomical situations. We can then proceed to consider life on Earth and the conditions that exist on other planets in our solar system. Our goal is to estimate the likelihood of life and of its development into technologically advanced civilizations throughout our galaxy and beyond. Could millions of civilizations exist among the stars, perhaps in contact with each other? If so, we seek to learn how we can join the galactic network and end forever our cosmic loneliness.

SUMMARY

Ever since scientists determined that the origin and evolution of life on Earth has proceeded through natural causes, we have attempted to generalize from our single example of life to speculate about the possibility of finding living creatures on other planets. We now stand at the threshold of the era in which we should be able to enter into communication with other civilizations, provided they are relatively numerous in our galaxy. In this effort, scientists will be guided by the approach that has proven so useful in the past, using the scientific method to refine our framework of knowledge, which has been repeatedly tested and modified by the results of numerous experiments. From such tests, we have learned much about the conditions under which life has developed on Earth, and in which it might be able to arise in other environments. As we follow scientific speculation to estimate the likelihood of other forms of life and other civilizations, we can grow in knowledge about ourselves, to appreciate better how our own existence fits into the structure and evolution of the universe.

QUESTIONS

1. Why has the geological record of the Earth's first billion years completely disappeared?

2. How can you explain the particular fascination that Mars has exerted on the human imagination over the past few hundred years? Consider both astronomical observations and astrological musings.

3. What are the "canals" of Mars? Do they really exist to carry water from the Martian polar caps to the warmer equatorial regions?

4. Compare the scientific approach with various other approaches—such as theological, spiritual, and astrological arguments—to estimate the probability of life elsewhere in the universe. Do you feel that the scientific attitude is basically different? Why?

5. Do you feel a sense of loneliness in contemplating the vast spaces that separate planets and stars from one another? How do you imagine that your great-grandparents felt when they thought about the planets and the stars?

6. Do you believe that human beings have evolved slowly from other forms of living creatures, or that humans have their origin in the relatively recent past—say, a few hundred thousand years ago? What arguments can be made for each point of view? How can they be tested?

FURTHER READING

Calvin, William. *How the Shaman Stole the Moon: In Search of Ancient Prophet-Scientists From Stonehenge to the Grand Canyon.* New York: Bantam Books, 1991.

Krupp, Edwin. *Beyond the Blue Horizon: Myths and Legends of the Sun, Moon, Stars, and Planets.* New York: Harper/Collins, 1991.

Krupp, Edwin. *In Search of Ancient Astronomies.* New York: Doubleday, 1978.

Ley, Willy. *Watchers of the Skies.* New York: Viking Press, 1963.

Lowell, Percival. *Mars as the Abode of Life.* New York: Macmillan, 1908.

Neyman, Jerzy (ed.). *The Heritage of Copernicus.* Cambridge, MA: MIT Press, 1974.

The Tyger

Tyger! Tyger! burning bright
In the forests of the night,
What immortal hand or eye
Could frame thy fearful symmetry?

In what distant deeps or skies
Burnt the fire of thine eyes?
On what wings dare he aspire?
What the hand dare seize the fire?

And what shoulder, and what art?
Could twist the sinews of thy heart?
And when thy heart began to beat,
What dread hand? And what dread feet?

What the hammer? What the chain?
In what furnace was thy brain?
What the anvil? What dread grasp
Dare its deadly terrors clasp?

When the stars threw down their spears,
And watered heaven with their tears,
Did he smile his work to see?
Did he who made the Lamb make thee?

Tyger! Tyger! burning bright
In the forests of the night,
What immortal hand or eye
Could frame thy fearful symmetry?

William Blake

PART TWO

The Universe at Large

The stars are threshed
And the souls are threshed from the husks.

—WILLIAM BLAKE

To organize our considerations about possible forms of life in the universe, we must learn how the universe itself has been organized through 15 billion years of cosmic history, starting with the immense clusters of galaxies that formed first and now extend through billions of light years to the visible horizon of the universe. From this study, we can learn how stars are born, live, and die; how they cluster into groups of various sizes, similar to our own Milky Way galaxy; and how the distribution of stars and the history of their explosions may determine the chance for life to originate around them. Cosmic evolution embraces the chemical history of matter, which apparently began primarily as protons, electrons, and helium nuclei, and only later produced a small fraction of elements such as carbon, nitrogen, and oxygen, essential to life on Earth but rarities in the cosmos. By following the history of the universe, we can understand what happens not only to the stars themselves but also to the raw material for life, which they continuously alter through their evolution.

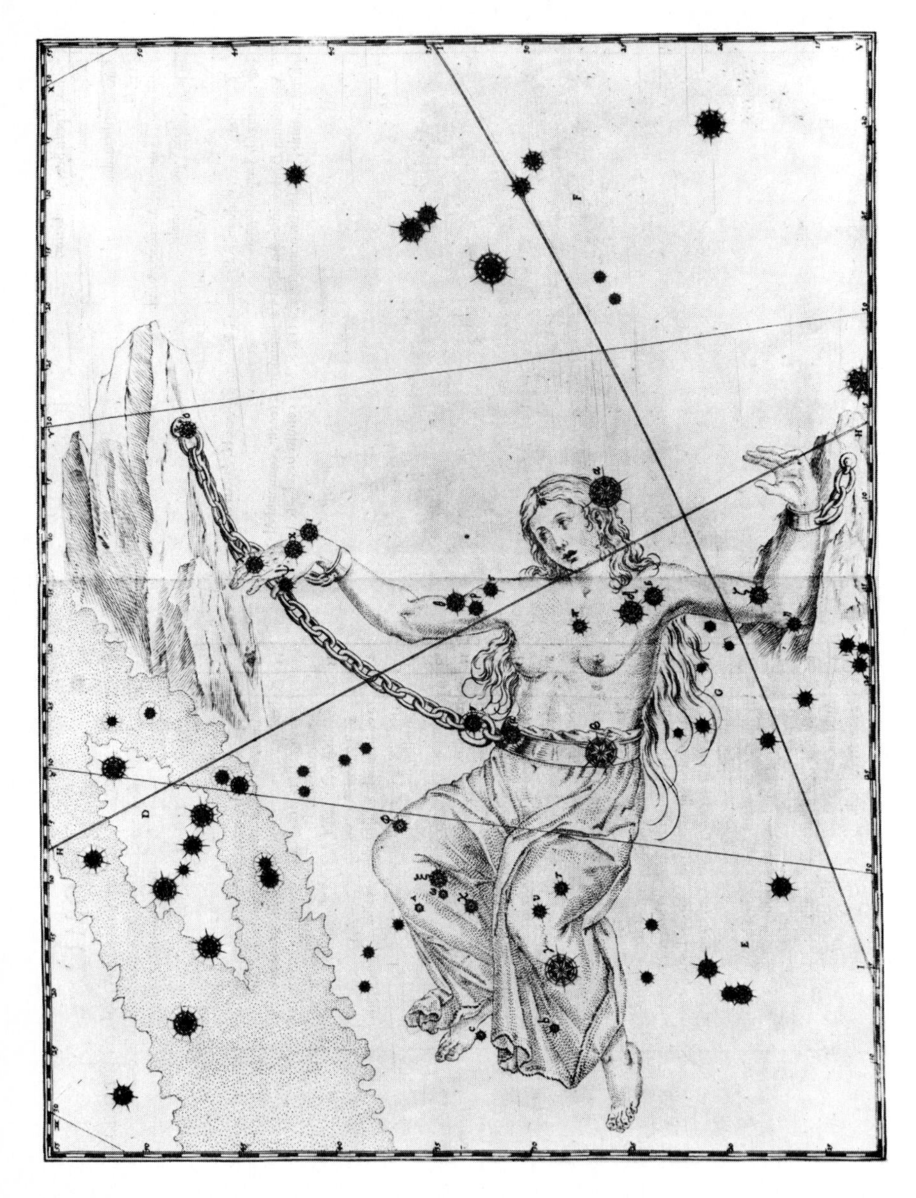

In 1603, Johann Bayer published the first modern star atlas, entitled *Uranometria*. This page shows Bayer's drawing of the constellation Andromeda. Although easily visible to the naked eye, the famed Andromeda galaxy (see Color Plate 1) does not appear on Bayer's map. The galaxy would form part of the chain below Andromeda's right elbow.

2

Space, Time, and the History of the Universe

Stand outside on a clear night, as far from the lights of civilization as you can, and look at the night sky: What do you see? What lies above you is the universe—all of space and everything in it, including all the other forms of life and intelligence that may exist. The vastness of the cosmos, which implies a corresponding minuscule importance for everything on Earth, has often seemed so threatening to human beings that they have gone to great lengths to show that things are not so lonely as they seem. But the progress of astronomy has left no escape from the fact that the cosmos contains unfathomable emptiness, along with objects that are so numerous and so varied that the human imagination may be temporarily staggered simply by contemplating them.

But in the vastness of space lie our hopes to find our cosmic neighbors. The same fact that makes the universe so threatening—its immense size—also implies that an immense number of worlds roughly similar to our own may well be sprinkled throughout the cosmos. But we should never forget that these two characteristics of immense size and tremendous numbers of objects emphasize one common, crucial fact: Distances between cosmic objects are enormous.

So look at the objects that spangle the night sky. What are they, and what possibilities do they offer for life in the universe? In answering this question, we summarize the results from three millennia of astronomical effort. Although it is an easy matter to name the types of objects that astronomers now identify in the universe, some respect is due the hard work and leaps of imagination that led us to our present state of knowledge.

The brightest and apparently largest objects in the sky are the sun and the moon. Each of them spans half a degree in angular size. The concept of "angular size" plays a crucial role in astronomy; it refers to the fraction of a complete circle that an object appears to cover when we see it in the sky. Using a measuring system that goes back to the ancient Babylonians, astronomers divide a complete circle into 360 degrees, more formally known as "degrees of arc," to remind us that degrees are units that measure angles (also called arcs in older geometry books). Each degree is subdivided into 60 minutes of arc, and each minute into 60 seconds of arc. Thus the sun and moon each have an angular size of 30 minutes of arc, or, if we prefer large numbers, 1800 seconds of arc.

The Distances to Astronomical Objects

From antiquity, people have seen that the sun and moon each cover the same angle on the sky. (Note that astronomers typically say "on the sky" rather than "in the sky"; by doing so, they remind us that they use a concept of the sky as projected onto an imaginary "celestial sphere," as if all the objects they see have the same distance from Earth. Wrong though this concept is, it proves quite useful for pointing telescopes in a given direction—and it reminds us that determining the distances to objects is a difficult task.) But in ancient times, no one had a good idea as to whether the sun lies farther away than the moon, or by how many times. What was known, because we see it on Earth repeatedly, is this: Any object's angular size decreases as its distance from the observer increases. If you look at a ship sailing away from you, the ship will appear to grow smaller—that is, its angular size will decrease—as it sails to greater distances. This principle of physics has become second nature to us; we use it repeatedly to estimate the distances to objects that we see.

But one great difficulty remains in applying the principle that angular size decreases with increasing distance. A large ship seen at a large distance has the same angular size as a small ship seen from a small distance. Therefore, if we do not know an object's true size, we cannot tell whether we are seeing a relatively small object close up, or a large object from a much greater distance. The sun and moon offer a perfect example: The sun in fact has a diameter 400 times the moon's diameter—1.4 million kilometers compared to 3476 kilometers. The fact that the sun and moon have the same angular size on the sky arises from cosmic coincidence: The sun lies just about 400 times farther away than the moon—150 million kilometers compared to 385,000 kilometers. The sun, with a diameter about 100 times that of Earth, lies so far away that it appears to us to be no larger than the moon, which has barely more than one-quarter of the Earth's diameter.

How did astronomers ever discover such facts? The answer takes us through 1500 years of intriguing speculation, which led to the modern concept of the solar system, centered on the sun and not on the Earth, with the sun's nine planets moving in orbit around it. Today astronomers can employ the finest geometric measurements yet made. They can even bounce radar waves from the sun, moon, and the nearer planets to measure their distances directly by timing how long it takes for the radar echo to return. As a result, they know the distance to the moon to a fraction of a centimeter, and the distances to the sun and the other planets to a few hundred meters. (This is a good thing, too, when astronomers must direct a spacecraft past one planet after another, as they did with the two *Voyager* spacecraft during the 1970s and 1980s.)

The basic configuration of the solar system was established during the sixteenth century, after the clergyman Nicolaus Copernicus revived an ancient Greek idea that all the planets orbit the sun. Today astronomers have added three more planets to the six that Copernicus knew (Fig. 2.1), and they know their distances and their motions in exquisite detail. Thanks to the automated spacecraft that have visited every planet except for Pluto, we also know the details of these orbiting worlds, and the satellites that accompany them, far better than we did a generation ago.

All of the planets orbit the sun in the same direction and in nearly the same plane. As a result, when we look at the objects in the skies of night, we find that the planets always appear in a well-defined band around the sky. This band includes the directions outward from the Earth that contain the planets' orbits around the sun (Fig. 2.2). In this band, and nowhere else

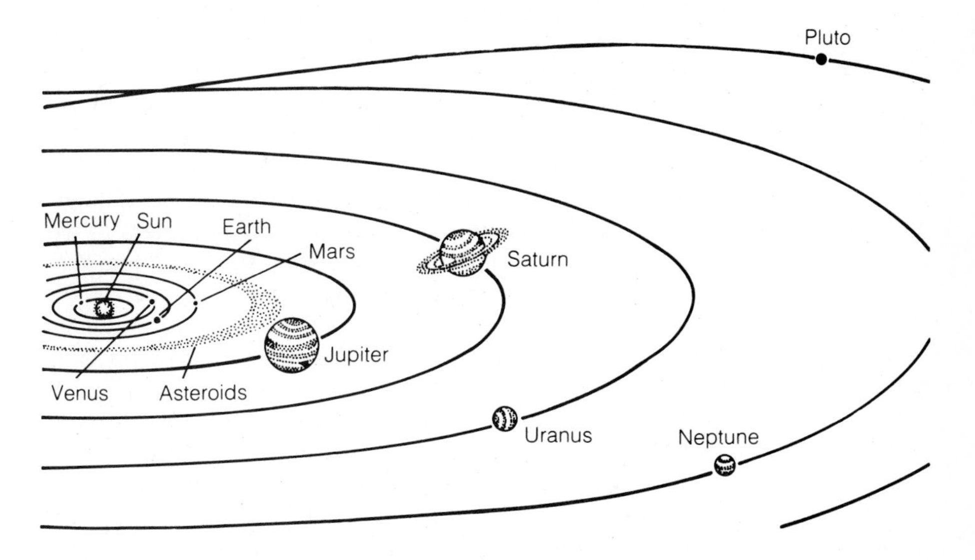

Figure 2.1 The solar system includes nine planets orbiting the sun, of which the Earth is the third from the sun. This diagram is not to scale.

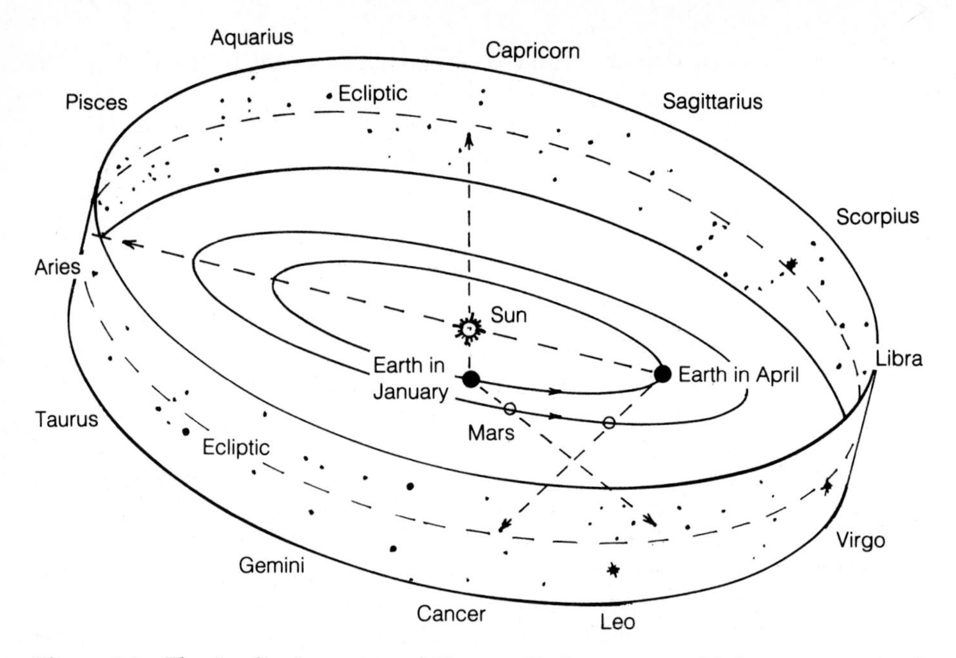

Figure 2.2 The "zodiac" consists of 12 constellations, most with human or animal names, forming a band around the entire sky. Within that band, the sun, moon, and all the planets are always to be found.

on the sky, you will find Mercury, Venus, Mars, Jupiter, Saturn, Uranus, Neptune, and Pluto.

Consider what this meant to those who first admired the sky. Uranus, Neptune, and Pluto are too faint to be seen with the unaided eye, but the five bright planets—Mercury, Venus, Mars, Jupiter, and Saturn—must have drawn great attention, reflected in the name of this class of objects. "Planet" is derived from the Greek word meaning "wanderer," which refers to the fact that these objects appear to wander among the stars (see again Fig. 2.2). Why? As the planets orbit the sun, their orientation with respect to the Earth changes. All the stars are so distant from us that even though they are moving in space, to the naked eye they appear to maintain exactly the same positions for many centuries. As a result, the Earth's daily rotation on its axis makes all the stars appear to move, but the stars maintain the same relation with respect to each other—the same patterns, which we call "constellations"—as the Earth's rotation appears to carry them across the sky from east to west. But the planets move through this pattern, changing their positions among the stars from week to week as they orbit the sun.

So ancient astronomers were drawn to the wandering planets and paid special attention to the part of the sky that always contained them. This band of the sky has the name *zodiac,* from the Greek word *zoe,* meaning "life," because the 12 constellations of the zodiac have (mostly) human and animal natures. The division of the zodiac into twelve constellations reflects the fact that the sun also appears to move around the zodiac. We now know that the Earth in fact orbits the sun once each year, but if we imagine ourselves to be at rest (as we inevitably do when we contemplate the night skies), it appears that the *sun* moves around the zodiac once every 12 months. Other cultures paid particular attention to the zodiac but were more impressed by the moon. Since the moon appears to move through the zodiac once every 27 1/3 days, if you want to mark out the moon's motion, you must divide the zodiac into 27 or 28 star groups, as the ancient Chinese astronomers did.

Today we know well that although the planets, like the sun and moon, all appear to move around the imaginary "celestial sphere," no such sphere exists. Instead, the planets have vastly different distances from the Earth, ranging from 40 million kilometers, when Venus approaches closest to the Earth, to Pluto's distance of 5.9 billion kilometers, more than 140 times larger! Table 11.1 provides the data of the planets' orbits, with their average distances from the sun given both in kilometers and in "astronomical units" (A.U.), with one A.U. equal to the average Earth-sun distance (149.6 million kilometers).

Vast though the distances to the sun's outer planets may be, these distances can still be measured by the signals sent from spacecraft back to Earth. But they pale into near-insignificance when we contemplate the starry realms around us. The closest stars to the sun lie hundreds of thousands of times farther away than the planets! How do we know this fact? For centuries, measurement of the distances to any stars beyond the sun lay beyond the abilities of astronomers: They knew the method but did not have the necessary equipment.

And what was the method to reveal the distances to the stars? These distance measurements rely on the "parallax effect," the simple fact of geometry that as an observer moves, nearby objects appear to shift their positions against a backdrop of much more distant objects (Fig. 2.3). Hence as the Earth orbits the sun, the closest stars should appear to change their positions, as seen against the background of stars much farther from Earth. If we can measure the amount of this change, called the "parallax shift," we can determine the distances to the nearby stars, because the greater the distance, the less will be the amount of the parallax shift.

In theory, then, the measurement of stellar distances by using the parallax effect should be easy. Once we know the Earth-sun distance, and once we can measure the amount of parallax shift that a nearby star exhibits as the Earth orbits the sun, we can use geometry to determine how many times the Earth-sun distance the star must be from us. Copernicus and his

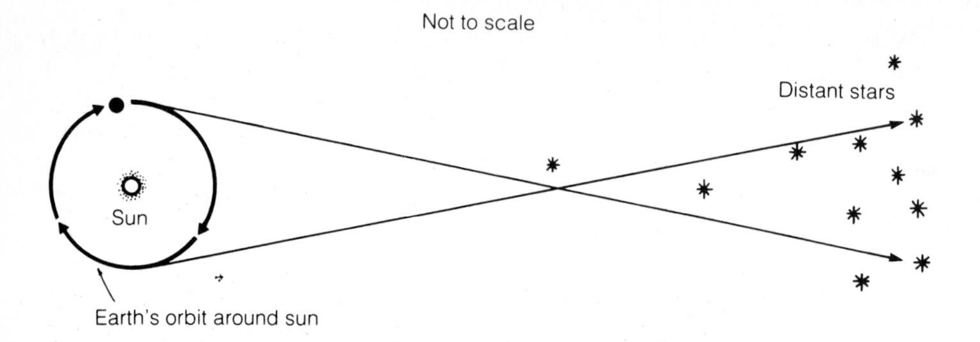

Figure 2.3 The Earth's orbital motion causes the nearer stars to appear to shift their positions against a backdrop of much more distant stars. Astronomers define one "parsec" as the distance a star must have in order for its back-and-forth parallax shift during the course of a year to equal one second of arc. More distant stars have smaller parallax shifts.

colleagues knew this quite well, and much of the scientific attack on Copernicus's sun-centered model of the solar system relied on the fact that no star showed any parallax shift! Copernicus replied, quite correctly, that even the closest stars to the solar system must be so far from us that the parallax effect was undetectable with the instruments of that day. In fact, the first parallax shifts were measured only during the 1830s, nearly three centuries after Copernicus published his epochal work on the solar system. But once astronomers had developed the technological ability to measure these tiny angles, they did so eagerly, and thereby established the basic ordering of distances to the stars around us. Astronomers found the parallax effect so useful that they created a new unit of distance, still in use today: the parsec.

One parsec (from parallax plus second of arc) is the distance that an object would have if its parallax shift equaled one second of arc (see again Fig. 2.3). In fact, no star lies close enough to the sun for parallax to shift its position by one second of arc. The closest stars, those in the Alpha Centauri system, have parallax shifts of three-quarters of a second of arc. Since the parallax decreases in proportion to the increase of distance, we know that the Alpha Centauri stars must lie at a distance of four-thirds of a parsec. And indeed in general, any star's distance, measured in parsecs, equals one over its parallax shift, measured in seconds of arc.

Today parsecs have replaced light years as astronomers' favored unit of distance measurement. One light year equals the distance that light travels in a year. Since the speed of light equals just about 300,000 kilometers per second, in a year's time (31.6 million seconds), light will travel nearly 10 trillion kilometers. One parsec—a unit based on a triangle with twice the

Earth-sun distance as one side and a vertex of one second of arc—turns out to equal 3.26 light years—a unit based on an entirely different concept, the distance that light travels in one year. Thus the Alpha Centauri stars, 4/3 parsec away, have distances of 3.26 × 4/3 = 4.3 light years from the solar system.

Light years have their particular usefulness: They remind us how long light has taken to reach us. All of astronomy is history, since we see objects not as they are but as they were when their light left them. It helps to remember that we multiply parsecs by a bit more than three to obtain the distance in light years—and thus the number of years that light has traveled on its way to us.

By now, astronomers have measured the parallax shifts of the few thousand stars that lie within a few dozen parsecs of the solar system. But these stars represent only the tiniest fraction of all the stars in our own Milky Way galaxy, the assemblage of stars, gas, and dust in which we live (see Color Plates 2–4). What prevents the determination of the parallax shifts, and thus the distances, of the vast majority of the stars in the sky? The answer lies above us, in the air that we breathe. Our atmosphere blurs our view of the cosmos, preventing astronomers from measuring tiny angles smaller than about 1/50 second of arc (and this is possible only by repeated observations under the finest atmospheric conditions). As a result, no star farther from us than about 50 parsecs can have its distance determined accurately by the parallax effect. Within 50 parsecs (163 light years) of the solar system are "only" about a hundred thousand stars—less than one-*millionth* of the total number of stars in the Milky Way! Yet these objects, which include such well-known stars as Sirius (2.5 parsecs away) and Vega (8 parsecs), have furnished us with our basic understanding of the distribution of stars in our galaxy.

For all the stars more distant than about 50 parsecs, astronomers must now use *indirect* methods to estimate distances; plans to obtain more accurate measurements from satellites above the atmosphere are just now being realized. The most important of the indirect methods to obtain stellar distances relies on a fundamental fact of astronomy—and of daily life: *The apparent brightness of any object decreases in proportion to one over the square of its distance from us.* This "inverse-square law of brightness" means that *if* we know that two stars have the same luminosity—that is, the stars produce the same amount of energy per second, and would therefore appear to us with the same brightness if their distances from us were identical—then the fainter star must be the more distant one. Furthermore, if one of the two stars has, for example, 25 times the apparent brightness of the other, then the fainter object must be five times farther away.

This method of comparing apparent brightnesses will work so long as we can be sure that the two stars *do* have the same luminosity (true brightness).

In practice, astronomers using this method to estimate distances know that the actual distance may vary from the estimate, often by 30 to 50 percent. Since stars vary widely in their luminosities, ranging all the way from amazingly luminous blue giants down to intrinsically faint red dwarfs, we may ask: How can we ever be sure that two stars have the same luminosity? Only then can we compare their apparent brightnesses to find the ratio of their distances from us. This question has an answer that deserves some attention.

The Spectra of Stars

The key to recognizing stars with the same luminosity lies in the stars' spectra—that is, in the distribution of their light at different frequencies and wavelengths.

We may picture light as streams of *photons,* massless particles that travel through space at the speed of light, close to 300,000 kilometers per second. One of the most important discoveries in physics, made during the early years of this century, was the recognition of the existence of photons. Each photon has a characteristic frequency, wavelength, and energy, all three of which are interdependent. To understand these quantities, it helps to picture the photon as a vibrating particle, in some sense quivering as it moves (see Fig. 2.4). The photon's frequency measures the number of vibrations per second, and its wavelength provides the distance the photon travels while it vibrates once. Since all photons travel at the same speed—called the velocity of light, "c"—those that vibrate more rapidly (higher-frequency photons) will travel shorter distances during each vibration (shorter wavelengths). Because the frequency is one over the period of vibration, the frequency and wavelength of any photon are *inversely* related: Long wavelengths correspond to low frequencies, and short wavelengths to high frequencies.

As was first realized by two great physicists, Max Planck and Albert Einstein, each photon carries an energy that varies in direct proportion to its frequency. This energy may be considered to be entirely kinetic energy (energy of motion). When a photon collides with an object, such as an atom, it can give all its energy to the object and then disappear. A photon with no energy simply does not exist, unlike an object with mass, which can exist even when it has zero kinetic energy.

Streams of photons are called *electromagnetic radiation,* sometimes shortened simply to "radiation." Among the various types of electromagnetic radiation, our eyes have evolved to detect photons of certain frequencies and wavelengths, which we call visible-light photons. Within the range that visible light covers, we perceive differences in frequency and wavelength as

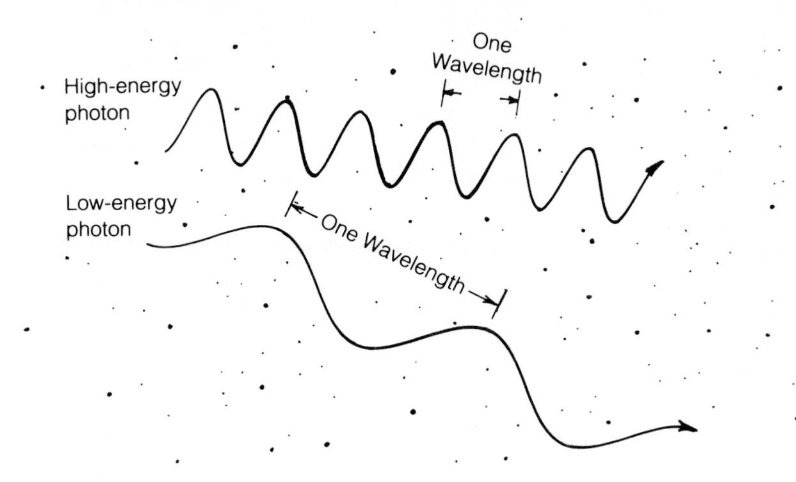

Figure 2.4 We can imagine a photon to be a vibrating particle, traveling at the speed of light and oscillating as it moves. The photon's *wavelength* equals the distance that it travels as it oscillates once, and its *frequency* equals the time it takes for each oscillation.

differences in color. Red-light photons have the lowest frequencies and longest wavelengths of all visible-light photons; violet-light photons have about twice the frequency and half the wavelength of red light.

But visible-light frequencies and wavelengths include only a tiny part of the entire *spectrum* or range of electromagnetic radiation. At shorter wavelengths and higher frequencies we find ultraviolet photons, and then (at still higher frequencies) x-ray and gamma-ray photons (see Fig. 2.5). In the other direction, photons with longer wavelengths and lower frequencies than visible light are classified as infrared, submillimeter, or radio photons (Fig. 2.5).

The light from stars such as our sun includes large numbers of visible-light photons. But the starlight differs in its details and provides astronomers with a sort of stellar fingerprint that allows them to classify stars in great detail. When astronomers employ "spectroscopy" to spread starlight into its various colors—that is, into the spectrum of its radiation—they see an alternating pattern of bright and dark bands. Today, they well know how to interpret such patterns: Stars tend to produce photons of all frequencies and wavelengths (the brighter parts of the pattern), but some of those frequencies and wavelengths are diminished, or even removed entirely (the darker parts of the pattern) as the photons pass through the outer layers of the star.

The reason that only certain frequencies and wavelengths are diminished or removed in the stars' outer layers rests on the way that photons

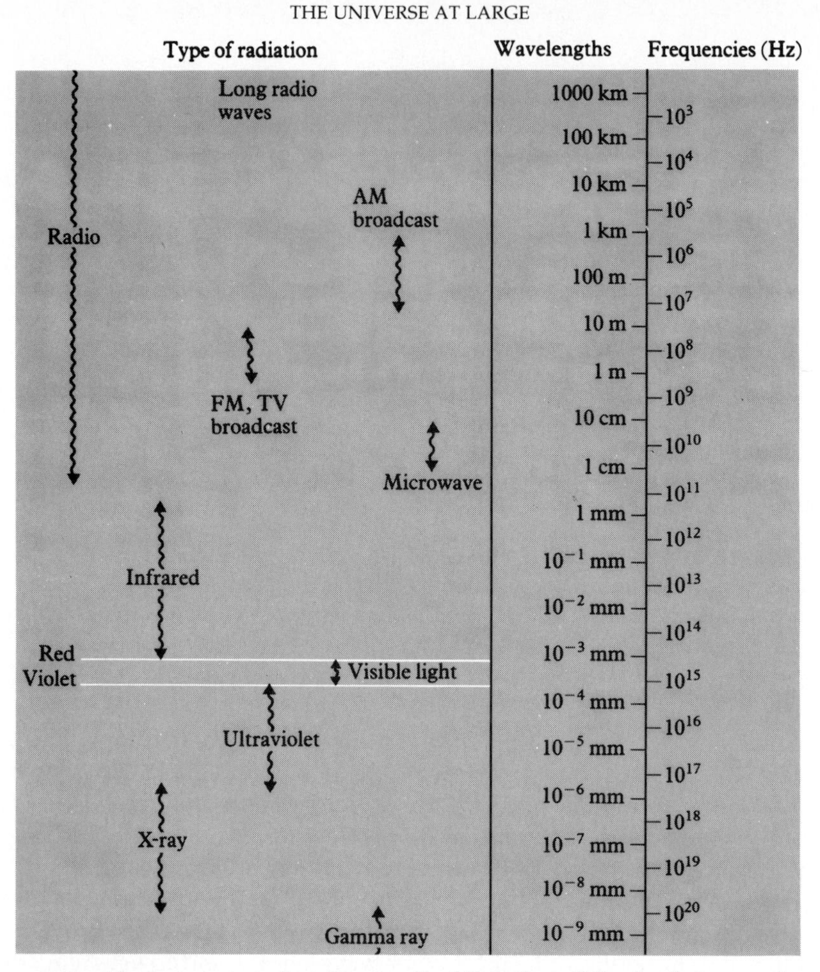

Figure 2.5 All photons travel at the speed of light but differ in their wavelengths and frequencies. Radio photons have the longest wavelengths and lowest frequencies, whereas gamma rays have the shortest wavelengths and highest frequencies. X-ray, ultraviolet, visible-light, and infrared photons are intermediate between these two extremes.

interact with atoms or molecules. We can picture an atom as a central nucleus, made of protons and neutrons, around which one or more electrons moves in orbit (Figs. 2.6 and 2.7). A molecule is a group of two or more atoms, and interacts with photons much as individual atoms do. When a photon encounters an atom, one of two events will occur: Either the photon will happen to carry just the right amount of energy to make an electron jump into a larger orbit, in which case the jump will occur and the photon will disappear, or else the photon's energy will not be matched to any such electron jump, in which case nothing at will happen, and the photon will pass by the atom with no interaction at all.

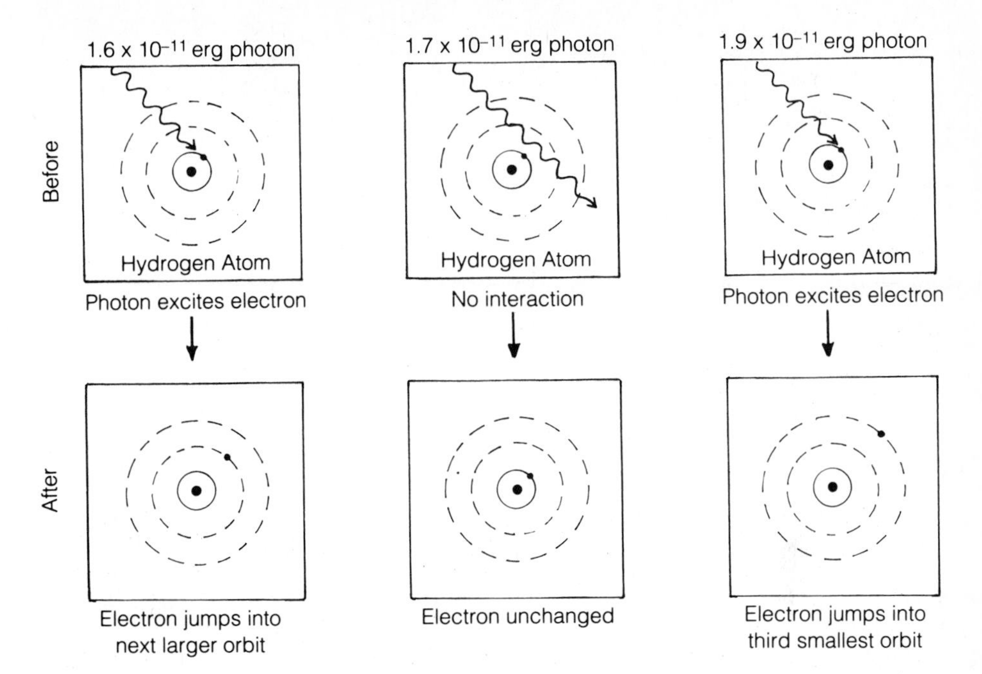

Figure 2.6 When a photon encounters an atom, it will do nothing to the atom (or the atom to it) unless it has the right energy to "excite" one of the atoms' electrons into a larger orbit—or to knock the electron completely loose from the atom. If one of these results occurs, the photon gives all its energy to the electron and disappears.

The reason that only certain photon energies produce an electron jump into a larger orbit is twofold. First, only certain orbits are possible for the electrons in an atom; the electron simply cannot move in the intermediate orbits. Since each orbit corresponds to a particular amount of energy for the atom, a jump to a larger orbit requires that the photon supply the difference in energy between the two orbits. Second, a photon cannot give *part* of its energy to an atom; it either gives all of it, and therefore disappears, or gives none and proceeds on its way unchanged (Fig. 2.6).

These two facts about photons and atoms imply that a set of atoms acts as a filter on streams of photons that pass through the atoms. The atoms remove photons of particular energies, and thus of particular frequencies and wavelengths—those that correspond to the energies needed to make the atoms' electrons jump into larger orbits. The more atoms of a particular type—that is, of a particular atomic *element*—that exist, the greater will be the fraction of photons of those particular energies that are removed by the atoms. If sufficiently large quantities of atoms of a particular type exist, *all* the photons of those frequencies and energies that interact with that type of atom will be removed from the beams of light.

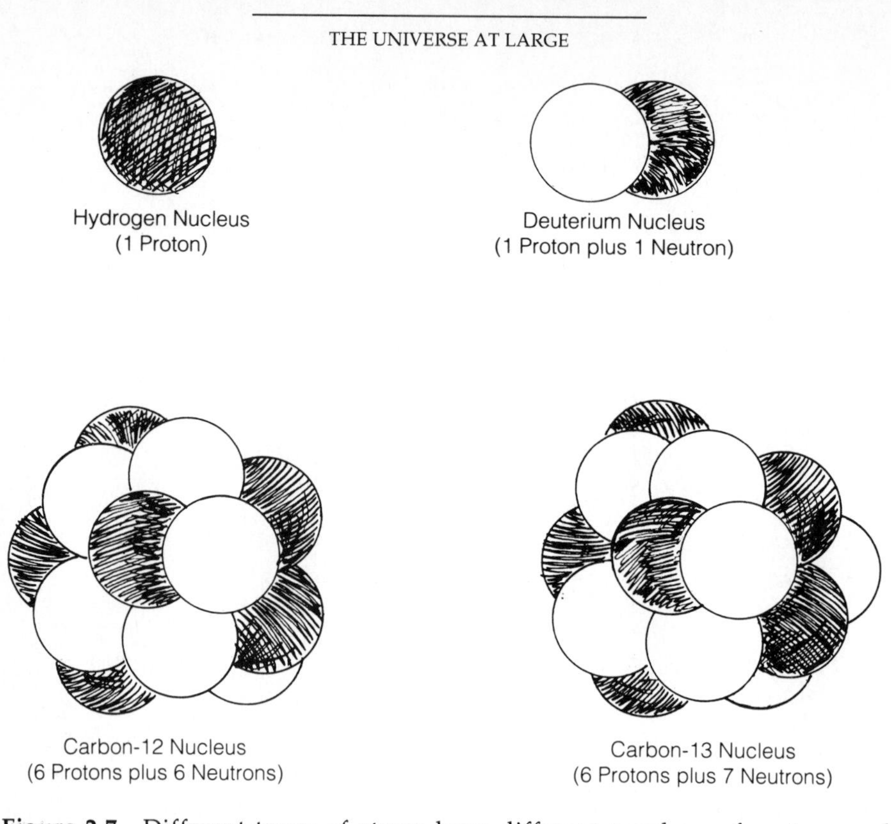

Figure 2.7 Different types of atoms have different numbers of protons and neutrons in their centers or nuclei. The number of electrons orbiting around the nuclei equals the number of protons in the nucleus. The electron orbits can have only certain specific sizes.

The process we have described produces a spectrum in which dark "absorption lines" appear at those frequencies and wavelengths filtered from the light. Through laboratory studies of just which types of atoms remove just which *frequencies* of light, astronomers can now recognize dozens of different elements by the "fingerprint" of their absorption lines. Furthermore, they can determine the amount of each type of atom by measuring the *strength* of the absorption lines (that is, the fraction of light removed) that it produces in the star's spectrum. Finally, comparison of the strengths of absorption lines produced by different elements reveals both the temperature and the densities (number of atoms per cubic centimeter) in the outer layers of the stars. The temperature and density affect the absorption lines in relatively subtle ways, but ones that astronomers have grown skilled at recognizing.

Thus through years of patient study of stellar spectra, astronomers can now determine not only what stars are made of but also how hot and how dense their outer layers are. They have also learned that two stars with the same mixture of elements, the same temperature, and the same density in their outer layers are almost certain to be identical, and thus to have the same luminosity (energy output per second). Astronomers know, for example, that the star Alpha Centauri A (the brightest of the three stars in the Alpha Centauri triple-star system) has a spectrum whose absorption lines almost perfectly match those in the spectrum of sunlight. Hence Alpha Centauri A must have almost the same luminosity as the sun. Because this star lies about 300,000 times farther from the Earth than the sun does, its apparent brightness is only about $1/(300{,}000)^2$, or one 90-billionth, of the sun's. Nevertheless the spectral fingerprints of stars reveal that the sun and Alpha Centauri A are near twins. If astronomers were to observe a third star with a nearly identical spectrum but an apparent brightness of only 1/10,000 of Alpha Centauri A's, they would be justified in assuming that this star has the same luminosity as Alpha Centauri A. Then its distance from us must be 100 times the distance to Alpha Centauri A, because 100 is the square root of 10,000.

The ability to recognize similar stars—stars with the same luminosity—at vastly different distances has unlocked the mystery of how the universe is arranged. Although the comparison of stellar spectra is not the only way to achieve this recognition, it is the prime method that astronomers use. The method of estimating distances by comparing apparent brightnesses works not only for individual stars but for entire *galaxies* as well. Suppose, for example, astronomers observe two galaxies that have similar shapes (for example, two spiral galaxies) and find that the spectra of light from the two galaxies nearly match one another. They then feel justified in assuming that the galaxies have nearly the same luminosity. In that case, the galaxy with the fainter apparent brightness must be more distant, and if it has only 1/100 the apparent brightness of the other galaxy, it must be 10 times farther away.

The Doppler Shift and the Expanding Universe

During the early years of the twentieth century, astronomers recognized that our Milky Way is simply one galaxy among millions, and that all of these other galaxies are enormously distant. The closest giant galaxy, the great spiral in Andromeda (Color Plate 1), has a distance of 2 million light years, and the closest large cluster of galaxies, called the Virgo Cluster, lies about 60 million light years away. The Milky Way and the Andromeda galaxy are the two largest galaxies in the "Local Group" of galaxies, which contains about two dozen members (Fig. 2.8). The Local Group itself forms an outlying part of the "Virgo Supercluster," which contains thousands of

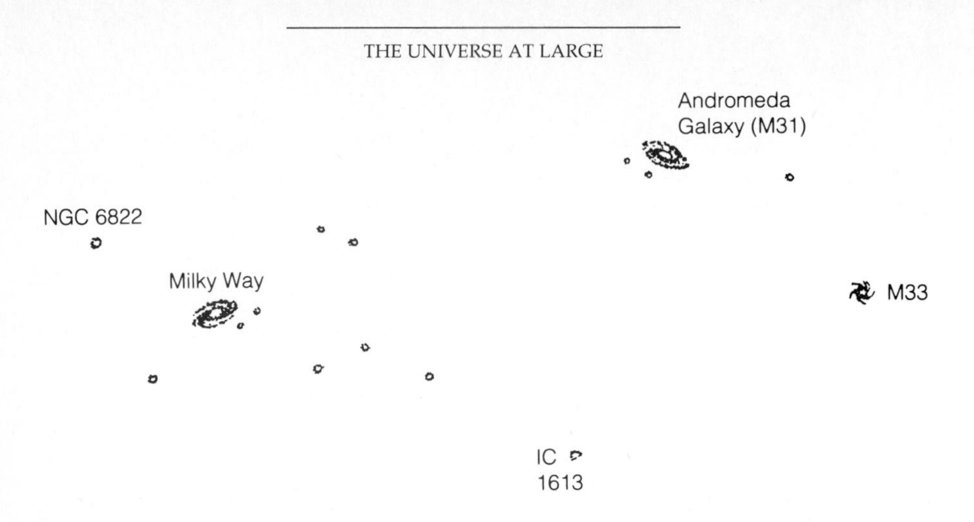

Figure 2.8 The Milky Way belongs to a small cluster of galaxies called the Local Group, with about two dozen members that span 600,000 parsecs of distance. The largest galaxies in the Local Group are the Milky Way, the Andromeda spiral galaxy, and a spiral galaxy in the constellation Triangulum called M33.

members in its inner regions (Fig. 2.9). Still farther from us than the Virgo Cluster (at the heart of the Virgo Supercluster) lie immense numbers of other galaxy clusters, some catalogued, most not, with estimated distances up to 10 billion light years from the Milky Way.

Estimating the distances to galaxies became possible during the 1920s through the efforts of Edwin Hubble, who found individual stars in the Andromeda galaxy whose apparent brightnesses could be compared with those of similar stars in the Milky Way. Hubble's work inspired further investigations of the distances to galaxies, which continue to this day. But even as those efforts began, Hubble made a startling discovery by combining his distance estimates with another key aspect of galaxies, revealed by the Doppler effect.

The Doppler effect, named after Johann Christian Doppler, the Austrian physicist who first analyzed it, describes the change in frequency and wavelength that arises when a source of waves is moving with respect to any observer of those waves. Doppler himself was interested in sound waves, but the same effect is at work for light waves. The Doppler effect describes this set of facts: If photons arrive from a source of light that is moving away from the observer, the motion continuously increases the distance that the photons must travel to reach the observer. As a result, each of the photons has a longer wavelength and a lower frequency than would be observed if no motion occurred (Fig. 2.10). Similarly, if a source of photons moves toward an observer, each successive photon has less distance to travel, and the observer sees photons with shorter wavelengths and higher frequencies than would be true without any motion. Furthermore, if the

Figure 2.9 The Local Group represents an outlying subregion of a great grouping of galaxies called the Virgo Supercluster, which contains several thousand member galaxies and has a diameter of 20 million parsecs. This photograph shows only the central regions of the Virgo Supercluster.

observer moves toward the source, the effect is the same as the one that would appear if the source were moving toward the observer; all that counts is the *relative* motion of the source with respect to the observer.

A key aspect of the Doppler effect is that the relative motion of source and observer *changes all the photon frequencies and wavelengths by the same fractional amount*. For relative motion at a speed v much less than the speed of light, c, the Doppler effect changes all the frequencies and wavelengths by the fractional amount v/c, so that more rapid motion produces a greater change in the frequencies and wavelengths. Furthermore, motion of the source away from the observer produces a "redshift" (a shift to longer wavelengths) that increases all the wavelengths by the fractional amount v/c and decreases all the frequencies by the same fraction. Motion of the source toward the observer, called a "blueshift," increases all the frequencies and decreases all the wavelengths by an amount equal to v/c times the original frequencies and wavelengths.

Because all the frequencies and wavelengths change by the same fractional amount, the Doppler effect preserves the *pattern* of a spectral fingerprint even as it "stretches" or "compresses" all the wavelengths and frequencies. This fact allows astronomers to recognize characteristic patterns in the spectra of the light from stars or galaxies that are moving toward us or away from us at a small fraction of the speed of light. In fact, once astronomers recognize a familiar spectral fingerprint in the light from an object, they can proceed to measure the amount by which the Doppler effect has changed the frequencies and wavelengths (called the "Doppler

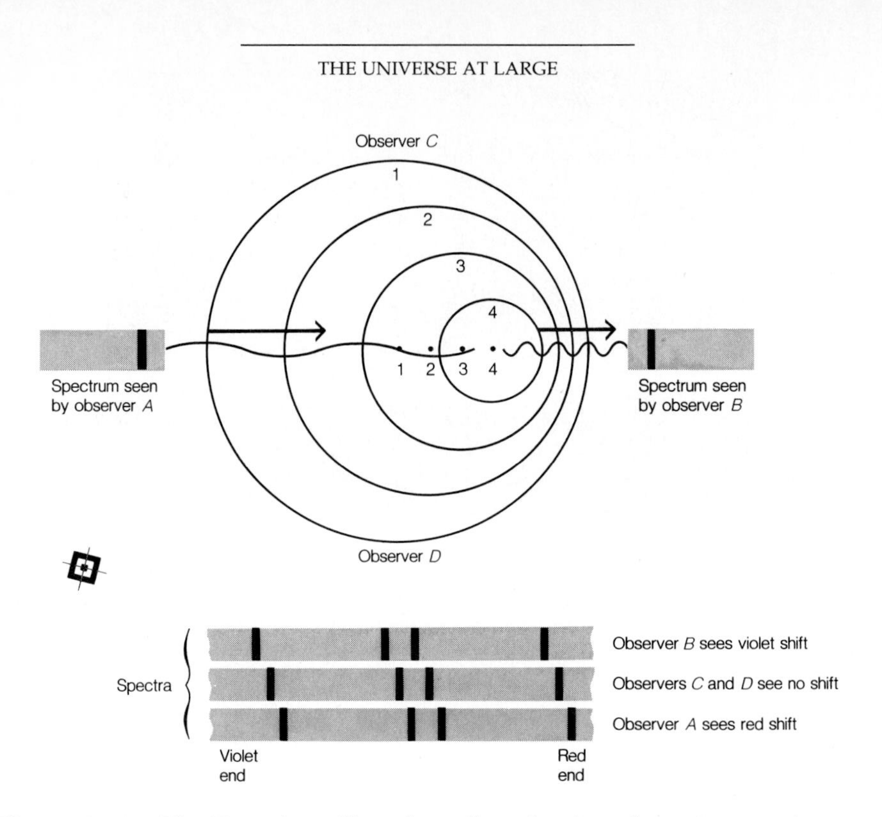

Figure 2.10 The Doppler effect describes the fact that photons from an approaching source are detected with higher energies, higher frequencies, and shorter wavelengths than photons arriving from the same type of source at rest with respect to the observer. Conversely, photons arriving from a receding source have lower energies, lower frequencies, and longer wavelengths than those from a stationary source.

shift"). They can then use this measurement to determine how rapidly the object is approaching us or receding from us (or how rapidly we are approaching or receding from it).

While Hubble was performing his pioneer estimation of the distances to galaxies, an astronomer named Vesto Slipher at the Lowell Observatory in Arizona was using the Doppler effect to determine how rapidly the brightest galaxies are moving toward us or away from us. In 1929, Hubble reached a startling conclusion by combining his distance estimates with Slipher's velocity estimates. Except for some of the closest galaxies, *all galaxies are receding from us, and with speeds that increase in proportion to the distances of the galaxies from us!* Today astronomers call this relationship between distances to galaxies and their speeds of recession from us "Hubble's Law."

What does Hubble's Law imply about the universe as a whole? Astronomers were quick to see that they could make one of two assumptions. Either our Milky Way has a special location in the universe—and this has seemed an inappropriate assumption ever since Copernicus dethroned the Earth as the center of the cosmos—or it does not, in which case we have a representative view of the universe. But if that is so, other observers throughout the universe should see galaxies receding *from them,* and at speeds that are proportional to the distances of the galaxies *from them.* The notion that all observers see galaxies receding from them corresponds to the concept that *the entire universe must be expanding.* In this case, the universe can be considered as analogous to the *surface* of an expanding balloon, on which dots represent galaxies (Fig. 2.11). When we blow up the balloon, every dot recedes from all the other dots, and at speeds proportional to the distances between dots.

But how can the entire universe resemble the skin of an expanding balloon? We must confront the fact that no one can have a true mental picture of the universe, since that would mean standing outside (even mentally) everything that exists. We must try to make do with analogies such as the balloon model for the expanding universe. Note that the balloon does have

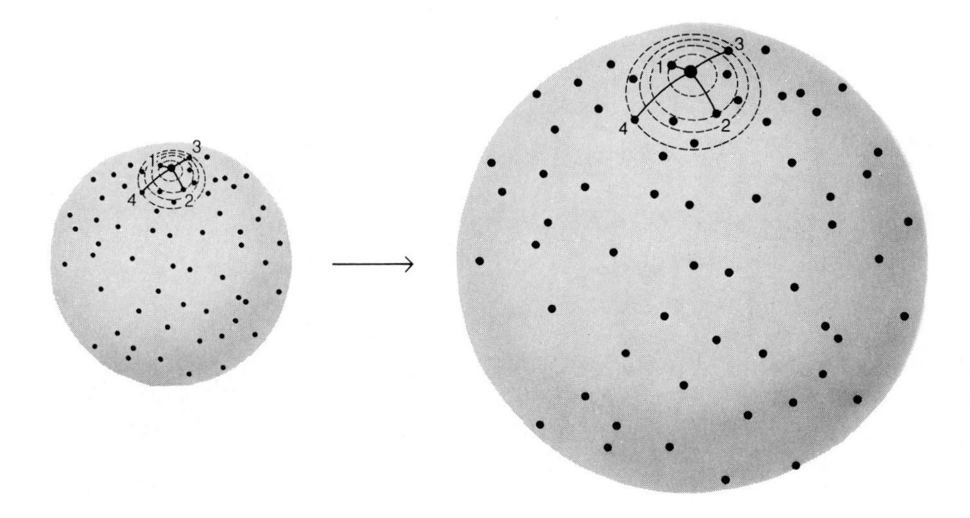

Figure 2.11 We can model the movements of galaxies in an expanding universe by imagining an expanding balloon with dots on its surface. The balloon's two-dimensional surface models all of three-dimensional space. As the balloon expands, every dot on its surface moves away from all other dots, and any two dots recede from one another with a speed that is proportional to the distance between them.

a center, but the center lies nowhere on the balloon's surface, with which we model *all of space.* On the surface, cosmic democracy reigns: No one dot can call itself *the* center of the expanding universe.[1]

The Big Bang

Hubble's work, coupled with the assumption that our view of the universe is representative, leads to the conclusion that clusters of galaxies are receding from one another everywhere. This straightforwardly implies that they used to be closer together. If we were to run a movie called *The History of the Universe* backward in our minds, we can see that the average density of matter was higher in past times. (An alternative theory of the universe, the "steady-state model," in which new matter appears as the universe expands, has been ruled out by observations of the cosmic background radiation described on page 41). If we proceed far enough back in time, we reach a moment when the universe had near-infinite density (Fig. 2.12). This moment, called the *big bang,* marks the start of the expansion, and for most purposes,

[1]The center appears in the "extra dimension" offered by the fact that we are using the two-dimensional surface of the balloon to make a model of three-dimensional space. Mathematically, the "center" of three-dimensional expanding space can be found in an extra, fourth dimension. However, so far as we can tell, "real" space has only the three dimensions that we know well.

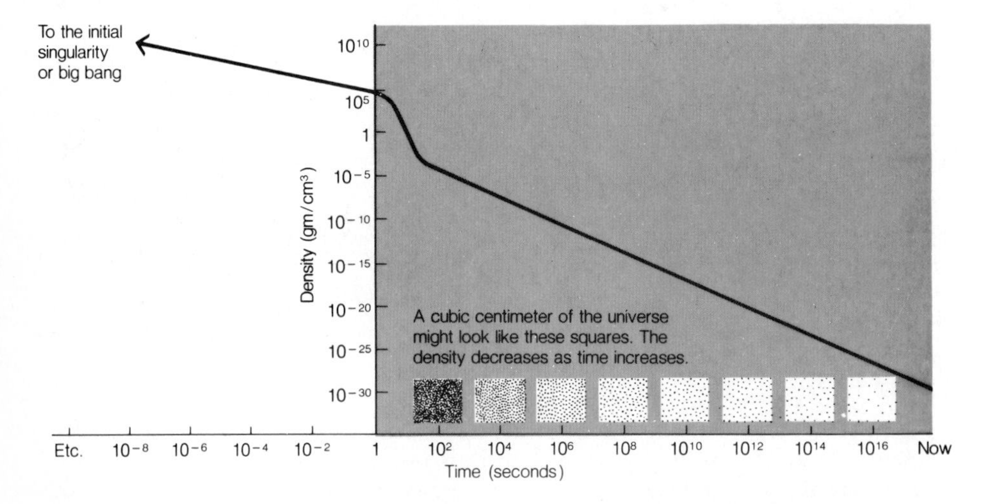

Figure 2.12 Ever since the big bang, the average density of matter has been decreasing, as more space has come into being but no significant new matter has been created.

the beginning of the universe. Astronomers' best calculations imply that the big bang occurred about 15 billion years ago, so that the Earth, which is nearly 5 billion years old, has about one-third of the age of the universe. We cannot now determine what the universe was up to before the big bang. It may have been contracting toward this moment, or it may not have existed at all.

But consider the universe as it was a few moments *after* the big bang. As Fig. 2.12 shows, in those moments the density of matter was enormous, as matter throughout the universe crowded together more densely than anything we see today. In those moments, no stars, galaxies, or planets existed; instead, the universe was a featureless, roiling broth of extremely hot matter, with elementary particles colliding to form new types of particles at every instant.

As the universe expanded, it cooled, much as gas escaping from a bottle cools as it enters a larger volume. The temperature dropped from trillions to billions to mere millions of degrees. Half an hour after the big bang, the basic mixture of particle types had emerged, because particles no longer collided with sufficient violence to generate new types of particles. So far as we can tell, the basic particle types were then—and are today—protons, neutrons, electrons, and helium nuclei, plus photons, neutrinos, and antineutrinos. The last three types of particles have no mass, though they can carry kinetic energy as they travel through space at the speed of light.

The helium nuclei produced in large quantities soon after the big bang each consist of two protons and two neutrons. All neutrons that had not formed part of a helium nucleus soon *decayed* (fell apart) into other types of particles, producing more protons, electrons, and antineutrinos. And what of nuclei more complex than helium? The moments soon after the big bang did not make them efficiently. All types of nuclei other than hydrogen nuclei (protons) and helium nuclei (two protons plus two neutrons) were *not* produced in significant quantities (see Table 2.1). Thus when we look for the source of nuclei such as carbon, nitrogen, and oxygen, so fundamental to life on Earth and quite likely fundamental to life anywhere in the universe, we must find a source other than the big bang itself. As explained in Chapter 6, we must look to the stars, and in particular to those few stars that explode violently at the ends of their lives, seeding the universe with nuclei heavier than hydrogen and helium.

The Cosmic Background of Photons

Photons have been the most abundant type of particle in the universe since a time soon after the big bang. Indeed, for every proton or electron in the universe today, there are 100 million photons. This enormous quantity

Table 2.1 Comparison of the Abundances of Different Elements with Models of Element Production After the Big Bang

Element	Number of Protons	Number of nuclei (per 10^{12} protons) produced during the first few minutes after the big bang*	Present number of atoms (per 10^{12} hydrogen atoms) in the solar system and in sunlike stars
Hydrogen	1	1,000,000,000,000	1,000,000,000,000
Helium	2	80,000,000,000	80,000,000,000
Carbon	6	1,600,000	370,000,000
Nitrogen	7	400,000	115,000,000
Oxygen	8	40,000	670,000,000
Neon	10	180	110,000,000
Sodium and all other heavier elements	11 or more	2500	140,000,000

*Different models of the early universe predict different numbers for the production of the nuclei heavier than hydrogen during the first few minutes after the big bang. The third column lists the maximum number of nuclei predicted under the most favorable conditions. Note that the elements heavier than helium are made only in tiny amounts during the first few minutes. The three elements between helium and carbon (lithium, beryllium, and boron) have tiny abundances, because they are easily changed into other elements through nuclear interactions.

of photons permeates all of space with electromagnetic radiation. Since this radiation provides a background to any other radiation we may detect—for example, the photons emitted by stars and galaxies—astronomers call it the *cosmic background radiation.*

With one important exception, the cosmic background radiation has been little altered since a time half an hour after the big bang, when particles ceased to collide with sufficient energy to form new types of particles. The exception arises from the expansion of the universe itself. This expansion has continuously reduced the energies and frequencies, and increased the wavelengths, of all the different types of photons in the cosmic background radiation, so that today each of these photons has far less energy than it did at the time when it was first created.

To see why this is so, consider the following fact. The photons in the cosmic background radiation were created 15 billion years ago. Hence any observer who detects these photons today observes photons that have traveled a distance of 15 billion light years. In view of the fact that more distant parts of the universe are receding more rapidly, we can calculate that any photon from such immense distances arrives from a source—the long-vanished early universe—that is moving away from us at speeds close to the speed of light. As a result, the photons in the cosmic background radiation arrive with tremendous redshifts, in comparison with the frequencies and wavelengths they would have if the universe had never expanded.

The existence of the cosmic background radiation (CBR for short) was predicted on theoretical grounds by George Gamow and his collaborators Ralph Alpher and Robert Herman during the late 1940s. In 1964, two radio astronomers at Bell Telephone Laboratories, Arno Penzias and Robert Wilson, discovered the CBR quite by accident, while testing a new radio antenna. The discovery of the cosmic background radiation, and its later confirmation in great detail (most recently from the COBE satellite that observed the radiation during 1989 and 1990) furnishes the most convincing evidence that the big bang actually occurred.

One of the best uses of the CBR is to prove that the expanding universe has no center, despite what our intuition may suggest. From where do the photons in the CBR arrive? From all directions, and in the same amounts from all directions. This shows that the early universe must have had the same conditions throughout, at least over the size ranges that the CBR samples. One of the greatest challenges of modern cosmology (the study of the universe as a whole) consists of explaining how the smooth and featureless early universe, which produced the smooth CBR that we observe 15 billion years later, managed to clump itself into stars and galaxies within a mere few billion years.

The Inflationary Universe

Astronomers have struggled for many years to explain how the universe developed its observed structure from an originally featureless hot broth of seething particles. They have attempted to calculate how the amount of matter they see in galaxies today managed to "pull itself together" by gravity, but they have consistently found that computer models of galaxy formation imply times that are too long (in comparison with the actual universe) to work. A recent new cosmological theory, called the inflationary universe, has added a new wrinkle to the astronomers' models, because it implies that *most of the universe consists of invisible matter of a completely unknown form.*

The inflationary theory grew from attempts to meld particle physics—the science of the smallest particles—with cosmology, the study of the universe at the largest size scales. Particle-physics experts saw that their theories predicted amazing behavior in the early universe—not during the first half hour after the big bang, but during the first 10^{-30} second! In that time, the theory says, what began as a submicroscopic bubble of "false vacuum" (we need not pause to consider what that might be) doubled in size, then doubled and doubled again, so that after 50 or 100 doublings, which took no more than 10^{-30} second, the bubble had become far larger than the visible universe today. Afterward, according to theory, this "inflationary" bubble behaved just like the "ordinary" expanding universe that we have described.

But the inflationary model of the earliest moments of the universe includes a prediction of just what the average density of the matter in the universe should be at all times. Today, 15 billion years after the big bang, that density should be approximately 10^{-29} gram per cubic centimeter. This does not sound like much, but it exceeds the density of the matter we *observe* by 50 or 100 times! In other words, if we accept the inflationary theory as valid, we must accept a universe in which "dark matter" provides nearly all of the matter in the universe, and "visible matter"—stars and galaxies made of stars—furnishes only 1 or 2 percent of the total.

The Dark Matter

Today, in the last decade of the twentieth century, cosmologists incline favorably toward the inflationary model of the early universe, because it seems to link what we know about subatomic particles with what we know about cosmology. They are quick to accept the difficulties of concluding that most of the universe consists of dark matter, but they point out that this dark matter offers possible advantages as well.

If the universe has a far greater density than is provided by the matter that shines, the additional matter could have helped to form galaxies, billions of years ago, since its gravity would add to the much smaller gravitational forces from ordinary matter. This argument seems promising, but it must be admitted that the best theories to date still have grave difficulties in explaining how the universe developed such enormous structures in only a few billion years. Among the largest of these structures are those revealed by modern maps of the distribution of galaxies, which include the "Great Wall" of galaxies, nearly 500 million light years across (Fig. 2.13).

Suppose that the inflationary theory is correct. What, then, forms the dark matter? No one knows, though speculation flows freely. Some cosmologists favor dark matter made of familiar types of particles, such as protons and helium nuclei, and predict that the universe consists mainly of Jupiter-sized objects, or burnt-out stars. Others favor more exotic objects, such as the "black holes" that arise when so much mass collects in so small a volume that nothing, not even light, can escape from it. Still others hypothesize entirely new types of particles, never detected on Earth or

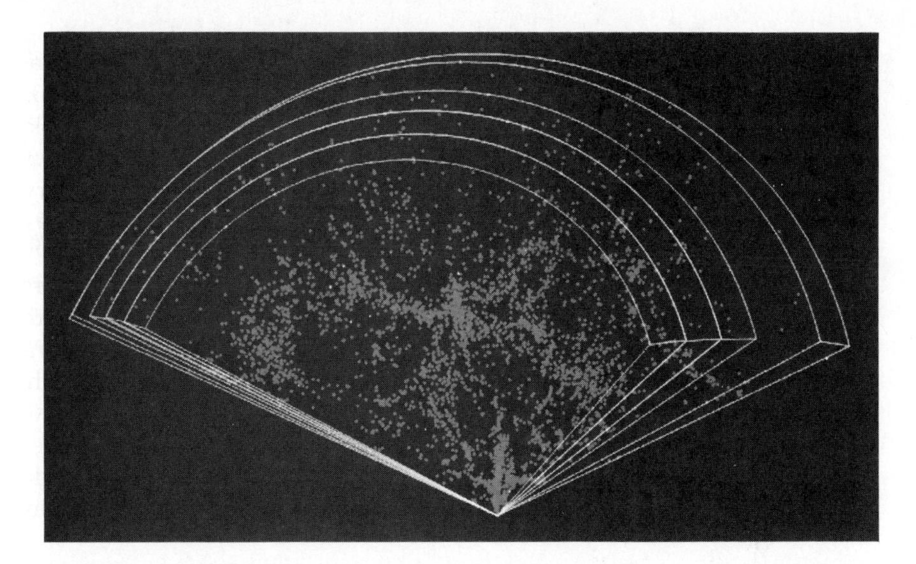

Figure 2.13 A map of the distribution of galaxies made by Margaret Geller and John Huchra shows slices of the sky extending out to a distance of about 150 million parsecs. The galaxies have arranged themselves into immense, loose structures such as the "great wall" that extends across the map (roughly horizontally) for half a billion parsecs.

elsewhere in the cosmos, that would nevertheless comprise most of the matter in the universe.

Years may pass before the mystery of dark matter yields even a partial solution. This could include the conclusion that we have highly overestimated the amount of dark matter, though we shall see in the next chapter that strong evidence exists for large amounts of dark matter in spiral galaxies such as our own Milky Way. For now, many find tantalizing the notion that most of the universe is "missing," that we have yet to determine the form assumed by most of the matter in the universe.

The implications of dark matter for life in the universe have yet to be explored in full. Certainly, since the amount of dark matter governs the eventual fate of the universe, its long-term effects on any form of life should be profound. Furthermore, once we learn the nature of the dark matter, we may obtain a clearer view of how the universe has evolved into stars and galaxies, which will tell us more about how we came to have so many potential sites for life (see Chapter 17). But since we know that stars and galaxies *do* exist by the billions upon billions, such information may prove more important to other fields of astronomy. For the investigation of life in the universe we shall seek to go first and foremost to the other possible places where life might exist, and to look around.

Is the Universe Finite or Infinite?

Two questions strongly related to the amount of dark matter in the universe are these: Will the universe expand forever? And is the universe—all of space and everything in it—finite or infinite?

We cannot now determine whether the universe spans a finite or an infinite extent, or whether it includes a finite or an infinite amount of matter. This problem has an intimate connection with the problem of whether or not the universe will expand forever, and both may be resolved within the next generation or so of astronomical effort. This effort hinges on attempts to test theories such as the currently fashionable inflationary model of the universe, which predicts that the universe *will* expand forever, and that it contains an enormous but finite amount of space and matter.

What does it mean to say that space in the universe could be finite? Could the universe have a boundary? Certainly not: The universe would then include the boundary and anything on the other side. Instead, in a finite universe, space itself curves so that all seemingly straight lines bend back to join themselves after a long but finite distance. (This kind of curvature has the technical name of positive curvature.) A straight-line rocket journey in a finite universe would eventually return us to our starting point.

Suppose that we ourselves had only two dimensions and spent our lives sliding around the surface of a smooth sphere (Fig. 2.14). When the subject of curved space came up, we would say, "I don't see how space could possibly curve back on itself," since we would be unable to conceive of all two-dimensional space (which would be all that was accessible to us) as a quantity with curvature. We might, however, discover this anyway: If we slid all the way around the sphere in a "straight line," we would return to our starting point, and we would know that something odd had occurred, even though we couldn't understand how we had returned. Despite the lack of comprehension, the fact of curvature would eventually leap out at us.

If you have difficulty in imagining a finite universe, consider the alternative possibility, an infinite universe. Just imagine space extending forever and ever and ever, totally without limit and containing a distribution of matter roughly similar to what we see around us. In that case, the universe would include an infinitely large number of stars, galaxies, planets, atoms, and particles. Consider what an infinite universe would contain. With an infinite set of situations, everything not forbidden would occur somewhere, not once but an infinite number of times! Other forms of life would duplicate our own, as well as every other conceivable possibility, over and over again in all variations, an infinite number of times for each individual possibility. This book would exist in every conceivable version, in every possible language (human or other), in every conceivable style of printing,

Figure 2.14 Flat creatures on the surface of a sphere might never intuitively "understand" that their space—the surface on which they live—is curved, but they could discover that fact by sliding all the way around it in a straight line and finding themselves back where they started.

photography, or paper. Sometimes the reader would have written it, sometimes not. Sometimes one word would be different from this text, sometimes two, sometimes more, and each possibility would be reality not just in one place, or in a few places, but in an infinite number of places.[2]

[2]The reader may think this is a joke, but it is not. No matter how improbable an event may be, in an infinite universe that event will occur not once, or a thousand times, but an infinite number of times. However, since we can observe, even in theory, only a finite part of the universe in a finite amount of time, we can see only an infinitesimal fraction of this infinite variety.

If we search for other forms of life, then we can rest assured that in an infinite universe, not only life but an infinite variety of life exists, of every possible function or appearance. A finite universe may seem less interesting, but since we know that the universe contains at least 100 billion galaxies, each with billions of stars, we still face an enormous array of possibilities in a finite universe.[3]

Will the Universe Expand Forever?

Since the universe is now expanding, we may conclude that either it will expand forever, or else the expansion will gradually cease, to be replaced by a universal contraction. The determination of which future—eternal expansion or cosmic recycling—lies in store for the universe presents astronomers with one of the most significant problems with which they grapple. The attack on this puzzle rests on the fact that if the universe ever starts to contract, it will do so because of gravity. Gravitational forces, so far as we know, provide the only chance to overcome the tendency to expand which the universe acquired at the moment of the big bang, and which has dominated the universe ever since. Every piece of matter in the universe attracts every other through gravitational forces, and these mutual attractions constantly act in a manner that opposes the universe's expansionist tendencies. As a result, the universe is no longer expanding as rapidly as it used to: The time for the distances between galaxy clusters to double has steadily increased. But will the distances ever stop increasing and start decreasing?

The question reduces itself to this: How can we test theories of the expanding universe, some of which predict eternal expansion while others include eventual contraction? All such theories predict a definite relationship between galaxy clusters' distances from us and their velocities of recession. The relationship starts out as one of direct proportion, but for galaxy clusters at truly enormous distances, more than 5 billion light years from us, each model diverges somewhat from the straight line of Hubble's Law (see Fig. 2.15). Thus if we can measure galaxies' recession velocities (a relatively easy task, thanks to the Doppler effect) and their distances (enor-

[3]Because only a finite amount of time has elapsed since the big bang, we can have interacted during that time with only a finite volume around us, no matter how large that volume may seem. At any time after the big bang, we can know about only those regions of the universe whose distance from us does not exceed the speed of light times the age of the universe. We therefore face the possibility that the universe may teem with life that lies too far from us to be detectable now, even if we could employ the theoretically best method, simply because those forms of life are too distant.

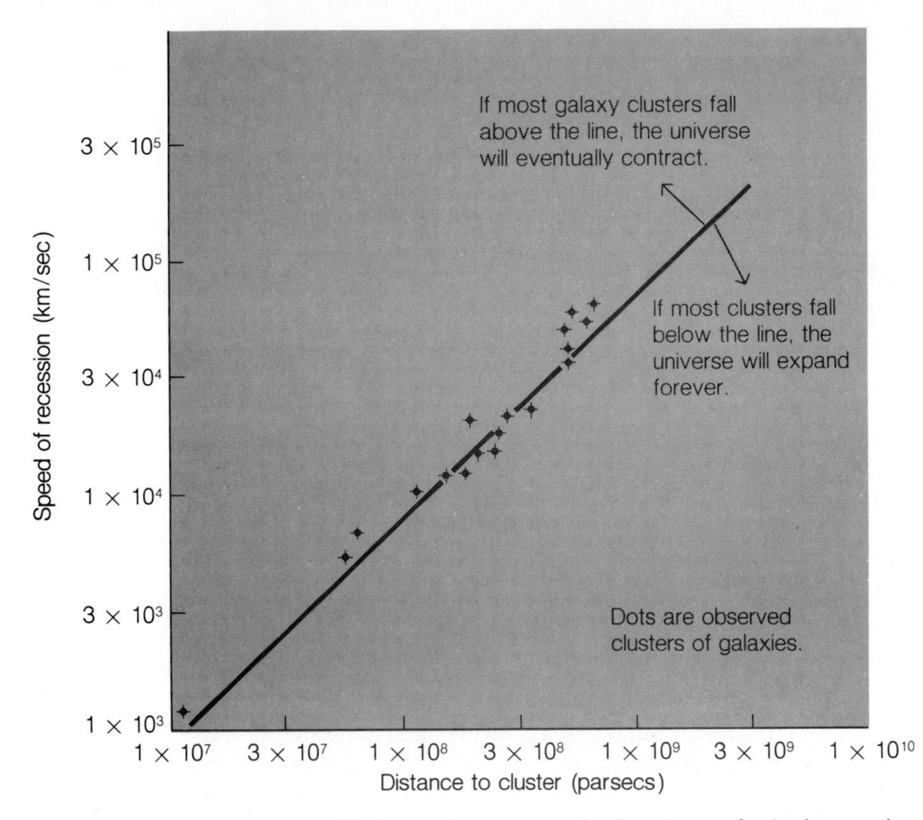

Figure 2.15 According to Hubble's law, a graph showing galaxies' recession velocities against their distances should reveal a straight-line relationship. But when we look at galaxies so far away that we are looking back a significant fraction of the time since the big bang, this is no longer true. The deviations from a straight-line relationship can in theory show whether the universe has enough matter eventually to contract, or whether it will expand forever. In practice, the results of these measurements are still too uncertain to reveal a definitive answer.

mously difficult) with sufficient accuracy, we can check which model, if any, of the expanding universe is correct.

Such an approach offers hope, but not for the immediate future. We may eventually be able to use large telescopes in orbit above the Earth's blurry atmosphere to determine the distances to galaxy clusters with much greater accuracy than we can today. Another line of attack exists to determine which model of the universe is correct: We can measure the average density of matter! If that density now exceeds a certain number, called the "critical density," then theory tells us that the universe must someday cease its expansion and begin to contract. If, instead, the average density of matter now falls below the critical value, then the universe will expand forever.[4]

If we could measure the actual value of the average density of matter in the universe, we would know its fate. We could also determine whether or not the inflationary model, which makes a completely definitive prediction of that density, is valid. The difficulty remains that of finding and measuring the dark matter, which we can do only by indirect means.

Quite possibly, one of these two lines of attack on determining the future of the universe will soon yield fruit. We may then discover whether the universe is finite or infinite, and whether or not it will expand forever.

SUMMARY

During this century, astronomers have learned how to estimate the distances to galaxies, and how to measure their velocities toward us or away form us by using the Doppler effect. These observations have revealed "Hubble's Law": Except for the galaxies that belong to our own Local Group, all clusters of galaxies are receding from us at speeds that increase in proportion to their distances from us. So long as we assume that we have a representative view of the universe, we must conclude that the entire universe is expanding. This expansion began with the "big bang," approximately 15 billion years ago.

The first minutes after the big bang saw a universe of enormously high densities and temperatures. In this cosmic furnace, the universe produced all varieties of elementary particles through tremendous numbers of high-energy collisions. After the first half hour, the basic mixture of particle types that we see today had been established. Photons emerged in huge quantities from the universe's first half hour, because many photon-generating collisions among particles had occurred. These photons then and now have filled the universe with cosmic background radiation. As the universe has expanded, each of these photons has steadily lost energy, as measured by any observer in the universe.

[4]As the universe expands, the average density of matter steadily decreases, but the "critical density" that separates eternal expansion from eventual contraction decreases in just the same way. Hence, if the actual density exceeds the critical density now, it will do so for all time. This is also the case if the actual density falls below the critical density.

Recent decades of observation have revealed that our Milky Way, and probably the entire universe, contains far more "dark matter" than visible matter. The nature of the dark matter remains completely unknown. Its existence has been deduced by the gravitational effect that it produces on the motions of stars moving in orbit around the centers of galaxies such as the Milky Way. The "inflationary" theory predicts that the dark matter should comprise about 99 percent of the mass in the universe.

If the universe ever ceases its expansion, it will do so because it has a sufficiently high density to overcome its current expansion through gravity. The inflationary theory predicts that the actual density of matter exactly equals the critical value of the density that separates eternal expansion from eventual contraction. In order to measure the actual density of matter, and thus to determine whether the universe will ever cease expanding, we must look back in time by observing ever more distant galaxies, whose light has taken still longer to reach us. These observations can reveal how the relationship between the distances and recession velocities of galaxies has changed with time since the big bang, and this will tell us how this relationship will change in the future. At the present time, direct observations of the average density of matter are extremely difficult because most of the matter resides in an invisible form. Furthermore, observations of distant galaxies cannot yet resolve the question of whether the universe will eventually contract.

QUESTIONS

1. Can you find situations on Earth in which the parallax shift plays a role in estimating distances? For which sorts of distances does this method prove most useful? Why does the parallax effect have less usefulness for greater distances?

2. Jupiter has five times the Earth's distance from the sun. Compare the sun's angular size and apparent brightness as seen from one of Jupiter's moons with these two quantities as seen from Earth. Why do angular size and apparent brightness change in different ratios as the distance from the object under observation changes?

3. How do you explain the fact that although the sun emits all types of electromagnetic radiation, including large amounts of ultraviolet, our eyes have evolved to detect only the type of electromagnetic radiation called visible light?

4. Suppose that a photon encounters a hydrogen atom whose electron occupies the atom's third-smallest orbit and knocks the electron into the fifth-smallest orbit. Does the atom gain energy or lose energy as a result of this transition? Explain your answer with reference to the kinetic energy of a photon.

5. If another photon can make a hydrogen atom's electron jump from the third-smallest into the fourth-smallest orbit, compare the wavelength of this photon with the photon described in Question 4. For both transitions (to the fourth-smallest and fifth-smallest orbits), explain what is likely to happen next to the hydrogen atom.

6. On Earth, we notice the Doppler effect for sound waves, which travel at about 1/1,000 the speed of light. Describe some of your experiences with the Doppler effect on the highways of America. Why does the difference in speeds make the Doppler effect easier to detect for sound waves?

7. When astronomers observe the photons that form the "cosmic microwave background," what photon source are they observing? Explain your answer in terms of locating the center of the expansion of the universe.

8. As the universe expands, where does new space come from? Does this question help to show that it is not easy to decide whether space is something—or simply the distance between objects? Does the notion that space is absolutely nothing contradict your intuition that space simply "sits there," unchanged and unchanging, no matter what happens in the universe?

9. How has the "dark matter" been detected, given that it is completely invisible? How does the amount of dark matter affect the future of the universe? Compare the expansion of the universe with the flight of a rocket from Earth, with the rocket launch analogous to the big bang. What forces are at work on the rocket and the matter in the universe? What would cause the rocket to move away from Earth, or the universe to expand, forever? What would cause the rocket to fall back to Earth, or for the universe to begin to contract?

10. The galaxies in the Coma cluster are receding from the Milky Way at 7000 kilometers per second, while the galaxies in the Corona Borealis cluster are receding at 21,000 kilometers per second. If the brightest galaxies in the two clusters have the same luminosity, how would their apparent brightnesses compare?

FURTHER READING

Bartusiak, Marcia. *Thursday's Universe.* New York: Times Books, 1986.

Bernstein, Jeremy. *Three Degrees Above Zero.* New York: Scribner, 1984.

Field, George, and Eric Chaisson. *The Invisible Universe.* Boston: Birkhäuser Books, 1985.

Gribbin, John. *In Search of the Big Bang.* New York: Bantam, 1986.

Harrison, Edward. *Cosmology: The Science of the Universe.* Cambridge: Cambridge University Press, 1981.

Krupp, Edwin. *Echoes of the Ancient Skies.* New York: New American Library, 1983.

Morrison, Philip, and Phylis Morrison and the Office of Charles and Ray Eames. *Powers of Ten: About the Relative Sizes of Things in the Universe.* New York: Scientific American Library, 1982.

Overbye, Dennis. *Lonely Hearts of the Cosmos.* New York: Harper Collins Books, 1991.

Trefil, James. *The Dark Side of the Universe.* New York: Scribner, 1988.

Tucker, Wallace, and Karen Tucker. *The Dark Matter.* New York: William Morrow, 1988.

Weinberg, Steven. *The First Three Minutes.* New York: Bantam, 1976.

COSMOLOGY MARCHES ON

3

Galaxies

Galaxies—each of them containing millions, billions, or (in extreme cases) trillions of individual stars, along with masses of gas and dust—form the basic structural units of the universe, and if we find life in the universe, we shall almost certainly find it within a galaxy. Our own Milky Way, a typical giant spiral galaxy, includes about 300 billion stars in its diameter of 30,000 parsecs (Color Plates 2–4). Within a few billion parsecs of our Milky Way—the region we may call the "known universe"—billions of individual galaxies exist, usually within clusters that each include thousands of members. These galaxies have apparently formed from what was once a featureless and nearly uniform medium, and they have done so during the 15 billion years since the big bang. The study of other galaxies provides some of the most exciting research in modern astronomy, because we still know so little about the ways that they formed, and how some of them produced such exotic objects as quasars—the most energetic of all objects—and giant "halos" of dark matter. In our search for life in the universe, the history of these galaxies and of the stars they contain has paramount importance, because in order for life to exist, it must find the opportunity to evolve under the fairly stable conditions these stars provide during this evolution.

Though the closest civilizations to our own lie (we hope!) within our own Milky Way, we must not overlook the fact that our galaxy represents a single grain of sand in the mainly uncharted ocean that we call the universe. To obtain a true perspective on the question of life in the universe, we should consider the numbers of different types of galaxies, and of the different kinds of stars and clouds of gas within them. This will reveal just how representative our own galaxy is, much as the study of our planetary neighbors in detail demonstrates how representative of planets our Earth may be.

Spiral Galaxies

We on Earth inhabit an outer corner of the Milky Way galaxy, a giant spiral of stars, gas, and dust that resembles the galaxy known to astronomers as NGC 6744 (Fig. 3.1). Since our present technological abilities do not allow us to travel outside our galaxy, we cannot yet obtain an overall view of the Milky Way like the one we have of NGC 6744. Instead, looking from within, we see the regions of our galaxy most crowded with stars as a diffuse band of light, a "milky way" in the heavens (Fig. 3.2). We can nevertheless attempt to map the structure of our galaxy from the inside, despite the difficulties of seeing the galactic forest among the starry trees around us. As a result of such efforts, astronomers can pinpoint the sun's location in the Milky Way: We are close to the plane of symmetry that divides the galaxy's "top" and "bottom" halves, but we orbit the galactic center at a distance of about 8000 parsecs, occupying an outer region of the Milky Way, a sort of galactic suburb with plenty of room among the stars.

Figure 3.1 The spiral galaxy NGC 6744, about 15 million parsecs away, shows complex "spiral arms," outlined by the youngest, hottest, most luminous stars and star-forming regions in the galaxy.

Figure 3.2 The "milky way" that we see in the sky consists of myriad stars that crowd the central plane of our spiral galaxy. Since the solar system also lies within this plane, when we look outward around the sky through the plane, we see far more stars than we do when we look in directions more or less perpendicular to the plane.

Until recently, astronomers had only one means of studying galaxies such as our own. They observed the *visible light* that galaxies produce and analyzed the spectrum of this light. Within the last few decades, however, other spectral regions—including radio waves, infrared, ultraviolet, and most recently, x-rays and gamma rays—have been added to the techniques of astronomical investigation. New observations have brought new discoveries; for example, x-ray observations allow us to penetrate much farther inside the Andromeda galaxy than we can with visible light (Fig. 3.3). Such new observations have greatly increased our understanding of the structure and evolution of galaxies—which naturally leads astronomers to speculate about how much remains to be discovered.

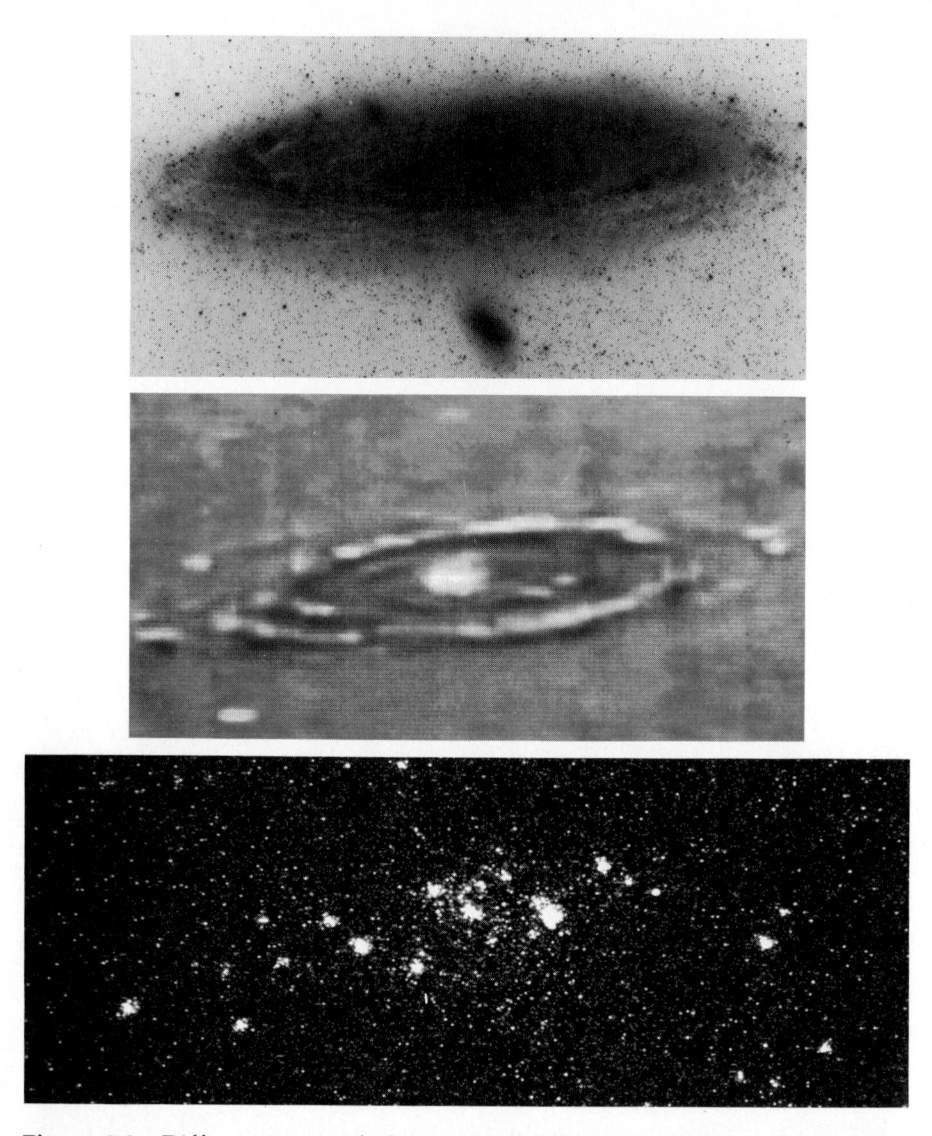

Figure 3.3 Different types of photons reveal distinctly different aspects of a particular object, since they typically arise from different processes. For example, an infrared image of the Andromeda galaxy (center) only roughly matches the visible-light image shown in the top panel (printed as a negative). An enlarged x-ray image of the innermost regions of the Andromeda galaxy (bottom) reveals x-ray sources that astronomers find do not match with any stars in the visible-light photograph (top). The sources of x rays are sites of high-energy activity close to the center of the galaxy.

The Structure of Spiral Galaxies

The structure of spiral galaxies such as the Milky Way exhibits two striking characteristics. First, most of the stars in a spiral galaxy are distributed throughout the *disk* of the galaxy, so that a spiral galaxy's shape is that of a flat plate, with a thickness barely equal to 1 percent of its diameter (Fig. 3.4). Second, within the disk, the most luminous stars are concentrated within the "spiral arms," as we saw in Fig. 3.1. The spiral arms are what make it a "spiral," but the fact that the galaxy has a highly flattened structure is even more important than the existence of the spiral arms. It is also significant that in addition to its disk of stars, each spiral galaxy has a "central bulge" or "nucleus" and a spherical "halo" of older stars. These stars apparently formed before the galaxy had finished its contraction into a disklike shape.

The apparent domination of a spiral galaxy's disk by the spiral arms looks more impressive than it actually is. The spiral arms contain the brightest, youngest stars, stars that have only recently (that is, a few tens of millions of years ago) begun to shine. Though they outline the spiral arms, these most luminous stars are useless as potential power sources for life on nearby planets, because the stars will burn themselves out in less than a hundred million years. This does not allow time for life to develop on the planets that might orbit these bright, evanescent beacons. The older stars in the disk of a spiral galaxy—our sun, for example—appear all through the disk, both between and within the spiral arms. Thus the disk actually has a fairly even density of stars, and the spiral structure is a sort of light frosting of young, hot, extremely luminous stars.

Why do the youngest, brightest stars appear only within the spiral arms of a spiral galaxy? These galaxies owe the existence of their spiral arms to a

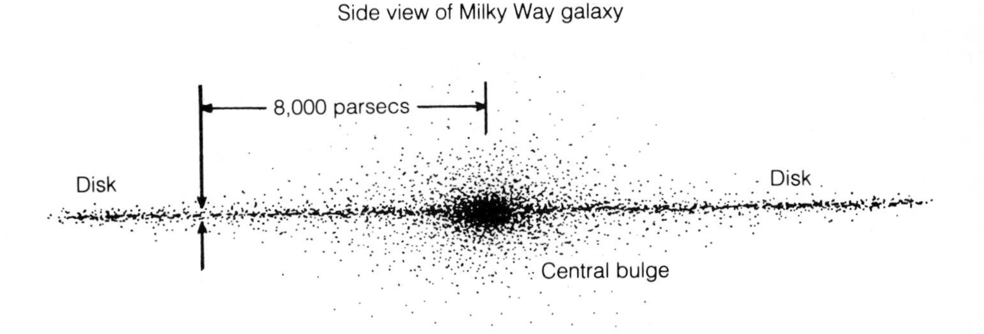

Figure 3.4 Most of the stars in a spiral galaxy such as our Milky Way lie within the thin galactic disk that surrounds the "central bulge" of stars. (See also Color Plates 1–4).

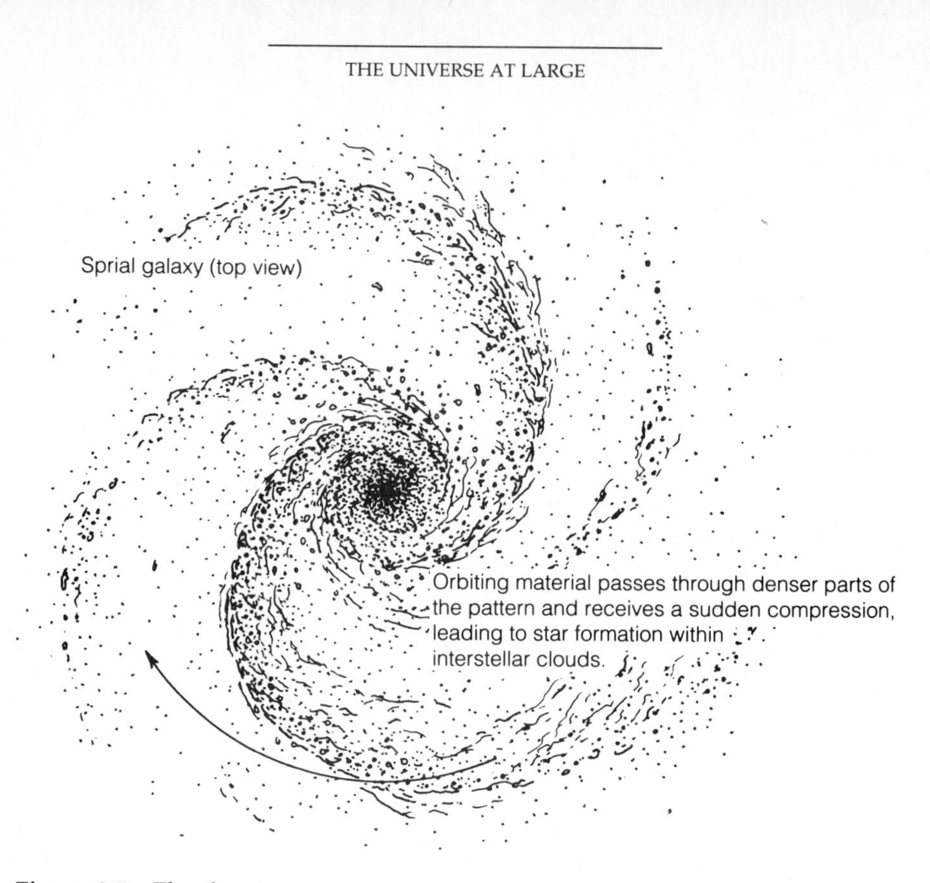

Figure 3.5 The density-wave pattern in a spiral galaxy consists of alternating regions of higher and lower density. The pattern rotates around the galactic center in the same direction as the stars orbit, but more slowly than any of the stars move in orbit.

rotating wave pattern that circles their centers (Fig. 3.5). The pattern consists of regions of alternating higher-density regions (the arms) and lower-density regions (interarm regions). This rotating pattern, in which the arms have a somewhat greater density of matter than the interarm regions, resembles the alternating ripples in water waves, except that we must replace the water by the stars plus the diffuse gas spread among the stars. Also, the water-wave pattern spreads outward from a disturbance in the water, whereas the density-wave pattern rotates around and around the galaxy, as Fig. 3.5 illustrates.

Even though the density of matter within the galaxy's spiral arms does not exceed the density outside the arms by much, this modest increase in density has important effects on the clouds of gas in the disk. The increase in density within the arms means that the average pressure must increase on the gas clouds that wander through interstellar space. When interstellar clouds enter the denser part of the wave pattern, the sudden increase in the pressure around them provokes the fragmentation of the clouds into small-

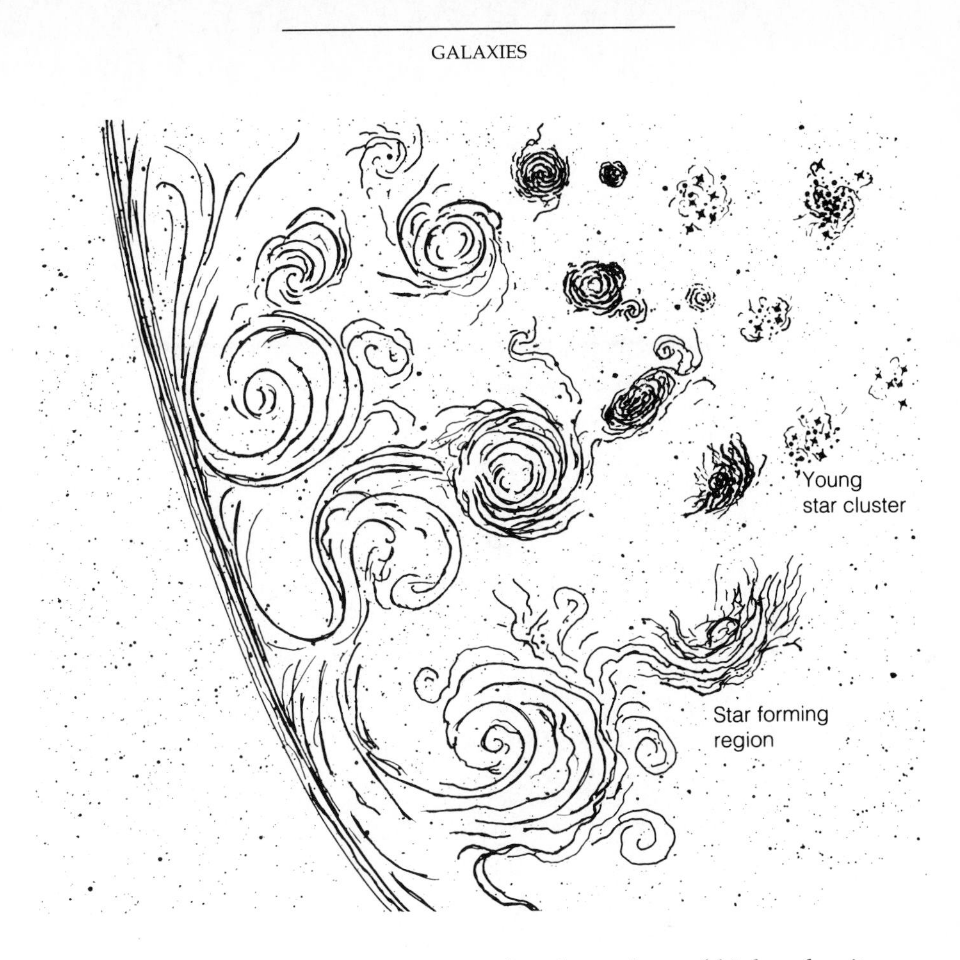

Figure 3.6 When orbiting matter overtakes the regions of higher density, gas clouds tend to be squeezed to a higher density, which triggers star formation in some of the clouds. This gives birth to the young stars and star-forming regions that are the mark of a galaxy's spiral arms.

er clumps of matter that can condense into stars through their own gravitation (see p. 81). As a result of this process, young stars are born inside spiral arms, which therefore provide a giant cosmic nursery. The youngest stars of all are closest to the leading boundary between the arm and interarm regions (Fig. 3.6).

Both the stars within a spiral galaxy and the gas clouds from which they form orbit the galactic center, much as the planets in our solar system orbit the sun (Fig. 3.7). Each of these stellar orbits reflects a balance for that particular star, between the combined gravitational pull from the material closer to the galactic center and the star's momentum, or tendency to move in space in a straight line. Most of the stars in a spiral galaxy have nearly

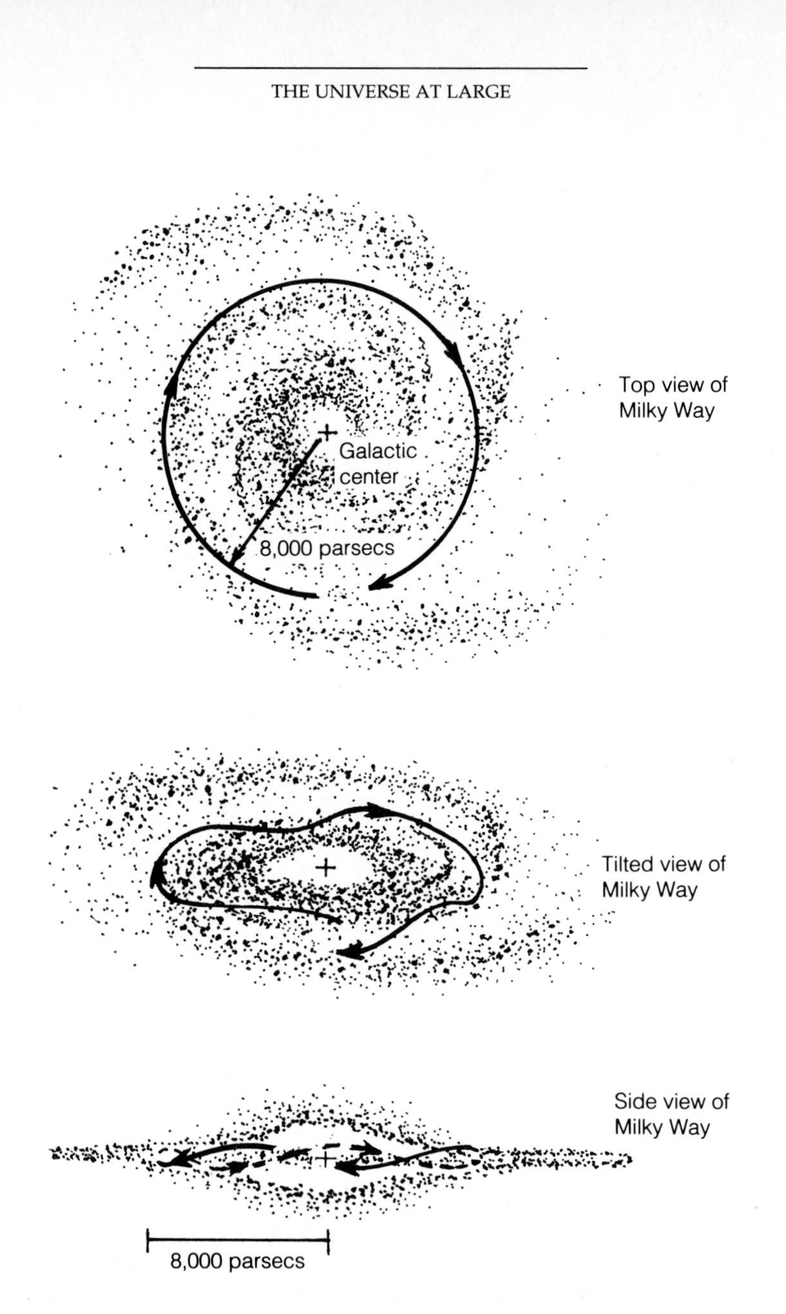

Figure 3.7 A typical star within the disk of a spiral galaxy, such as our own sun, moves in a nearly circular orbit around the galactic center, bobbing up and down several times as it does so. Each up-and-down bob carries the star only a few dozen parsecs above and below the galaxy's median plane, while the basic orbit has a diameter of many thousand parsecs and requires hundreds of millions of years to complete.

circular orbits around the center, as our sun does in the Milky Way. The sun takes 240 million years to complete one trip around the Milky Way. As it does so, like most other stars, the sun bobs up and down, reaching a maximum distance of about 80 parsecs above and below the galaxy's median plane (see again Fig. 3.7).

Dark Matter in Spiral Galaxies

During the past two decades, studies of the motions of stars in giant spiral galaxies such as our Milky Way have revealed a startling fact: The bulk of the mass of these galaxies lies beyond the halo of older stars, and we simply do not know what form this mass takes (Fig. 3.8)! Through basic detective work, astronomers have deduced that most of the matter in spiral galaxies resides not in stars but rather in *dark matter* of unknown form. The evidence

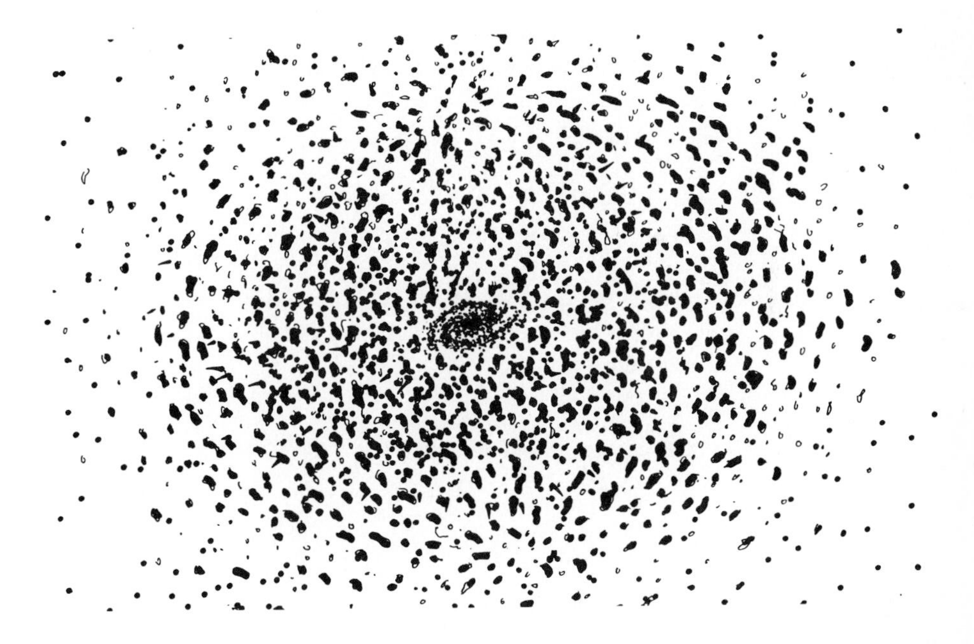

Figure 3.8 This diagram schematically indicates the existence of a halo of "dark matter" that surrounds the Milky Way and similar galaxies. Since the dark matter cannot be seen—and since we don't know what comprises the dark matter—any such diagram must be taken with a large dose of imagination. It would be more accurate to draw blank space around the visible galaxy at the center!

for this dark matter consists of the gravitational force that the dark matter exerts upon the matter that astronomers *can* see.

How does such astronomical detection proceed? First, we study the Milky Way and other giant spiral galaxies in detail, attempting to determine how rapidly the stars at various distances from the centers of these systems are moving in orbit (Fig. 3.9). Such studies were performed during the 1970s by Vera Rubin and her collaborators, and they revealed a startling fact. The orbital speeds at large distances from the center do not decrease with increasing distance, as astronomers expected. Their expectation relied on the fact that the amount of visible matter—the matter in stars—clearly diminishes sharply at distances greater than about 5000 to 15000 parsecs from the center. But the study of stellar motions revealed quite a different distribution of matter: To explain the observed motions of stars, most of the galaxy's matter must be located *outside* the visible disk and halo of stars in a spiral galaxy. The conclusion follows that most of a giant spiral galaxy—as much as 90 or 95 percent of its total mass—lies not in the stars but in dark matter surrounding nearly all the visible stars and star-forming regions.

The existence of enormous amounts of dark matter in giant spiral galaxies has now become an accepted fact among experts. This result does not prove that the entire universe consists mainly of dark matter, though it

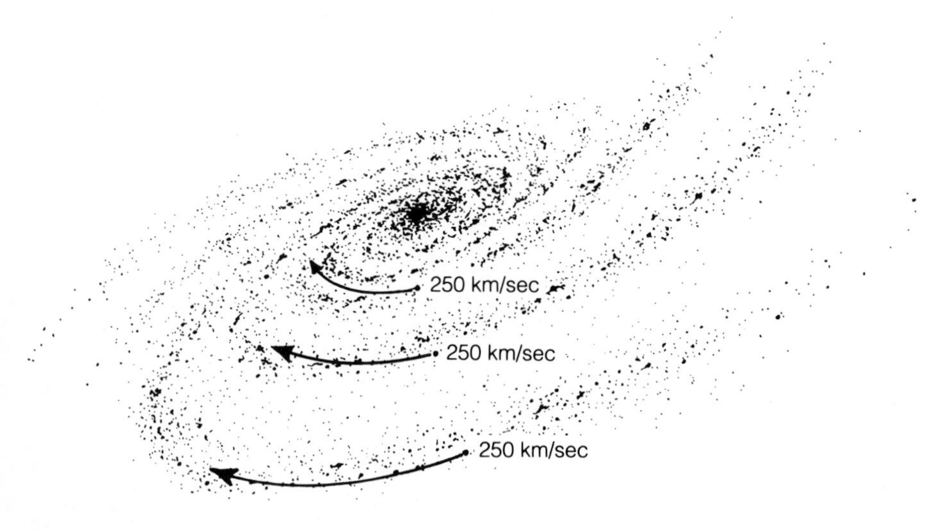

Figure 3.9 From studies of the speeds at which stars from the center of the galaxy orbit the center, astronomers have deduced the existence of the dark matter halo surrounding giant spiral galaxies such as our Milky Way. The fact that the velocities remain constant at large distances from the center reveals the presence of enormous amounts of dark matter. Such dark matter halos contain from 10 to 100 times more mass than the total contained in the galaxy's stars.

strongly suggests such a conclusion. Nor does it prove that the dark matter has sufficient abundance to ensure that the universe will eventually contract. In order for this to occur, dark matter must form not simply 90 percent or even 95 percent of the mass in the universe. Eventual contraction requires that fully 99 percent of all matter be invisible dark matter, whose form astronomers are now striving to determine. Hence the results from spiral galaxies, though fantastic in their own right, are simply tantalizing and not conclusive in what they tell us about the universe as a whole.

Elliptical Galaxies

Elliptical galaxies are the other major type of galaxies. Unlike spirals, elliptical galaxies have no flattened disk of stars; instead, their stars spread into an almost spherical configuration, sometimes elongated into an ellipsoid (see Fig. 3.10). Because elliptical galaxies have no complex structural pattern comparable to spiral arms, one elliptical tends to resemble another. Elliptical galaxies are nearly as numerous as spirals, and a giant elliptical can contain as many stars as the largest spiral galaxy, or even more. Since the motions of stars in an elliptical galaxy are more chaotic than those in a spiral galaxy, astronomers cannot deduce so easily how much dark matter elliptical galaxies contain.

Elliptical and spiral galaxies differ in another way that is significant in the search for life in the universe: the amount of interstellar gas and dust that they contain. Elliptical galaxies contain almost no interstellar gas and

Figure 3.10 A giant elliptical galaxy such as M49 shown here contains several hundred billion stars, as a giant spiral galaxy does, and has about the same size as a giant spiral—but quite a different shape. The stars in an elliptical galaxy all orbit the center, but they do not all move in nearly the same plane, as the stars in a spiral galaxy do.

Figure 3.11 Between the stars in a spiral galaxy are large amounts of interstellar gas and dust, often concentrated into "clouds" of much higher density than average. The gas (mostly hydrogen and helium) is transparent, but the dust grains absorb visible light and produce the dark lane of absorbing matter visible in this edge-on photograph of the spiral galaxy NGC4565.

dust, whereas the disk of a spiral galaxy contains significant amounts—a few percent by mass (Fig. 3.11).

Interstellar gas and dust are important as the raw material from which stars are born. Equally important in the search for life, this interstellar matter becomes gradually enriched in elements other than hydrogen and helium by material that is ejected from exploding stars (see Chapter 6). The fact that elliptical galaxies contain little or no interstellar matter implies that they ceased to form stars long ago. Supporting this conclusion, astronomers find that ellipticals contain no stars younger than a few billion years, whereas spirals have stars only a few million years old, and others that are still being born or that will be born in the future. Although the origin of life may follow the birth of a star (and of any planets around it) by a long time (see Chapter 8), we may nevertheless draw a general conclusion: Elliptical galaxies are unlikely to include many places where life has recently appeared. In contrast, spiral galaxies should offer a much wider variety of sites, ranging from planetary systems many billion years old, comparable to our five-billion-year-old solar system, down to planets that are forming even now. This means that our Milky Way, a typical giant spiral, should offer a host of different places, loaded with different amounts of the key elements for life, that await investigation by a civilization capable of exploring the galaxy.

When astronomers study the motions of individual galaxies within a cluster, they find a striking result, analogous to the result of studying the motions of stars within a galaxy. Galaxy clusters apparently contain far more dark matter than the matter we can account for in stars. We know this because the cluster members move more rapidly than would be expected if stars provided the bulk of the matter. If large amounts of dark matter did not exist, the galaxies would have escaped from the cluster during the time since they formed, or would now be moving more slowly. Thus galaxy clusters provide evidence for enormous amounts of dark matter, though we do not yet know whether most of this dark matter lies *within* the individual galaxies or *among* them—that is, spread out diffusely within the cluster of galaxies.

Irregular Galaxies

Besides spirals and ellipticals, a third type of galaxy exists, the irregular galaxy. Irregular galaxies are unlike either spirals or ellipticals, which together account for 90 percent of all the galaxies that we see. Irregulars have neither the disklike flattening of spirals nor the smooth ellipsoidal shape of ellipticals. Two well-known irregular galaxies appear in Fig. 3.12: the

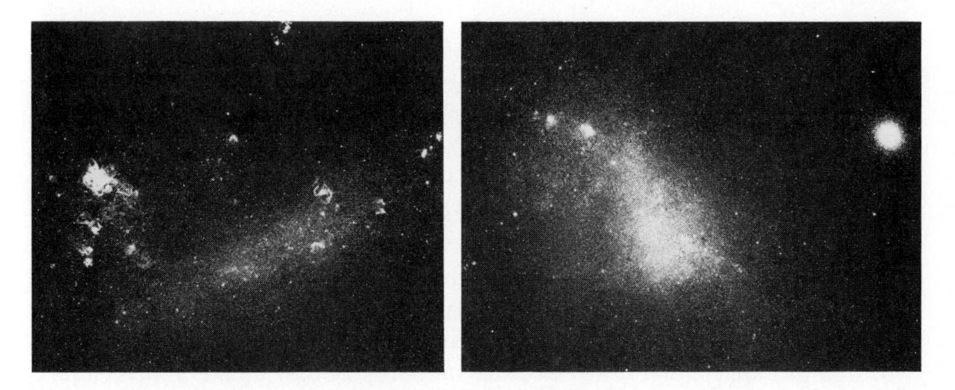

Figure 3.12 The Milky Way's two closest neighbors are satellite galaxies called the Large Magellanic Cloud (left) and the Small Magellanic Cloud (right), two irregular galaxies that each contain 5 to 10 billion stars.

satellites of our own galaxy called the Large and Small Magellanic Clouds. Irregular galaxies have the highest fraction of the galaxy's mass in the form of gas and dust strewn among their stars—as much as 20 to 50 percent of the total, as compared with a few percent within the disk of a spiral galaxy and less than 1 percent in an elliptical galaxy. Since irregular galaxies therefore have the largest amount of star formation occurring within them, astronomers study them in order to learn more about the processes that formed all the stars that shine in the skies of night.

The Formation of Galaxies

Astronomers think that all of these galaxies, except perhaps for a few of the irregulars, have existed for billions of years. Galaxy formation apparently began soon—less than a few billion years—after the big bang, 15 to 20 billion years ago. The details of the galaxy-formation process remain a matter of debate among astronomers. It does seem evident that protogalaxies (galaxies in formation) were larger than galaxies are now, and contracted as the result of the self-gravitation of the matter in each protogalaxy. Astronomers have found many examples of colliding galaxies, and it is possible that such collisions may accelerate the rate of star formation as gas clouds collide. Some theories suggest that elliptical galaxies are not formed directly, but rather emerge from collisions between spiral galaxies. Much more research must be done before we can answer the simple question, Just how did galaxies form?

Star Clusters

Whatever the details of galaxy formation may be, the first objects to form as a protogalaxy contracted toward its present sizes were its "globular star clusters," groups of many thousands of stars only a few parsecs in diameter (Fig. 3.13). Globular clusters are subunits that belong to galaxies, but because they began to form when the protogalaxy was larger than the galaxy it became, many of these clusters orbit the galaxy's center on highly elongated trajectories, sometimes reaching enormous distances from the center (Fig. 3.14). The gas in the protoclusters that became globular clusters must have undergone further fragmentation into individual protostars (stars in formation), since today we see only stars, and no interstellar gas, within a globular cluster. Most of the stars in globular clusters rank among the oldest known stars, and none of them is as young as the bright stars that outline the arms of a spiral galaxy. These stars are relatively "poor" in elements heavier than helium, another sign of their great ages (see page 140).

Figure 3.13 The globular cluster M13, located about 8000 parsecs away in the direction of the constellation Hercules, contains about half a million stars and has a diameter of about 10 parsecs.

Still, many of these stars have greater luminosities than the sun. If we lived on a planet that orbited a star in a globular cluster, the concentration of stars would provide us with dozens of stellar neighbors, each shining as brightly as the full moon on Earth!

Within the flat disk of spiral galaxies, we find a different type of star cluster, the "open" or "galactic" cluster. Open clusters, about the same size as globular clusters (a few parsecs in diameter) contain only a few hundred stars, or at best a few thousand, rather than the hundreds of thousands or millions of stars that fill a globular cluster. The fact that open clusters are found only in the disk of a spiral galaxy shows that these clusters formed only after the galaxy contracted to its disklike shape. (Elliptical galaxies contain no open clusters, although they have many globular clusters within and around them.) Open clusters do not persist as clusters for the full lifetime of a galaxy (about 10 billion years, so far, in the case of the Milky Way). Unlike globular clusters, which contain enough mass to hold themselves together as separate units for billions of years, open clusters contain too few stars to remain compact over the entire life of a galaxy. After a dozen or so galactic rotations, each of which takes a couple of hundred

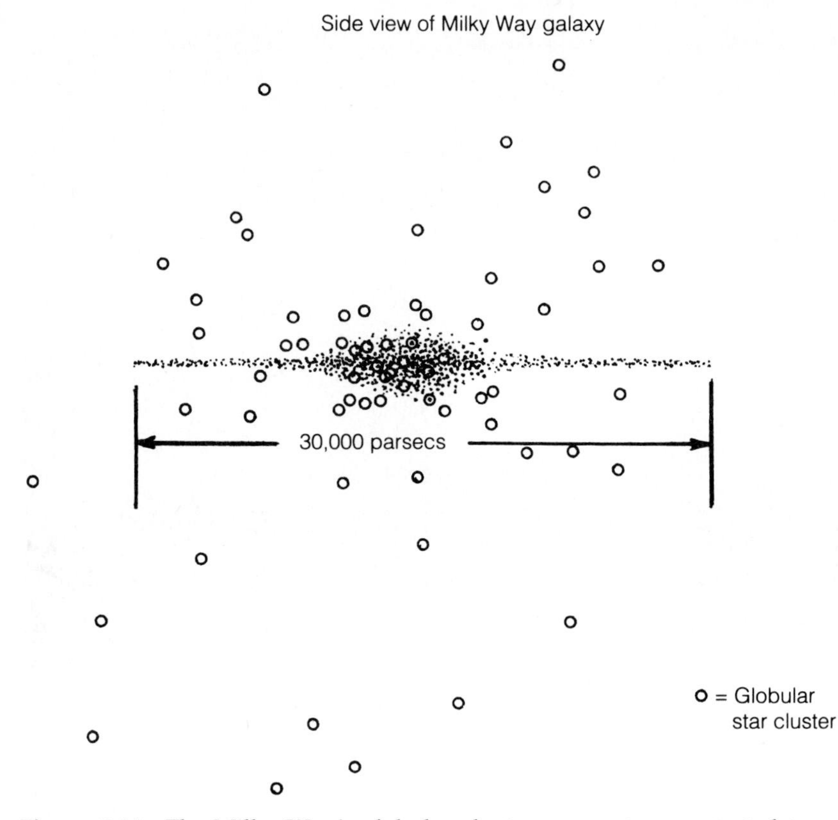

Figure 3.14 The Milky Way's globular clusters are not concentrated toward the plane of the galaxy. Instead, they appear primarily in the central bulge and in a spherical distribution that extends far above and below the Milky Way's median plane.

million years, the stars in an open cluster will disperse among the other stars in the disk of the galaxy. Our sun, for instance, probably formed along with several hundred other stars as part of a loose association, but we have no idea where in the galactic disk the sun's brothers and sisters may be now. They are certainly much farther from us than our closest stellar neighbors.

A typical open cluster that now exists, such as the Pleiades (Fig. 3.15 and Color Plate 6), has an age measured in tens or hundreds of millions of years, and shows some remnants of the gas and dust from which the individual stars have condensed. Since we think that life requires at least a few hundred million, if not billions, of years for its origin and development, young open star clusters do not appear to be good places to search for extraterrestrial life. All the stars in the Pleiades are, we think, far too young

Figure 3.15 The Pleiades are an open or galactic cluster in the constellation Taurus, about 120 parsecs from the solar system, containing only about a hundred stars and spanning a region about 5 parsecs across.

for life to have begun on any planets that may orbit around them. Our hopes for finding life rest not with the young, bright members of the stellar population but with the quieter, older stars (but not so old that they lack heavy elements) that predominate in a galaxy such as our own.

Radio Galaxies

Among the millions upon millions of galaxies accessible to our telescopes, we find a few exceptional ones that produce great quantities of radio waves. Since stars emit only a small fraction of their nuclear-generated energy as radio waves, and since galaxies consist mainly of stars, in these cases we must be dealing with peculiar galaxies indeed. Astronomers give the name "radio galaxies" to those galaxies that radiate as much (or even more) energy per second in radio photons as they do in visible-light photons. To see how exceptional this is, compare a radio galaxy with the Milky Way, a typical giant spiral galaxy, which produces only one-millionth as much radio energy as visible-light energy each second.

Radio galaxies apparently owe their tremendous outflow of radio photons to violent events occurring inside them. By studying the details of the photons emitted by these galaxies, astronomers have concluded that the radio waves from most radio galaxies arise from the "synchrotron process." This name, bestowed in honor of the particle accelerators where scientists first observed the process in detail, describes the fact that photons are produced whenever charged particles move at nearly the speed of light in a magnetic field and accelerate, that is, change their velocity, or their direction of motion, or both. All four requirements—charged particles, velocities close to the speed of light, magnetic fields, and acceleration—must be present. If they are, then photons are produced, and produced with a characteristic spectral distribution that can be recognized, even billions of parsecs away.

The production of photons through this synchrotron process robs the charged particles of some energy. Without an additional energy input, the charged particles will soon slow down and will cease to produce photons by the synchrotron process. Hence some sort of violence within radio galaxies has accelerated the particles to these enormous velocities (and continues to do so), even though we cannot specify the cause of this violence.

The radio photons from radio galaxies typically arise from large regions far from the center of the galaxy, and indeed often beyond the galaxy's visible stars (Fig. 3.16). The most widely accepted theory of radio galaxies suggests that the centers of the galaxies are places where particles receive tremendous accelerations. Somehow, certain peculiar galaxies shoot fast-moving particles in opposite directions. As the particles move outward, they encounter intergalactic gas. To clear this gas out with fast-moving particles becomes progressively more difficult, so a "snowplow" effect slows the particles, which tend to pile up near the outer edge of the clouds. The amount of radio emission therefore also peaks at the edges of "radio clouds" such as those shown in Fig. 3.16.

Quasars

Quasars, or quasistellar radio sources, appear to be the most distant objects known and the most powerful sources of photon emission. Quasistellar radio sources owe their name to the fact that most of them emit large amounts of radio waves, and the first ones to be found emit simply stupendous amounts of energy. When astronomers located these radio sources accurately on visible-light photographs, they found points of light similar to

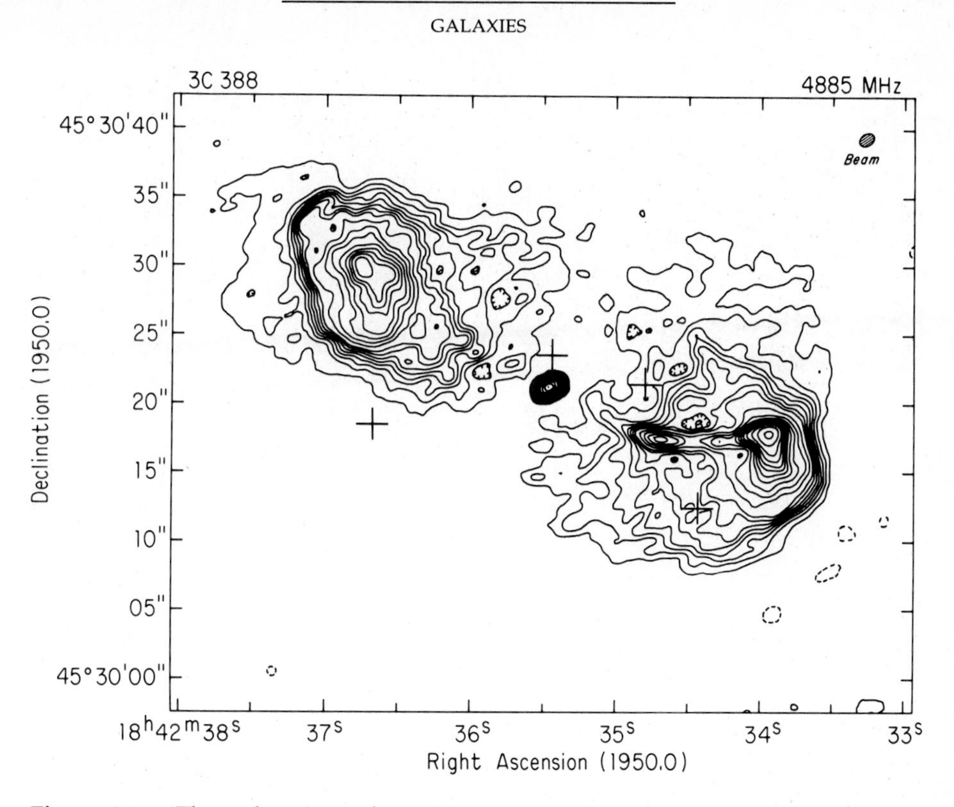

Figure 3.16 The radio waves from the radio galaxy 3C 388 arise primarily from two enormous regions, symmetrically located on either side of the visible galaxy, but hundreds of times larger.

stars (Fig. 3.17). The light from these quasistellar images, spread into its spectrum of colors, astonished astronomers by showing none of the spectral features familiar to them from their study of stars. After a period of confusion, the apparent explanation dawned on the astronomers Maarten Schmidt and Jesse Greenstein: The light from the first two quasars to be discovered had such large redshifts that familiar spectral features had been shifted all the way from the yellow region of the spectrum into the red. The redshifts for these two quasars, interpreted as the result of the Doppler effect, revealed recession velocities equal to 15 and 30 percent of the speed of light!

By using Hubble's Law, astronomers could calculate that these velocities imply distances to the two quasars of 900 million and 1.8 billion parsecs. This put the quasars as far away as the most distant galaxies, yet the apparent brightnesses of the quasars equaled those of some large galaxies 100 times closer to us. This implies that the quasars have 10,000 times the

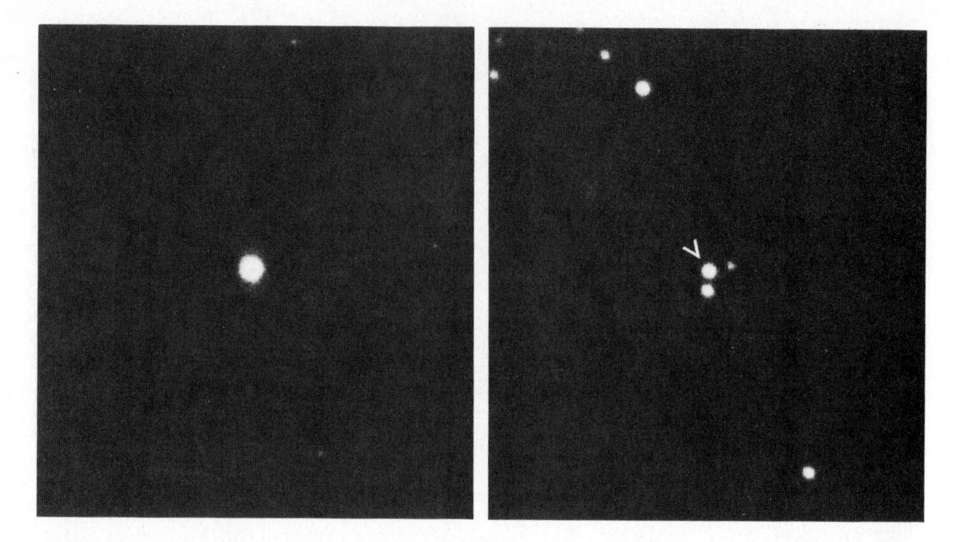

Figure 3.17 Quasars such as 3C 48 (left) and 3C 147 (right) appear as pointlike objects. In fact they are not stars but immensely energetic objects, hundreds of millions or billions of parsecs away.

luminosity of large galaxies! By now, quasars have been detected that are receding from us at more than 90 percent of the speed of light, and have distances greater than 12 billion light years. Even more mysterious, it seems clear that quasars are smaller than galaxies. Quasars usually appear to be point sources of light, while even the most distant of galaxies appear as fuzzy, extended blobs of light (Fig. 3.18). More complex arguments have convinced astronomers that most quasars have diameters no larger than the distance from the sun to Alpha Centauri!

Astronomers have now detected infrared emission and x rays from quasars, and they have found that some quasars emit even more energy per second in these types of photons than they do in visible light or radio waves. When we add all the observed forms of photon energy, we find that some quasars produce 100,000 times more energy per second than a giant galaxy does, provided that our estimate of the quasars' distances from us is correct.

No one knows for certain how quasars manage to emit so much energy from such small volumes, or why their tremendous energy output varies in a relatively brief time. Quasars may represent some early stage in the formation of galaxies, since their enormous distances from us imply that we see them as they were billions of years ago. The most popular models for quasars involve one of the most impressive results of astronomical theorizing: an "accretion disk" of matter spiraling into a "supermassive black hole."

Figure 3.18 Even a distant galaxy appears as a fuzzy blob rather than as a single point of light. This cluster of galaxies lies about 1 billion parsecs away, and we see the galaxies as they were more than 3 billion years ago. All the objects in the photograph that are not perfectly round points of light are galaxies in this cluster.

Accretion Disks and Supermassive Black Holes

To astronomers, an "accretion disk" consists of matter moving in ever-tightening orbits around a strong source of gravity. A "black hole" is an object with such strong gravity that nothing—not even light or any other type of massless particle—can escape from within a certain critical distance of the black hole's center. That distance equals 3 kilometers times the mass of the black hole in units of the sun's mass. An object with a mass equal to the sun's will become a black hole if all the mass somehow packs itself within 3 kilometers of the center. Then nothing can escape from within that distance, though material farther away has at least a theoretical chance of escaping before it passes inside the critical distance.

Black holes with masses comparable to the sun's may be one endpoint of stars' lives, as we discuss in Chapter 6. But for quasars we must imagine still more fantastic objects. Consider not one solar mass but *one billion solar masses* of matter. If this matter concentrates within a radius of 3 billion kilometers, it must become a black hole—a "supermassive black hole," in astronomers' parlance. Then nothing can escape once it approaches within 3 billion kilometers of the black hole's center (Fig. 3.19).

Now 3 billion kilometers approximately equals the radius of Uranus's orbit around the sun. In other words, a supermassive black hole has about the same size as the solar system (which contains only one solar mass—the sun's—rather than a billion). Such an object will attract matter toward it, but the matter falling into the black hole will most likely not fall straight inward. Instead, the matter will swirl in spiral orbits, ever closer to the black hole, before it finally passes inside the critical distance and is seen no more.

But before that happens, the infalling matter forms an "accretion disk" around the black hole, and the accretion disk may be not merely visible but spectacularly so. As the matter spirals inward, it moves more and more

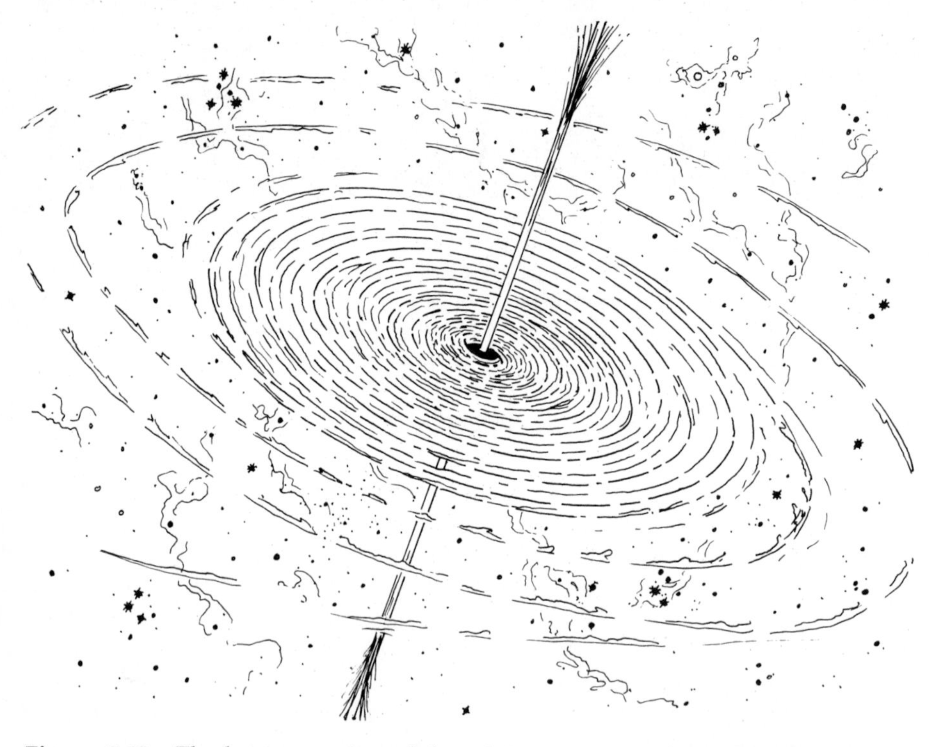

Figure 3.19 The best current model explaining quasars hypothesizes a supermassive black hole, around which a disk of material moves in spiral orbits, eventually falling into the black hole. Within this "accretion disk," material heated by collisions reaches enormous temperatures and emits great amounts of x rays, ultraviolet, visible light, infrared, and radio waves.

rapidly, and the density of matter rises higher and higher as it approaches closer to the critical distance. Collisions among particles in the swirling matter will occur at enormous velocity. These collisions heat the matter in the accretion disk and make it glow, most strongly in its inner regions (but not at its actual center, which remains black).

Quasars may well turn out to be accretion disks of matter spiraling into supermassive black holes. They may represent an early stage in the lifetime of a giant galaxy, in which a supermassive black hole has formed and the matter nearby, pulled inward by gravitation, shines brightly on its way to oblivion. This stage may end when the black hole has consumed all the matter relatively close to it, leaving only material far from the black hole, moving in "safe" orbits that can persist for billions of years. If this model proves correct, then quasars should die down once they exhaust the supply of "food" for their black holes. Every giant galaxy, including our own, may contain a once-active supermassive black hole at its center, relatively starved of the matter that made its accretion disk a quasar billions of years ago.

Could Quasars Be Intergalactic Beacons?

If we let our imaginations roam still more freely, we might imagine that quasars would make excellent beacons, with which a civilization could signal its existence across billions of light years of distance. But we have no evidence that this is the case, and we must take note of the fact that a quasar would be an immensely wasteful way (in energy terms) to attract attention. As we will discuss in Chapter 18, radio waves of a carefully bounded set of wavelengths and frequencies offer a way to attract attention, and to communicate information, at a tiny fraction of the energy expenditure of a quasar. That is why we use radio and television broadcasting at specific wavelengths on Earth, rather than employing signal fires or simply trying to drown out other radio signals at all wavelengths. For now, it seems safe to conclude that quasars represent a natural rather than an artificial astronomical puzzle—a conclusion that grows much stronger from the fact that we cannot begin to understand how a civilization would make a quasar.

SUMMARY

Fifteen billion years after the big bang, matter no longer spreads evenly through space, as it did in the early universe. Instead it appears in clumps called stars, which themselves clump into galaxies and galaxy clusters.

Such clumps have apparently bound themselves together through their self-gravitational forces early in the history of the universe. They then contracted toward their present sizes as the rest of the universe continued its expansion.

Spiral galaxies, which number about half of all galaxies, show a disk-like distribution of matter, in which the youngest, brightest stars outline the distinctive spiral arms. Elliptical galaxies, the second major galaxy type, are far less flattened than spirals and have apparently turned all of their original gas and dust into stars. Spiral galaxies still have a few percent of the mass in the disk in interstellar gas, while irregular galaxies, with no recognizable structure, have 20 to 50 percent of their mass in interstellar matter. We now know that spiral galaxies, and quite likely elliptical galaxies as well, have enormous halos of "dark matter" surrounding their visible distributions of stars.

In spiral and elliptical galaxies, and in irregular galaxies as well, a small fraction of the stars belong to compact "star clusters." A spiral galaxy contains two distinctly different types of star clusters. Globular star clusters represent the first parts of the galaxy to form as the protogalaxy contracted; their immense, elongated orbits can carry them much farther from a galaxy's center than most stars ever go. In contrast, open clusters are found only in spirals and irregulars, not in elliptical galaxies, and always appear in spiral galaxies near the plane of symmetry. Open clusters usually consist of young stars, such as those in the Pleiades, and have only a few hundred members rather than the million or so stars in a globular cluster.

Quasars—quasistellar radio sources—appear as points of light on photographs, in contrast with the fuzzy blobs that represent galaxies; they produce vast quantities of infrared and visible light, and of x rays, in addition to their radio emission. The visible-light spectra of quasars show the largest redshifts yet observed. If these redshifts arise from the expansion of the universe, then quasars must be the most distant objects detected so far, as well as the most powerful sources of photon emission. A quasar may draw its immense luminosity from matter falling into a "supermassive black hole" at its center, with the matter heating through high-energy collisions as it spirals inward towards oblivion.

QUESTIONS

1. Why do galaxies cluster together instead of appearing to be strewn evenly through space? How does this contrast with the overall appearance of the universe in the years immediately following the big bang?

2. In what ways do spiral and elliptical galaxies resemble one another? What are the chief points of difference between these two galaxy types?

3. Why do the youngest stars in a spiral galaxy always appear in the galaxies' spiral arms? If these stars have lifetimes of only a few million years, how does the spiral pattern persist through hundreds of millions, even billions of years?

4. What are the differences between the two major types of star clusters in the Milky Way galaxy? Do similar star clusters appear in elliptical galaxies?

5. The galaxies M87 and NGC 7793 have almost the same apparent brightness, but M87 has almost four times the distance of NGC 7793. How do the luminosities of the two galaxies compare?

6. Consider the light reaching Earth from the Orion Nebula (1600 light years away), the center of the Milky Way (25,000 light years away), the Andromeda galaxy (2 million light years away), and the quasar 3C 9 (8 billion light years away). What was going on in our vicinity at the time that the light left these four objects?

7. A snowflake falling to Earth carries a kinetic energy of about one erg. In contrast, each square meter of a radio antenna pointed at a quasar receives an energy of about one-billionth of an erg per second. How large an antenna would you need to build in order to receive one-thousandth of the energy of a snowflake each second from the quasar? With an antenna 100 meters across, how long would you need to wait to collect the energy of a falling snowflake?

8. The quasar 3C 147 shows a redshift in its visible-light spectrum of 0.55; that is, all the wavelengths in the spectrum are 55 percent longer than they would be in a terrestrial laboratory. How have the frequencies and energies of the light in the quasar's spectrum changed?

9. The mathematical relationship between the observed photon energy and the original photon energy for the Doppler effect is

$$\frac{\text{Observed energy}}{\text{Original energy}} = \sqrt{\frac{1-(v/c)}{1+(v/c)}}$$

in which v is the velocity of recession and c is the velocity of light. What recession velocity does the quasar 3C 147 have, if we assume that its redshift arises from the Doppler effect?

FURTHER READING

Disney, Michael. *The Hidden Universe.* New York: Macmillan, 1984.

Ferris, Timothy. *Galaxies.* New York: Harrison House, 1987.

Friedman, Herbert. *The Astronomer's Universe.* New York: Norton, 1990.

Gale, George. "The Anthropic Principle." *Scientific American* (December 1981).

Hartmann, William and Ron Miller. *Cycles of Fire.* New York: Workman Press, 1987.

Hodge, Paul. *Galaxies.* Cambridge, Mass.: Harvard University Press, 1986.

Preston, Richard. *First Light: The Search for the Edge of the Universe.* New York: Atlantic Monthly Press, 1987.

Rubin, Vera. "Dark Matter in Spiral Galaxies." *Scientific American* (June 1983).

Trefil, James. *The Dark Side of the Universe.* New York: Scribner, 1988.

Tucker, Wallace, and Karen Tucker. *The Cosmic Inquirers.* Cambridge, Mass.: Harvard University Press, 1986.

Tucker, Wallace, and Riccardo Giacconi. *The X-Ray Universe.* Cambridge, Mass.: Harvard University Press, 1985.

4

Interstellar Gas and Dust

Strewn among the stars in spiral and irregular galaxies lie vast quantities of gas and dust, often clumped into interstellar clouds. All the stars and planets once formed from such clouds. Even today, stars still form within our galaxy; astronomers have detected the infrared emission from stars in their final formation stages, and from disks of material around such stars that may be in the process of forming planets (Fig. 4.1). The great gas clouds,

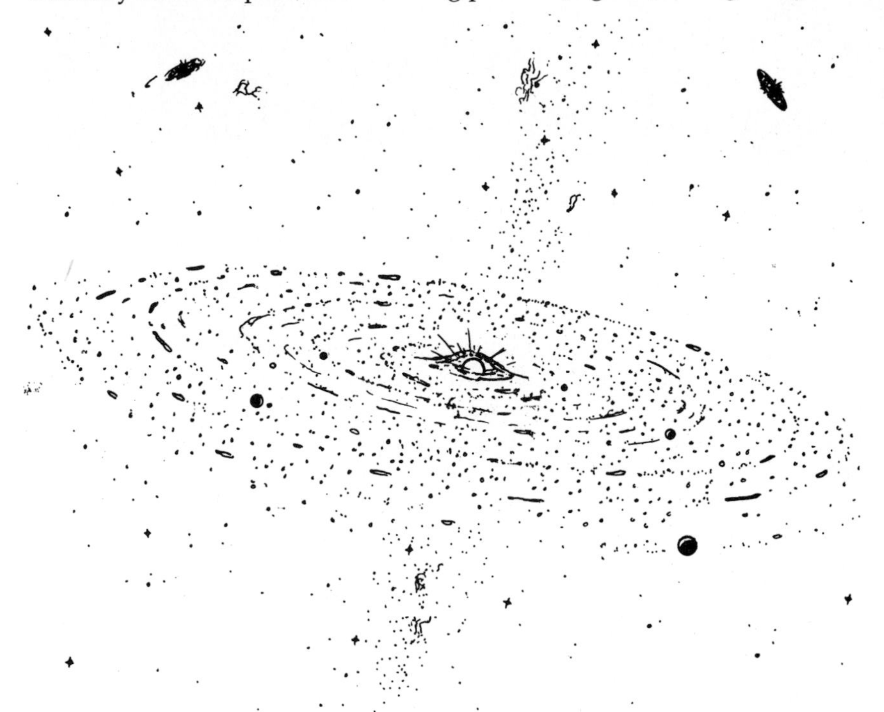

Figure 4.1 Planets such as those in the solar system apparently formed from a flattened disk of gas and dust in orbit around a star-in-formation, as individual clumps within the disk grew larger when more particles stuck to them. This process may have produced many planetary systems throughout the Milky Way galaxy.

Figure 4.2 The Lagoon Nebula is a star-forming region relatively close to the solar system. There, some 1200 parsecs away, a cluster of young, hot, luminous stars have formed within the past million years or so.

lit from within by new stars, represent "cosmic nurseries" within which the youngest stars have just begun to shine (Fig. 4.2 and Color Plate 5).

Within the past two decades, astronomers have discovered that interstellar clouds contain many types of complex molecules, which must have assembled within these clouds, as similar molecules once formed on the primitive Earth. Some astronomers have even suggested that life itself began in interstellar clouds, later to migrate onto planetary surfaces. Although this last hypothesis has not received much acceptance, all astronomers agree that the interstellar gas and dust represent the best place to study both the birth of stars and the formation of a wide variety of molecules in space.

Probing the Interstellar Medium

Astronomers first discovered the existence of interstellar matter by observing the "milky way," the band of diffuse light that circles the sky (see again Fig. 3.2). This glow of starlight is the combined emission from millions of stars in our galaxy, each too distant to be seen as an individual star without telescopic aid.

After Galileo showed that the milky way consists of stars, later astronomers realized that our solar system lies within a system of stars, the Milky Way galaxy: We see the galaxy's central plane as a band of light. At certain points in the band, most clearly in the constellation Cygnus, the band seems to split in two. The galaxy does not divide at these points; instead, tiny dust particles that absorb starlight produce the apparent split by blocking our view of the stars that happen to lie behind the obscuring matter. Thus astronomers learned to "see" interstellar particles by observing the effects of their absorption, which produces an absence of light from parts of the milky way (Fig. 4.3). Today, thanks to the results from Infrared Astronomy Satellite (IRAS), astronomers can map the dust in our galaxy by observing the infrared emission produced by the dust grains (Fig. 4.4 and Color Plate 3).

More detailed studies of this absorption of starlight have allowed us to distinguish the effects of dust particles, each consisting of millions of atoms, from the effects of individual atoms or molecules. Dust grains absorb all colors of starlight, though they absorb blue light more efficiently than red light; atoms and molecules can absorb only a few particular frequencies or colors. By analyzing the spectrum of starlight that has passed through clouds of interstellar matter, we can therefore distinguish the different kinds of atoms and molecules in the clouds. We can also determine both the numbers of the different types of particles and the general properties of the interstellar dust grains.

Some clouds of interstellar gas and dust do not happen to lie between ourselves and bright stars, so we cannot detect them easily by their absorption of starlight. Other clouds are so dense that no starlight can penetrate them, no matter how bright the stars behind them may be. To study these clouds, astronomers use the fact that interstellar matter can itself emit various kinds of photons, so the clouds can be observed directly. For example, interstellar gas that lies close to young, hot stars will be lit by the starlight energy that the gas absorbs, even though we cannot see the stars themselves. Still more important, many kinds of molecules, as well as hydrogen atoms, can emit radio waves without any energy input from nearby stars.

The first example—the interstellar gas around young, hot stars—produces the giant, glowing gas clouds.[1] In these clouds, the atoms of gas have all been ionized—stripped of one or more electrons—by the intense flow of

[1] These clouds are called "H II regions." The abbreviation "H II" represents ionized hydrogen—that is, a collection of hydrogen atoms that each have had their electron knocked loose and consist of bare protons. Neutral or un-ionized hydrogen carries the abbreviation "H I."

Figure 4.3 In the North American Nebula, dust grains concentrated in the "Gulf of Mexico" absorb the light from more distant stars. Similar concentrations of dust produce "dark lanes" in the "milky way."

ultraviolet light from the hot stars that the clouds contain (Fig. 4.5). As the electrons recombine with the ions, each atom can emit one or more photons of visible light when the recombining electron jumps from a larger orbit into

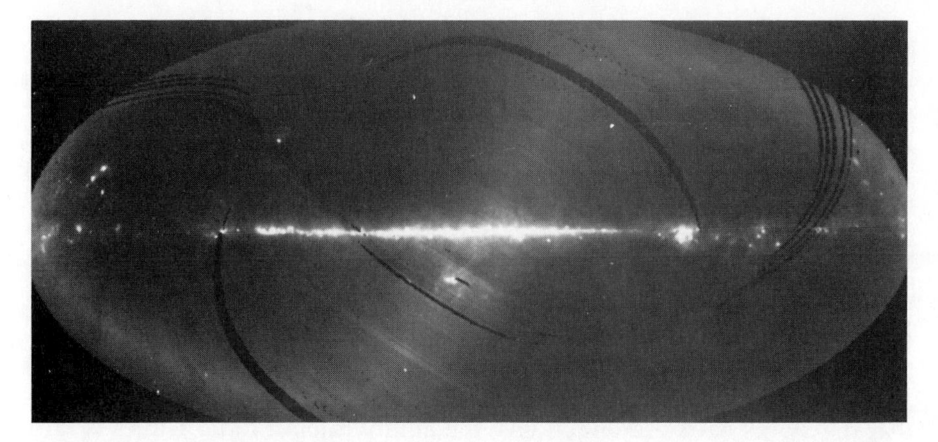

Figure 4.4 In 1983, the Infrared Astronomy Satellite mapped the Milky Way in four infrared wavelengths. This map shows the view all around the sky as seen from Earth, less only a few swaths that the mapping project missed. Note the concentration of infrared-emitting regions toward the plane of the Milky Way.

Figure 4.5 Young, hot, luminous stars radiate mostly ultraviolet, along with copious amounts of visible light. The ultraviolet radiation ionizes the gas around them, producing what astronomers call an H II region, in this case the Rosette Nebula.

a smaller one. As a result, the entire cloud of gas, whose atoms are repeatedly ionized, recombined, and ionized again, glows with the photons emitted as part of the recombination process. The stars that power this entire cycle emit most of their photons with ultraviolet energies, which in fact are just the energies needed to ionize hydrogen and other common atoms in the gas around the star. The recombination process thus produces visible-light photons from the original, ionizing energy of the ultraviolet photons from the star.

Radio Waves from Interstellar Hydrogen Atoms

Hydrogen atoms far from any star can be detected by the radio waves that the atoms emit. Among atoms, the ability to emit radio waves is rare. Remarkably enough, hydrogen atoms, the most abundant atoms in the universe, possess this ability. Hydrogen atoms in our galaxy, each with one proton and one electron, can emit radio waves because the two particles in each atom resemble tiny, spinning magnets (Fig. 4.6). The rules of atomic physics dictate that these spinning magnets can have spins that are either parallel or antiparallel to one another.

When the proton and electron spins are parallel, the atom has a bit more energy than in the case of antiparallel spins. A parallel-spin atom can therefore flip the electron spin into the antiparallel position as shown in Fig. 4.6. As the atom does this, it emits a radio-wave photon, whose small energy equals the difference in energy between the two spin configurations. This photon's frequency will always be 1420 megahertz, perhaps the most significant frequency that the universe produces; the corresponding wavelength of the radio emission equals 21.1 centimeters.

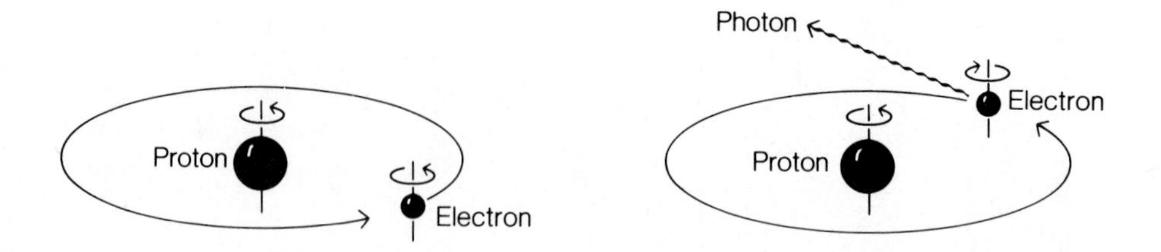

Figure 4.6 The electron that orbits the proton in a hydrogen atom can have its spin in one of only two possible orientations: either parallel to the proton's spin, or antiparallel to that spin, meaning that the spins are in opposite directions. When the spin orientation flips from the parallel into the antiparallel position, a photon appears with a frequency of 1420 MHz and a wavelength of 21.1 centimeters.

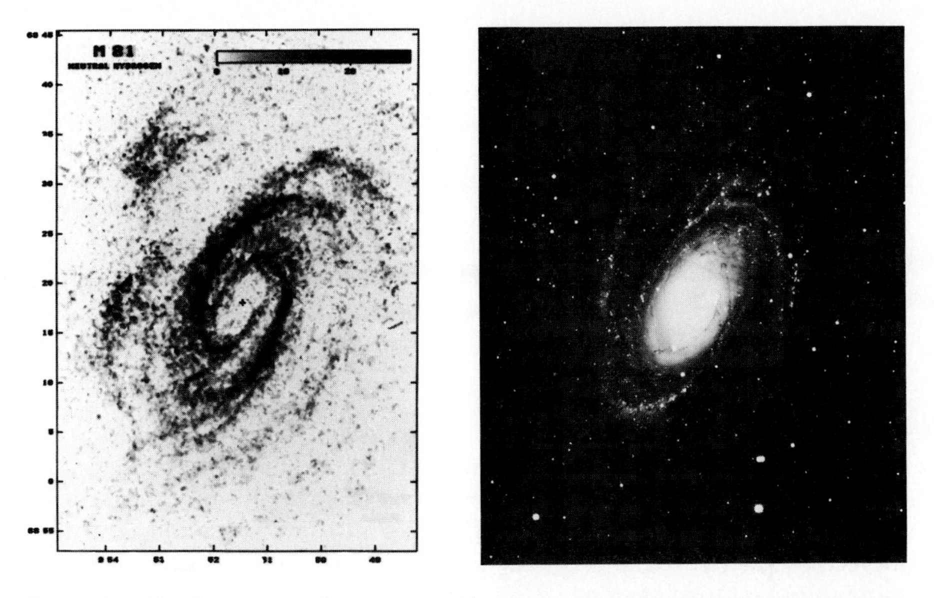

Figure 4.7 By observing radio waves with a frequency of 1420 MHz, astronomers can map the distribution of hydrogen atoms in a spiral galaxy such as M81 (left). Comparison with a visible-light photograph (right) shows that the galaxy's central regions are relatively devoid of hydrogen atoms.

In interstellar space, hydrogen atoms by the trillions upon trillions emit radio waves of 1420-megahertz frequency. Once an atom has reached the antiparallel-spin state, mild collisions between atoms can bump the electron's spin back into the parallel position. The atom can then once again emit a 1420-megahertz photon. With carefully designed antennas and receivers, astronomers can determine the number of hydrogen atoms in a given direction simply by measuring the intensity of the 1420-megahertz radio waves from that direction. In addition, they can estimate the distances to various groups of hydrogen atoms by observing the Doppler effect in the frequency of the radio emission. The amount of change in the frequency can be related to the atoms' velocities with respect to the solar system, and thus to their distances from us, because we know the pattern that the different parts of the Milky Way follow as the galaxy rotates. By observing 1420-megahertz radio waves, we can map the distribution of hydrogen atoms in our own galaxy and also in other spiral galaxies (Fig. 4.7).

Interstellar Dust Grains

Since hydrogen is the most abundant element in all galaxies, maps of hydrogen atoms already tell us a great deal about the distribution of interstellar matter. We should also keep track, however, of the interstellar dust grains. Dust grains are specks of matter that each include a few million atoms, with diameters of a few millionths of a centimeter. These grains apparently form in the atmospheres of cool stars, which expel them into interstellar space as the stars age. The dust grains consist mostly of silicon, carbon, and oxygen atoms, perhaps with an outer mantle of hydrogen and water molecules. Their total mass is far less than that of the atoms and molecules in the interstellar gas, but they play a key role in the formation of hydrogen molecules (see below).

Interstellar Molecules

Within the interstellar medium, molecules that have formed from two or more atoms have a special interest to the search for life, because the initial stages in forming living creatures must begin with the formation of molecules. In interstellar clouds of gas and dust, the most abundant molecules are hydrogen (H_2), each made from two hydrogen atoms, the most abundant type of atom.

The formation of hydrogen molecules occurs by a process that may seem roundabout but nonetheless describes how hydrogen atoms can pair to make hydrogen molecules. Individual hydrogen atoms that hit dust grains within the interstellar clouds will tend to stick for a while and will wander over the surface of the grain (Fig. 4.8). During this time, the atoms can combine with other hydrogen atoms that have also stuck to the grain. The resulting molecule will most likely pop off the grain as part of the formation process. This process, according to astronomers' calculations, works well for hydrogen molecules, but other types of molecules will not appear in significant numbers as a result of such grain-surface reactions.

In contrast, other common interstellar molecules, such as carbon monoxide (CO), ammonia (NH_3), and formaldehyde (H_2CO) have apparently formed in a different way: First two atoms combined with each other, then a third atom joined the first two, and so forth, without any sticking to the surfaces of interstellar dust grains (Fig. 4.9). To form molecules in this way, however, requires a fairly large density of matter—that is, a fairly large number of atoms per cubic centimeter. Otherwise, collisions among atoms will be so rare that only a relatively few molecules can form, even during the billions of years that an interstellar cloud of gas and dust may exist.

When we examine the distribution of interstellar matter within the Milky Way galaxy, our hypothesis about the formation of molecules finds confirmation. Within the general pancake-shaped arrangement of matter, interstellar gas and dust tend to collect into clouds, regions where the density of matter far exceeds the average density. Taking the galaxy as a whole,

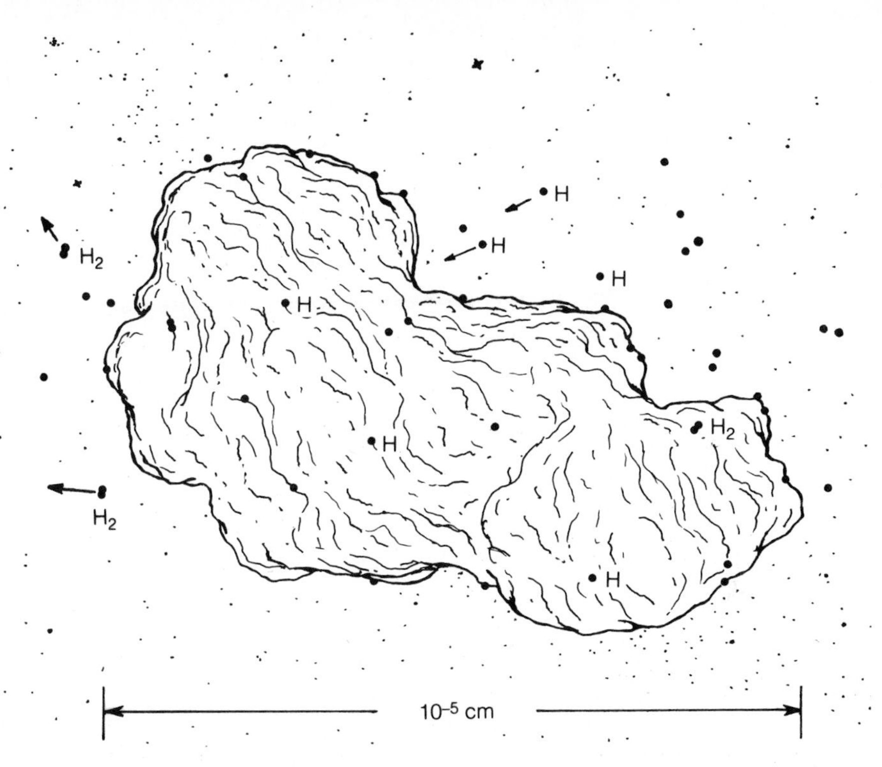

Figure 4.8 The chief process by which hydrogen molecules (H_2) form in interstellar space involves dust grains, which typically include a million or so atoms. Individual hydrogen atoms (H) stick to the surface of the dust grain and wander over it. When one of them encounters another H atom, the two atoms bond together to form a molecule, which leaves the surface of the dust grain.

the interstellar matter averages only about one atom per cubic centimeter, while the interstellar clouds each contain at least 10 times this density of matter. But an important difference exists between two types of interstellar clouds, the diffuse clouds and the dense, or molecular, clouds.

Diffuse or atomic interstellar clouds have densities of matter that range up to a few hundred atoms per cubic centimeter. Such clouds typically contain dust grains as well, whose total mass is about 1 percent of the mass of the atoms. The diffuse clouds contain many hydrogen molecules, but the number of hydrogen atoms far exceeds that of hydrogen molecules. Some of these diffuse clouds are the interstellar clouds revealed by the absorption of starlight that we described on page 83. By carefully measuring the absorption produced by particular sorts of atoms, we can determine that these clouds have temperatures between 40 and 250 K (the symbol K stands for degrees above absolute zero), colder than Earth's surface but well above

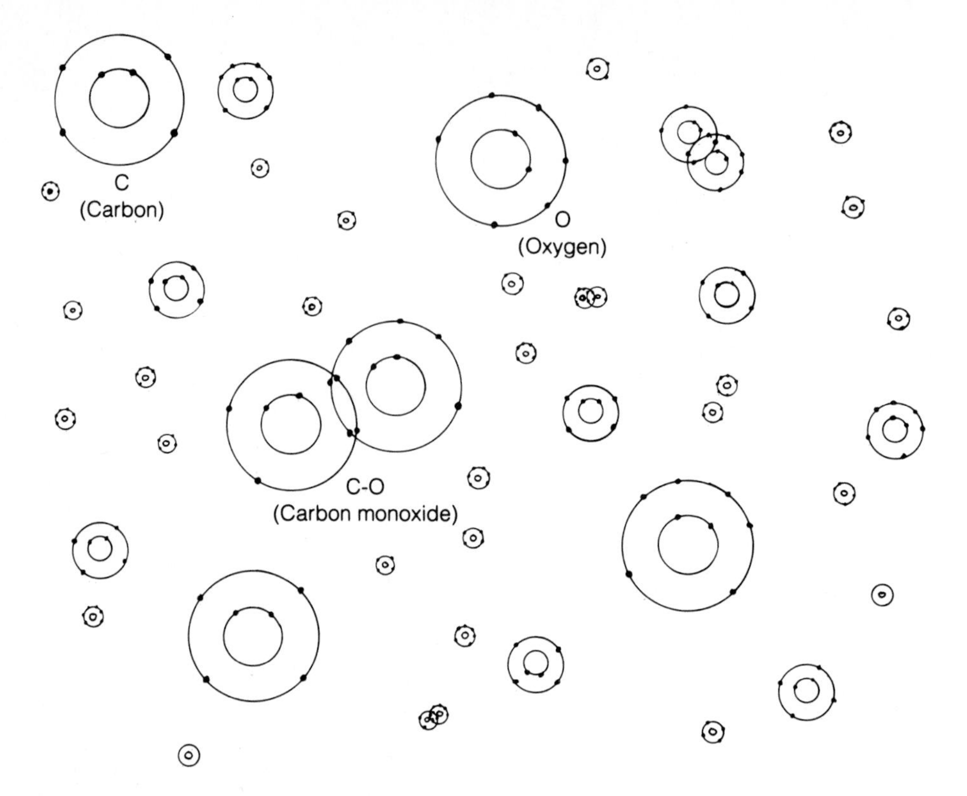

Figure 4.9 Most molecules other than hydrogen form in the "gas phase"—that is, without the intervention of interstellar dust. Instead, the atoms manage to stick together as molecules when they collide.

the zero temperature that would occur if the matter in the clouds did not receive some photon energy from other regions of space.

Molecular Clouds

The second type of interstellar cloud, the "dense" or "molecular" cloud, presents a different mixture of constituents. As their name implies, these clouds are rich in molecules, which have been able to form more readily because of the greater density of matter in such a cloud. Each molecular cloud has a mass of at least a few thousand solar masses, and often hundreds of thousands or millions of times the sun's mass, so the masses of these clouds far exceed the masses of diffuse interstellar clouds. However, because the matter in molecular clouds has a greater density than that in diffuse clouds, the dense clouds do not occupy volumes that are greater than those of diffuse clouds (Fig. 4.10). Within a molecular cloud, the density averages many thousands of particles per cubic centimeter, but the cloud includes clumps where the density rises to about 1 million particles per cubic centimeter. (For comparison, the air we breathe contains about 3×10^{19} molecules per cubic centimeter.) Thus the gas in molecular clouds is thousands of times denser than the gas in a diffuse interstellar cloud. Within molecular clouds, the temperature ranges from about 30 to 100K, somewhat lower than the temperature within atomic clouds.

The greater density of matter within a molecular cloud gives atoms a much better chance to collide with one another to form molecules, and it is precisely in these parts of interstellar space that we find molecules in great abundance. The molecular clouds have therefore passed through the first simple step on the road from atoms to life: They have already manufactured a great variety of molecules from individual atoms.

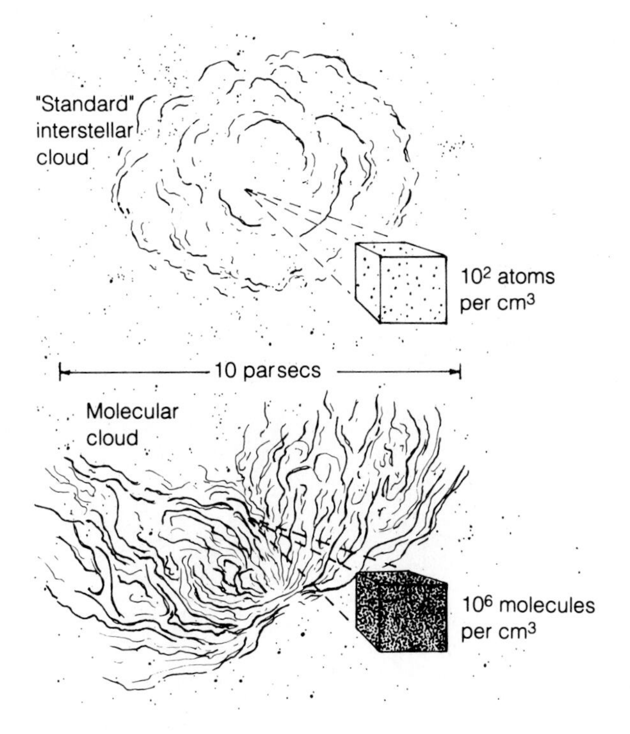

Figure 4.10 Clouds of interstellar material are either molecular clouds or atomic clouds. An atomic cloud consists mainly of hydrogen atoms, with a density of only a few hundred atoms per cubic centimeter. In contrast, within a molecular cloud nearly all of the hydrogen atoms have paired together in molecules, and the average density of molecules equals a few hundred thousand per cubic centimeter.

Let us take a look at a typical dense cloud, the closest and best-studied example, the Orion molecular cloud. This concentration of gas and dust in the direction of the constellation Orion has a million times the mass of the sun. Most of this matter has a temperature just a few tens of degrees above absolute zero, but in a small part of this giant cloud complex, the part we call the Orion Nebula, the density of matter has grown so large that stars have condensed and have recently begun to shine (Fig. 4.11 and Color Plate 5). We see the star-forming region only because it lies on *our* side of the much larger Orion molecular cloud. If the star-forming region lay within the molecular cloud, or on its far side, we could never see the newborn stars and the gas they illuminate, because the dust grains in the molecular cloud would absorb all the young starlight headed in our direction. The stars inside the Orion Nebula have ages of no more than a few hundred thousand years. Thus they are all far younger than a typical star, such as our sun, whose age is measured in billions of years.

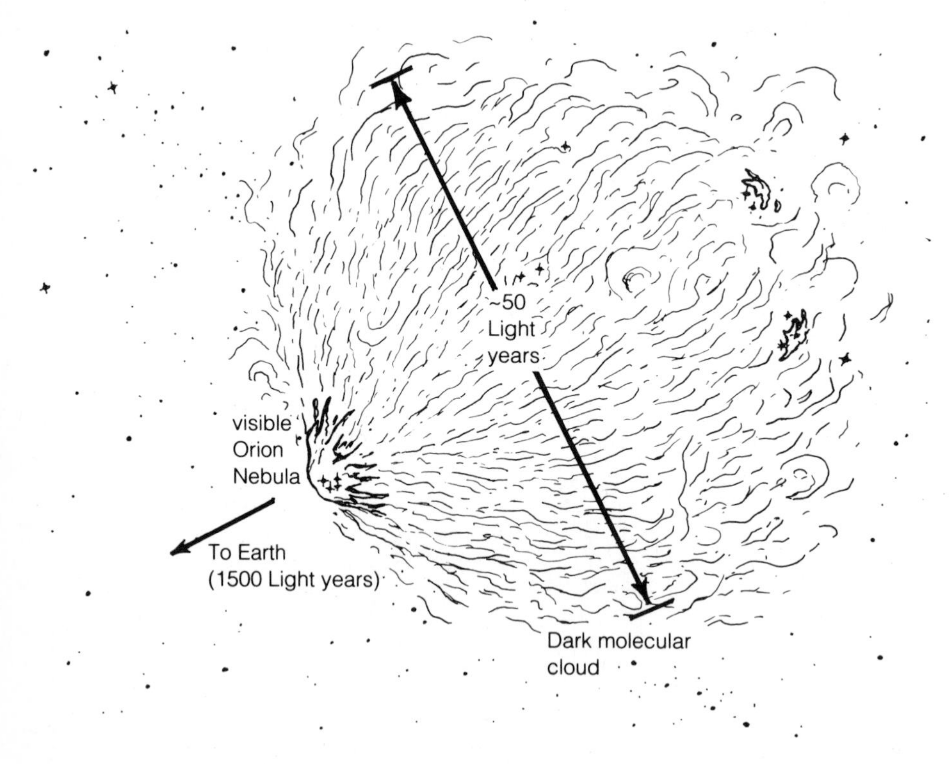

Figure 4.11 The Orion Nebula lies within a much larger "giant molecular cloud," most of which has yet to form stars. Most of the giant molecular cloud lies on the far side of the Orion Nebula, which forms a "blister" on part of the cloud's surface.

The molecular cloud in Orion is the closest of a few thousand similar regions that dot the disk of the Milky Way. These relatively dense clouds, each of them capable of forming hundreds of thousands or millions of stars, collectively contain a few percent of the mass of our galaxy. They serve both as the reservoir of material for future star formation, and as a reminder that the bulk of the matter in the galactic disk has already formed stars.

The Different Types of Molecules in Molecular Clouds

The Orion molecular cloud contains both an active star-forming region—the Orion Nebula—and also regions where stars have not yet formed, but where they are now forming or are about to form during the next few tens of thousands, or hundreds of thousands, of years. These future star-forming regions have densities of many millions of particles per cubic centimeter, far more favorable to the formation of molecules than the much lower densities of matter in diffuse interstellar clouds. Within the molecular clouds of the Orion complex, astronomers have founds dozens of different types of molecules, ranging from small molecules such as carbon monoxide (CO) and cyanogen (CN) up to molecules as large as ethyl alcohol (CH_3CH_2OH) and even larger.

The larger molecules draw our attention in the search for life, though they are not as abundant as the small ones, because they take us farther down the road to the complex molecules found in living organisms. With molecules such as methylamine (CH_3NH_2), we seem well on our way to forming the simplest amino acids (see page 162). Amino acids, the basic structural units in protein molecules, are certainly not themselves alive; nor has anyone found amino acids in dense interstellar clouds (though in fact the search for amino acids has barely begun). The discovery of molecules made of as many as a dozen atoms in a single dense interstellar cloud does, however, suggest that amino acids might well have formed there.

The simplest amino acid, glycine, contains 10 atoms; the next simplest, alanine, has 13 atoms; the other amino acids have from 14 to 26 atoms (see page 167). Almost all these atoms are either hydrogen, carbon, nitrogen, or oxygen, with an occasional sulfur atom in some of the amino-acid molecules. Dense molecular clouds, such as those at the center of our galaxy and in a relatively nearby cloud called the "Taurus dark cloud," provide the prime candidate areas in the search for interstellar amino acids. If amino-acid molecules do indeed exist in interstellar molecular clouds, then we may well expect that amino acids have managed to form under a variety of conditions throughout our galaxy, and in other galaxies too.

Table 4.1 lists the more than 90 different types of molecules that have been detected, so far, within dense interstellar clouds. These molecules consist

TABLE 4.1 Identified Interstellar Molecules Classified by Number of Atoms per Molecule

2 Atoms	3 Atoms	4 Atoms	5 Atoms
H_2	H_2O	NH_3	SiH_4
OH	H_2S	H_3O^+(?)	CH_4
SO	SO_2	H_2CO	CHOOH
SO^+(?)	HN_2^+	H_2CS	H(C≡C)CN
SiO	HNO(?)	HNCO	CH_2NH
SiS	H_2D^+(?)	HNCS	NH_2CN
NO	HCN	CCCN	H_2CCO
NS	HNC	HCO_2^+	C_4H
HCl	HCO	CCCH	C_3H_2
NaCl	HCO^+	c-CCCH	CH_2CN
KCl	HOC^+	CCCO	C_5
AlCl	OCS	CCCS	SiC_4
AlF	CCH	HCCH	H_2CCC
PN	HCS^+	$HCNH^+$	HCCCO(?)
CH	CCO(?)	C_2H_2	
CH^+	CCS		
CN	C_3		
CO	SiC_2		
CS			
C_2			
SiC			
CP			

6 Atoms	7 Atoms	8 Atoms
CH_3OH	CH_3CHO	$CHOOCH_3$
NH_2CHO	CH_3NH_2	H_3C(C≡C)CN
CH_3CN	CH_3CCH	
CH_3NC	CH_2CHCN	
CH_3SH	$H(C≡C)_2CN$	
C_5H	CH_3CCN	
HC_2CHO	C_6H	
C_5H	CH_2	

9 Atoms	10 Atoms	11 Atoms	13 Atoms
CH_3CH_2OH	$CH_3(C{\equiv}C)_2CN$(?)	$H(C{\equiv}C)_4CN$	$H(C{\equiv}C)_5CN$(?)
$(CH_3)_2O$	$(CH_3)_2CO$(?)		
CH_3CH_2CN			
$H(C{\equiv}C)_3CN$			
$H(C{\equiv}C)_2CH_3$			

Note: A question mark after the symbol for the molecule indicates that its identification has not been confirmed.

primarily of the most abundant elements in the universe (with the exception of helium and neon, which do not combine with other atoms easily)—hydrogen, oxygen, carbon, and nitrogen, plus silicon, sulfur, phosphorus, and other heavy elements. Notice, in particular, the large number of molecules that contain one or more *carbon* atoms. Carbon atoms form the "backbone" of our molecular structure and hence the key to life on Earth. The importance of carbon atoms comes from each atom's ability to combine with as many as four other atoms and to form long chains in which a string of carbon atoms supports the entire molecule. The existence of still larger molecules built on chains of carbon atoms seems a fairly sure bet in dense interstellar clouds, but the possibility of much larger molecules, those with many dozens, or many hundreds, of atoms, remains a question for open speculation.

The discovery of these 90-odd varieties of simple molecules in dense interstellar clouds has important implications in our search for life. First, these molecules have formed under conditions quite different from those on planetary surfaces, yet in interstellar clouds we find the same sorts of molecules as those that we believe to have existed on the surface of our planet early in its history.

The fact that these molecules can assemble themselves in clouds which, even though we call them "dense," are far more rarefied than our atmosphere shows that we may expect to find these types of molecules widely distributed in the cosmos, since many more favorable sites should exist. We might conclude straightaway that we would restrict ourselves unnecessarily if we look for life only on the surfaces of planets. Should we consider the possibility of life in dense interstellar clouds?

Second, we notice that different types of molecules found in dense interstellar clouds have different degrees of relevance to life on Earth. Molecules such as methylamine (CH_3NH_2) have an intimate connection with the kinds of molecules found in terrestrial organisms, whereas molecules such as sulfur dioxide (SO_2) have far less relevance to the kind of life we find on Earth.

We shall follow the implications of the distinction between life-involved and nonlife molecules in Chapters 8 and 11. For now, we shall pursue the implications of the existence of so many types of molecules in dense interstellar clouds.

First, nearly all of the molecules listed in Table 4.1 have been detected in the interstellar medium during only the past 25 years, and half of them during the past decade. We may therefore reasonably assume that many more molecular types await our discovery. No doubt exists that the simple molecules are likely to be far more abundant than the more complex molecules; thus, for example, hydrogen molecules outnumber all other types combined by a factor of more than 1000. The fact, however, that molecules are constantly forming (and coming apart) in interstellar clouds—the continuing "chemical evolution" of molecular clouds—shows that chemical reactions pervade our entire galaxy, even if they occur primarily in localized regions called dense interstellar clouds or molecular clouds.

Second, we must admit that we have no knowledge of how far this chemical evolution has gone in molecular clouds. What sorts of truly complex molecules may have formed there we do not know. Molecules with greater numbers of atoms become progressively more difficult to detect, especially if the abundance of these molecules falls below that of the less complex molecules. The chemical evolution in dense clouds, such as those in the Orion molecular complex or at the center of our galaxy, may have proceeded to form much larger molecules than those listed in Table 4.1.

Third, we do not know how the existence of interstellar molecules relates to the existence of molecules on planetary surfaces. Our own planet, for example, has a great variety of molecular types on its surface, some made by nonbiological processes during the Earth's 4.5-billion-year history, others made by the early stages of what we call life, still others made by human beings in complicated chemical reactions, unlikely to occur elsewhere without planned intervention into natural events. The question of which molecules may have existed on the Earth before life developed remains incompletely answered.

Did Life Begin in Interstellar Clouds?

Two well-known astronomers, Fred Hoyle and Chandra Wickramasinghe, have suggested that complex organic molecules could form in great quantities in dense interstellar clouds. The typical sorts of molecules that Hoyle and Wickramasinghe have in mind are polysaccharides, long-chain molecules made mostly of carbon, oxygen, and hydrogen atoms. The best-known examples of polysaccharides are cellulose molecules, the principal structural molecule in plants. Hoyle and Wickramasinghe suggest that interstellar clouds are loaded with cellulose, and that we should not ignore the possibility that life has begun in these clouds. But more important still from a human perspective, these astronomers believe that comets, the most primitive objects in the solar system, have preserved complex organic molecules such as polysaccharides formed in interstellar clouds, and that comets may even contain living cells and viruses. In other words, Hoyle and Wickramasinghe assign the origin of life to the cometary lumps of gas, ice, and dust that condensed in molecular clouds such as those in the Orion complex.

We shall discuss comets in Chapter 11, but we can pause now to see the implications of this theory, if it were true. (Most astronomers do not think that this theory has much merit as a description of reality.) If comets do contain primitive forms of life, then the possible interaction of comets such as Halley's comet with our Earth would become extremely important. Comets could seed planets with life as they pass by, and every close pass of a comet would carry the possibility of further seeding. Fred Hoyle has even suggested that outbreaks of epidemics, such as influenza and smallpox, are the result of such close encounters with a comet, and that the age-old tradition that comets bring bad luck has its origins in similar epidemics.

Although the theories proposed by Hoyle and Wickramasinghe have not been accepted by astronomers or biologists, they serve to remind us that interstellar molecules may have a direct bearing on the origin of life. Dense interstellar clouds not only make molecules, but give birth to stars and planets (see below). The molecules from dense interstellar clouds that were directly incorporated in planets could hardly have survived the formation process; instead, the molecules almost certainly broke apart into their constituent atoms at that time. But comets, frozen lumps of old interstellar matter, could preserve the molecules that formed in dense clouds and might later have deposited some of these molecules on the surfaces of planets after they had formed. We shall discuss this idea in Chapters 8 and 11.

If the latter hypothesis is correct, it provides a strong argument in favor of the idea that life should generally be about the same throughout our galaxy, since life should then have arisen from much the same kinds of molecules made in similar molecular clouds. If, however, the opposing hypothesis—that the molecules used by living organisms were assembled on Earth rather than brought from outside—turns out to be correct, then we should expect to find a greater diversity of types of life from planet to planet, since each planet would provide a specialized set of conditions within which life could begin. But we should note that the molecules listed in Table 4.1 are extremely small and simple in comparison with biological molecules such as proteins and DNA, which we discuss in Chapters 7–9. If interstellar clouds produce only the molecules listed in Table 4.1 and similar ones, they do not advance much along the road to life. Our understanding of the chemistry in dense interstellar clouds remains uncertain. Larger and more complex molecules may well await discovery. We do know that stars and planets form in these clouds. Just which compounds form there as well, and whether they reach the surfaces of planets intact, are topics of ongoing research, to which we shall return in Chapter 11.

The Birth of Stars

From the viewpoint of most astronomers, the most significant role of dense molecular clouds in the universe consists of giving birth to stars. Starbirth is a process that is shrouded in mystery twice over. Firstly, the initial stages of the formation of a star occur long before the star begins to shine by nuclear fusion, or even to heat itself significantly by its own contraction (see Chapter 5). Secondly, even when a "protostar"—a star in formation—does begin to emit large amounts of energy, dust distributed throughout the star-forming region blocks the radiation from the protostar, so that we cannot hope to see anything until and unless the veil of dusty material lifts. Hence astronomers to this day must guess at (or calculate) the initial stages of star formation, unable to tell much by direct observation.

In the final stages of star formation, however, many newborn stars can be recognized by the situation called "bipolar outflow." This term refers to gas flowing outward in two opposite directions (see Fig. 4.12). These outflowing jets, traveling at tens of kilometers per second, are driven by the heating of the material close to a young star, and they have been observed both in visible light and in radio waves. If the material around a newly formed star had exactly the same distribution in all directions, bipolar outflows would probably not occur, but because the gas around the protostar does not have such perfect symmetry, the jets tend to "punch through" the surrounding material preferentially in certain directions. We do not yet know whether the jets are a more or less continuous phenomenon around newborn stars, or whether they come and go intermittently (that is, on time scales of centuries or millennia) as the material around the young star adjusts to the new source of heat.

How Many Stars Form with Planets Around Them?

As we shall discuss in Chapter 16, one of the most favorable locations for the origin and development of life in the universe consists of planets in orbit around stars that last for billions of years. Though this may seem a prejudiced statement—since the only type of life that we know inhabits just such a site—arguments based on the fundamental chemistry of living organisms do point to planets as the first places to look for life elsewhere. Hence the fraction of all stars that form with *planets* around them plays a crucial role in assessing the likelihood of extraterrestrial life.

Figure 4.12 Radio astronomers have detected many cases of "bipolar outflow," in which streams of gas are moving away from an object within the Milky Way in opposite directions. Bipolar outflows are apparently stars in the final stages of formation, in which the protostar has acquired a disk of material around it, and matter squirts outward above and below the disk as most of the matter moves inward to form the star.

And what is that fraction? Astronomers cannot tell us for certain, because they cannot hope to make direct observations of planets around even the closest stars to the sun. But tantalizing evidence does exist that a significant minority of stars—and perhaps even as many as half of all stars—give birth to orbiting worlds as they form.

That evidence is indirect and draws on several types of observations. One strand of evidence deals with precise measurements of the motions of stars, which could reveal the effects of a planet's gravitational tug, first in one direction and then in the other, as it orbits. These observations suggest that some of the closest stars do have companions with masses several times the mass of Jupiter, the sun's largest planet. Another strand of evidence measures the *motions* of stars by using the Doppler effect, again seeking the effects of a planet's gravitational pull, first in one direction and then in another. But the strongest evidence consists of direct observations that reveal *disks of material* in orbit around young stars. Though this material has not yet formed planets, astronomers think that it occupies the type of situation in which planets may eventually form.

These "protoplanetary disks" can be identified by the radiation they emit. Since the material comprising them has relatively low temperatures, this radiation is mostly long-wavelength infrared and submillimeter. Newly developed techniques have allowed astronomers to make improved observations of the cool material around stars. They have established that a large fraction of stars younger than half a billion years do have matter moving in orbit, and that the matter farther from the center moves more slowly. If and when this matter clumps together to form planets, as happened in the solar system 4.5 billion years ago, it will become completely invisible to our best techniques today, because most of the matter will be packed away *inside* a planet. Thus we have a better chance to see preplanetary material than planets themselves, and that is just what astronomers believe they have identified in some cases.

Of course, nothing can compare with the direct observation of actual planets—worlds of their own, which may orbit other stars in the Milky Way by the billions. We may hope to make that discovery during the next decade. For now, we must regard the question of how many worlds await life, or have given birth to it, as unanswered.

SUMMARY

Spiral galaxies such as our own Milky Way contain clouds of gas and dust, often with masses a million times that of the sun. Astronomers can investigate the properties of these interstellar clouds by studying the radio waves that certain atoms and molecules emit, as well as the photons of particular energies that are absorbed when starlight passes through an interstellar cloud. Through radio and visible-light studies of the interstellar medium, astronomers have found that the less dense clouds contain few molecules, having their gas (mostly hydrogen and helium) instead in the form of atoms. Young, hot stars within a cloud of gas will ionize most of the atoms to produce the H II regions that shine brightly as the atoms temporarily recombine.

In dense interstellar clouds, more than 90 types of molecules have been discovered, ranging from hydrogen (H_2), the simplest and by far the most abundant, to molecules as complex as HC_9N, which contains a chain-like structure of nine carbon atoms. These more complex molecules in many cases resemble the basic building blocks of living matter on Earth. The fact that such molecules appear in "dense" interstellar clouds, which are far less dense than our atmosphere, tells us that molecules of at least this complexity seem to form naturally in relatively difficult situations. This conclusion in turn suggests that the basic molecules required for life may be widely distributed in interstellar space as well as on planets.

QUESTIONS

1. On a clear night, find a place to look at the sky that is free of city lights and spot the band of light called the "milky way." What does this "milky way" consist of? Why does it circle the entire sky, as seen from Earth? Is it accurate to say that if you could see the entire "milky way" in the sky, you could see the entire Milky Way galaxy? Why or why not?

2. How are molecular clouds distinguished from other interstellar clouds? What is the importance of these clouds for star formation? What is their importance in speculation about the origin of life?

3. Would you expect the formation of molecules on Earth to be more difficult or less difficult than their formation in interstellar clouds? Why?

4. The ability of a given receiver of photons to resolve small angles on the sky varies in proportion to the diameter of the receiving antenna divided by the wavelength of the photons that it detects. How does the human eye, with an antenna diameter of 3 mm, compare with the Arecibo antenna, with a diameter of 300 m, if the human eye observes visible-light photons of 5×10^{-5} cm wavelength, and the Arecibo telescope observes radio photons of 50 cm wavelength? How large would the Arecibo telescope have to be to achieve the same angular resolution, when observing photons of 5×10^{-5} cm wavelength?

5. The ability of any antenna to gather photons for detailed study varies in proportion to the area of the antenna. If the effective diameter of the Arecibo antenna exceeds that of the human eye by 100,000 times, how many times more photons can the Arecibo antenna detect each second?

6. Interstellar dust scatters photons of all visible-light frequencies, but the dust scatters blue-light photons more effectively than red-light photons. Will light from a star that passes through a thin cloud of interstellar dust emerge redder or bluer than before its passage? Why?

FURTHER READING

Audouze, Jean, and Guy Israel, eds. *The Cambridge Atlas of Astronomy.* New York and London: Cambridge University Press, 1983.

Goldsmith, Donald. *The Astronomers.* New York: St. Martin's Press, 1991.

Lovell, Bernard. *The Jodrell Bank Telescopes.* Oxford and New York: Oxford University Press, 1985.

Sagan, Carl. *Cosmos.* New York: Random House, 1980.

Spitzer, Lyman *Searching Between the Stars.* New Haven: Yale University Press, 1982.

Verschuur, Gerrit. *The Invisible Universe Revealed.* New York: Springer-Verlag, 1987.

5

Energy Liberation in Stars

Stars are the basic units of the visible universe, giant balls of gas that are capable of shining for millions or billions of years. The sole example of life that we know today exists on a planet that nestles close to its parent star. From this star the Earth receives nearly all of the energy available to any forms of life on it. Though our constant sun has shone for about 5 billion years, and will shine in much the same way for another 5 billion, we must never forget that our star, like all others, will some day fade into obscurity. Thus in the extremely long view, no form of life can depend on a star for its existence, no matter how much life may owe to that star for its origin and apparent success.

Life on Earth clearly depends on the sun, a representative star whose light and heat arise from the same process that powers all stars: nuclear fusion (Fig. 5.1). When we examine the universe to assess the chances of life elsewhere, our attention is drawn directly to the stars, or more accurately to any planets that may orbit the sun's neighbors. This is so because astronomers believe that planets offer the likeliest places for life to develop in the universe—but only because stars keep them warm.

So far as life is concerned, the most important characteristics of a star are the star's luminosity, or intrinsic brightness, its stability, and its total lifetime. The first characteristic has much to say about the chances of life arising on a planet, while the second and third determine the hazards and the length of time that any forms of life that do arise on that planet will encounter as they try to survive and to evolve. Many stars have luminosities too low to keep their planets warm, if the planets orbit at distances like those of the planets from the sun in our solar system. A few stars vary so erratically in luminosity that life would have a hard time surviving. By a significant irony of stellar evolution, the stars with the greatest luminosities have the shortest lifetimes. These stars burn themselves out long before life would have a chance to evolve on planets in orbit around them.

When we study the luminosities and lifetimes of various stars, we conclude that our hopes for finding life on planets around them rest with the average stars, those sufficiently luminous to keep planets warm but not so

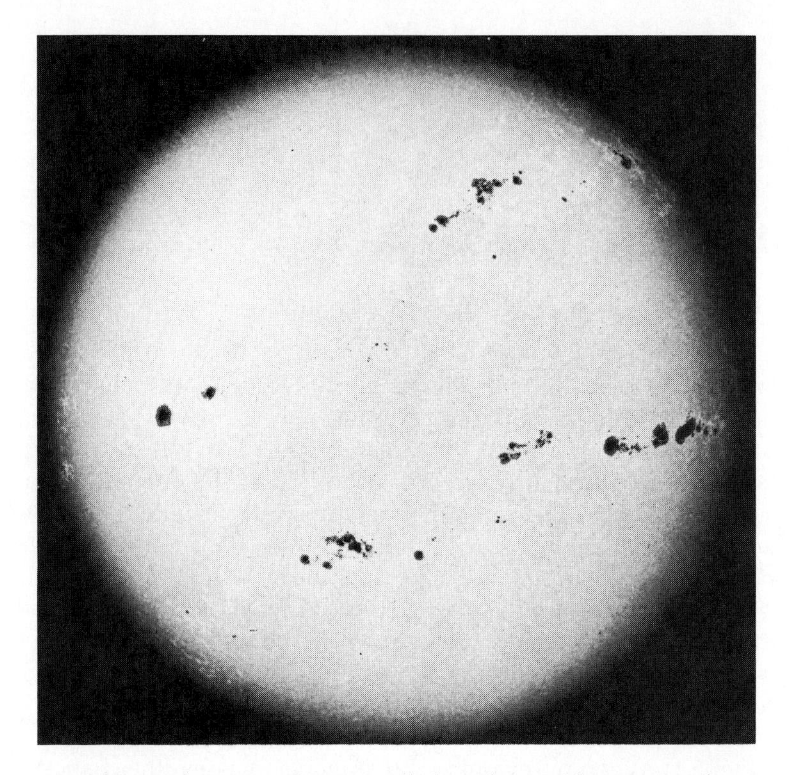

Figure 5.1 The sun has a seething, mottled surface, with occasional, temporary "sunspots"—regions that are about 1500 degrees cooler than the average surface temperature of 5800 K.

luminous that they burn out before life can originate or evolve in a significant way. Our own sun furnishes an example of such a representative star: More luminous than 95 percent of all stars, it will last for a total lifetime of 10 billion years. Half this time has already passed, during which life on Earth has appeared, evolving into a fantastic complexity that continues to increase.

Stellar Lifetimes

Why do some stars continue to shine for billions of years, while others have their energy-liberating lifetimes measured in mere millions of years? Why do some stars explode violently as supernovae, while most stars simply fade calmly into white dwarf obscurity? Through generations of patient observations of stars, and through calculations that attempt to determine what happens deep inside stellar interiors, astronomers have found the answers to these questions.

The most far-reaching result of this research into stellar structure and evolution connects the total lifetime of any star with the star's mass: Stars with higher masses burn themselves out far more rapidly than stars with lower masses do. Large-mass stars turn out to be spectacular fireworks, Roman candles with lifetimes only one-one-thousandth of the sun's 10 billion years. These high-mass stars, which last for only a few million or perhaps a few hundred million years, are prime candidates for supernova explosions, violent catastrophes that end the stars' lives with a bang, not a whimper (Fig. 5.2).

To understand why stars of different masses have widely different lifetimes, we must understand how all stars liberate energy through nuclear fusion reactions. We must also consider the possibilities open to a star as it runs out of its basic nuclear fuel, which will show that most of the elements present in the Earth and in our bodies were made in fiery stellar furnaces that later exploded to seed their ashes through the universe.

Figure 5.2 A supernova explosion (arrow) marks the end of one of a minority of stars, which blast their outer layers into space to seed their galaxies with the elements they have made during their stellar lives. For a few weeks, the supernova can shine with the luminosity of a billion ordinary stars, rivaling an entire galaxy.

How Stars Liberate Energy

All stars that shine owe their energy output to the fact that they turn energy of mass into kinetic energy deep in their interiors. Stars perform this conversion through processes of *nuclear fusion*. Indeed, the conversion of energy of mass into energy of motion (kinetic energy) through nuclear fusion is the fundamental way that new kinetic energy appears in the universe (Fig. 5.3). All other energy-converting processes, such as chemical reactions among different types of atoms and molecules, pale to insignificance in comparison with nuclear fusion.

The conversion of one form of energy into another through nuclear fusion draws on Albert Einstein's most famous equation, $E = mc^2$. This equation specifies the amount of energy of mass, E, that any object with mass m contains: We multiply the mass by the square of the speed of light, c. Every

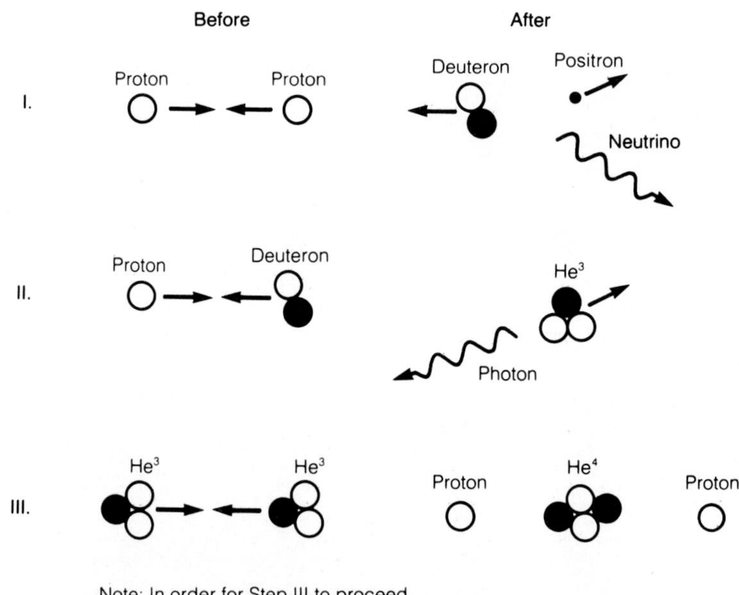

Figure 5.3 The proton-proton cycle of nuclear fusion consists of three steps, in each of which some energy of mass becomes kinetic energy (energy of motion). The first step fuses two protons to produce a deuteron (^2H nucleus), a positron, and a neutrino. The second step fuses a proton with a deuteron to produce a nucleus of helium-3 (^3He) and a photon. In the third step, two helium-3 nuclei fuse to produce a helium-4 (^4He) nucleus, two protons, and more kinetic energy.

object with nonzero mass therefore has an equivalent energy of mass, *m* times c^2. The energy of mass can, at least in theory, be converted into kinetic energy. For example, the energy of mass in a 5-gram nickel, 4.5×10^{21} ergs, equals the amount of kinetic energy consumed in the United States each minute.[1] Hence we could supply the energy needs of the United States with just half a million nickels each year, if we could find a simple way to convert energy of mass into kinetic energy with complete efficiency.

On Earth, this problem remains unsolved. Even in the universe at large we rarely find complete efficiency in converting energy of mass into kinetic energy. Deep inside countless trillions of stars, however, where the temperature rises to tens of millions of degrees, nature has created natural nuclear-fusion reactors, shielded by thousands of kilometers of the stars' outer layers. Even though the conversion of energy of mass into energy of motion proceeds with only 1 percent efficiency in these natural fusion reactors, the resulting energy produces the starlight that sprinkles the sky as far as we can see. One of these reactors, inside our sun, provides the basic energy source that allows life to flourish on Earth.

The Proton-Proton Cycle

To know stars, we must know the secrets of their energy production. Thanks to the work of generations of nuclear physicists and astrophysicists, we can look (mentally!) into the hearts of stars with surprising clarity. We know the fundamental series of nuclear-fusion reactions in most stellar interiors. That series is called the "proton-proton cycle," so named because in the first of the three nuclear reactions in the cycle, two protons collide.

In order for nuclear fusion to work, such a collision between two protons must occur with sufficient energy—that is, with sufficiently high speeds for the two protons—that the two protons actually fuse together instead of bouncing off each other. In this fusion, the two protons vanish, and three new particles appear in their place: a *deuteron,* a *positron,* and a *neutrino* (Fig. 5.3.)

We can think of a deuteron as a proton and neutron combined into one nucleus. However, a key fact about the deuteron is that a deuteron has *less* mass than two protons do, and still less mass than a proton mass plus a neutron mass. Somehow the act of binding particles together through strong forces causes the mass of the deuteron to be less than the mass we would expect from the fact that a deuteron is basically a proton bound to a neutron. In fact, the sum of a deuteron's mass, a positron's mass, and a neutrino's

[1] One erg is about the energy of a mosquito in flight.

mass (zero) amounts to less mass than the mass of two protons. Here is the key to nuclear fusion: Some mass disappears when the protons fuse. The missing mass amounts to $m = 7 \times 10^{-28}$ gram. According to Einstein's equation, this corresponds to an energy equal to $m \times c^2 = 6.3 \times 10^{-7}$ erg. Precisely this amount of energy appears as new kinetic energy, which adds to the total kinetic energy of the particles, as shown in Fig. 5.3.

The first step of the proton-proton cycle, in which two protons (hydrogen nuclei) fuse together, typifies the way that stars work: Some energy of mass disappears in the fusion process and becomes new kinetic energy. This newly liberated kinetic energy adds to the total of the particles' kinetic energy, so that the particles that emerge from the fusion reaction have a combined kinetic energy greater than that of the original colliding particles.

How does the new kinetic energy pass outward from the center of the star, where it is made by nuclear fusion? Enormous numbers of collisions among the particles that surround the nuclear-fusing region share this kinetic energy among the particles immediately around the center. Other collisions transfer the energy outward. Thus the newly liberated energy of motion spreads outward as heat (that is, the rapid movement of particles) among the particles in the interior.

Although some photons are produced in nuclear-fusion reactions, most of the photons in stellar interiors arise simply because stellar interiors are hot. Any object at a temperature above absolute zero radiates photons, and the hotter it is, the more photons it radiates each second. The photons from each star's interior eventually emerge (in altered form) from the star's surface in the form of ultraviolet, visible, and infrared light.

The positron (also called an antielectron) produced in the first step of the proton-proton cycle will soon meet an electron and the two particles will mutually annihilate. This annihilation of the positron and electron turns all of the energy of mass of both particles into the kinetic energy of the photons, neutrinos, and antineutrinos that emerge from the annihilation. The neutrinos that appear in the first step of the cycle, and in electron-positron annihilations, can, amazingly enough, escape directly from the center of the star! Neutrinos are so reluctant to interact with matter that most of them can pass through hundreds of thousands of kilometers of matter in a straight line, as easily as visible-light photons pass through air.[2] In contrast, the kinetic energy liberated by the proton-proton cycle takes about a million years to diffuse outward to the star's surface.

[2] The same property of neutrinos—their unwillingness to interact with matter—makes them extremely difficult to detect on Earth. Modern detectors use enormous quantities of fluid held in a giant tank a kilometer underground, in order to screen out the effects of other particles that would mimic the result of a neutrino collision. With this apparatus, physicists can detect about one neutrino per day from the core of the sun.

In the second step of the proton-proton cycle, a proton collides with a deuteron. The two particles fuse together to produce a nucleus of helium-3 (^{3}He) and a photon (see again Fig. 5.3). Once again, some energy of mass disappears during the fusion process, and the total kinetic energy increases by the same amount.

The third and final step of the proton-proton cycle liberates the greatest amount of kinetic energy. In this step, two nuclei of ^{3}He collide, and this collision produces a nucleus of helium-4 (^{4}He) and two protons. The total mass before the collision again exceeds the total mass after the collision; once again, the decrease in the total energy of mass matches the increase in the total kinetic energy.

The three steps of the proton-proton cycle are shown in Fig. 5.3. Because each nucleus of ^{3}He comes from the fusion of a proton and a deuteron, the first and second steps of the cycle must each occur twice for the third step to occur once. The three steps in the cycle, along with the positron-electron annihilations, liberate a grand total of 4.25×10^{-5} erg of kinetic energy. This does not seem like much energy, since even a bumblebee uses ten million times more energy each second as it moves. Inside a star such as our sun, however, about 10^{38} proton-proton fusions occur each second, so that a grand total of 4×10^{33} ergs of kinetic energy appears each second, 10,000 times more energy than the human race has consumed during the past 5000 years!

Strewn throughout the universe, in star after star, controlled nuclear fusion reactors exist, as perfect as nature can make them, in which the rate of energy liberation hardly varies over most of the stars' lifetimes. The tremendous power of the proton-proton cycle arises from the enormous number of individual fusion reactions that occur, and this number in turn arises from the tremendous abundance of protons (hydrogen nuclei) in every star.[3] Simply stated, the fusion of the most abundant nuclei (hydrogen) into the next most abundant (helium) turns energy of mass into kinetic energy, the fundamental way in which new kinetic energy appears in the universe.

The liberation of energy through nuclear fusion inside stars provides a good way to see the interplay of the four types of forces that exist in the universe. "Gravitational forces" hold the entire star together. "Strong nuclear forces" hold nuclei together and are responsible for the fusion reactions among nuclei, but they act only over extremely small distances. "Weak nuclear forces" likewise act only over distances no larger than the sizes of atomic nuclei; they modify the effects of strong forces in subtle ways, and

[3] A minority of stars, those with the largest masses, fuse protons into helium nuclei not through the reactions of the proton-proton cycle but rather through a different set of reactions called the carbon cycle. Both sets of nuclear reactions, however, have the same net result: Four protons fuse into one helium nucleus, liberating kinetic energy as a result of the fusion process.

they are responsible for the decays of isolated neutrons. At distances much larger than the diameter of an atomic nucleus, "electromagnetic forces" predominate. They cause the positively charged nuclei to repel one another but keep the negatively charged electrons in orbit around the nucleus in an atom.

The Importance of Temperature Inside Stars

For the proton-proton cycle of nuclear fusion to occur, protons must fuse together in the first step of the cycle. Since each proton has a positive electric charge, electromagnetic forces between the two protons produce a mutual repulsion. Only if the two protons can approach within about 10^{-13} centimeter can they fuse together through the effect of strong forces. But how can the protons ever achieve this close approach, since they repel one another?

Temperature provides the answer, because the temperature in degrees Kelvin measures the average energy of motion per particle.[4] At low temperatures, particles have little kinetic energy and low velocities. At higher temperatures, each particle's kinetic energy, and therefore its velocity, must increase. Protons need enormous velocities to overcome their mutual electromagnetic repulsion and fuse together. For this reason, the proton-proton cycle of fusion reactions cannot begin inside stars until the temperature reaches about 10 million K.

How do these enormous temperatures arise in stellar interiors? The temperature rises because of the gravitational forces that hold the star together. Gravity pulls each part of the star toward every other part, and the total result makes each part of the star feel an attractive pull from the star's center (Fig. 5.4).

In a contracting protostar, the strength of this gravitational attraction increases as the particles approach one another, because gravitational forces vary in proportion to 1 over the square of the distance. As the particles are more highly squeezed together, they collide more often and move more rapidly. Squeezing particles into a smaller volume always tends to increase the average kinetic energy per particle. If we pump air into a bicycle tire, for example, we find that the air inside the pump (and the pump itself) grow warmer from the fact that we compress the air as we pump it. Stars contain so much matter that each star, although completely gaseous, holds together solely by gravitation. Likewise, gravity provides the force of compression as a protostar contracts.

[4] The absolute or Kelvin scale of temperatures begins at absolute zero, the coldest possible temperature. On this scale, water freezes at 273.16 K and boils at 373.16 K; the temperature steps are the same size as those in the familiar centigrade (Celsius) scale.

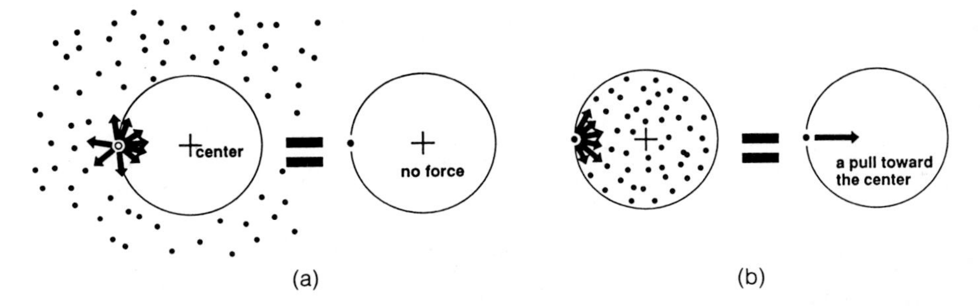

Figure 5.4 Inside a star, each part of the star feels a gravitational pull toward the star's center. The gravitational pulls from the parts of the star at a greater distance from the center (outside the circle in the diagram) cancel one another to produce a zero net force, but the gravitational pulls from the parts of the star closer to the center (inside the circle) all combine to produce a net force toward the center.

Inside such a contracting protostar, the increasing gravitational forces produce an increase in the gas pressure and gas temperature during the millions of years of slow contraction. Even though protostars radiate away some of the additional heat that arises from their contraction, the temperature inside a protostar continues to rise—first to thousands of degrees at the center, then to hundreds of thousands, finally to millions of degrees—as the protostar contracts to an ever-smaller size. Finally, when the temperature at the center of the protostar reaches about 10 million K, the proton-proton cycle of nuclear reactions begins liberating kinetic energy from energy of mass. This release of kinetic energy halts the contraction by opposing the star's self-gravitation. Perversely, therefore, nuclear fusion "cools" the situation by providing a way to oppose an otherwise ever-increasing contraction and heating.

Only at temperatures of tens of millions of degrees can protons start to fuse together despite their mutual electromagnetic repulsion, and only then can kinetic energy be created from energy of mass. The newly liberated kinetic energy pushes outward, opposing the inward pull of gravity (Fig. 5.5). This opposition allows the protostar to cease its contraction and to become a star, capable of supporting itself against its self-gravitation by liberating new kinetic energy, energy that eventually flows from the star's center to its surface, from which it escapes outward into space.

The Struggle Between Gravity and Pressure

Every star represents a cosmic battleground, a place where opposing forces continuously clash. One of these forces is gravity, the pull of every piece of a star for every other piece, which attracts all parts of the star

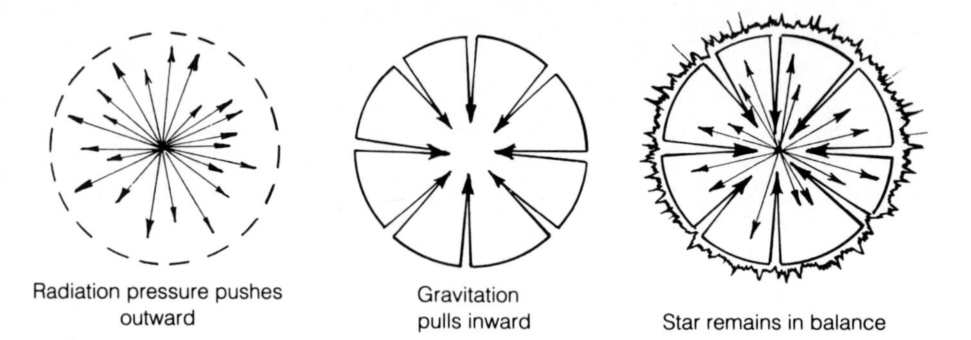

Figure 5.5 Energy released through nuclear fusion at a star's center tends to expand the star. The star's self-gravitation, however, opposes this tendency and keeps the star from exploding.

toward its center. The other force arises from the pressure difference between the center of the star, where the gas pressure is high, and the regions far from the center, where the gas pressure is much lower. Since gas pressure depends on the product of the temperature and the number of particles per unit volume, we can easily see why the pressure rises much higher near the star's center, where both the temperature and the gas density have much larger values than they do near the star's surface. This pressure difference tends to expand—even to explode—the star, just as the high pressure created by a bomb blast leads to an explosion into the low-pressure surrounding gas. And indeed without a star's own gravity, every star would immediately explode, shooting the many-million-degree gas at its center at enormous velocities into interstellar space.

But gravity makes the difference. The invisible hand of gravitation holds the star together, keeping a lid on the tendency to explode. Gravity allows a near-perfect balance to reign between the two conflicting tendencies. Without gravity, the star would explode, but without its tendency to explode—that is, without the high pressure at its center—the star would collapse under its own gravitational forces. Because both gravity and the pressure differences exist within a star, every bit of the star, from its nuclear-fusing core to its more modestly seething outer layers, can persist in suspense, held between gravity's pull to the central regions and the outward thrust from the higher pressure in the star's interior.

The Influence of Mass upon Stellar Lifetimes

All stars balance gravity and pressure differences. The details of this balance distinguish one star from another. More massive stars have higher central temperatures, and therefore shorter lifetimes, than less massive stars do. A star with 10 times the sun's mass, for example, will last for only about 100 million years, instead of the sun's 10 billion years, before using up its stock of protons and exhausting the star's potential for any further nuclear fusion. Such a star, bright though it may be while it lasts, does not furnish a good environment for life, because we think (using our Earth as a guide) that life needs billions of years to evolve from the first primitive organisms to an advanced civilization.

Why do more massive stars liberate kinetic energy at such enormous rates? Their central temperatures are higher, so protons and other nuclei fuse together much more rapidly. If we double the temperature at the center of a star—say, from 10 million to 20 million K—the number of nuclear-fusion reactions and therefore the rate of energy liberation will increase, not by a factor of two but by a factor of 50. The particles' increased ability to overcome their mutual repulsion produces a vast increase in their rate of fusion, and thus in the star's ability to turn energy of mass into kinetic energy.

And why do more massive stars have higher temperatures at their centers? Because the total weight of the overlying layers presses down more firmly. The Earth's atmosphere, for example, exerts 1 kilogram of pressure per square centimeter on any object at the Earth's surface. This pressure arises from the weight of the atmosphere, held down by the Earth's gravity and compressed into a much smaller volume than it would occupy if the Earth's gravity were much less. If we doubled the total amount of gas in our atmosphere, we would increase the pressure at the Earth's surface, because more gas would weigh down on each square centimeter.

Similarly, if we double the mass of a star, we increase the mass pulled toward the star's center by its self-gravitation. The increased tendency to contract will cause an increase in the temperature because of the extra pressure that the extra tendency to compress exerts within the gas. In response to this additional tendency to compress, the more massive star liberates more kinetic energy each second. Each star has just the "right" central temperature for its individual mass. The rate of energy liberation in the star's center (governed by the central temperature) provides just the amount of energy of motion needed for each star to resist its own self-gravitation. Table 5.1 shows the central temperatures that have been calculated to exist inside stars of various masses, along with the stars' rates of energy liberation and their expected lifetimes. The table shows that the less massive, far less energy-liberating stars will outlast their more massive, energy-prodigious cousins by tremendous margins.

Even though massive stars begin with a greater supply of hydrogen nuclei, these stars liberate kinetic energy at such large rates that they manage to burn through their fuel reserves at relatively great speeds. Roughly speaking, the rate at which a star liberates kinetic energy from energy of

TABLE 5.1 Estimated Masses, Central Temperatures, Luminosities, and Lifetimes of Certain Stars

Star	Estimated Mass (Sun = 1)	Central Temperature (millions of Kelvins)	Rate of Energy Liberation (Sun = 1)	Main-Sequence Lifetime* (Billions of Years)
Rigel	10	30	50,000	0.002
Sirius	2.3	20	23	1
Procyon	1.8	18	7.6	2.4
Alpha Centauri A	1.1	15	1.5	7
Sun	1.0	13	1.0	10
61 Cygni A	0.6	8	0.08	80
Proxima Centauri	0.1	6	0.00006	16,000

* The main-sequence lifetime is the most stable period of a star's life (see p. 119).

mass will vary as the *cube* of the star's mass, so a 10-solar-mass star will turn energy of mass into kinetic energy at least 1000 times more rapidly than the sun. Such a star, even with 10 times the sun's original supply of protons, can last for less than a hundredth of the sun's lifetime. In general, *stars' lifetimes vary in proportion to the stars' masses divided by their rates of energy liberation.* Their lifetimes therefore vary approximately as 1 over the *square* of the stars' masses, so a star with twice the sun's mass should have a lifetime one-quarter of the sun's, whereas a star with half the sun's mass should have a total lifetime four times longer than that of the sun.

The total lifetime of a star, and particularly the length of time that the star can radiate energy at a steady rate, has great importance in the search for life. Only those stars whose lifetimes exceed a few billion years are likely to permit intelligent life to develop on any planets they may have. This means that high-mass, high-luminosity stars are poor sites for life in their planetary systems.

Life on Earth required billions of years to evolve beyond the stage of single-celled, primitive organisms. If our planet's history represents any sort of guide to life in the universe, we must look to the lower-mass stars, stars with less than about 1.5 times the sun's mass, to guarantee the long lifetimes needed for life to develop. Luckily for us in the search for life, such stars are the majority: They far outnumber the more spectacular high-mass, high-luminosity but short-lived stars that form most of the stars we can see in the night skies.

When we examine the stars, we cannot detect their masses or their central temperatures directly. The largest telescopes can hardly show stars other than the sun as anything but points of light—clear proof of the enormous distances to even the nearest stars. What astronomers can determine with relative ease are the stars' surface temperatures and apparent brightnesses. By using these two measured quantities, astronomers have developed a fruitful science of classifying the various types of stars.

The Types of Stars

To determine the temperatures of the surface layers of stars, astronomers use spectrographs to spread the starlight into the spectrum of the light's various frequencies or wavelengths. Fig. 5.6 shows the spectrum of the light from our sun, by far the best-studied star. In this figure, the frequency (and thus the photon energy) decreases from top to bottom and also from left to right. The brightness at a particular location in the spectrum shows how many photons of that particular frequency or wavelength are present in the sun's spectrum of visible light. At certain frequencies, the sun emits few photons or none at all. These particular types of photons have

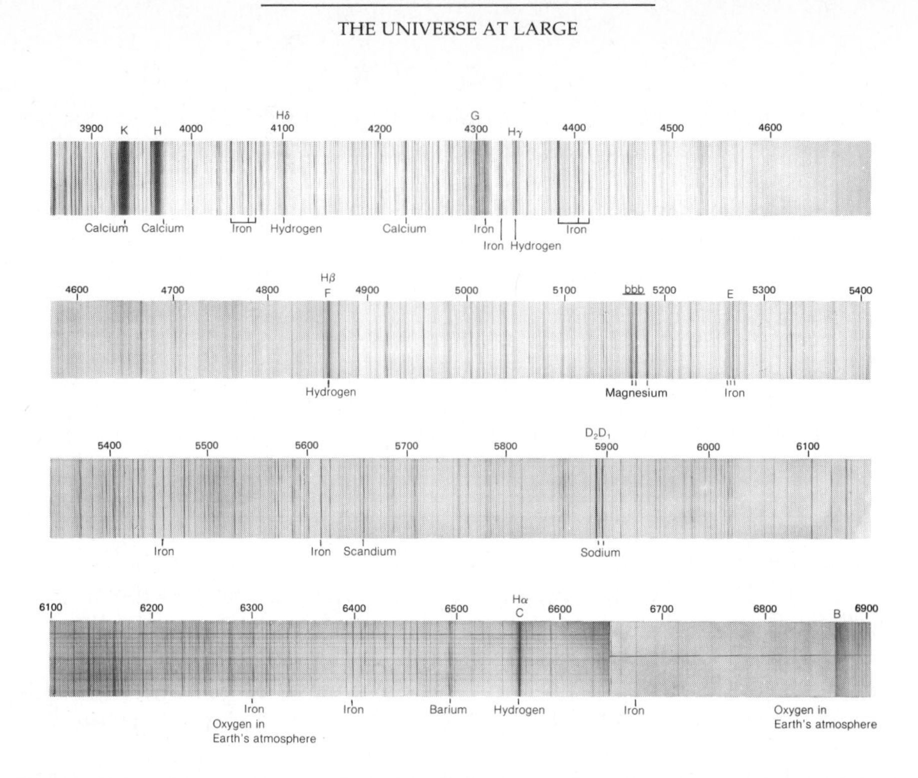

Figure 5.6 The spectrum of visible light from the sun shows a host of dark absorption lines, named "Fraunhofer lines" after their first investigator, that arise from different types of atoms, ions, and molecules in the sun's outer layers. The most prominent of these lines (top left) arise from calcium ions. In this spectrum, the wavelengths are designated in angstroms (units of 10^{-10} meter).

been absorbed in the sun's surface layers, as the light from below (which contains photons of all the frequencies of visible light) escapes into space.

The temperature of the sun's surface layers determines both the types of atoms that will be ionized and the degree of the atoms' ionization. Since the presence or absence of an absorption from a given type of atom or ion is determined by the temperature, temperature has a direct influence on the spectrum of the light that emerges from the sun or from any other star.

Astronomers use their knowledge of stellar spectra to classify stars on the basis of their surface temperatures. These spectral types, shown in Table 5.2 and in Fig. 5.7, were named before astronomers recognized that the differences in stars' spectra arise primarily from differences in the stars' surface temperatures. For this reason, the names of the spectral types show a notable lack of consistency: The hottest stars are type O; next are type B, then types A, F, G, K, and M. The O stars have surface temperatures of 30,000 K or more, 10 times those of the M stars. In addition to these basic

TABLE 5.2 Surface Temperatures and Spectral Features of Different Types of Stars

Spectral Type	Surface Temperature (Kelvins)	Chief Spectral Features
O	30,000	Absorption lines from ionized helium; weak hydrogen absorption lines
B	20,000	Absorption lines from non-ionized helium; stronger hydrogen absorption lines
A	10,000	Hydrogen absorption lines are most prominent spectral feature; still some absorption from non-ionized helium atoms
F	7,000	Hydrogen absorption lines still dominate spectrum; absorption lines from "heavy" elements are also present
G	5,500	Hydrogen absorption lines weak; many absorption lines from once-ionized and non-ionized "heavy" elements
K	4,000	No hydrogen lines visible; increasing number of absorption lines from non-ionized "heavy" elements
M	3,000	Many absorption lines from non-ionized atoms and from simple molecules

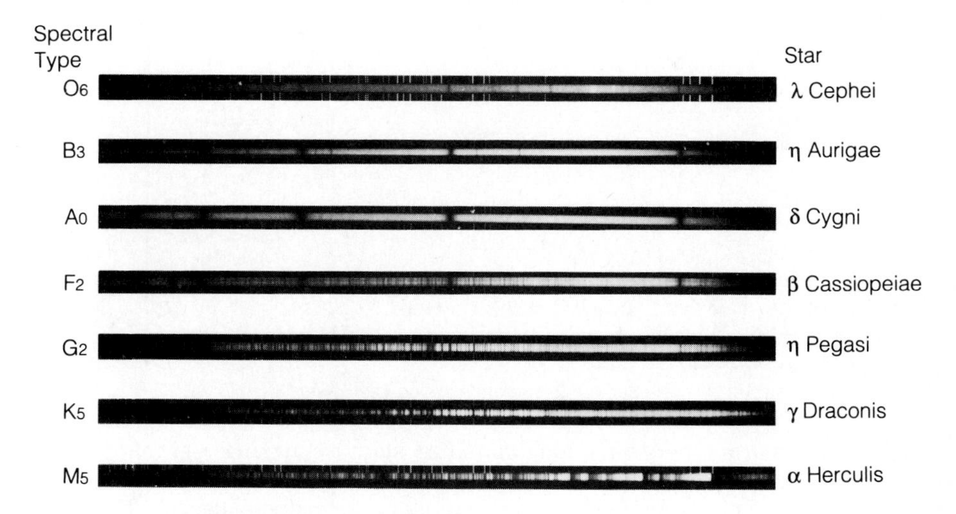

Figure 5.7 Stellar spectra are arranged into different classes that show different patterns of absorption lines. The stars with the hottest surface layers (top) show absorption lines from once-ionized helium, whereas those with the coolest surface layers (bottom) show spectra dominated by different types of molecules.

categories, astronomers subdivide each class of stars by placing a number from 0 to 9 after the letter, so that, for example, a G9 star has a slightly greater surface temperature than a K0 star. A popular mnemonic for recalling the spectral types from hottest to coolest is "Oh, Be A Fine Girl, Kiss Me," in which the initial letters of the words give the spectral types.

The Temperature-Luminosity Diagram

When we look at the luminosities of stars together with the stars' surface temperatures, a remarkable pattern emerges. Figure 5.8 shows a graph of the luminosities and the surface temperatures of a representative group of stars,

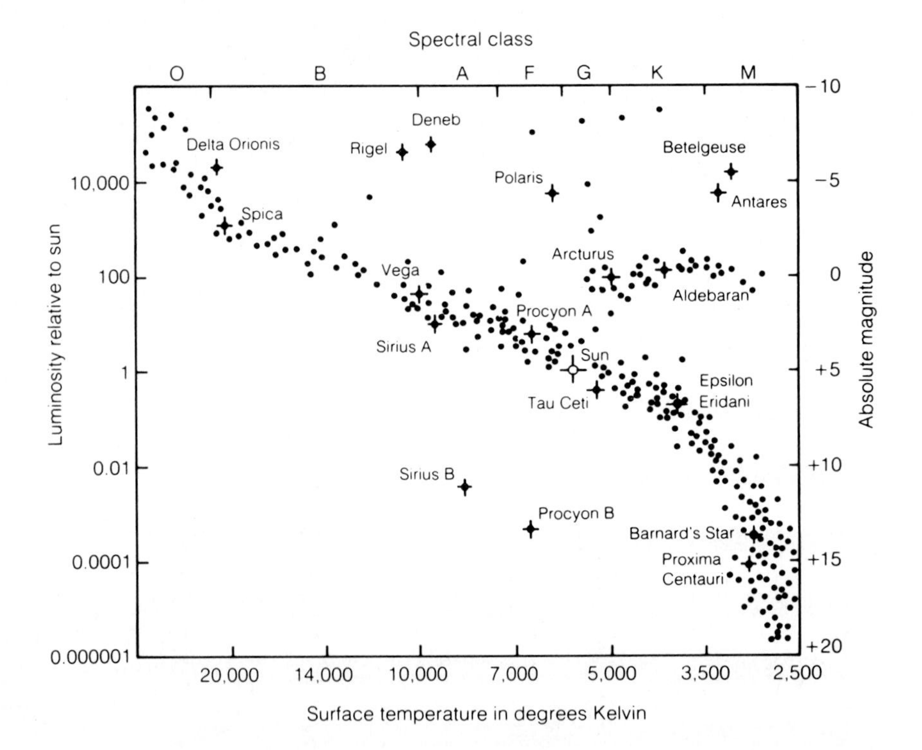

Figure 5.8 Astronomers have long classified stars by making a graph of their surface temperatures and luminosities. This classification system reveals that most stars' characteristics place them somewhere along a "main sequence." The chief exceptions are "red giants," such as Antares, Betelgeuse, Arcturus, and Aldebaran, which have higher luminosities than main-sequence stars of the same surface temperature, and "white dwarfs," such as Sirius B and Procyon B, with far lower luminosities than main-sequence stars of the same surface temperature.

including some of the best-known ones. The temperature-luminosity diagram is also called the Hertzsprung-Russell (H-R) diagram in honor of its originators. Figure 5.8 shows that most stars have surface temperatures and luminosities that place them along a particular swath of the graph, called the "main sequence." Stars on the main sequence have surface temperatures that range from a few thousand degrees to 30,000 K or more. These stars show a clear correlation between their surface temperatures and luminosities: *Stars with higher surface temperatures have much greater luminosities.*

This result could be expected if all of the stars on the main sequence have roughly the same size. The hotter stars radiate much more energy per second from each square centimeter of their surfaces, and the number of square centimeters on each star's surface would be about the same. (In fact, the hotter stars also tend to be larger than the cooler ones on the main sequence, which makes the hotter stars even more luminous in comparison to the cooler ones.)

The main sequence includes most of the stars that we can observe. *All stars spend the major part of their energy-liberating lifetimes on the main sequence.* The stars whose surface temperatures and true brightnesses place them somewhere on the main sequence are turning energy of mass into kinetic energy at a steady rate. These stars can hold themselves to an almost constant size as they liberate kinetic energy, thanks to the balance between their own self-gravitation and the outward push of the liberated energy of motion. Each star's mass determines its position on the main sequence: High-mass stars have high surface temperatures and large luminosities; low-mass stars have low surface temperatures and small luminosities (see again Fig. 5.8).

During the contraction of the protostars, the stars did not occupy the same points in the temperature-luminosity diagram that they do now (Fig. 5.9). Since these stars began to liberate kinetic energy from energy of mass, however, they have maintained nearly the same surface temperatures and the same luminosities, which imply almost the same positions on the main sequence. In short, stars evolve onto the main sequence, and later evolve off it, but they remain in almost the same position on the main sequence throughout most of their lifetimes, while energy liberation is occurring steadily in their interiors.

Red Giants and White Dwarfs

But what about the stars that do *not* appear along the main sequence on the temperature-luminosity diagram? Such stars, as we shall discuss in the next chapter, have completed the stage of steady liberation of kinetic energy. These stars have moved farther toward the end of their lives than the

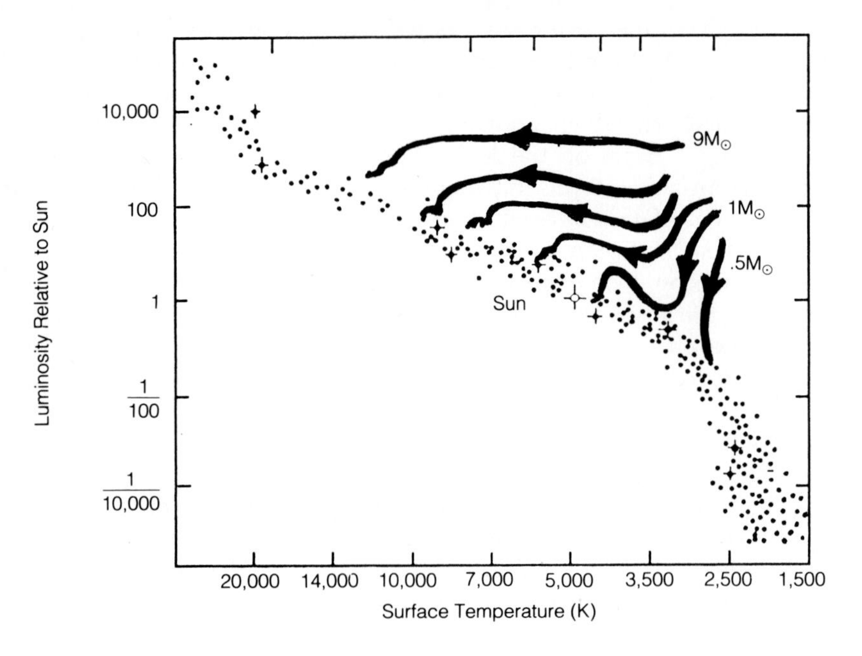

Figure 5.9 As a protostar shrinks to become a star, it undergoes a brief phase of high luminosity, during which the star's contraction provides most of its energy output. As the star approaches the main sequence on the temperature-luminosity diagram, its position on that sequence is determined by the star's *mass:* The more massive a main-sequence star, the higher its surface temperature and luminosity will be.

main-sequence stars, and appear either above the main sequence (the "red giants") or below it (the "white dwarfs").

Red-giant stars have relatively low surface temperatures (2000 to 6000 K), but great luminosities. For the luminosity of a low-temperature star to be large, the star itself must be enormous, since a low-temperature surface will radiate relatively little energy per second from each square centimeter. We therefore require a red giant to have an enormous surface area and thus an enormous radius. The largest such stars, the "red supergiants," have sizes thousands of times larger than the sun's radius, large enough to contain the *orbits* of the Earth and Mars! Since these stars have masses of no more than a few dozen times the mass of the sun, however, they must be extremely tenuous; they consist mainly of a gauzy near-vacuum that surrounds a smaller, dense core.

White dwarf stars, on the other hand, have relatively high surface temperatures (5000 to 15,000 K) but extremely small true brightnesses. Therefore the white dwarf stars must have relatively tiny surface areas and radii (Fig. 5.10). Most white dwarfs have sizes similar to the Earth's, yet contain almost as much mass as the sun! Since the sun's radius exceeds the Earth's by 100 times, the sun's volume is 1 million times that of the Earth. A white dwarf that has the mass of the sun within a volume as small as the Earth must have a density of matter 1 million times the density of matter in the sun. A cupful of white-dwarf matter would weigh 1000 tons at the Earth's surface!

Most of the matter in any star on the main sequence consists of hydrogen nuclei (protons), with helium nuclei providing about 25 percent of the total mass. All the elements heavier than hydrogen and helium form no more than 1 or 2 percent of a star's mass. This fact should not surprise us when we consider that the big bang left behind mostly hydrogen and helium, and that most of the nuclear fusion within a star produces helium nuclei from hydrogen nuclei. In fact, the question that must be answered turns out to be: Why do stars contain as much as 1 or 2 percent of their mass in the form of heavy elements—carbon, nitrogen, oxygen, neon, and so forth? How were these elements produced? The answer to these questions lies in the final stage of stars' lifetimes, when the familiar ways of liberating kinetic

• White dwarf (10^4 km diameter)

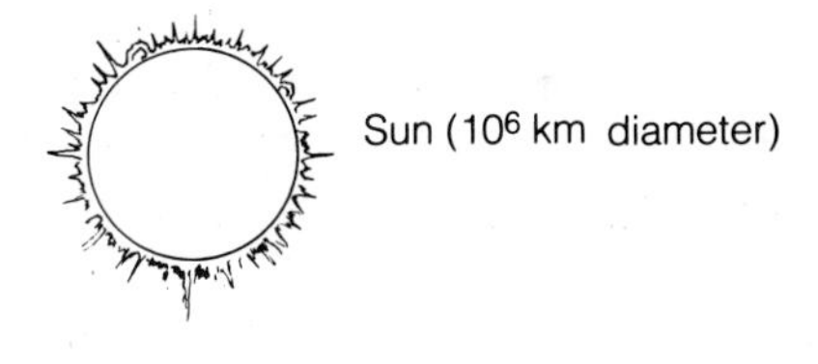

Arcturus (10^7 km diameter)

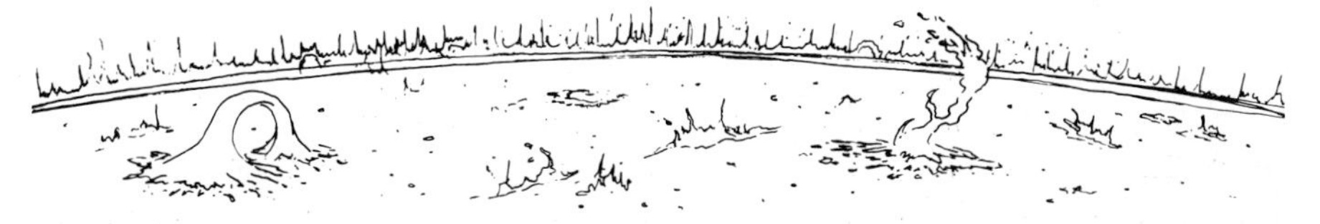

Figure 5.10 A red-giant star (lower) may have two or three times the sun's mass but a radius more than a hundred times the sun's, giving it an average density only one-millionth of the sun's 1 gram per cubic centimeter. In contrast, a white dwarf has a radius close to the Earth's, only one one-hundredth of the sun's, and therefore has a density of matter about a million times *greater* than that of the sun.

energy become no longer available. In the next chapter we shall see what happens inside stars that have exhausted their basic energy-liberating processes and thus have passed through middle age to enter a glorious senility.

SUMMARY

Stars shine by turning energy of mass into kinetic energy, in accord with Einstein's formula $E = mc^2$. Stars achieve this transformation by using a series of nuclear-fusion reactions in which nuclei—the centers of atoms—collide at such high energy that they fuse together, turning themselves into other types of nuclei. In most stars, this occurs through a series of nuclear reactions called the proton-proton cycle. In this cycle, four protons combine to form one helium nucleus, with the release of some kinetic energy as the energy of mass decreases slightly. Though each reaction releases only a tiny amount of new kinetic energy, huge numbers of reactions occur within the star's hottest, innermost regions each second.

This newly made kinetic energy spreads among all the particles inside the star through collisions. This gives the particles enough kinetic energy to resist the star's tendency to collapse under its own self-gravity. Because the outward flow of liberated kinetic energy balances the inward pull of self-gravitation, most stars can regulate their rate of energy liberation accurately. Stars that achieve a steady rate in this manner appear on the main sequence of a graph of stars' luminosities and surface temperatures. The stars with larger masses have higher surface temperatures and much greater luminosities than small-mass stars. As a result, the large-mass stars burn out their supply of protons more rapidly than do low-mass stars, by fusing them into helium nuclei to release kinetic energy.

QUESTIONS

1. What does it mean to say that stars "liberate" energy? How do they turn energy of mass into energy of motion (kinetic energy)? Which part of a star has a temperature high enough for this process to occur, and why is this region favored for nuclear fusion?

2. On a clear night, take a star chart and identify some of the brightest stars in the sky, such as Vega, Deneb, and Altair in the fall; Sirius, Betelgeuse, Rigel, Capella, and Aldebaran in the winter; or Regulus, Arcturus, and Spica in the spring. Can you see the colors of these stars, if only faintly? What

property of the star does the color help to show, for use in an attempt to locate these stars on the temperature-luminosity diagram? What data are you lacking in order to determine the stars' actual positions on the diagram?

3. Which part of the proton-proton cycle produces a particle of antimatter? What happens to this antiparticle?

4. Why does a protostar cease to contract? Is it fair to say that the onset of nuclear fusion keeps the star from growing hotter? Why? Why doesn't the start of nuclear fusion blow a star apart from the effect of newly released kinetic energy?

5. Why do more massive stars exhaust their nuclear fuel in less time than less massive stars?

6. Alpha Centauri and Capella are both G-type stars with roughly the same apparent brightness and surface temperature, but Capella has about 10 times Alpha Centauri's distance. By how many times does Capella's luminosity exceed Alpha Centauri's? What sort of a star is Capella?

7. Why do most stars lie close to the main-sequence line in the temperature-luminosity diagram? Which stars do not lie along the main sequence?

8. Why do stars of different masses occupy different positions on the main sequence?

9. Antares and Barnard's star (the fourth closest star to the sun) have the same surface temperature, but the surface area of Antares exceeds that of Barnard's star by 100 million times. How do the luminosities of these two stars compare?

FURTHER READING

Bernstein, Jeremy. *Hans Bethe: Prophet of Energy.* New York: Basic Books, 1980.

Gamow, George. *A Star Called the Sun.* New York: Viking, 1964.

Haramundanis, Katharine, ed. *Cecilia Payne-Gaposchkin: An Autobiography and Other Recollections.* London and New York: Cambridge University Press, 1984.

Jastrow, Robert. *Red Giants and White Dwarfs,* 3rd ed. New York: Warner, 1990.

Maffei, Paolo. *When the Sun Dies.* Cambridge, Mass.: MIT Press, 1982.

Malin, David, and Paul Murdin. *Colours of the Stars.* New York and London: Cambridge University Press, 1984.

Wentzel, Donat. *The Restless Sun.* Washington: Smithsonian Institution Press, 1989.

6

How Stars End Their Lives

If our own existence on Earth provides a guide to the cosmos, stars are the source of the energy that maintains life on planets in orbit around them. Therefore, both for ourselves and in our search for life, we have more than a passing interest in the fates of stars as they begin to exhaust their nuclear fuel. When the sun runs out of protons in its nuclear-fusing core, the era that we now enjoy because of its present, steady energy output will be gone forever. But this is not the only way in which stellar aging and death can affect us. Even now we are living by and through the ashes of stars that have long since exhausted their supplies of protons. Every molecule in our bodies contains matter from exploded stars. The air we breathe, the ground on which we live, the seas around us, and the flowers that bloom in the spring all came from the cinders of long-vanished stars that first collapsed, then erupted, as they exhausted all the processes that could liberate more kinetic energy in their interiors.

Nuclear Fuel Consumption in Stars

Because a star's self-gravitation never ceases, a star will inevitably contract if it can neither liberate sufficient kinetic energy each second to overcome its self-gravitation nor find some other means to support itself against its own gravitational force. Once most of the protons in a star's central regions have been fused into helium nuclei, the star cannot liberate much more energy from the fusion of protons. Helium nuclei provide a natural back-up supply of "fuel" for additional fusion reactions, but with a catch: Helium-4 (^{4}He) nuclei are extremely reluctant to fuse together. To make ^{4}He nuclei fuse requires a temperature not of millions of degrees K, but rather of hundreds of millions of degrees K. Only at these enormous temperatures can ^{4}He nuclei overcome their mutual repulsion and approach closely enough in collisions to fuse together.

These fusion reactions, called the "triple-alpha process" after the old name ("alpha particle") for a ^{4}He nucleus, first make two ^{4}He nuclei fuse

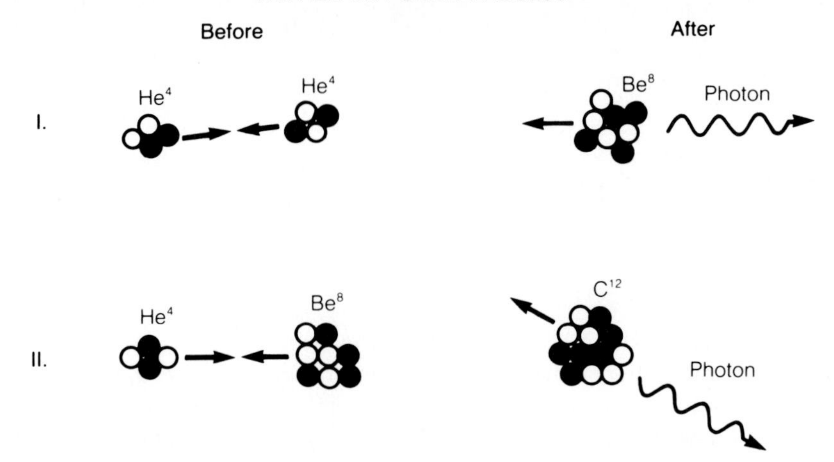

Figure 6.1 At temperatures of about 200 million K, nuclei of helium-4 (^{4}He) will fuse together. This fusion produces a nucleus of beryllium-8 (^{8}Be). If another helium-4 nucleus fuses with the beryllium-8 nucleus before it decays, the second fusion will produce a nucleus of carbon-12 (^{12}C) and a photon.

into a nucleus of beryllium-8 (^{8}Be) plus a photon. Soon afterward, the ^{8}Be nucleus will fuse with another ^{4}He nucleus to produce a nucleus of carbon-12 (^{12}C) and another photon (Fig. 6.1). The combination of the two reactions fuses three ^{4}He nuclei into a ^{12}C nucleus plus two photons, and liberates 1.2×10^{-6} erg of kinetic energy from energy of mass. Notice that this amount of energy is only one-quarter of the 4.2×10^{-5} erg of energy liberated in the proton-proton cycle: The star can liberate far less kinetic energy from the fusion of ^{4}He nuclei than from the fusion of protons. Since each ^{4}He nucleus was made from four protons, the disparity in energy-liberating ability between hydrogen and helium nuclei is even greater than it first seems.

The Evolution of Stars

Consider what happens as a star exhausts the protons in its center through fusion reactions. First, the dwindling supply of protons must continue to supply enough energy to keep the star from collapsing. The star can do this by contracting its central core: Its self-gravitation dominates the kinetic energy released by nuclear fusion, and the core grows somewhat smaller. This contraction raises the temperature in the core, causing the remaining

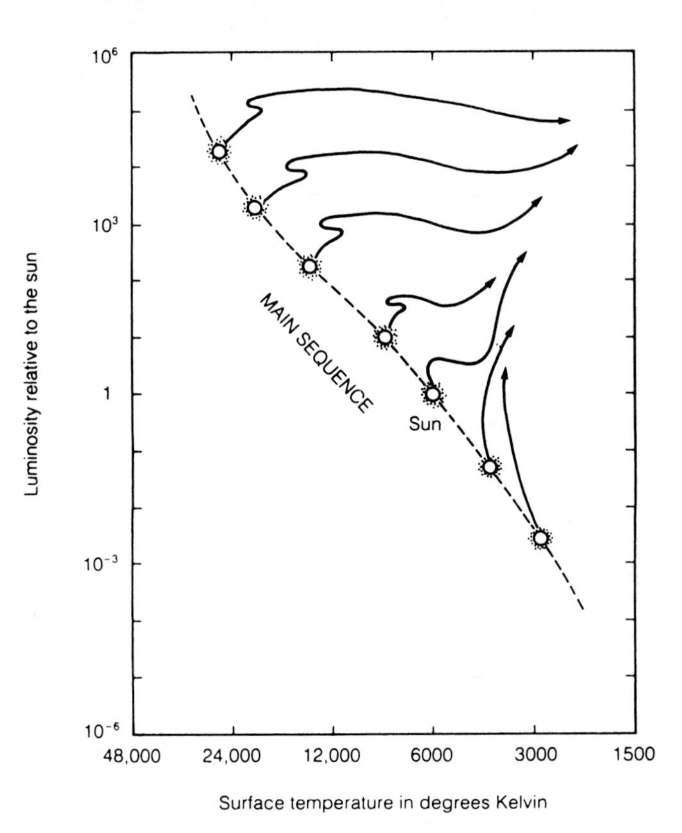

Figure 6.2 As a star ends its main-sequence lifetime, it grows more luminous. As its outer layers expand, they cool, taking the star into the red-giant region at the top right of the temperature-luminosity diagram.

protons to fuse together more rapidly, with a greater rate of energy liberation and a greater rate of exhaustion of the proton supply. The star's core behaves like a driver who reacts to a dwindling fuel supply by pressing down harder on the gas pedal. The star's central regions, which contain about half the star's mass, keep on contracting and growing hotter, and becoming almost totally depleted of protons, until the helium nuclei themselves start to fuse together because the temperature in the core has risen to 100 million K.

The onset of nuclear-fusion reactions among the helium nuclei provides a sudden rush of kinetic energy that briefly makes the star much brighter. This "helium flash" marks the high point of a star's luminosity; afterward, the slide towards obscurity continues.[1] Figure 6.2 shows the changes in the

[1] The helium flash takes less than a second in the star's core, but for the energy to diffuse outward takes much longer. Thus if we were lucky enough to find a star during the period immediately following the core's helium flash, we would see a brightening for several days.

luminosities and surface temperatures of stars at various points along the main sequence. As most of the protons in the stars' inner regions fuse into helium nuclei, the stars leave the main-sequence locations on the graph that typify steady proton fusion. During this phase, the stars' inner cores become almost pure helium, while protons continue to fuse in a shell surrounding the central core (Fig. 6.3). Some stars actually increase their luminosities while contracting their cores; they all raise their central temperatures to fuse the remaining protons more and more rapidly in thin shells of material surrounding their pure-helium cores.

The helium flash liberates so much kinetic energy in such a short time that the core expands slightly. This in turn lowers the temperature and the rate of helium fusion. Nonetheless, the helium fusion reactions continue, and the shrinking core continues to liberate kinetic energy from a steadily diminishing supply of helium nuclei. Because the rate of fusion reactions depends on the temperature in an extreme manner, the star can, for a brief time, liberate more kinetic energy per second from a depleted stock of nuclei. As the star does so, the increased amount of newly liberated kinetic energy pushes outward on the star's outer layers, expanding them beyond their original size. The expansion cools the gases, so the star creates a cool, highly extended outer "envelope" at the same time that its core grows extremely dense and hot.

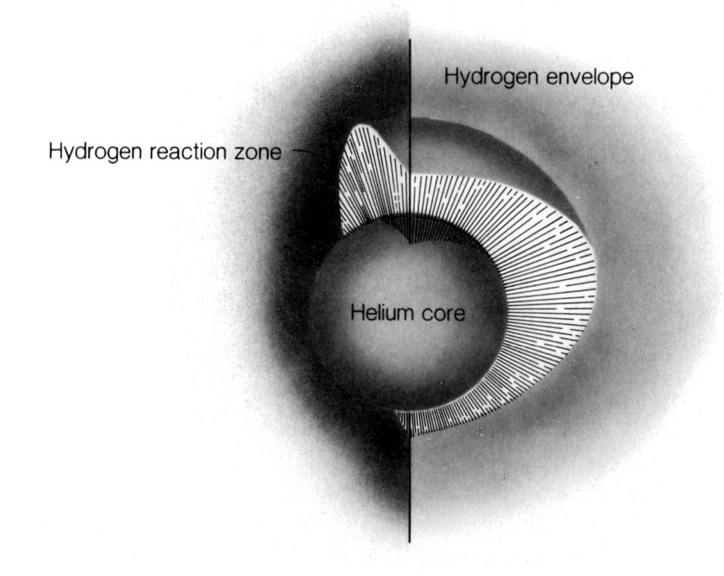

Figure 6.3 After a star has fused nearly all the hydrogen nuclei (protons) in its core, the fusion of protons continues in a shell of material surrounding the nearly pure-helium core.

The star thus becomes a red giant, cooler and more rarefied on the outside, but hotter and denser in its core, than a main-sequence star. Eventually the star will expel its outermost layers completely into space, exposing the hotter subsurface layers beneath. Then for a time, the star may shine with a high luminosity and a high surface temperature—so large, in fact, that the star emits most of its energy in the ultraviolet part of the spectrum, invisible to the human eye. The ejected outer layers may, however, trap some of the star's ultraviolet photons, which will ionize the atoms in the expanding shell around the star. As the atoms recombine, electron jumps into smaller orbits can produce visible-light photons, so we see a "planetary nebula" from the star's discarded mantle (Fig. 6.4). Each planetary nebula radiates energy that comes from a nearly invisible (to us) star that occupies the center of a spherical shell of gas expelled from that star.

How long do the various phases of a star's evolution take? Suppose that we analogize the entire lifetime of a star with a human lifetime (70

Figure 6.4 A planetary nebula arises in shells of material expelled from a star's outer layers during its red-giant phase. As the outer layers evaporate, the hot surface they expose radiates ultraviolet photons that ionize most of the hydrogen atoms in the surrounding shell. Once these ions recombine with electrons, the newly formed atoms radiate visible light as the electrons jump into smaller orbits.

years). Then the initial contraction and the protostar phase represent the star's childhood (say 15 years), while the final stages of contraction and the onset of energy liberation pass much more rapidly (1 year). The star's main-sequence lifetime lasts for perhaps 50 years, and the evolution from the main sequence up to the time of the helium flash takes just a year more. The helium flash itself must be reproduced by a billionth of a second (in real life it takes less than a single second), whereas the red-giant phase that follows lasts for perhaps three years. The subsequent planetary nebula phase and all that follows will last less than a year before the star either explodes or becomes a white dwarf. We use the analogy with a human lifetime, rather than naming specific numbers of years in a star's true history, because stars of different masses require quite different times for each phase of their lifetimes. Thus, for example, the sun will spend 10 billion years on the main sequence and 100 million years from there to the helium flash, whereas a star with five solar masses will spend just 70 million years on the main sequence and 5 million years from there to the helium flash. These times, estimated from detailed calculations, underscore the key fact that *most of the star's lifetime passes on the main sequence.*

We have followed a star's evolution from the onset of proton-proton reactions through the stages of helium-to-carbon fusion. What happens as a star exhausts its helium nuclei by fusing them into carbon nuclei? Each star will follow one of two basically different life stories, depending on the mass that the star has. Low-mass stars will become white dwarfs, while high-mass stars will not and are likely candidates for supernova explosions.

The Exclusion Principle in White Dwarf Stars

White dwarf stars are the tremendously compressed centers of once-normal stars that have shed their outer layers to expose their inner cores, now made mostly of carbon nuclei and electrons. What distinguishes white dwarfs from other stars, and supports the white dwarfs against their self-gravitation, is the fact that the electrons in white dwarf stars are supported by the *exclusion principle.* This phrase is the scientific name for something that seems impossible to believe: Certain types of particles (in particular, electrons, protons, and neutrons) simply cannot be squeezed together indefinitely. They impose their resistance to further compression once the spaces within "normal" matter (atoms) have been "squeezed out." This resistance—the exclusion principle—acts in addition to, and much more powerfully than, the general repulsion that electrons exert on one another through electromagnetic forces. Actually, the exclusion principle plays a role in the structure of all familiar matter, because it governs how many electrons can fit into orbits around an atomic nucleus.

Within aging stars, the exclusion principle becomes important only when the density of matter exceeds about 1 million grams per cubic centimeter, a million times the density of water. At lower densities, the exclusion principle plays no role in the structure of a star's interior. But consider a star that keeps contracting its central regions, thus raising its central temperature. Once most of the helium in the star's core has become carbon nuclei, the carbon nuclei could potentially fuse into heavier nuclei, liberating still more kinetic energy. However, if the star's interior has contracted so far that the matter reaches a density of 1 million grams per cubic centimeter, such carbon fusion will not occur. Why not? The electrons, obeying the exclusion principle, will refuse to compress to a greater density. If the electrons resist further compression, so do the carbon nuclei: The electromagnetic attractive forces between the negatively charged electrons and the positively charged nuclei effectively lock the nuclei in position, so that they cannot react to their mutual gravitational pull toward the center of the star.

The fact that the electrons in a white dwarf hold the carbon nuclei apart rather than allowing them to rush toward the star's center prevents the nuclei from fusing together, since they can never collide with speed sufficient to overcome their mutual electromagnetic repulsion. Main-sequence stars have a density of matter in their centers that is lower than that of white dwarf stars, and in the interiors of main-sequence stars, head-on, high-velocity collisions maintain the cycle of proton-proton fusion reactions. In contrast, white dwarf stars have extremely high densities but no possibility of fusion reactions, because the nuclei can never collide with enough kinetic energy to fuse the nuclei together.

In a white dwarf—the Earth-sized remnant core of a once-ordinary main-sequence star—heat energy passes outwards by conduction from the interior and radiates into space, but the star liberates no additional kinetic energy. Strangely enough, because white dwarfs radiate relatively little energy (their luminosities reach at most one-thousandth of the sun's), they can shine for billions of years, growing slowly but steadily dimmer as they expend the stored energy from their earlier years (Fig. 6.5).

These white dwarfs would provide a stable energy source for any living creatures that orbited close around them, content with a tiny fraction of a percent of the sun's energy output. When we consider, however, the red giant and helium-flash phases that preceded the white dwarf, not to mention the planetary-nebula expulsion of the star's outer layers, we may easily conclude that life might not reach this stable configuration. Any civilization on a planet orbiting an aging star would have to deal with tremendous changes in the star's energy output. Only if a civilization foresaw what would happen to their parent star, then managed to ride out the ups and downs of the red-giant phase, and finally moved close to the white dwarf

Figure 6.5 The white dwarf companion to the nearby star Sirius, called Sirius B, has about the same mass as Sirius A, but shines with about one-millionth the luminosity of its main-sequence companion, which has not yet evolved into its white dwarf phase. This highly overexposed photograph barely reveals the white dwarf in the glare of Sirius A.

remnant, would the possibility of "lowered expectation" living become a reasonable strategy for the next billion years.

The best-known white dwarf is Sirius B, the companion to the brightest star in our night skies. These two stars orbit their common center of mass with an orbital period of about 50 years. Packed to about a million times the density of water, the matter in Sirius B totals a mass almost equal to the sun's within a diameter a bit larger than the Earth's. Since we believe that the star that has become Sirius B was born at the same time as Sirius A, and since we know that more massive stars age more rapidly, it seems likely that Sirius B once had more mass than Sirius A (2.3 solar masses), but lost most of its mass during its red-giant phase before settling into white-dwarf obscurity.

Supernova Explosions

So much for the majority of stars—the numerous but modest ones that end as white dwarfs. What of the second path of stellar evolution, the one that produces supernovae like those that made the elements within us? Why do some stars explode, seeding the universe with useful nuclei, rather than fade away as compact dwarfs? The fact that a small proportion of stars explode at the end of their lives allows us to exist and to ponder their nature; we are fortunate, however, that our own sun is not one of the fast-evolving, soon-to-explode supernovae that prove so essential for the existence of life.

Stars end their lives differently because they have different masses, which in turn imply different densities at the stars' centers. More massive stars have lower central densities, because the more massive stars can reach a given central temperature with a lower density of matter in their cores. Why? Because the more massive stars have more matter pressing down on their centers, weighing more heavily on the matter there. The material at the center responds to this greater weight by raising its temperature without requiring so large a density of particles to produce a given temperature.

This relative ease of producing high temperatures gives high-mass stars higher central temperatures than low-mass stars. In low-mass stars, the lesser gravitational force cannot raise the star's central pressure and temperature as easily, so the star must achieve this result (in order to balance the gravitational force) by squeezing matter more tightly in the star's interior. Less massive stars therefore have a greater density of matter at their centers than more massive stars do at the same point in their evolution.

In the less massive stars (masses less than about 1.4 times the sun's mass), the exclusion principle becomes important by the time the stars' helium nuclei have fused to form carbon nuclei. All white dwarf stars, so far as we can tell, have masses less than 1.4 times that of the sun, and most have masses that are less than the sun's.

The more massive stars never become so dense in their interiors, even after helium-to-carbon fusion, that the exclusion principle prevents further nuclear fusion. These stars can proceed to fuse carbon nuclei into heavier nuclei. Such fusion reactions liberate a bit more kinetic energy, but each successive type of reaction liberates less and less energy. If two ^{12}C nuclei fuse together to form a nucleus of magnesium-24 (^{24}Mg), the fusion process liberates only one-tenth as much kinetic energy as the fusion processes in the proton-proton cycle. But the star must continue to liberate kinetic energy as rapidly as always or face ever-increasing contraction. In fact, massive stars pass through a series of fusion reactions that yield less and less kinetic energy, producing all the elements from carbon to iron before they completely exhaust their ability to liberate more kinetic energy through nuclear fusion.

Iron marks the end of the line, because to make nuclei heavier than iron-56 (^{56}Fe, with 26 protons and 30 neutrons per nucleus), we must *add* kinetic energy. In other words, the fusion reactions that form the nuclei *less* massive than iron result in less energy of mass and more kinetic energy, whereas fusion reactions that make nuclei *more* massive than iron finish with more energy of mass and less kinetic energy.[2] A star that has fused most of the nuclei in its core into ^{56}Fe has no way to find a new, even if less

[2] Since the fusion of nuclei up to and including iron-56 liberates kinetic energy, we must supply kinetic energy to reverse the process and break apart these lighter nuclei. In contrast, since making nuclei more complex than iron requires kinetic energy, we can gain kinetic energy when these nuclei split apart (for example, from the fission of uranium nuclei).

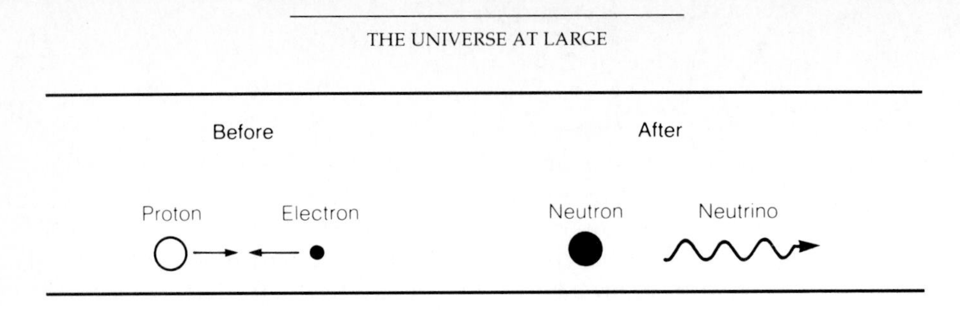

Figure 6.6 At the enormous densities of matter that exist within a collapsing stellar core, weak forces allow electrons to fuse with protons. This never occurs at "ordinary" densities of matter. The "weak reactions" produce neutrons and neutrinos from the protons and electrons.

effective, fusion reaction to liberate more kinetic energy. Such a star, if it has not yet achieved a central density high enough for the exclusion principle to be significant, has exhausted all means of supporting its core against gravitational collapse.

So the core collapses under its own gravitation, shrinking to a thousandth of its initial size in less than a second. Why doesn't the exclusion principle prevent this collapse? The reason is that at the fantastically high temperatures inside the collapsing core, the electrons fuse together with protons broken from the colliding nuclei. As the core of such a star collapses, the collapse itself provides the energy needed to break the iron nuclei apart into protons and neutrons as the nuclei collide with each other. The protons will then fuse with electrons to produce neutrons and neutrinos, in a reaction that is almost the opposite of the decay of individual neutrons (Fig. 6.6).[3]

Such fusion reactions between protons and electrons are weak reactions, governed by weak forces. Since electrons do not feel the strong forces that bind protons and neutrons together, they do not ordinarily fuse with nuclei as protons and neutrons do. Furthermore, electromagnetic forces, which do attract the oppositely charged electrons and nuclei, always get the electrons into orbits *around* the nuclei, and never make them fuse together. This leaves weak forces as the means of producing the fusion of electrons and protons, since gravitational forces are insignificant at the size level of elementary particles. As their name implies, weak forces do not usually have much success at producing such fusion. If, however, the temperature in stellar cores rises to billions of degrees and the density rises to hundreds

[3] Owing to the effect of weak forces, any isolated neutron will fall apart (decay) after about 10 minutes into a proton, an electron, and an antineutrino. If we imagine the exact reversal of this process, we must bring together a proton, an electron, and an antineutrino to form a neutron. A near-duplicate of this reverse process brings a proton and an electron together to make a neutron and a neutrino.

of thousands of grams per cubic centimeter (still not high enough for the exclusion principle to have an important effect on the electrons), then the electrons will indeed begin to fuse with the protons. As the electrons disappear, they take with them the star's chances of using the exclusion principle's effect upon electrons to hold itself up. The fusion of protons with electrons allows the star's core to continue its sudden, violent collapse—which itself provides the temperatures and densities high enough to make the protons and electrons fuse together.

The Second Type of Supernova

We have described the origin of what astronomers call a "Type II supernova"—a massive star whose core collapses once it has become mostly iron nuclei. Another type of supernova exists, called the "Type I supernova." These supernovae, intriguingly, are believed to arise when a white dwarf belongs to a double-star system and receives matter from its companion, presumably when the companion star swells through its red-giant phase. If the infalling matter is relatively rich in protons (as are the outer layers of a red giant), then the matter can accumulate on the surface of the white dwarf and suddenly detonate, once the temperature rises to the point that protons can fuse together. This "nuclear conflagration" repeats, on a white dwarf's surface, what occurs in the cores of normal stars. Because the nuclear fire has nothing to contain it, the detonation can ignite all at once and can blow apart the white dwarf and the material that has accumulated upon it.

So far as astronomers can tell, about half of the supernovae that occur are Type I, white-dwarf, supernovae, and the other half are Type II, core-collapse supernovae. The two types of explosions arise in different situations, though both occur near the ends of the lives of extraordinary stars or star systems. From the point of view of their effect on the universe, however, the two types of supernovae have a striking similarity: They made us what we are today.

The Production of Heavy Elements in Supernovae

The collapse of the core of a massive star, or the "nuclear conflagration" on the surface of a white dwarf, takes no longer than a second or so of time. This is an important second for the star and for the rest of the universe. During the collapse of a massive star's core, fusion reactions by the trillions occur as the nuclei smash into one another. Such collisions mostly have the effect of stripping the nuclei (iron-56 and similar nuclear species) into their constituent protons and neutrons. In addition, however, such collisions can produce small amounts of nuclei heavier than iron, such as mercury, silver, lead, gold, platinum, and uranium. In effect, the star's collapse uses the last energy source available—the energy that held up the core—to make these heavy elements.

In a similar way, the nuclear fusion that occurs on the surface of a white dwarf that accumulates matter from its companion star can also produce relatively small amounts of the nuclei of heavier elements. Both types of supernovae contribute to the cosmic supply of heavy elements. (In addition, a small contribution arises from aging stars, which manufacture some heavy elements as byproducts of their basic mechanisms of nuclear fusion, and gently expel these elements into space as they expand as red giants.)

When we look at the elements that form the Earth and comprise living organisms, we find an important difference in their abundances. Nuclei of the types that were made during the millions of years before a massive star's core could collapse—such as carbon, nitrogen, oxygen, aluminum, silicon, and iron—are far more abundant than those nuclei made during the single second of collapse, such as molybdenum, silver, platinum, gold, and mercury (Fig. 6.7). Living organisms, in fact, consist almost exclusively of elements lighter than iron—that is, of the more abundant elements, made during the long periods before stellar collapse. Trace amounts of the elements heavier than iron do, however, play a role in most living creatures on Earth. In a similar fashion, more than 99.9 percent of the Earth consists of elements no heavier than iron, but the relatively tiny amounts of silver, gold, and uranium made during the final second of a star's life have played a role in human history, partly because of their very scarcity, the result of the short amount of time during which they were made.

How do these elements blast outward from the collapsing star? In addition to the production of heavy nuclei, the rush of fusion reactions inside the collapsing star creates enormous numbers of neutrons, the result of proton-electron collisions at tremendous energies. These neutrons quickly form a core called a "neutron star," a compact object only a few kilometers in diameter made almost entirely of neutrons (Fig. 6.8). Like electrons, neutrons obey the exclusion principle and cannot be squeezed past a certain density much higher than the corresponding density for electrons. As the neutrons refuse to be squeezed further, matter from the outer layers falls onto the surface of the newly made neutron star and "squeezes" it to a slightly smaller size. The "squeezing" of the neutron star produces an outward "bounce," like that of a squeezed rubber ball. The "bounce" starts an outward explosion of the outer layers that produces a tremendous outpouring of energy, a last gasp of the star that can light up the sky. This sudden outburst signals that the collapse of the star's inner regions, yielding an incredibly dense neutron-star core, has occurred (Fig. 6.9).

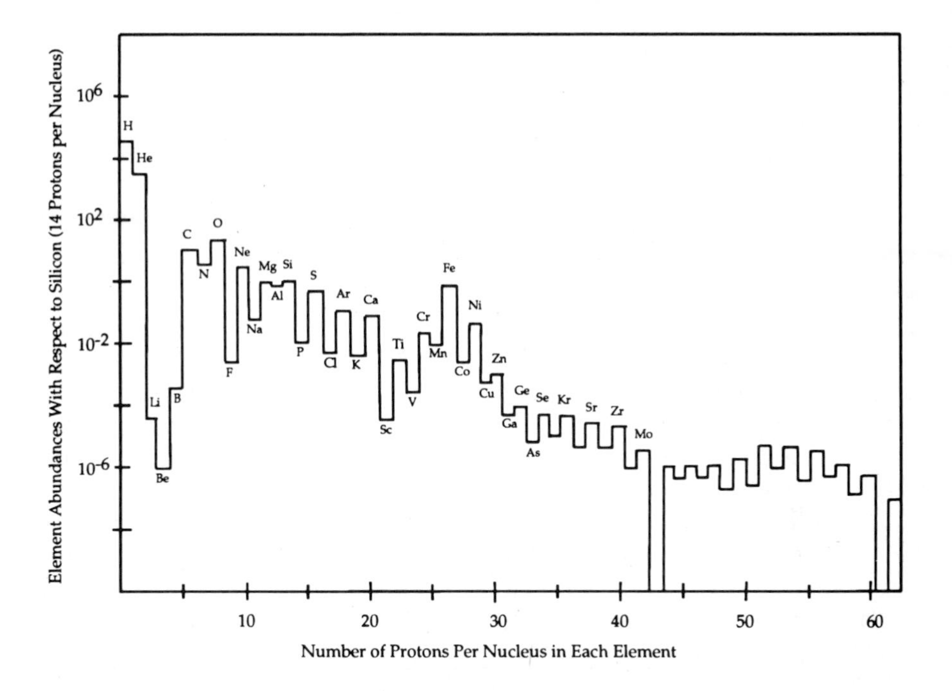

Figure 6.7 The abundances of the different elements show a steep decline from the two lightest and most abundant, hydrogen and helium, toward heavier elements. Note that the abundance scale is logarithmic, proceeding in powers of ten. Note also that starting with carbon, the elements with an even number of protons tend to be more abundant than neighboring elements with an odd number of protons, the result of the fact that many elements form from the fusion of ^{4}He nuclei.

Such supernova explosions would be fatal for any life that inhabited planets around the star. Even if we suppose that living creatures on these planets had managed to survive the pre-supernova evolution of the star, when the star radiated energy at a greater and greater rate, the final outburst would surpass all preparation. Most supernovae release hundreds of millions or even billions of times more energy each second than the sun does, and they do so for several weeks or months. The effect on a planet like the Earth would exceed what would happen here if the sun were suddenly brought to the moon's distance from us. In fact, this comparison is far too mild; it would be more accurate to compare the effect of a supernova's energy release on nearby planets to the experience of a hydrogen bomb close up.

From a distance, supernova explosions are spectacular but safe. A large spiral galaxy—our own Milky Way, for example—has a supernova about

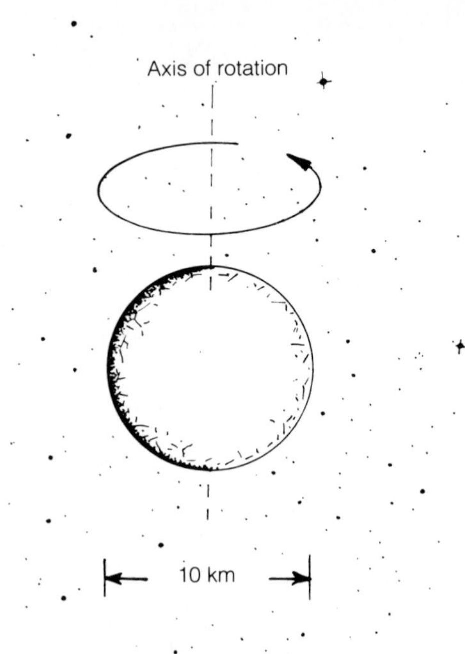

Figure 6.8 A neutron star, with a 10-kilometer radius and containing more mass than the sun, amounts to a single crystalline structure, made almost entirely of neutrons, which are held in place by the exclusion principle.

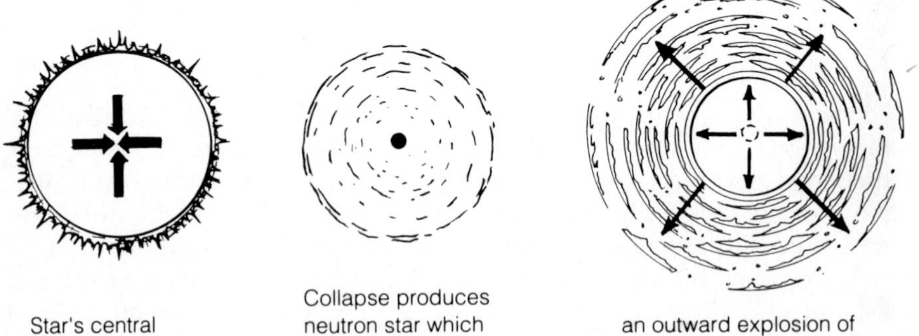

Figure 6.9 One type of supernova is thought to arise when a star's core collapses, then squeezes itself so tightly that it "bounces" outward. The bounce starts a shock wave that spreads outward at greater and greater velocities as it encounters progressively less dense material, and blasts the star's outermost layers into space at nearly the speed of light.

every 50 or 100 years. The most spectacular "nearby" supernova of this century appeared on February 23, 1987, in the Large Magellanic Cloud, the closest galaxy to the Milky Way (Fig. 6.10). Astronomers then see a new star appear where they could see no star before, since the pre-supernova star typically has too little brightness to be visible. The supernova, however, can be a billion times brighter than the star that preceded it, brighter than an entire medium-sized galaxy (see Fig. 5.2)! Within a few months, it fades to invisibility. The total energy output during this explosion, immense though it seems, is less than the energy steadily radiated by the star during the hundreds of millions or billions of years preceding this spectacular event.

Astronomers last observed supernovae within our own galaxy in 1572 and 1604, and Chinese astronomical records of supernovae go back three millennia. The supernova remnant shown in Fig. 6.11 arose from an exploding star that appeared in our skies in the year 1054, and passed completely unrecorded in western Europe, though it was noted by Chinese and Islamic

Figure 6.10 In 1987, a supernova (lower center) appeared in the Large Magellanic Cloud, some 50,000 parsecs from the Milky Way. This supernova, which had actually exploded about 165,000 years ago, was the closest supernova to be seen since the year 1604.

Figure 6.11 The Crab Nebula, about 2200 parsecs away in the direction of the constellation Taurus, is the remnant of a star that was seen to explode in the year 1054 but had actually exploded some 7000 years before that.

scholars. We may expect to see another supernova in our galaxy soon, since one explodes each 50 to 100 years, on the average. In view of the fact that we are probably thousands of light years from the next one that we shall see—as, indeed, we were from the last three—we may be almost positive that the star responsible for it has already exploded, but its light has not yet reached the Earth. Every year brings the discovery of several new supernovae in relatively nearby galaxies.

In a galaxy such as our own, the first generation of stars formed from the matter that emerged from the big bang—matter that was almost entirely hydrogen and helium nuclei, plus electrons, neutrinos, and antineutrinos. Less than one-millionth of the matter made soon after the big bang consisted of nuclei heavier than helium. These first stars therefore contained relatively few of the elements heavier than helium. By now, these stars have all vanished. The oldest stars that we see now, called Population II stars, must have formed somewhat later, as a second generation of stars. Population II stars include smaller amounts of the "heavier" elements (carbon, nitrogen, oxygen, neon, and so forth) than the fraction (1 percent by mass) that characterizes stars like our sun. Stars that are sunlike in their composition are called Population I stars; they must have formed *after* the Population II stars did. It appears likely that a high proportion of the early stars had relatively high masses, several times the mass of our sun. These stars therefore evolved quickly (in less than a few billion years) through their energy-liberating lifetimes and exploded as supernovae, flinging nuclei of the heavier elements (all the way from carbon through uranium) out into the rest of the galaxy.

The third and following generations of stars, which include most of the stars now in our galaxy, resemble our sun in the fraction of heavy elements they contain. Approximately 1 or 2 percent (by mass) of these stars consists of nuclei heavier than helium, formed in other stars and seeded through space when the stars exploded. Planets such as the Earth represent a collection of these cinders from burnt-out stars, from which the two most abundant elements in the universe—hydrogen and helium—have almost entirely evaporated, leaving a residue of "heavy" elements, many of which are essential to life.

Cosmic Rays

As if this were not enough, supernovae seem to have made another contribution to the evolution of life. We have seen that the outer layers of a supernova explode into space at enormous velocities. Because the presupernova star had progressively less density of matter at greater distances from its center, the blast wave produced by the catastrophic infall and subsequent rebound will encounter progressively less matter to accelerate as it moves outward. As a result, the matter in the outermost layers of a supernova acquires fantastic speeds, 99.9 percent of the speed of light and even more. These layers quickly turn into individual cosmic-ray particles—electrons, protons, helium nuclei, and heavier nuclei traveling at nearly the speed of light. Such cosmic rays (actually fast-moving elementary particles) permeate our galaxy, and they arrive at the top of the Earth's atmosphere by the billions every second. Luckily for us, our atmosphere prevents most of the cosmic-ray particles from reaching the Earth's surface, thus saving us from a deadly rain that would tend to destroy life as we know it within several generations. Some of these cosmic rays can penetrate molecular clouds in interstellar space, where they ionize hydrogen atoms to begin the process of forming molecules shown in Fig. 4.8.

The cosmic-ray particles that do reach the Earth's surface may play a vital role in our existence. Such particles, along with similar fast-moving particles from radioactive minerals (also the result of supernova explosions!) are one cause of mutations, sudden changes in individual genes from one generation of life to the next. These changes allow for the basic pattern of evolution, the difference in reproductive success among members of the same species. Most mutations reduce an individual's ability to survive and to reproduce, so the characteristics typical of such mutations quickly disappear from the general population, but successful individuals owe their success in part to the specific mutations that comprise their genes. If cosmic-ray particles do somehow induce mutations, then supernova explosions provide not only the raw material of which we are made, but also some of the changes that allow biological evolution to occur.

Could a supernova destroy life on Earth? Possibly. If it exploded among any of the hundred thousand closest stars—that is, among the stars that lie within about 200 light years of the solar system—life on Earth might be completely extinguished (at least in the forms that we know now) by the high-energy cosmic-ray particles from the explosion. Among the 100,000 closest stars, we expect one supernova to explode in a time scale of roughly a billion years—about the same interval of time that life has existed on Earth. In the long run, supernova explosions might be one of the factors that limit the longevity of a civilization on a particular planet, though an advanced civilization might learn how to predict supernovae and could therefore protect themselves in ways unknown to us now.

Pulsars: Cosmic Lighthouses?

In 1972 and 1973 two spacecraft—*Pioneer 10* and *Pioneer 11*—left Cape Kennedy, accelerated away from Earth, and after almost two years of travel coasted by the planet Jupiter. Two decades after their launch, each of these spacecraft has covered about 9 billion kilometers and is farther from the sun than any of its known planets. Together with the two Voyager spacecraft (see page 350), these are the objects that humans have flung most deeply into space. After 100,000 years, these spacecraft will have covered a distance equal to that from the sun to the nearest star, though they are not directed toward Alpha Centauri but simply out into the depths of interstellar space.

Each of the two Pioneer spacecraft carries a gold-anodized plaque, 15 by 25 centimeters, with an intriguing picture etched on it (Fig. 6.12). The right-hand portion of the figure, which drew the most attention on Earth at the time of launch, shows a man and a woman in front of a stylized drawing of the spacecraft (to give the correct sizes). The top of the drawing shows the spin flip of a hydrogen atom (see page 86), while the bottom represents schematically the path of the spacecraft out from the Earth, the third planet from the sun.

And what of the left side of the picture, a sort of spider-web arrangement of lines with dashes along them? This diagram represents the best way that astronomers Carl Sagan and Frank Drake could think of to show our sun's location in the Milky Way galaxy, in case some far-distant civilization might find the plaque after millions of years and wonder from what outlying corner of the Milky Way it might have come. Each of the lines in

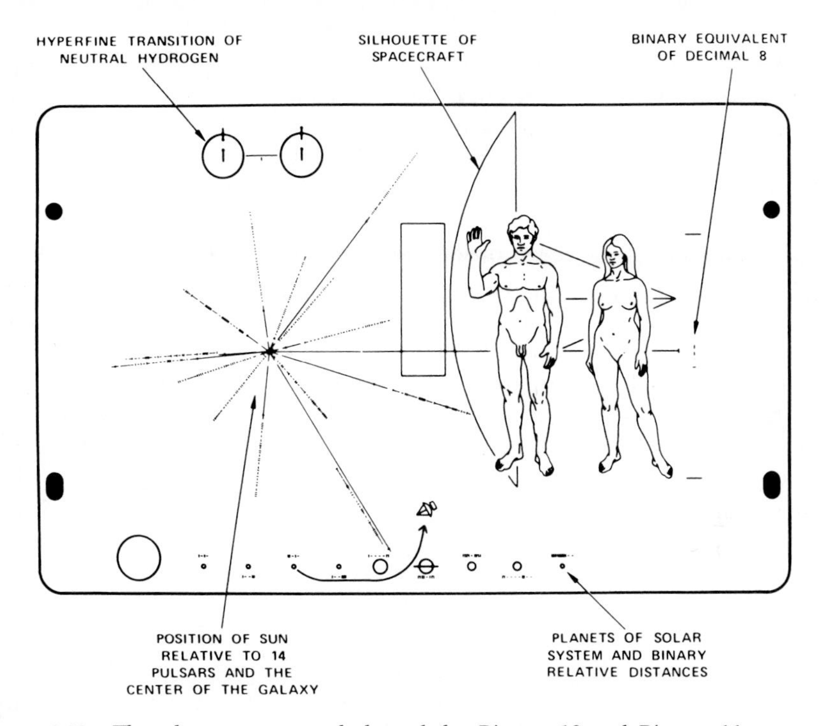

Figure 6.12 The plaques mounted aboard the *Pioneer 10* and *Pioneer 11* spacecraft, the first to leave the solar system, include a map locating the solar system (left) and a picture of two humans in scale with the spacecraft (right). It is not totally clear whether any beings finding the spacecraft could tell the map from the picture, and reading the map might be difficult: It locates the solar system by showing its position with respect to relatively luminous pulsars, whose periods are shown in terms of the frequency of the spin-flip radiation from hydrogen atoms.

the spider-web figure shows the direction from the solar system to a "pulsar," a strange sort of cosmic object that emits pulses of radio waves, and sometimes of light, at precise intervals. The exact intervals between pulses are written in binary notation along each of the 14 lines, using the spin-flip frequency of hydrogen atoms (hence the top part of the drawing) to define the basic interval of time.

Even though the Pioneer plaques will only be messages to ourselves for at least the next few thousand centuries, they do show how pulsars can serve as cosmic beacons, lighthouses that differ from one another in the interval between successive flashes. Pulsars may prove extremely useful when one civilization wants to tell another its precise location in space, because they each have a characteristic interval between flashes, recognizable at great distances throughout the galaxy.

What produces these precisely timed beacons, the pulsars? Once again, we meet supernova explosions in a role important to the search for life. Pulsars arise from neutron stars, the collapsed inner remnants of stars that have exploded. The collapse squeezes the core so tightly that it turns entirely into neutrons. In addition, the core acquires a rapid rate of spin and an enormously strong magnetic field. Both of these characteristics arise from the fact that a formerly large object (the core before its collapse) has been compressed into a much smaller volume.[4]

Near the surfaces of neutron stars, some charged particles constantly appear, as neutrons decay into protons, electrons, and antineutrinos. (In the star's interior such decays are quickly compensated by the additional fusion of protons and electrons into neutrons and neutrinos.) The charged particles are rapidly accelerated almost to the speed of light by the rotating magnetic field.

As a result of the tremendous increase in its rate of rotation and the strength of its magnetic field, the newly formed neutron star arising from the core's collapse is a super-dense, fast-spinning magnet, whipping the field around and around many times each second (Fig. 6.13). Any charged

[4] The compression increases both the rate of spin and the strength of the magnetic field in proportion to the square of the factor by which the core has shrunk in size. For example, if the core were rotating once each month before its collapse, and shrank from a radius of, say, 40,000 kilometers to the radius of a neutron star, only 20 kilometers across, it would spin 4 million times more rapidly. (2000 squared equals 4 million).

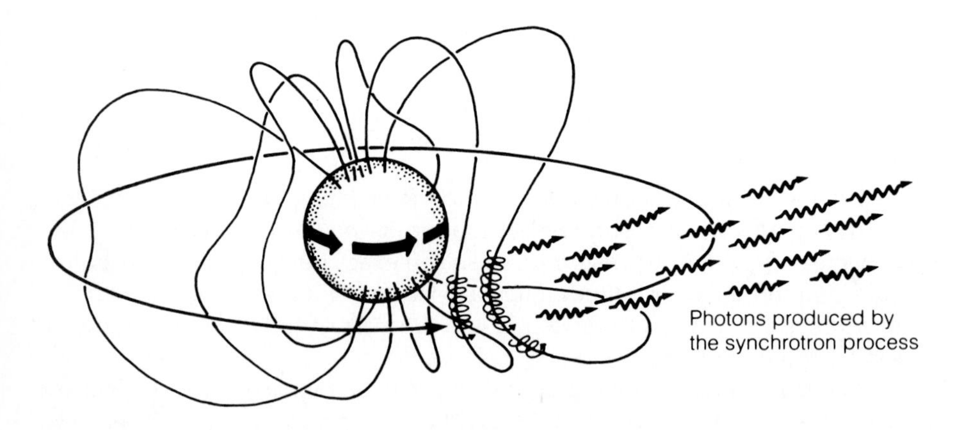

Figure 6.13 Radio and visible-light emission from the vicinity of a rapidly rotating neutron star arises as the magnetic field surrounding the neutron star whips through surrounding material at enormous velocities, dragging charged particles along with it. The particles accelerated to nearly the speed of light generate "synchrotron radiation" as they move through the magnetic field.

particles that remain near the star's surface will be accelerated by the motion of the magnetic field and will spiral along the field lines. Some of them will eventually escape into space with quite impressive energies, adding to the cosmic ray flux from the original supernova explosion. Even more important, the charged particles accelerated by the rotating magnet will emit photons by the synchrotron emission process we described on page 72. These are the photons detected at visible light and radio frequencies that reveal the pulsar's existence.

This radiation process decreases the star's rotational kinetic energy by transferring this energy first to the accelerated particles and then to the synchrotron photons. As a result, the neutron star's kinetic energy must decrease, so the neutron star gradually slows its rotation from hundreds or thousands to 10 rotations per second, then to four, to two, and to even fewer. This slowing down occurs at an extremely leisurely rate, perhaps by one part in a thousand in the rate of rotation each year.

Each pulsar, of which a few hundred have been found so far, produces pulses of photons, rapidly spaced bursts that recur with remarkable regularity anywhere from once every four seconds (for the slowest) to 641 times per second (for the fastest). These photon pulses are usually detected at radio frequencies, but two of the best-studied pulsars show photon bursts at gamma-ray, x-ray, and visible-light frequencies, timed in precise synchrony with their radio pulses.

According to the best theories now available, the reason that pulsars emit photons in pulses rather than at a steady rate stems from the fact that the pulsars' magnetic fields do not align perfectly with the pulsars' rotation axes (Fig. 6.14). Photons emitted by the synchrotron process tend to appear preferentially away from the direction of the magnetic field. Thus, as the neutron stars spin, we see an effect like that of a lighthouse beacon, and we sometimes get a full blast of photons, sometimes only a few, with the pattern repeating over and over.

The photons from pulsars arrive with great but not total regularity, because the neutron star's rate of rotation steadily decreases, as described above. For example, at the center of the Crab Nebula, the supernova remnant left behind by the outburst seen in the year 1054 (Fig. 6.11), we find a pulsar that blinks on and off 33 times each second, one of the most rapid spin rates of all known pulsars (Fig. 6.15). Owing to the rapidity of the pulsar's rotation, we may conclude that this pulsar must be young—a conclusion supported by our prior knowledge of the fact that just over 900 years have passed since it first appeared. Precise timing reveals that the period between pulsar flashes is increasing: Each year, the pulse interval increases by about one–hundred-thousandth of a second.

In 1967, when astronomers detected the first pulsar, they passed through a phase of wondering whether they had not in fact found the first

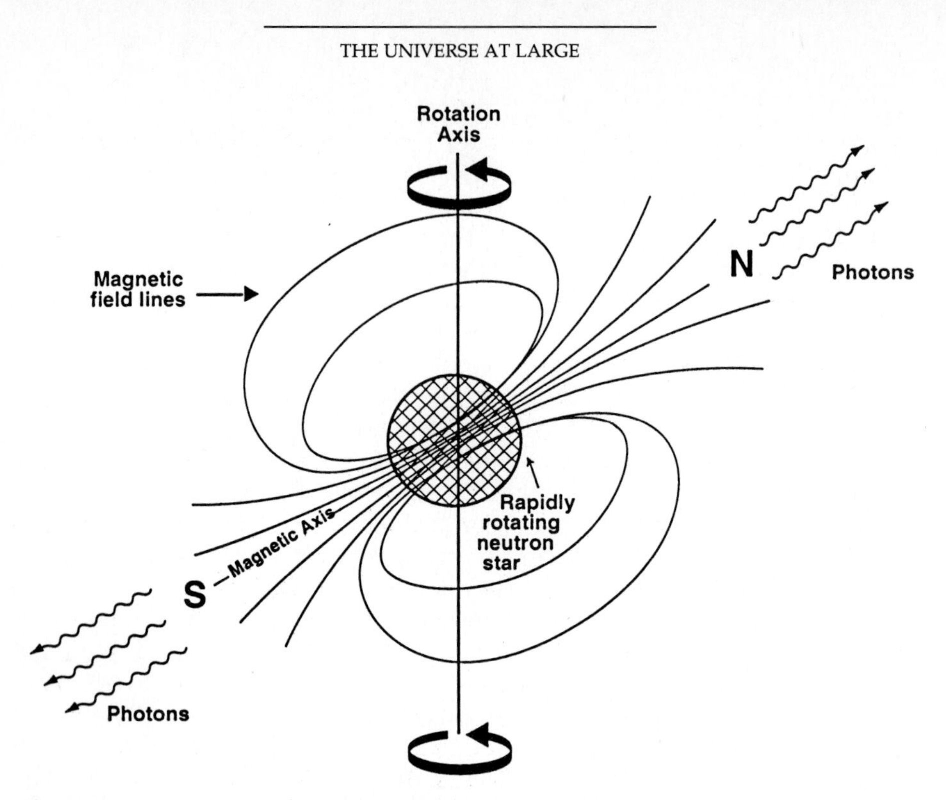

Figure 6.14 In many pulsars, the neutron star's rotation axis apparently does not coincide with the line joining the north and south poles of its magnetic field. As a result, the neutron star's rotation alternately exposes regions of greater and lesser radiation through the synchrotron process, producing a regularly spaced pulsation.

artificial interstellar beacon, another civilization's equivalent of our lighthouses. Each terrestrial lighthouse has its rate of sweep set at a precise interval of time, so that sailors can determine which lighthouse they see simply by timing the interval between flashes. Pulsars would serve admirably in the same manner, as the Pioneer plaques demonstrate, but they seem to be completely natural, though impressive, cosmic timekeepers.

Since most pulsars slow down only gradually, another civilization could recognize the pulsars described on the Pioneer plaque even though the time intervals had lengthened somewhat during many thousands of years. A comparison of the time intervals given for the pulsars on the plaque with those observed when the plaque was found would reveal how much the pulsars had slowed down. Any advanced civilization would be able to duplicate our ability to observe the rates of slowdown over even a few years. Thus it could determine just when the spacecraft had been launched. The plaque tells not only where it came from but when it left as well.

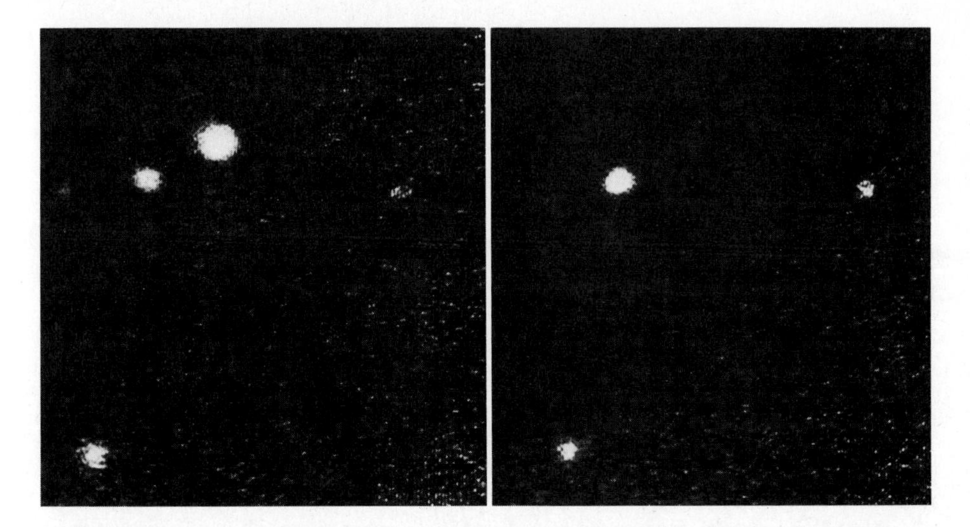

Figure 6.15 Close to the center of the Crab Nebula, a pulsar flashes on and off in visible light and radio waves 33 times per second. The left-hand photograph was made by taking repeated exposures during the "on" part of the pulsar's cycle, while the right-hand photograph combines exposures made during the "off" phase.

The vastness of interstellar space makes it highly unlikely that the Pioneer plaques will ever be found and interpreted by another civilization. If this were to occur, exploding stars would have played crucial roles in identifying our location (through the pulsars that some of them leave behind) as well as creating the key elements from which life on Earth—and the plaques themselves!—are made. Meanwhile, the plaques form a fascinating kind of "space art," one that might appeal to Earthdwellers but not, so far as we now know, to anyone else. If we ever hope to put on an art exhibition for another set of sentient beings—or they for us—we ought to know something about their biological evolution. For that we must start with the composition and history of the one form of life that we do know, life on Earth.

SUMMARY

As stars grow older, they eventually exhaust the supply of protons at their centers. As a result, each star loses its former ability to liberate kinetic energy from energy of mass. For a time, a star can compensate for the increasing shortage of protons by contracting its central regions, raising the temperature there to fuse its remaining protons at an ever-increasing rate.

The additional kinetic energy liberated in this way expands the star's outer layers, cooling them slightly in the process and producing a red-giant star that has a huge outer surface and a small, rapidly fusing inner core.

A red-giant star will eventually contract its core enough to start fusing the helium nuclei into carbon nuclei. This helium flash temporarily expands the core, but as the helium nuclei all fuse into carbon, the core again grows denser and denser as the star tries to maintain itself against self-induced gravitational collapse.

Smaller stars will stop contracting their cores after they have fused helium nuclei into carbon nuclei. The star's interior will be so dense at this point that the exclusion principle keeps the electrons from packing any tighter. The electrons hold the nuclei from further compression through electromagnetic forces, so the exclusion principle can support the entire star against its own gravitation. After the star's outer layers evaporate, the inner core persists as a white dwarf star that begins a long, slow fade into dimness.

No white dwarf, however, can exist if its mass exceeds a value of about 1.4 times the sun's mass. Stars with more mass are likely to produce supernova explosions as they age. Supernovae of this type arise from the fact that more massive stars will not become so dense in their centers as less massive stars do when they fuse helium nuclei into carbon nuclei. Instead, these stars will fuse carbon into heavier and heavier nuclei to keep on liberating kinetic energy, but once the star's core has become mostly iron nuclei, no more energy-liberating reactions exist, so the star's core collapses.

The collapse produces a tiny, incredibly dense neutron star, which bounces outward to start an explosion of the star's outer layers. This explosion produces an exceedingly bright object, a "supernova," visible for a few weeks or months. Such supernovae seed heavier nuclei throughout the galaxy in which they are located, and they probably produce most of the cosmic rays—nuclei and electrons traveling at nearly the speed of light. Cosmic rays may be responsible for some of the mutations that allow evolution on Earth to proceed.

Nature has another way to produce supernovae. The second type of supernova arises in white dwarf stars that move in orbit with relatively nearby companion stars. If the white dwarf's surface accumulates enough mass, it can detonate that new material, rich in protons that can fuse together, in a sudden explosion that matches the outburst of a high-mass star whose core collapses.

Neutron stars are the collapsed cores of stars, made of neutrons and only a few kilometers in diameter. At birth, they are rapidly rotating, highly magnetic objects. These spinning magnets form pulsars, sources of photon pulses that repeat with impressive regularity, and the photons themselves originate from charged particles accelerated almost to the speed of light by the neutron star's rapidly rotating magnetic field.

QUESTIONS

1. What happens to the central region of a star as most of the protons there fuse into helium nuclei? What happens to the outer layers of the star?

2. What keeps white dwarf stars from collapsing? How does the exclusion principle keep all the carbon nuclei in a white dwarf star from rushing to the star's center?

3. Do nuclear fusion reactions occur inside a white dwarf star? What makes a white dwarf keep on shining?

4. Which stars do not end their lives as white dwarfs? Do these stars pass through their main-sequence lifetimes more rapidly or more slowly than other stars do? Why?

5. What happens to the stars that do not become white dwarfs? What importance does their fate have for the formation of later generations of stars?

6. The supernova that appeared in our galaxy in the year 1604 had a distance from us of about 5000 parsecs. In approximately what year did the explosion actually occur?

7. The supernova that appeared in the Andromeda galaxy in 1885 had a distance of 600,000 parsecs, about 100 times the distance of the 1604 supernova. In what years had the two supernovae actually exploded?

8. Why do Population I stars contain a larger fraction of their mass in the form of heavy elements than Population II stars do? Why do we distinguish heavy elements from light elements? To which population does the sun belong?

9. What are cosmic rays? Which part of a supernova may have produced them?

FURTHER READING

Clark, David. *Superstars.* New York: Crown, 1985.

Goldsmith, Donald. *Supernova! The Exploding Star of 1987.* New York: St. Martin's Press, 1989.

Greenstein, George. *Frozen Star: Of Pulsars, Black Holes, and the Fate of Stars.* New York: New American Library, 1983.

Kaufmann, William. *Black Holes and Warped Spacetime.* New York: Freeman, 1979.

Kirshner, Robert. "Supernova: Death of a Star." *National Geographic* (May 1988).

Marschall, Laurence. *The Supernova Story.* New York: Plenum Press, 1989.

Thorne, Kip. "The Search for Black Holes." *Scientific American,* (December 1974).

Tucker, Wallace, and Riccardo Giacconi. *The X-Ray Universe.* Cambridge, Mass.: Harvard University Press, 1985.

Had I the heavens' embroidered cloths,
Enwrought with golden and silver light,
The blue and the dim and the dark cloths
Of night and light and the half-light,
I would spread the cloths under your feet:
But I, being poor, have only my dreams,
I have spread my dreams under your feet,
Tread softly because you tread on my dreams

—W. B. Yeats

PART THREE

Life

The force that through the green fuse drives the flower
Drives my green age

—DYLAN THOMAS

In all the cosmos, we know of only one form of life: life on Earth. The living systems on our planet—united in their origin and genetic organization—provide us with a single example of extraordinary complexity that nonetheless relies on a basic structure of impressive simplicity. The complexity arises from billions of years of evolution; the simplicity from the limited number of types of atoms and molecules that dominate all life on Earth. Both the complexity and the simplicity deserve close examination when we seek to estimate how often life might have arisen elsewhere, and how life might have evolved in situations other than our own. In this effort, we must rely on a double extrapolation: first to determine the past course of evolution on our own planet, then to apply this knowledge to the imperfectly known situations that exist throughout the universe. Although this extrapolation remains unavoidable, we can rely on well-tested principles of science as we form our ideas about the origin and evolution of life on Earth and the possibilities for life in the universe at large.

Shiva, the god of destruction and rebirth, appears here trampling upon Asura, the demon of ignorance. In the Hindu religion, Shiva is often called the Bright One or Happy One, because death represents a transition to new life.

7

The Nature of Life on Earth

Now that we have obtained a general understanding of the various components of the universe, we could start to examine the cosmos for the best places to find life, and then proceed to study this remarkable phenomenon. Our knowledge of life on Earth must be used to direct our search, but as we review this knowledge, we shall quickly recognize a basic limitation in the use of terrestrial life as a guide to the cosmos. For all its amazing diversity, life on Earth has a fundamental unity at the molecular level—a unity that clearly marks it as a single example. Thus we cannot begin our study of life in the universe with a comparative approach, as we have done so successfully with stars and galaxies. Instead, we must proceed as best we can from life as we know it, trying to estimate how life begins, why life is what it is and where it is, and how often the conditions that lead to the origin and development of life occur in our galaxy and in the universe at large.

As we study these problems, we must be aware that many characteristics of life on Earth, especially the most visible ones, such as the size and weight of living creatures, are undoubtedly the product of the specific environments offered by the particular places where life has developed. We shall be more interested in the "fundamental" properties of life—properties we can identify for use in our efforts to determine the probability that life has arisen many times in the universe, rather than being confined to the single example we know on Earth.

What Is Life?

Since life seems to be a property of matter, we can attempt to investigate life as we might study another property of matter, such as magnetism. What is the mysterious something that distinguishes "living" matter from all other combinations of atoms and molecules? Can we describe life in terms of conventional physics and chemistry, or does this approach miss some special insight that would enable us to solve the mystery of life? Does there exist a "vital force," an essence that can make ordinary matter come alive?

Instead of struggling with the philosophical implications of these questions, we shall use a descriptive approach, seeking to identify the characteristics that distinguish life from inanimate matter. *The most distinguishing characteristics of matter that we call "alive" are the abilities to reproduce and to evolve.* A flame can ignite other flames (thus "reproducing itself"), but it cannot evolve into other sorts of flames, such as iron rust. But we have extensive evidence from fossils that life has been able to transform itself continuously, producing the abundant examples of living matter that surround us, from whales and redwoods to amoebae and blue-green bacteria. To understand how these transformations could occur, and to gain an appreciation of the unity that underlies the apparent diversity of life, we must examine both the types of atoms and molecules that form living terrestrial organisms, and also the basic chemical reactions that enable matter to remain alive.

We can begin by asking, What is life made of? This apparently simple question immediately leads to a useful perspective: Life, as we know it, is made from a recipe dominated by a remarkably small number of ingredients. Eighty-five stable elements exist in nature, ranging from hydrogen (the lightest) to uranium (the heaviest). Just four of these elements—hydrogen, oxygen, carbon, and nitrogen—comprise more than 95 percent (by weight) of the matter that we call alive! These four elements are the most abundant elements in the universe, except for the inert gases (helium and neon), which do not form chemical compounds. In striking contrast, the four most abundant elements on Earth are silicon, iron, magnesium, and oxygen. In other words, the composition of living matter resembles the composition of the stars more closely than the composition of the planet on which we find ourselves (see Table 7.1).[1]

The large amount of hydrogen and oxygen in living organisms follows naturally from the high percentage of water that all life contains, and this percentage seems reasonable because water is so abundant at the Earth's surface. But carbon and nitrogen, though abundant in stars, are relatively rare on Earth, far less abundant than silicon or iron, for example. The concentration of carbon and nitrogen in living matter demands an additional explanation. This explanation hinges on the chemical properties of carbon and nitrogen atoms.

The unique ability of carbon atoms to form complex molecules, able to store the information necessary for the continuation of life, clearly makes carbon an important element for biology. But complexity in molecules has no value if the molecules are either unstable or too stable to undergo reactions

[1] Does this explain why so many cultures believe that their ancestors descended from the sky and that they will return to the stars when they die? Do we see here a longing of our atoms to return to their primordial wombs? (!)

TABLE 7.1 Element Abundances in Different Situations*

Sun		Earth		Earth's Crust	
Hydrogen	90.99%	Oxygen	50%	Oxygen	47%
Helium	8.87	Iron	17	Silicon	28
Oxygen	0.078	Silicon	14	Aluminum	8.1
Carbon	0.033	Magnesium	14	Iron	5.0
Neon	0.011	Sulfur	1.6	Calcium	3.6
Nitrogen	0.010	Nickel	1.1	Sodium	2.8
Magnesium	0.004	Aluminum	1.1	Potassium	2.6
Silicon	0.003	Calcium	0.74	Magnesium	2.1
Iron	0.003	Sodium	0.66	Titanium	0.44
Sulfur	0.002	Chromium	0.13	Hydrogen	0.14
Argon	0.0003	Phosphorus	0.08	Phosphorus	0.10
Aluminum	0.0003			Manganese	0.10
Calcium	0.0002			Fluorine	0.063
Sodium	0.0002			Strontium	0.038
Nickel	0.0002			Sulfur	0.026
Chromium	0.00003				
Phosphorus	0.00003				

Earth's Atmosphere		Bacteria		Human Beings	
Nitrogen	78%	Hydrogen	63%	Hydrogen	61%
Oxygen	21	Oxygen	29	Oxygen	26
Argon	0.93	Carbon	6.4	Carbon	10.5
Carbon**	0.03	Nitrogen	1.4	Nitrogen	2.4
Neon	0.0018	Phosphorus	0.12	Calcium	0.23
Helium	0.00052	Sulfur	0.06	Phosphorus	0.13
				Sulfur	0.13

* Abundances are given as the percent of the total number of atoms, rounded to two significant figures (after work by C.W. Allen, B. Mason, E. Anders, and N. Grevesse).

** In the Earth's atmosphere, 99.9% of the carbon is in carbon dioxide molecules and 0.1% in methane molecules.

with other molecules. Nitrogen and oxygen atoms each have the unusual ability of being able to share more than one electron with a carbon atom. This allows nitrogen and oxygen to form chemical bonds with carbon that are strong but breakable. The possibility of such chemical bonds, together with the ability of carbon atoms to link with one another in many ways, allows carbon, nitrogen, and oxygen to form an extraordinary variety of stable molecules. Nitrogen has the additional ability to form a stable gas that aids in cycling this element between organisms and their environment. Oxygen atoms can easily combine with other atoms and molecules in chemical compounds that release energy as they form (Fig. 7.1). Given the unusual chemical properties of carbon, oxygen, and nitrogen, we may conclude that these atoms, together with hydrogen, dominate life not by chance, but because of the special nature of these particular atomic elements. We shall return to this issue in Chapter 10.

Aside from reminding us of our kinship with the stars, the coincidence of elemental abundances between the universe and life on Earth has reassuring implications in our search for life elsewhere in the universe. If life on Earth were made primarily from such rare elements as hafnium and holmium, we might well conclude that our kind of life should be exceedingly rare. But in actuality, the *elemental composition* of life as we know it poses no barrier to the prevalence of life in the universe.

Let us continue with the list of ingredients that form life. If we add calcium and phosphorus to our four basic elements, we can account for 98.6

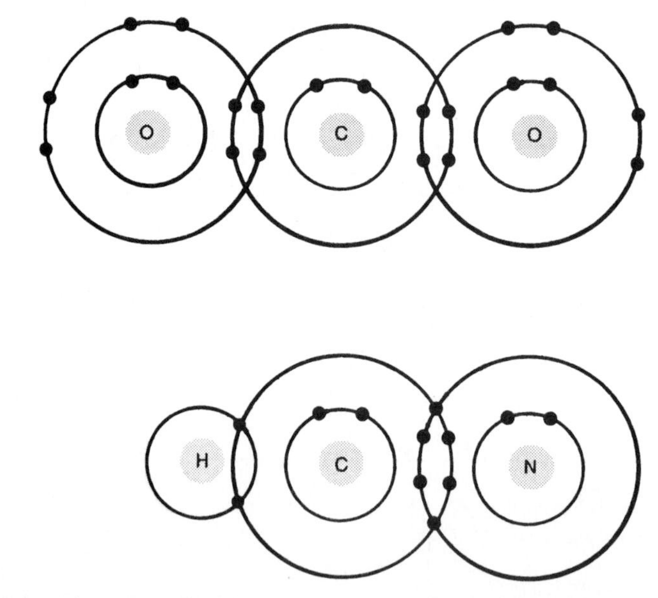

Figure 7.1 A carbon atom has four electrons in its outer electron shell, where they can be shared with other atoms. Some of the different ways in which carbon can form bonds with other elements include (a) sharing two electrons with each of two atoms of oxygen to make carbon dioxide (CO_2) and (b) sharing three electrons with a nitrogen atom and one electron with a hydrogen atom to make hydrogen cyanide (HCN).

percent of living matter (by weight). The remaining 1.4 percent usually consists of chlorine, sulfur, potassium, sodium, magnesium, iodine, and iron, plus tiny amounts of biological "trace elements," which include manganese, molybdenum, silicon, fluorine, copper, and zinc. This completes the list of elements in a typical terrestrial organism, but great variations in the abundance of minor elements exist from one organism to another, and even the composition of a single cell within an organism can change with time. We can point, however, to a single important relationship: The concentrations of trace elements in bacteria, fungi, plants, and land animals show a strong correlation with the concentrations of these elements in sea water. This correlation not only suggests that life began in the seas, but also indicates that life as we know it indeed arose on Earth rather than arriving here from some other environment (for example, from the surface of Mars, or from an interstellar cloud of gas and dust), where the relative abundances of trace elements could be quite noticeably different.

The elements that add to the basic four can therefore tell us much about living matter. Some of the trace elements, rare as they are in living creatures, are vital to the organism that contains them. Copper appears with great enrichment in certain marine animals, such as oysters and octopi. In these animals, hemocyanin (made with copper) plays the same role that hemoglobin (made with iron) does in vertebrate mammals, such as mice and men or wombats and women, namely, the transport of oxygen through the circulation of blood. Zinc is a necessary component of insulin, while manganese is an essential ingredient in some enzymes. Certain kinds of southwestern "locoweed," a plant that induces hallucinations in animals that eat it, contain as much as 1.5 percent by weight of the rare element selenium. We should remember that even though four elements provide the bulk of all living creatures, as many as three dozen additional elements may be essential to such complicated organisms as human beings. This fact strongly suggests that life has evolved in complex interactions with its environment.

Biologically Important Compounds

The list of elements in living creatures does not tell us what life is, any more than the list of ingredients describes a cake. We must know how the elements fit together into simple molecules, and how simple molecules join to form more complex molecules, in order to see how living organisms work.

When we examine various organisms at the molecular level, we find that *most life forms consist of a small number of types of rather simple molecules.* Scientists use the term *monomer* to describe any of several types of small molecules that can join together to become components of larger, more complex molecules called *polymers.* Among the most important monomers are the *amino acids* that form proteins; other well-known types of monomers are sugars, fatty acids, and nucleotides. Of particular interest in the search for life is this fact: Some simple monomers come in two forms, which we can characterize as left-handed and right-handed (Fig. 7.2). The two forms are identical in their atomic composition but differ in the way that the atoms that form the monomers fit together, so that one is the mirror image of the other. The left- and right-handed molecules are like the two gloves in a pair, similar but not identical. The interesting fact about these molecules is this: Except for the components of some proteins found in the cell walls of certain bacteria, *all the amino-acid monomers found in life on Earth are of the left-handed, never the right-handed, variety.* This distinctive property of life on Earth apparently arose by chance. The use of just one of the two possible forms increases the efficiency of chemical reactions required to sustain life, but either structure would serve this purpose. This is an example of the selectivity of life at the molecular level, and provides a useful test for distinguishing biogenic from nonbiogenic compounds. The latter contain equal amounts of left-handed and right-handed amino acids; the former just the left-handed variety.

The ways in which individual types of monomers form larger molecules ultimately distinguish matter that is alive from the inanimate matter that forms the bulk of the universe. One way to think about this distinction is to realize that the molecules in living organisms must be able to store and

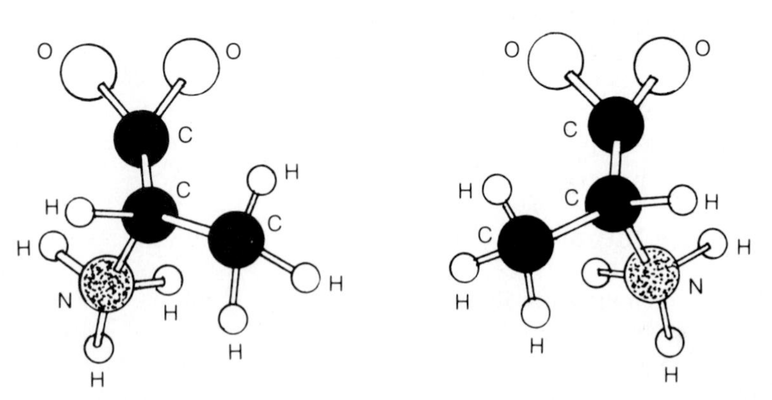

Figure 7.2 This drawing schematically represents a typical amino acid. Except for glycine, each amino acid, such as alanine, which is shown here, may appear in either one of two forms. There are the left-handed (*levo-*, or L) and right-handed (*dextro-*, or D) versions of the same chemical compound.

transmit enormous amounts of information. This storage and transfer can be accomplished by varying both the *composition* (types of atoms) and the *structure* of the molecules.

Not surprisingly, we find that living matter on Earth consists mainly of long, chainlike molecules—monomers strung together into polymers—in which *a given pattern repeats over and over again*, sometimes with small variations (Fig. 7.3). In these polymers, ringlike structures and side chains often occur, and the chains themselves sometimes fold into elaborate, highly complex, but extremely specific shapes. The ability to assume specific shapes, along with the ability to change shape at appropriate times, allows some polymers to act as "catalysts," sites where chemical reactions can occur much more rapidly than they could otherwise. These catalysts are called "enzymes."

The monomers can link together in many different ways to form an immense family of chemical compounds. An average protein molecule—the quintessential organic polymer—consists of a few hundred amino-acid monomers. Each type of protein differs from the others in the types of amino acids it contains, and in the order in which the amino acids are strung together to form the polymer chain. But from a tremendous number of *possible* amino acids, *only 20 commonly occur in life as we know it* (see Table 7.2). Nevertheless, an average protein molecule, containing perhaps 100 amino acids, could be made in at least 20^{100} different ways. This enormous number, far greater than the number of atoms in our galaxy, implies a correspondingly super-astronomical variety of possible proteins. Yet most living organisms make and use fewer than 100,000 types of protein molecules.

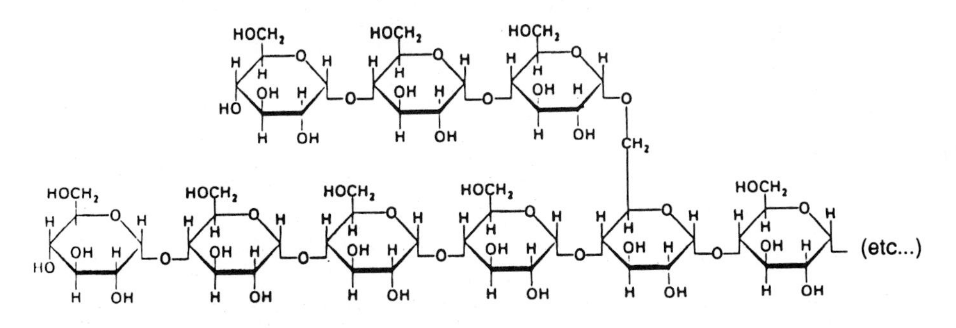

Figure 7.3 Glycogen, the chief carbohydrate used to store energy in animals, is a polymer composed of a long, branched chain of glucose molecules, each of which contains 22 atoms; in the figure, the hexagons each represent six carbon atoms linked together in a ringlike structure. A "polymeric chain" made of such glucose monomers can repeat the monomers thousands of times to make an enormous molecule of glycogen.

TABLE 7.2 The 20 Amino Acids Found in Living Organisms

Amino Acid*	Chemical Formula	Number of Atoms
L-Alanine	$C_3H_7O_2N$	13
L-Arginine	$C_6H_{15}O_2N_4$	27
L-Asparagine	$C_4H_8O_3N_2$	17
L-Aspartic Acid	$C_4H_6O_4N$	15
L-Cysteine	$C_3H_7O_2NS$	14
L-Glutamic Acid	$C_5H_8O_4N$	18
L-Glutamine	$C_5H_{10}O_3N_2$	20
Glycine	$C_2H_5O_2N$	10
L-Histidine	$C_6H_9O_2N_3$	20
L-Isoleucine	$C_6H_{13}O_2N$	22
L-Leucine	$C_6H_{13}O_2N$	22
L-Lysine	$C_6H_{15}O_2N_2$	25
L-Methionine	$C_5H_{11}O_2NS$	20
L-Phenylalanine	$C_9H_{11}O_2N$	23
L-Proline	$C_5H_9O_2N$	17
L-Serine	$C_3H_7O_3N$	14
L-Threonine	$C_4H_9O_3N$	17
L-Tryptophan	$C_{11}H_{12}O_2N_2$	27
L-Tyrosine	$C_9H_{11}O_3N$	24
L-Valine	$C_5H_{11}O_2N$	19

*For those amino acids that have both a left-handed (L) and a right-handed (D) form, we have indicated that only the left-handed member of these stereoisomer pairs appears in living organisms. Only glycine, the simplest of the amino acids, has no L and D forms, and thus requires no L or D designation. See page 167 for diagrams of these molecules.

In other words, *life shows an extraordinary selectivity in the kinds of molecules that it uses.* The ability to form highly specific compounds, and to reject a far greater number of molecular arrangements, provides one of the defining characteristics of life as we know it.

In all of these complex molecules, carbon is the element that allows the structure to exist; without carbon, life as we know it would not occur. This is so because carbon has the unique ability to form large, intricate molecules with a wide variety of elements, a point that we shall examine further in Chapter 10. The variety and complexity of molecular structure must exist in order to store, and to transmit, the information that allows one configuration of matter to reproduce itself, or to choose and to produce only the compounds that an organism needs in order to remain alive.

The Capacity to Reproduce

The most basic property of life, the essence of its existence and persistence, rests in life's ability to reproduce itself. Cells divide; plants make seeds that grow into new plants; birds and reptiles lay eggs; most mammals give birth to live babies. Despite the diversity that we see, in all organisms reproduction at the *molecular* level follows the same basic plan: a certain kind of long, skinny polymer called DNA (deoxyribonucleic acid) governs the process of reproduction. Furthermore, DNA molecules, together with their close relatives RNA (ribonucleic acid) tell the new organism, as well as the old one, how to function. The general principles by which DNA and RNA molecules work deserve attention. Although we do not expect identical molecules to appear in other forms of life that may exist on other worlds, we may well expect that molecules with similar functions are present. We ought therefore to understand how DNA and RNA molecules can do so many things for life on Earth.

DNA molecules store genetic information that tells the next generation of organisms how to carry out metabolism, to grow, and to reproduce. Each time that a cell divides to form "daughter cells," whether from a simple bacterium or from a tiny part of a complex animal, each new "daughter" cell must receive a complete copy of the DNA in the original cell. Only then can the daughter function as a reproduction of the original cell. But how does the cell make a perfect copy of the DNA molecule? As it turns out, the means of this replication is "built in" to the basic structure of DNA.

DNA has two strands, wound around each other like the handrails of a spiral staircase: the famed double helix of molecular biology. The monomers that bond together to form each strand of DNA are called *nucleotides.* Each nucleotide consists of a sugar (called deoxyribose), a phosphate molecule (PO_4), and one of four possible carbohydrates called "nitrogenous bases." Many other bases could possibly be used, but life has again shown its selectivity in using just these four in DNA molecules. These four bases are adenine (A), guanine (G), cytosine (C), and thymine (T) (Fig. 7.4).

In all the DNA monomers, the sugar and phosphate components are identical; only the bases vary. The monomers connect to one another to form the "backbone" of DNA, the two strands of the double helix. The bases appear inside these two strands, where they play the key role of joining the two strands: A base from one strand links to a base from the other strand. *It is crucial to note that the base pairs cannot link together at random.* Instead, adenine (A) pairs only with thymine (T) to form one kind of link, and guanine (G) pairs only with cytosine (C) to form the other kind (Fig. 7.5).

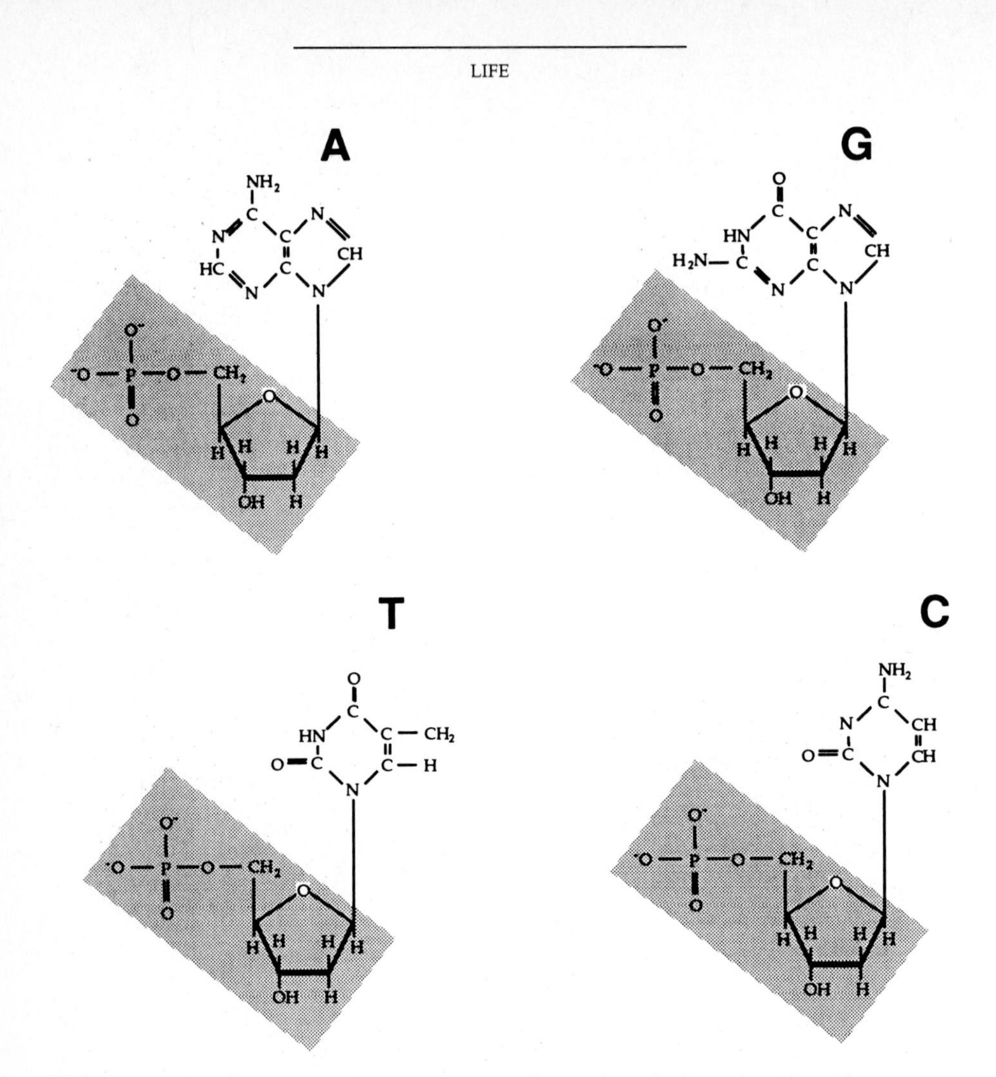

Figure 7.4 The four DNA monomers are collectively called nucleotides. The sugar and phosphate portions of a nucleotide (shaded) are identical, while each nucleotide has a different "base": **A** indicates adenine; **G,** guanine; **T,** thymine; **C,** cytosine.

Further Functions of DNA

In addition to reproducing themselves and guiding the reproduction of entire organisms, DNA molecules also govern the formation or "synthesis" of protein molecules. Proteins are the workhorses of the living cell. Some are structural; they help to hold things together. Others, the enzymes, are sophisticated catalysts; they cause specific chemical reactions necessary for growth and metabolism to occur at the proper rate and at the correct time and place.

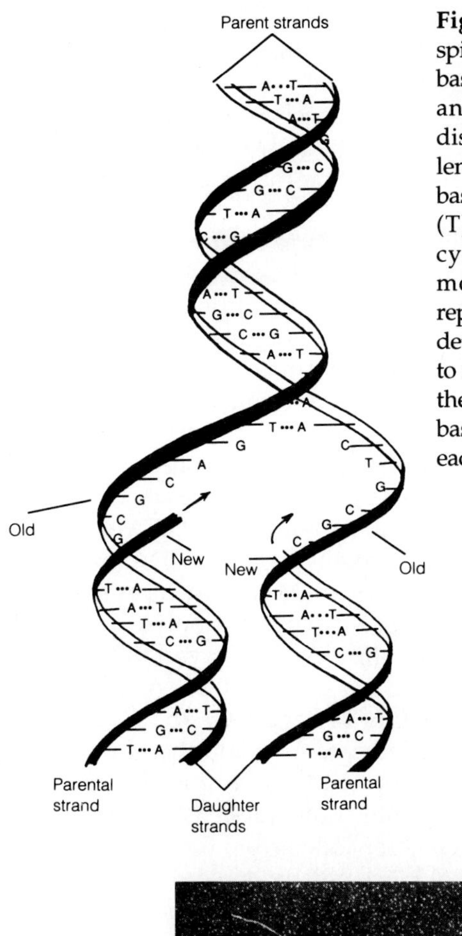

Figure 7.5 DNA molecules consist of two spirals, linked by pairs of nitrogen-containing bases. The long strands, which consist of sugar and phosphate molecules, keep the same distance from one another throughout the length of the molecule. Of the four types of bases, adenine (A) can pair only with thymine (T), and guanine (G) can pair only with cytosine (C). Therefore, when the DNA molecule divides lengthwise during replication, each separate half of the molecule determines precisely which bases can link up to reconstitute the missing strand. In this way, the genetic code, carried by the sequence of bases along each strand, can be reproduced in each of the two "daughter" molecules.

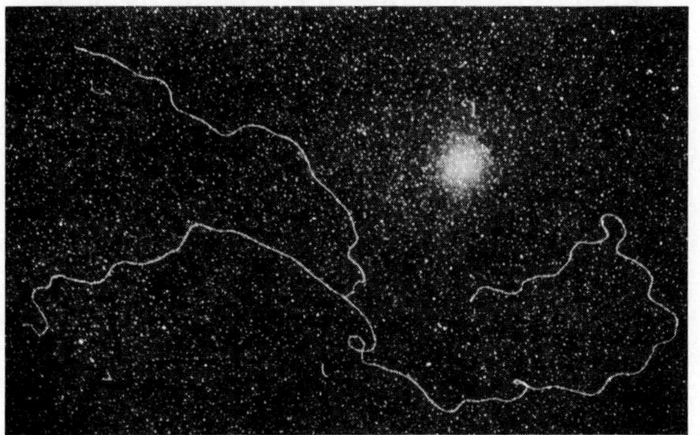

Figure 7.6 An electron micrograph of a DNA molecule shows the double-stranded spiral in the process of replication. As the molecule duplicates itself, a region with two double helices appears, until eventually two complete double spirals form from one original molecule.

Proteins are polymers made of amino-acid monomers. As we saw above, only 20 different amino acids are used by life forms on Earth (Fig. 7.7). Each of the amino acids contains an identical set of atoms that can be joined together to form a uniform protein backbone. As with the four DNA monomers, each of the 20 amino acids has its own unique "signature" group of atoms that determines the properties that the amino acid can contribute to a protein molecule.

The final, folded, three-dimensional structure of the protein chain determines how it will function. But it is the *sequence* in which the amino acids are strung together along the chain that establishes *what three-dimensional structure they will choose.* This sequence of amino acids is determined by the sequence of bases along a strand of DNA. The DNA instructions for making proteins are "written" in a format called the "genetic code." The sequence of bases along a strand of DNA determines the message encoded by that stretch of DNA, much as the sequence of dots and dashes establishes the message encoded in a portion of Morse code.

How can the sequence of bases along the strands of a DNA molecule dictate the sequence of amino acids in a protein molecule? Suppose that *each* of the "letters" (bases) of the DNA alphabet were to represent one amino acid. Then the four types of bases could represent only four amino acids in the DNA message. This wouldn't allow for much variety in life! If each amino acid were encoded by a unique *pair* of bases (AA, AG, GA, AC, CA, AT, TA, GT, TG, GC, CG, GG, TC, CT, TT, and CC), the code could specify 16 different amino acids. The two-letter code would therefore not be sufficient to represent all 20 of the different amino acids that life on Earth uses. However, it is worth noting that quite some variety can be derived from 15 amino acids (reserving one two-letter code group for "stop"). So far as we can tell, a complex, informational biochemistry could develop with a two-letter code. Perhaps life on some other planet has developed just such a system.

But if not one, nor two, but *three* bases are used to code each amino acid, we obtain 64 possibilities, more than the 20 we need. Careful experiments have shown that DNA *does* use just such a triplet sequence as the code that specifies how proteins will be made. The additional possibilities allow not only for "stop" but also for a molecular redundancy, with several code groups specifying the same amino acid. This redundancy prevents most mistakes, analogous to typographical errors in a book, from having harmful effects. Using this three-letter code, a sequence of 100 to 500 triplets of bases along a strand of DNA can specify a protein. If this sequence encodes a product of significance to the organism, the sequence is called a *gene.*

How does this affect us? The protein-synthesizing machinery of the cell does not read the DNA message directly. Instead, the cell makes "working

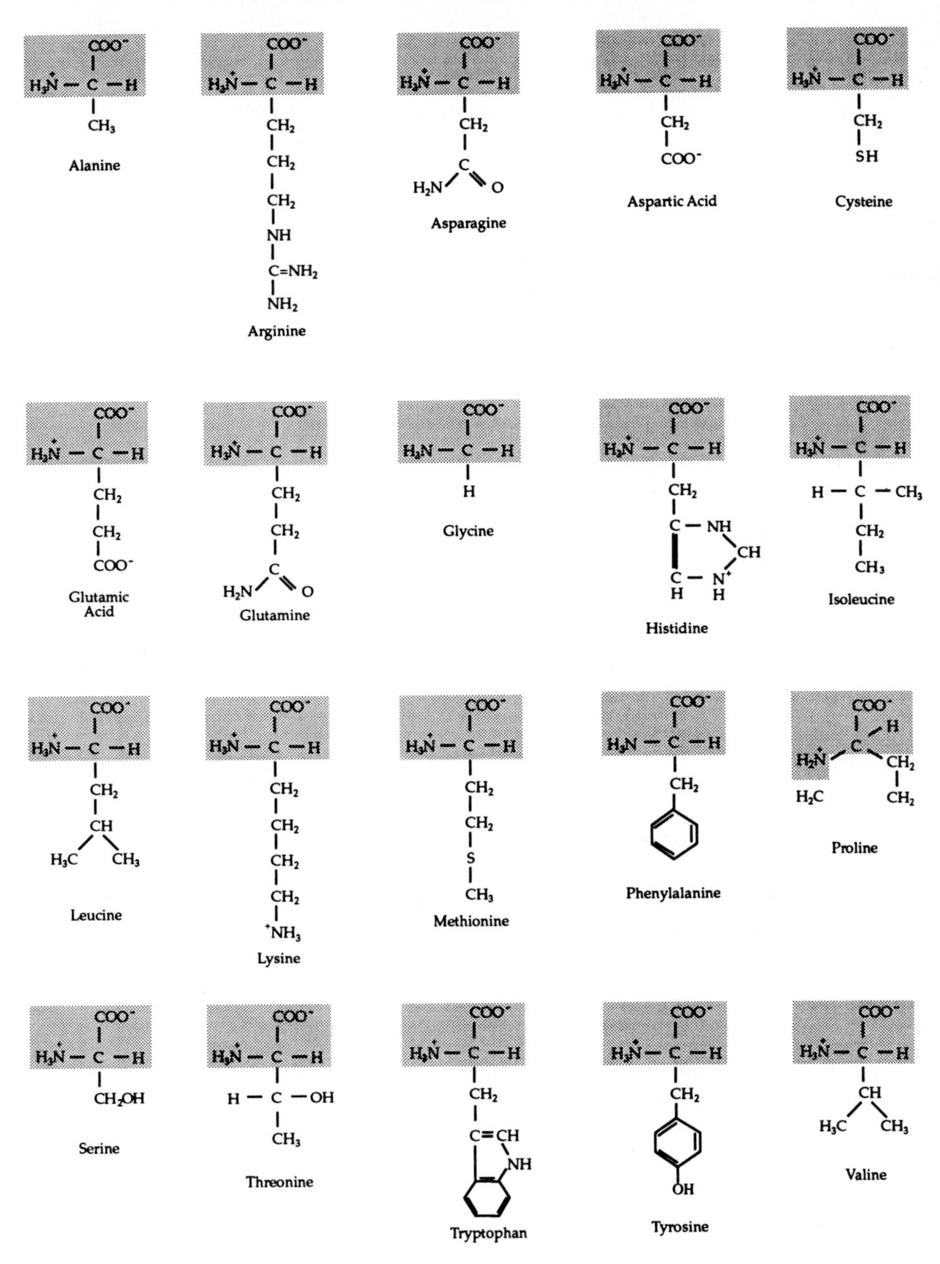

Figure 7.7 The 20 amino acids that are commonly used to make proteins by life on Earth are shown diagrammatically here (compare Table 7.2 and Figure 7.2). The shaded portions of each amino acid are common to the set, whereas the unshaded parts reveal the differences that distinguish one amino acid from another.

copies" of the genes it needs at a particular moment. These working copies are not DNA but the closely related molecule RNA. In RNA, uracil (abbreviated U) is used in place of DNA's thymine (T). Since the cell actually reads RNA when translating code into amino-acid language, the biologists who deciphered the "code book" of life wrote it out in RNA language (Table 7.3).

Impressive as the nucleic acids DNA and RNA may seem, they are not all-powerful. DNA itself is not "alive" in the strict sense, for the replication

TABLE 7.3 The Genetic Code

First Letter	Second Letter: U		C		A		G		Third Letter
U	UUU	leu	UCU	ser	UAU	tyr	UGU	cys	U
U	UUC		UCC		UAC		UGC		C
U	UUA	leu	UCA		UAA	stop	UGA	stop	A
U	UUG		UCG		UAG	stop	UGG	trp	G
C	CUA	leu	CCU	pro	CAU	his	CGU	arg	U
C	CUC		CCC		CAC		CGC		C
C	CUA		CCA		CAA	gin	CGA		A
C	CUG		CCG		CAG		CGG		G
A	AUU	ile	ACU	thr	AAU	asn	AGU	ser	U
A	AUC		ACC		AAC		AGC		C
A	AUA		ACA		AAA	lys	AGA	arg	A
A	AUG	met	ACG		AAG		AGG		G
G	GUU	val	GCU	ala	GAU	asp	GGU	gly	U
G	GUC		GCC		GAC		GGC		C
G	GUA		GCA		GAA	glu	GGA		A
G	GUG		GCG		GAG		GGG		G

Abbreviations:

ala = alanine	gln = glutamine	leu = leucine	ser = serine
arg = arginine	glu = glutamic acid	lys = lysine	thr = threonine
asn = asparagine	gly = glycine	met = methionine	trp = tryptophan
asp = aspartic acid	his = histidine	phy = phenylalanine	tyr = tyrosine
cys = cysteine	ile = isoleucine	pro = proline	val = valine

of DNA molecules requires a complex and highly specialized environment. At the present time, such an environment occurs naturally only within living cells. An example of this limitation appears in the case of "viruses." Viruses consist of a nucleic acid surrounded by a coat of proteins, and they cannot reproduce outside of cells. (If only they could, we would be spared many unpleasant diseases, for it is the viruses' requirement for additional material to reproduce that troubles virus-laden cells.) DNA molecules need special enzymes to help them uncoil and to determine the part of the information they encode that will be read into an RNA message.

We can now summarize certain unifying facts about life on Earth. All life as we know it contains DNA and its close relative RNA. These polymers provide the basis for replication, and replication can occur only within cells. Hence, the smallest, simplest systems that live—that can grow and reproduce—are cells. All forms of life contain proteins, polymers made from amino-acid monomers in response to the information carried by the nucleic acids. Proteins are responsible for much of the structure, and for most of the functions, of living cells. Other polymers serve as food (carbohydrates), can store and transport energy (fats), and form the basic components of the membranes that organize molecules into cells (lipids).

Evolution and the Arrow of Time

In order to follow the changes in life through billions of years, we shall look more closely at the property of life that we have called most distinctive: the capacity to reproduce and to evolve. At the molecular level, life's ability to reproduce begins with the replication of DNA, during which two new spirals are created that are exact replicas of the original molecule.

Sometimes a change in the sequence of nucleotide bases, called a *mutation,* occurs in the DNA polymer. Such changes arise basically at random, sometimes from the impact on the DNA molecules of high-energy gamma rays or of cosmic-ray particles, or from exposure to various chemical agents called mutagens, or even from rare errors made by the cell's own DNA-copying machinery. We do not know which of these processes predominates in causing mutations throughout the history of life on Earth. When a mutation arises in a part of the DNA where information for a protein is encoded, it can cause a different, "incorrect" amino acid to be inserted into the protein under construction.

Many mutations are neither helpful nor hurtful. They are simply called "neutral" mutations. If the new protein does its cellular job poorly, the organism may be less fit for survival and reproduction, or it might not survive at all. On other occasions, mutations may actually change a protein in such a way that it does its job *better* than the original protein did. The lucky organism with this mutation would have some advantage over its fellows—perhaps it can replicate its DNA more quickly, swim more rapidly, sense food more efficiently, have more brightly colored feathers, or smell predators at a greater distance.

Such advantages would give the organism greater reproductive success than its fellows: Because they are healthier, more resistant to cold, prettier, better at escaping from predators, or superior for some other reasons, the more fit organisms will (by definition) produce more surviving offspring than their less fit relatives. As a result, organisms carrying favorable mutations will, over time, come to predominate in a population. "Differential reproduction"—the greater or lesser success that organisms achieve in producing offspring—lies at the heart of the process called *natural selection.* Differential reproduction determines whether a given mutation becomes established, or "fixed" in the general population. Thus natural selection, operating through differential reproductive success, causes the characteristics of a species gradually to change when advantageous, or "adaptive," mutations sweep through the population. In this way, differential reproduction allows one species to evolve into a new species (Color Plate 8).

Sometimes groups within a species become isolated from each other for many generations. When this happens, different mutations, appearing at random, become fixed in the separated populations. Gradually, the populations differ more and more from each other. When populations differ significantly (usually, when they can no longer interbreed to produce fertile offspring) we call them two separate species. In 1839, Charles Darwin noticed such changes among bird species in the Galápagos Islands. Darwin studied populations of birds resembling, but not identical to, those that he knew well from his native England. He found different species and subspecies of finches on the different islands that he visited, and his speculations about how the differences arose led to his theory of evolution, first published in 1859. In this epochal work, Darwin identified what we have called natural selection (differential reproductive success) as the driving force that makes new species on Earth.

Biological evolution seems to violate our common sense awareness that in general, *disorder tends to increase as time passes.* All around us, we see configurations of matter move from order to disorder, from improbable states to more probable ones. Paint weathers; rocks crumble; iron rusts; wood decays; stars radiate away their energy. But here on Earth, life continues to combine elements into specific molecules and monomers into lengthy polymers, making ever-greater complexity and order from simplicity and disorder. A single DNA molecule, which may be a million times longer than it is wide, represents an exceedingly nonrandom bit of matter. Such molecules store the tremendous amounts of information needed to carry out life's activities, information that can be preserved undamaged through thousands of replications. Hence, a great degree of order is required to keep matter alive. This order not only persists but has actually *increased* as life has evolved to ever more complex forms on Earth.

How can we explain this apparent contradiction—the maintenance, and even the increase, in the order and complexity of life in a universe that is inexorably evolving toward increasing disorder? The resolution lies in a consideration of the total system, of which life is just one part. Life on Earth does not form a closed system. Instead, life can maintain its highly improbable configuration only at the expense of its environment; that is, life can become highly organized only by increasing the disorganization of its surroundings. The disorder of the total system *increases,* while the disorder of living creatures within it *decreases.* Here "disorder" refers not to pollution but to the way that life acquires the energy it needs.

Energy

The various chemical reactions that occur in living organisms require some source of energy. Billions of years ago, living organisms may have drawn some of this energy from the heat of the Earth itself, as a few organisms in hot springs and near thermal vents in the ocean depths still do. But the dominant source of energy throughout the lifetime of our planet has been the *sun.* Most of the forms of life on Earth have linked their destiny to the energy arising from thermonuclear fusion deep within our star.

How does this use of solar energy proceed? During the past few billion years, many forms of life have developed the ability to obtain energy directly from sunlight through the extraordinary process called *photosynthesis.* In photosynthesis, sunlight provides the energy to convert carbon dioxide and water into complex organic compounds containing many carbon atoms. These organic molecules store the energy generated by the sun in the chemical bonds that hold their atoms together. The process of making the organic molecules releases oxygen as a by-product (Fig. 7.8). (Some bacteria can perform photosynthesis using sulfur rather than oxygen, which was probably the way this process originated, as we shall see in Chapter 8.) We can represent a typical example of photosynthesis by the following equation:

6 carbon dioxide molecules
+6 water molecules
+SUNLIGHT ENERGY = 1 organic molecule
+
6 oxygen molecules

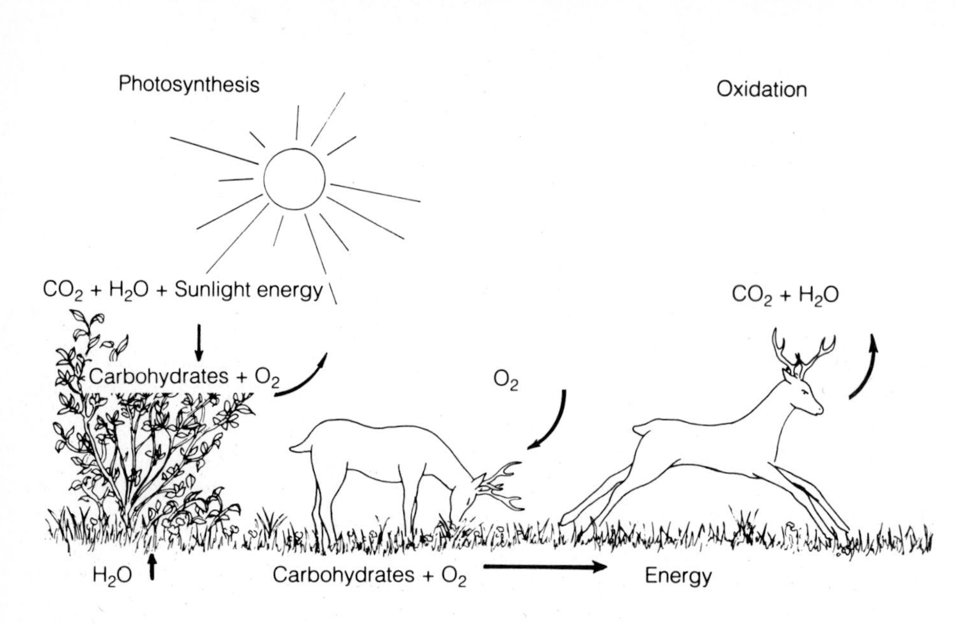

Figure 7.8 In plant photosynthesis, sunlight provides the energy to combine carbon dioxide and water molecules, forming carbohydrate molecules with the release of oxygen. The carbohydrates store energy in their chemical bonds. This energy can be released upon oxidation, which occurs, for example, during combustion or metabolism. In these processes, carbohydrate molecules are broken down again by reacting with oxygen molecules, returning CO_2 and H_2O to the environment.

If we prefer chemical symbols, the equation of photosynthesis is as follows:

$$6\,CO_2 + 6\,H_2O \xrightarrow{\text{light}} C_6H_{12}O_6 + 6\,O_2$$

As we pass from the left to the right side of this equation, we follow photon energy from the sun being converted into chemical energy, energy stored in the bonds of the organic molecules. On Earth, this activity now occurs predominantly in photosynthetic bacteria and in the "chloroplasts" contained in the cells of plants, allowing these organisms to store energy for future use while releasing oxygen into the atmosphere (see again Fig. 7.8).

Some forms of life—humans and other animals, for example—simply feast on the energy that plants have stored. This provides animals with energy in a process that is complex in structure but simple in operation. We eat, but we do not become what we eat (despite popular sayings to the contrary). Our bodies assimilate the "dead" matter we ingest and use its constituents to remain alive; in this sense, the "dead" food regains "life." This property of living organisms provides an elegant solution to an ancient riddle: How can you unscramble a scrambled egg? Answer: Feed it to a chicken! Like all other living creatures, we constantly exchange the matter within ourselves for new atoms and new molecules, yet we remain much the same.

A similar pattern holds true for life itself. We tend to be struck by the fragility of life, for we are surrounded by organisms busily passing through cycles of birth, reproduction, and death. But if we take a longer view, we can easily be impressed that *the exceedingly improbable state of matter that we call life has remarkable continuity*. Life has persisted on Earth for at least 3.5 billion years, a span of time greater than the lifetimes of many stars. How has life managed to do this? There are many answers to this question, but *the key to continued survival is a steady source of energy*. The development of photosynthesis, the connection life has formed with the sun, has assured terrestrial life a chance to survive on cosmic time scales.

During life's existence on Earth, living organisms have changed the face of our planet. Consider the most important side effect of plant photosynthesis: the release of oxygen molecules (O_2). These molecules enter the Earth's atmosphere, where they offer an additional source of energy to any form of life that can use oxygen in "oxidation," the process of combining oxygen with other molecules. We can see from this fact another reason why animals can be called parasitic: Without plants and the oxygen they release, animals could not develop, breathe, or eat. We may puzzle over the order of the chicken and the egg, but we do know that *oxygen-producing organisms appeared before oxygen-using animals* as life evolved on Earth.

Look once again at the equation for photosynthesis. If we reverse the process, and pass from right to left in the equation, we describe oxidation or respiration, in which energy is liberated rather than stored. The released energy may appear in the form of work, heat, or even light. Familiar examples of oxidation are the burning of plants, wood, coal, or oil, which liberate solar energy stored by photosynthesis. The fact that large deposits of energy-rich organic material exist on our planet indicates that photosynthesis has dominated respiration over past ages; that is, the net direction of the arrow in our equation has been just the way it is drawn, from left to right. Human activities, however, often require exactly the opposite goal, so we are rapidly depleting the richest deposits of energy stored through photosynthesis: petroleum, natural gas, and coal.

The Unity of Life

We have now identified the *chemical elements* of which life consists, the principal *monomers* (small molecules) formed from these elements, the most important *polymers* made of these monomers, and some of the basic *chemical reactions* that occur in any configuration of matter that we call alive. We have seen that despite the extraordinary diversity of life on our planet, an underlying unity appears at the molecular level. Just 20 amino acids out of the thousands that *could* exist are used to make proteins, and these same 20 appear in all forms of life. The different types of life all use the same genetic code, with the same four molecular bases, taken three at a time in a code that specifies particular amino acids. Although organisms differ in the sequence of base pairs—the fact that causes them to be different types of organisms—the *code* remains nearly identical among all organisms.

The unity of life's chemistry suggests that all the life we see around us descended from the same line of early organisms. This conclusion in turn leads to the expectation that life on some other planet could be quite different from life on Earth, even if that life employs the same elements that we do. We might expect extraterrestrial life to use a different set of amino acids, for example, or different types of bases in its equivalent of DNA. But this alien life might, of course, differ in still more fundamental ways from ours, perhaps by using types of organic molecules completely different from ours for the processes of replication, information storage, and energy transport.

We would have a far better ability to assess these possibilities if we understood the essence of life: What happens to matter to bring it to the level of complexity where reproduction occurs? To this question we still have no answer. No one has yet made a self-replicating polymer from simple monomers without using products furnished by living cells. So how did nonliving matter assemble itself into a self-replicating configuration? We shall deal at some length with the fundamental question of life's origin in the next chapter. Unfortunately, we shall find it extremely difficult to reconstruct the first steps toward life from the available evidence. If we discovered another kind of life somewhere else in the universe, a fossil record of such life, or a planet on which the first natural experiments in life's origin were still occurring, we could certainly learn much more than we know now.

SUMMARY

The elemental composition of life on Earth resembles that of the sun and other stars much more closely than the composition of the matter that forms the Earth. We may conclude from this fact, and from more detailed studies of the ways that life behaves, that the four basic elements that appear in living organisms—carbon, oxygen, hydrogen, and nitrogen—have a special role that could make them essential for life throughout the universe.

We must, however, be cautious in the conclusions we draw from our single example, life on Earth. We refer to all of life as a single example because all life on Earth is based on the same chemical system. In all terrestrial organisms, the double-stranded spiral molecules of DNA carry the genetic information that determines what the next generation will look like, as well as the additional information needed to tell the various parts of an organism how to function. When DNA molecules divide, each half of the double spiral can regenerate the missing half from smaller component molecules, for two reasons: Only certain molecules can connect with those exposed by the splitting, and the order in which they connect is uniquely specified by the "parent" molecules.

We can define living organisms through their capacity to reproduce themselves and to evolve, but the basic activity of an individual organism—when it is not reproducing itself—consists of continuing its metabolic activities through the direct or indirect use of the radiant energy provided by sunlight. Plants can take direct advantage of sunlight through the chemical reactions involved in photosynthesis, which stores radiant energy in the chemical bonds of carbohydrate molecules formed from water and carbon dioxide. Animals (and some rare plants) depend on this stored energy, which they release through the digestion and assimilation of plant matter, or of other animals that have eaten plants. In addition, the oxygen molecules released in photosynthesis provide another important contribution from plants: the metabolic processes in animals consist of reactions of these oxygen molecules with organic matter from plants.

In other words, plants make carbohydrates and oxygen (more precisely, they release oxygen molecules into the atmosphere as part of their photosynthetic reactions), while animals burn carbohydrates and oxygen. On Earth, we still find a net surplus of plant-made carbohydrates, along with the hydrocarbons (natural gas, petroleum, and coal) that originated as decaying plant or animal matter. We have come to understand and appreciate the cycle of life, in which photosynthesizers such as plants had to precede animals, and through which we humans find the food, oxygen, and hydrocarbon fuels we consider essential. What remains unknown, at least in any true detail, is the exact way in which the microorganisms that began the great chain of life on Earth emerged from some mixture of a small number of simple molecules.

QUESTIONS

1. How would you define "life"? Can you see any exceptions you would like to make to this definition? For example, would robot machines, programmed to make other robots, be "alive" by your definition? Why or why not?

2. Why do we say that the composition of living organisms resembles the composition of the stars? What does this imply for theories of how life might have originated throughout the universe?

3. Why is carbon such an important element in all living creatures on Earth?

4. What are polymers? In what form do smaller molecules, the sort we call monomers, become strung together: in straight chains, in rings, or in more complicated structures?

5. How do DNA molecules replicate, or reproduce themselves?

6. What other key functions besides governing the reproductive process do DNA molecules regulate?

7. What do protein molecules do in living systems? Carbohydrate molecules? Molecules called "lipids" and "fats"?

8. How do animals use the sunlight energy that plants have stored through photosynthesis? Describe the chemical reactions that first stored the energy, and later, in almost the reverse reaction, released it.

9. Where did oil, coal, and natural gas acquire the energy that they contain ready to be released upon burning?

10. Would we expect another form of life, in some other planetary system, to be made of about the same elements as life on Earth? Would we expect such life to use the same kinds of amino acid molecules as we do? Should we expect such life to have developed DNA molecules like those in living organisms on Earth? Why or why not?

FURTHER READING

Asimov, Isaac. *The Genetic Code.* New York: Signet Books, 1962.

Beadle, George, and Muriel Beadle. *The Language of Life.* New York: Doubleday, 1966.

The Biosphere. San Francisco: W. H. Freeman, 1970.

Dawkins, Richard. *The Selfish Gene.* Oxford: Oxford University Press, 1976.

Eiseley, Loren. *The Immense Journey*. New York: Vintage Books, 1946.

Goldsmith, Donald. *The Quest for Extraterrestrial Life: A Book of Readings.* Mill Valley, CA: University Science Books, 1980.

Thomas, Lewis. *The Lives of a Cell.* New York: Viking, 1974.

Watson, James. *The Double Helix.* New York: Signet Books, 1968.

8

The Origin of Life

We know that life existed on Earth as long as 3.5 billion years ago, because we can find the fossil evidence left by microorganisms in rocks that old. But these organisms were already well-advanced, living in colonies and performing photosynthesis. They must have been preceded by still simpler forms, which in turn followed the remarkable transformation from inanimate matter to matter that we could recognize as being alive. When and how did that transformation occur? Did it occur only once or many times on Earth?

Unfortunately, the fossil record cannot help us here. Because of erosion and the motions of the Earth's crust, the geological record of the first billion years of this planet, including evidence of the impact craters that doubtless then covered the Earth's surface (as they still cover Venus, Mars, and the moon), has vanished forever, dragged down into the mantle. The crustal motions are driven by convection currents within the Earth's mantle, which derive their energy from heat liberated by radioactive minerals deep in the Earth's interior (Fig. 8.1). Not only has the primitive crust vanished; even the present configuration of the continents is a relatively recent phenomenon, the result of the slow movements of the crustal plates. Just 200 million years ago the continents were much closer together than they are today (Fig. 8.2). Geologists can reconstruct most of the Earth's history from the record in the rocks. But they cannot reach back through time more than 3.8 billion years, to provide an understanding of what Earth was like when life began.

We can, however, reconstruct some of the Earth's early history by using our increasing knowledge of the way the planets formed. The "primitive Earth" that we then visualize is a rocky spheroid, of the same size and composition as the Earth today but with a greater rate of heat energy released from radioactive minerals, some of which have now "decayed" and no longer generate heat. The most profound differences—and the most important ones for any organisms on its surface—between the Earth today and the Earth of 4 billion years ago consist of changes in the Earth's *atmosphere.*

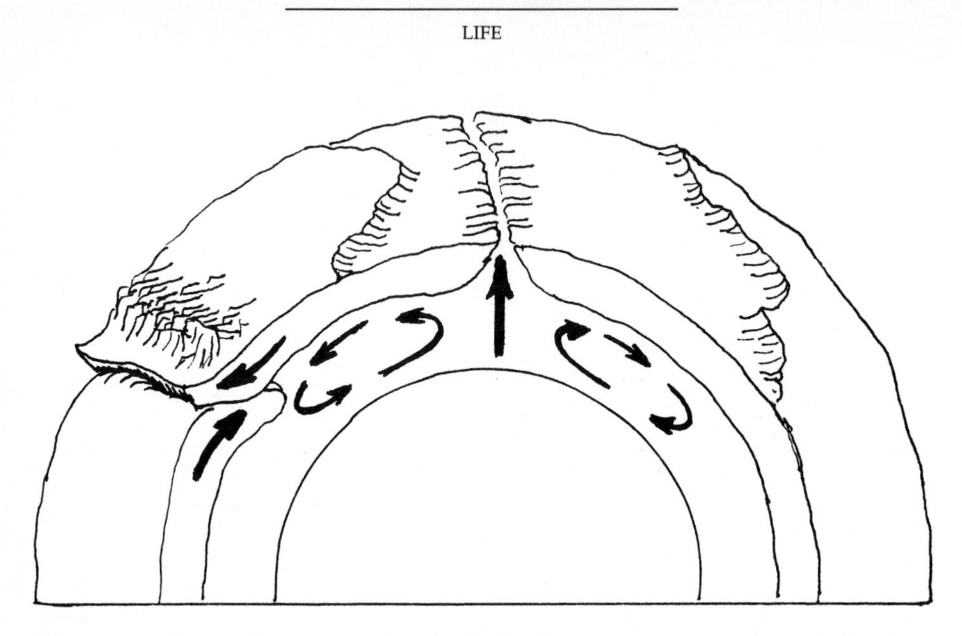

Figure 8.1 Convection currents inside the Earth transport energy through the slow movements of mantle rock in much the same way (but far more slowly) that hot air rises and cool air descends in the atmosphere. At the Earth's surface, these currents cause crustal spreading that can move the continents while bringing new rock to the surface.

These changes are more difficult to reconstruct, but we must try to do so, since they hold the key to the origin of life on our planet.

Our quest may be compared to the building of the first transcontinental railroad in the United States. To forge this great connecting link, teams worked from the West and from the East, meeting in Utah, where a golden spike was driven to commemorate the event. In our quest, we must work *backward* through time by using the history recorded in the Earth's rocks, and must also work *forward* in time from the formation of the Earth by using all the available evidence about the evolution of planets. This evidence comes both from our theories of planetary formation and also from studies of other objects in the solar system that have preserved a record of the first billion years of their history. These two approaches must join at a point close to the time when life began. Unfortunately, the point of joining is not so well-defined as those of two sets of railroad tracks; it is as if the tracks disappear into a fog at the California border and do not reappear until we reach Kansas or Colorado. We know that they must connect, but we don't yet know where or how. We are still waiting for the golden spike that will commemorate the meeting of planetary history and the history of life.

Figure 8.2 Two hundred million years ago, most of Earth's land areas were concentrated into a single continent, Pangaea, allowing living species to diffuse more easily than is true now. This figure shows both the land-rich (left) and land-poor (right) hemispheres of the Earth at that time. The continents continue to move; for example, the Pacific plate will carry Los Angeles past San Francisco on its way to Alaska in about 10 million years.

How the Earth Got Its Atmosphere

The Earth apparently grew to its present size through the "accretion," or sticking together, of smaller particles within the "solar nebula," the dense interstellar cloud from which the solar system formed (see Chapter 4). All the planets, satellites, asteroids, and comets that now orbit the sun passed through this accretion process, 4.5 billion years ago, to reach the sizes they have today. Because hydrogen formed the most abundant element within the solar nebula, many scientists once believed that hydrogen must have formed a significant part of the Earth's primitive atmosphere. When large amounts of free hydrogen atoms and molecules are ready to combine with other atoms and molecules, we have what chemists call "reducing" conditions. For many years, scientists concluded that the Earth's primitive atmosphere must have been highly reducing, that is, laden with hydrogen and hydrogen-rich molecules: methane, ammonia, and water vapor. With this composition, the Earth's primitive atmosphere would have resembled the atmospheres of the giant planets Jupiter and Saturn, where primitive conditions have persisted down to the present day. This model of our original atmosphere had widespread acceptance for many years, but it requires some adjustment in the light of new discoveries.

Through studies of the abundances of the "inert gases"—neon, argon, krypton, and xenon—that form a tiny part of our atmosphere, and through careful investigation of the oldest rocks, scientists have found evidence that the Earth probably never had an atmosphere captured from the cloud of dust and gas that formed the planets. The process of accretion that built the planets may have occurred in such a way that the icy and rocky material richest in "volatile elements" joined the Earth last (Fig. 8.3). The volatile elements, the lightest and most easily vaporized, include hydrogen, carbon, nitrogen, and oxygen. These elements form most of the molecules in our present atmosphere and in life itself. This late-accreting material must have been similar to some of the meteorites and comets that we find in our solar system today (see Chapter 11). Comets in particular, being so rich in

Figure 8.3 The material richest in the volatile elements—hydrogen, carbon, nitrogen, and oxygen—joined the Earth last as the proto-Earth grew into the present Earth through the process of accretion. The earlier stages of accretion occurred at higher temperatures than the later stages, since they involved the impact of larger objects.

"volatiles" (easily vaporized elements and compounds), must have played an important role in contributing these substances to the Earth.

According to this new understanding of our planet's formation, the volatile matter that permeated the outer layers of Earth (and of the other inner planets) would have included only a tiny abundance of free hydrogen atoms and of hydrogen molecules (H_2) that were not bound into larger molecules. Most of the hydrogen would have been part of molecules such as water and organic compounds, formed by chemical reactions within interstellar clouds or within the solar nebula itself.

Heated by its collision with the forming Earth, the late-accreting material would have vaporized most of its volatiles to produce the Earth's original atmosphere. Any hydrogen atoms and hydrogen molecules that were present must have escaped from Earth during its first few hundred million years, because our planet had insufficient gravitational force to retain them. Hydrogen bound into heavier molecules such as water vapor would have remained. Even this hydrogen was not completely safe. Ultraviolet radiation from the young sun would have broken some of the heavier molecules apart. This process, called "photodissociation," produced molecular fragments that could combine with one another to form new compounds. Whenever one of these hydrogen-containing compounds was struck by an ultraviolet photon, or was broken apart by some other source of energy (lightning, for example), individual hydrogen atoms and molecules had a good chance of being released to react with other compounds or to escape from the planet.

When we add to this picture the chemical reactions occurring between the atmospheric gases and the primitive crust of the Earth, the net result would have been a mildly reducing primitive atmosphere, made mostly of carbon monoxide (CO), carbon dioxide (CO_2), nitrogen (N_2), and water (H_2O), with only a small amount of continuously escaping H and H_2. A similar situation should exist on any Earth-sized planet orbiting relatively close to its central star. The atmosphere on such planets will undergo a steady progression from more reducing to less reducing to "oxidizing" (oxygen-rich) conditions whether or not life originates. We can see this outcome in our own solar system on Mars and Venus.

The Evolution of the Atmosphere

Once the bulk of the hydrogen had escaped from the primitive atmosphere, a significant change occurred: The atmosphere could produce enough "ozone" to shield the Earth's surface against ultraviolet light from the sun. Ozone molecules (O_3) each consist of three oxygen atoms, while ordinary oxygen molecules (O_2) are pairs of oxygen atoms. When ultraviolet photons strike molecules of water vapor (H_2O), they break the molecules apart into hydrogen and oxygen atoms. So long as hydrogen molecules were abundant in the atmosphere, any free oxygen atoms soon recombined with hydrogen to form water vapor and could therefore not form oxygen or ozone molecules. But once hydrogen evaporated from Earth, ozone and oxygen molecules *did* form—in abundances that were small but nevertheless sufficient to prevent ultraviolet photons from reaching the Earth's surface. Ozone molecules have such a great ability to absorb ultraviolet that even a small concentration of ozone protects us against the deadly (to us!) ultraviolet radiation from the sun.

We must therefore add to the water vapor, carbon dioxide, nitrogen, and carbon monoxide in the Earth's evolving atmosphere a small amount of oxygen (less than 1 percent), and a far smaller, though crucial, amount of ozone. We still seem nowhere near the Earth's present atmosphere, which consists mainly of nitrogen and oxygen molecules, with small amounts of water vapor, only traces of carbon dioxide, and next to no carbon monoxide. What happened to produce these changes?

There is a basic answer: life! Life on Earth has provided almost all of the oxygen in our atmosphere. It has helped to remove the carbon dioxide while retaining most of the nitrogen. Let us therefore follow the history of each of the basic components of our atmosphere to see what happened during the past 4 billion years to establish the present atmospheric composition.

First of all, the water vapor produced in such abundance early on still forms part of the atmosphere. A cycle of evaporation and rainfall, followed by water runoff into the seas, keeps a small fraction of the water in the form of atmospheric vapor. We see this vapor when it condenses on a cold surface, or as rain clouds or snow. But most of the Earth's water resides in liquid form within the great reservoirs that cover 71 percent of our planet's surface. These oceans, which distinguish Earth from all other planets in the solar system, have probably been in existence from the earliest period of the Earth's history, from the time when the surface was first cool enough to allow them to condense. As we shall see when we study Venus and Mars, the persistence of liquid water on the surface of a planet—an essential condition for the origin and continuation of life as we know it—requires rather special conditions, which the Earth just happens to provide.

The Effects of Life on the Atmosphere

Carbon dioxide (CO_2) may once have dominated the composition of our atmosphere; it is the second most abundant volatile on Earth, after water (H_2O). Today most of the carbon dioxide on Earth has become locked up in chemical compounds in abundant types of rocks, principally in "calcium carbonates" (calcium-oxygen compounds) such as chalk and limestone. These calcium-carbonate rocks form on ocean bottoms, using the carbon dioxide dissolved in sea water. As this carbon dioxide disappears into rock deposits, more of this gas dissolves in the seas, and the process might continue until the atmosphere contained little or no carbon dioxide. Although calcium carbonates will form even if life does not exist, the presence of living marine creatures greatly accelerates the formation process. We see their shells in the limestones they helped to produce. Billions of years of both biological and nonbiological formation of calcium carbonates have led to the disappearance of nearly all the carbon dioxide that has been produced over geologic time. A small amount remains because volcanic eruptions, the weathering of rocks, animal respiration, and the decay of organic matter all return carbon dioxide to the atmosphere.

Atmospheric oxygen, as we have seen, existed only in small amounts before life began to flourish. Green-plant photosynthesis has greatly increased the abundance of oxygen, so that 21 percent of our atmosphere now consists of this gas. Fossil records show that rocks formed and buried more than about 2.5 billion years ago are "suboxidized" (formed under oxygen-poor conditions). Fully oxidized rocks have been made only during the past 2.5 billion years. Confirming the evidence for the time of rapid oxygen release are the layered rocks called "stromatolites," rocks made from colonies of blue-green bacteria living at the boundaries between water and rock sediments (Fig. 8.4). The blue-green bacteria that formed stromatolites were among the first to release oxygen into the atmosphere. Ancient stromatolites occur in the fossil record at many places on Earth at times only 2.0 to 2.3 billion years ago. Before this they are rare. We may therefore hypothesize that oxygen-releasing organisms such as the bacteria that formed stromatolites appeared in large numbers at a time somewhat less than 2.5 billion years ago—just over halfway back through the Earth's history. Large plants—trees, flowers, and grasses—go back no more than 600 million years in time, so we have blue-green bacteria to thank for the basic oxygen enrichment of the atmosphere (Color Plate 7).

The nitrogen that now forms 78 percent of our atmosphere came from "outgassing" (venting into the atmosphere from subsurface rock), but the existence of life on Earth helps to explain the *persistence* of nitrogen in the atmosphere. Each time that lightning discharges occur, the electrical currents cause some of the atmospheric nitrogen to combine with oxygen to form nitric oxides. Rain then washes the nitric oxides into the soil and oceans. If the Earth lacked life, only the slow weathering of the rocks would return this nitrogen to the atmosphere. In fact, however, "denitrifying bacteria," tiny organisms that break apart nitric oxides, live in the soil in

Figure 8.4 These modern stromatolite beds on the western coast of Australia are exceptional at the present time, but they typify the conditions under which stromatolites have been made for billions of years on this planet.

immense numbers. These denitrifying bacteria play an important role in recycling the nitrogen into the atmosphere once again.

When photosynthesis began to liberate large amounts of oxygen, this highly reactive element tended to combine with every available chemical compound that was not "fully oxidized" (carrying as many oxygen atoms as it could hold). In this way, carbon monoxide that once formed an important component of the Earth's atmosphere long ago combined with oxygen to form carbon dioxide.

Argon, the last important component of the present atmosphere, forms about 1 percent of our air. Here we have one gas not affected by life, for argon is an inert element that does not form chemical compounds. Over 99 percent of the argon in our atmosphere was generated by the decay of small amounts of radioactive potassium in the Earth's crust. The erosion and melting of this potassium-containing rock then liberated the argon, so that it could continuously enter the atmosphere through the same "outgassing" process that had earlier contributed the nitrogen and other gases trapped in the material that accreted to form the Earth (Fig. 8.5). We therefore expect to find some argon on any rocky planet in any planetary system, provided that the planet is massive enough to prevent this gas from escaping into space, and is sufficiently old to have released significant amounts of argon by radioactive decay.

Figure 8.5 Volcanoes such as this one in Hawaii release new gases into our atmosphere even today. The release of gases from impacting comets and meteorites, as well as from molten magma, gave the Earth its primitive atmosphere.

The preceding paragraphs summarize how the Earth's original atmosphere changed by solar ultraviolet radiation, by chemical reactions, by outgassing, by the escape of hydrogen, and by the origin and evolution of life. As a result of these effects, the carbon dioxide, nitrogen, hydrogen, and carbon monoxide of the primitive atmosphere were replaced by the nitrogen and oxygen that we now breathe. Atmospheric water vapor—along with the oceans that produced it!—has probably been present throughout the Earth's history.

We must now attempt to follow the ways in which life has developed beneath and within the Earth's changing atmosphere, whose largest changes have apparently been caused by the activities of life itself. We have stressed the essential unity of life at the molecular level, and have suggested that this unity indicates a descent from a common ancestor. The next chapter follows this descent in detail, but we can recognize immediately a key fact: *The complexity of life has increased with time.* The earliest rocks that indicate the presence of life contain evidence only of microbial life, and older rocks show no evidence of life at all. The absolute and relative ages of these rocks are determined by measuring the amounts of parent and daughter elements in a radioactive decay process, similar to the one we described as producing most of the argon in the air we breathe from the potassium in the rocks we tread. Thus there is no doubt about the great age of the Earth. A planet only a few million years old would not have generated nearly as much argon as the Earth has. Similarly, the evolution of life from simple to complex forms is not a hypothesis; it is an inescapable fact. But how did this process begin? How did atoms and molecules, on the surface and in the atmosphere of Earth, ever combine into assemblages that are alive?

Early Ideas about the Origin of Life

If we examine how various cultures have dealt with the question of life's origin, we find that human intuition usually invokes outside intervention, a reasonable extrapolation of our observations of how new things appear: Some force or causative agent "makes it happen." The hidden assumption in applying the rule of causation to life is that something alive had to exist before life could appear on our planet. This is consistent with simple observation of how new life appears on Earth today. Thus it is not surprising that most religions invoke outside intervention, usually in the form of a divine and omnipotent being, to explain the appearance of life.

Some scientists, still impressed with the idea that life could never have arisen spontaneously on Earth, have suggested that life appears everywhere in the universe. This notion of "panspermia," developed in 1901 by the Swedish chemist Svante Arrhenius, suggests that life floats as spores through the interstellar medium and occasionally comes to rest on a planet, where replication and evolution can occur. As we have learned more about the hazards of space travel, the hypothesis of panspermia has become less and less believable. The combined effects of high-energy ultraviolet radiation, x rays, and cosmic-ray particles in interstellar space would probably prove lethal over the length of the trip, tens of millions of years or more for random wandering from one planetary system to the next. Furthermore, the panspermia hypothesis does not confront the question we are ultimately trying to answer. If the Earth was impregnated by spores from outer space, where did the spores come from?

The Chemical Evolution Model for the Origin of Life

How *did* life arise? A wide variety of laboratory experiments have shown that in an atmosphere containing free oxygen, the chemical compounds that form all living systems cannot arise from spontaneous chemical reactions. If the transition from inanimate to living matter happened on

Earth, conditions must have been quite different at the time that this change occurred. This conclusion is entirely consistent with the deductions that we can make from both the fossil record (our path backward in time) and from our reconstruction of the Earth's earliest history (the path forward from planet formation).

The preceding paragraph carries a sobering message, to which we shall return in Chapter 17: *If all living organisms had ever been completely eradicated during the roughly 4 billion years since life arose, life would not have begun again.* Thus if some global event such as an extreme ice age, a nearby supernova, or an impact with a large asteroid had destroyed life on Earth, our planet would have remained completely barren, its empty oceans lapping the uninhabited land beneath an azure sky unstirred by flight. We must keep such a possibility in mind when we attempt to extrapolate from the history of life on Earth to possible life on planets circling other stars.

The traditional scenario for the early Earth, in which the Earth's primitive atmosphere consisted of methane, ammonia, and hydrogen molecules, with water able to condense, provides an environment in which the production of organic (carbon-containing) molecules is very easy. The less reducing environment that now seems more likely for the early Earth also allows organic compounds to be formed, though not as easily. The energy needed to drive the chemical reactions that make organic molecules could have arisen from many sources, but especially from the sun. Solar ultraviolet would have reached the surface of the primitive Earth before the free hydrogen had escaped, because no ozone existed to act as a shield against this short-wavelength radiation. Such high-energy photons would have been the primary energy source, but additional energy could have come from lightning, from local geothermal sources (hot springs and volcanoes), from radioactive elements in rocks, from impacts by meteorites and sub-atomic particles, and even from shock waves generated by thunder or by ocean surf (Fig. 8.6).

It may seem ironic that solar ultraviolet radiation helped life begin, since today ultraviolet destroys many molecules essential to life. This apparent paradox can be resolved by noticing that organic molecules formed in the atmosphere could be washed out into pools and ponds by rainfall, or the molecules could have formed at the interfaces between water, land, and air, from which they could enter deeper water. Just a few meters of clear water or a thin layer of other organic compounds can provide adequate screening from solar ultraviolet.

The traditional scientific picture of life's formation on the primitive Earth involves the spontaneous formation of complex molecules from atmospheric gases, and the molecules' later concentration in pools of water. Some of the chemical reactions could have been catalyzed, or speeded up, by soil materials with suitably active surfaces. The runoff of rainwater would have helped bring the product molecules into ponds, lakes, and tide

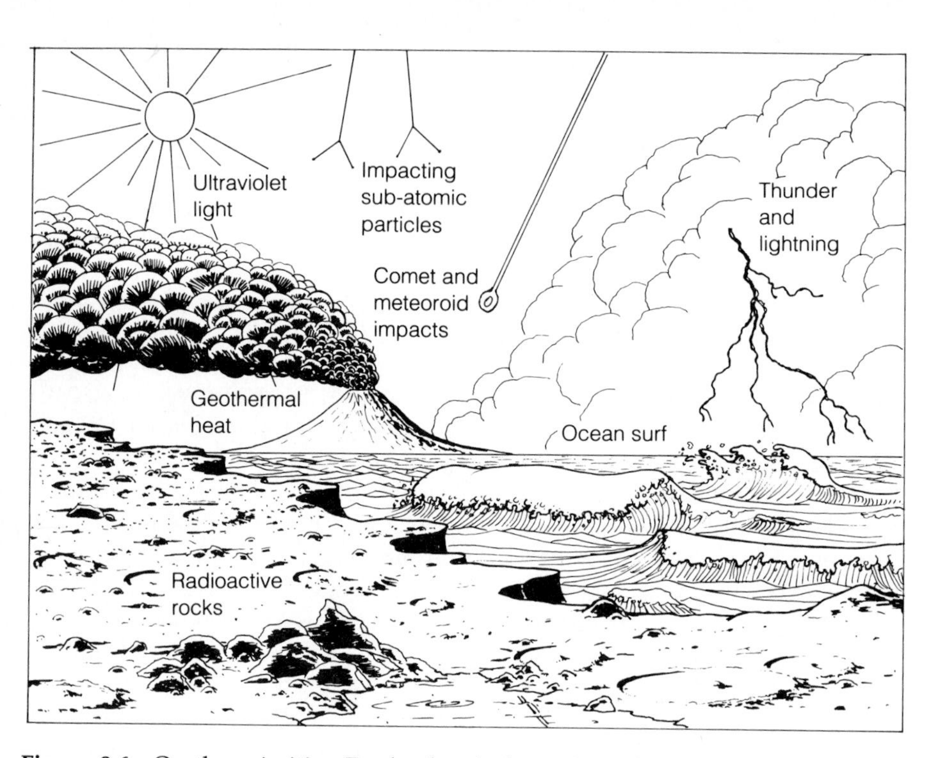

Figure 8.6 On the primitive Earth, chemical reactions that require energy could have obtained this energy from many diverse sources: the planet's magnetosphere; ultraviolet radiation from the sun; lightning discharges; localized geothermal heat sources; radioactivity; meteorite impacts, or even the shock waves produced by ocean surf.

pools at the edges of oceans. These pools and ponds would have become rich in organic compounds and dissolved minerals. They probably served as the nurturing, protective environments in which molecules interacted at random, until some of the reactions produced compounds that could serve as guides in the formation of other molecules. Grains of clay at the edges of these ponds could also have helped to organize small molecules into larger ones (see page 198).

This tide pool mixture of life-producing ingredients had a vague preview in Charles Darwin's conception of "some warm little pond" in which life could have begun. Later, the British biologist J. B. S. Haldane and the Soviet biologist Alexander Oparin independently developed this model in detail, characterizing the primitive oceans of Earth as a dilute organic broth, a "primordial soup" in which more complex molecules could assemble.

In outline, the concept of life arising in primordial soup suggests that natural conditions on the primitive Earth would lead to (1) the production of simple organic molecules, (2) the selection of pathways to complexity that avoid chemical diversity, (3) the development of specific forms of chemical activity that start the machinery of life, and (4) the appearance of membranes that prevent the dispersion of the chemicals that make all this occur. But how good is this concept? How can we test it?

An Experimental Test of the Primordial-Soup Model

The first part of this model—the hypothesis that organic molecules were produced naturally on the early Earth—was tested experimentally by Stanley

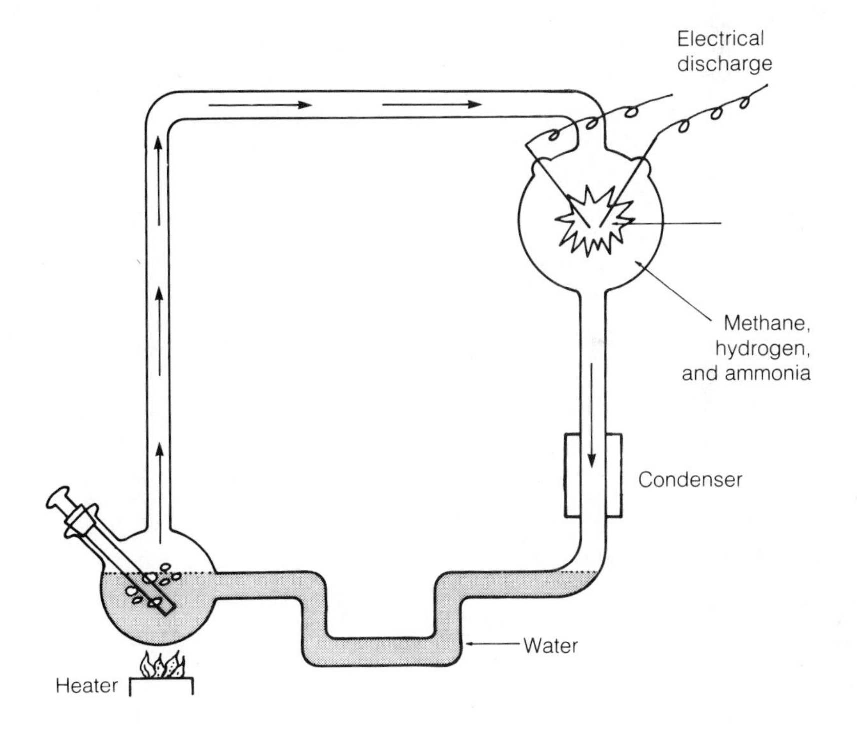

Figure 8.7 A schematic diagram of the experimental equipment used by Stanley Miller and Harold Urey shows the 5-liter flask that contained water, methane, hydrogen, and ammonia to simulate their concept of the Earth's primitive atmosphere and oceans, plus the electrical discharge apparatus that released energy into this mixture.

Miller and Harold Urey in 1953. Figure 8.7 shows a schematic representation of their classic experiment. Here the atmosphere of the primitive Earth is represented by a mixture of methane, ammonia, hydrogen, and water vapor, which resides in the upper flask. Energy arrives from an electric discharge, which could be replaced by a source of ultraviolet photons (or by shock waves). The water in the lower flask models a pool of water on the Earth's surface. The heater and condenser provide circulation of water vapor through the system in the same way that water evaporates from pools, moves into the atmosphere, and then condenses as rain, falling back to Earth along with the chemicals formed in the atmosphere. The mechanical aspects of this experiment seem to represent the conditions of the primitive Earth quite well, although we now have doubts about the mixture of gases used to represent the early atmosphere, and the flask is no substitute for the mineral-rich surface of our planet.

What happens when we turn on the energy supply and let the experiment run? After a few days of continuous operation, organic molecules appear in the water-containing flask. Chemical analysis of this mixture provides exciting news: In addition to a large amount of unidentified organic "gunk," amino acids have formed, including glycine and alanine, as well as sugars, the nucleotide bases, and numerous other compounds of interest.

We can illustrate what has occurred with the following chemical formula for the production of glycine:

1 ammonia molecule
+
2 methane molecules
+
2 water molecules
+
ENERGY

=

1 glycine molecule
+
5 hydrogen molecules

In symbolic language,

$$NH_3 + 2CH_4 + 2H_2O \xrightarrow{\text{Energy}} C_2H_5O_2N + 5\,H_2$$

The formation of other amino acids occurs in a similar manner in the experiment we have described. When we recall that one of the major activities of life as we know it consists of the manufacture of proteins from amino acids, we can see that the production of these acids represents a significant step toward living matter, and a general confirmation of the idea that the basic substances of which life is made could form under at least one set of possible primitive conditions. A further illustration of the ease with which these compounds can form under natural conditions comes from the discovery of amino acids in meteorites (see Chapter 11).

In contemporary life on Earth, of course, amino acids do not come from meteorites or from laboratories duplicating the Miller-Urey experiment. Instead, they are made by living systems, as we can easily demonstrate by investigating the symmetry of these important molecules. The amino acids found in the soup that arises in the Miller-Urey experiment (and those found in meteorites) include equal amounts of right-handed and left-handed molecules. As we have seen (page 160), the proteins in living systems contain amino acids exclusively of the left-handed variety. Life on Earth apparently selected one of the two possible "handedness" forms at the time of its origin and has maintained this exclusive choice ever since. As we pointed out previously, this selection confers an advantage, since it allows more precise programming of the structures of more complex molecules made from these amino acid units.

Thus life elsewhere, if it uses amino acids, may also have chosen one of the two configurations. Here we have a good test for distinguishing between biologically produced molecules and those that formed spontaneously. The latter will always occur in equal mixtures of both configurations, while the former should show a marked preference for one over the other, as is certainly true on Earth.

Did Life Really Originate in This Manner?

Let's return from chemistry laboratories to the primitive Earth. We have identified a way in which the building blocks of proteins could form. What about the components of lipids, carbohydrates, and the nucleic acids?

We find that some of the simplest organic acids formed directly in the Miller-Urey experiment: acetic acid, formic acid, and proprionic acid. Organic acids help to form fatty acids (lipids), so once again we feel some confidence that we are on the right track. But after these triumphs, the picture becomes less clear.

From our discussion of DNA in the last chapter, we know that each nucleotide contains a sugar (deoxyribose) together with one of four bases (adenine, guanine, thymine, or cvtosine) plus a phosphate (phosphoric acid). RNA uses uracil instead of thymine. Not all of these substances arise in the Miller-Urey experiment, but the experiment can produce formaldehyde, cyanoacetylene, hydrogen cyanide, and urea. These compounds take us part of the way to our destination. If we heat formaldehyde in alkaline solutions, or in the presence of clays, we produce sugars. We thus form carbohydrates, and in the process develop a pathway to produce one of the key structures in nucleic acids. As John Oró demonstrated in a different laboratory experiment, further reactions of hydrogen cyanide can produce adenine, while urea and cyanoacetylene can react to form cytosine. Phosphates could presumably come from rock weathering, with subsequent water runoff into collecting areas. We observe these processes occurring even today.

So far, so good; but no one has yet found a way to make all of the essential building blocks from this mixture of compounds. We may simply be ignorant about the appropriate chemical reactions, or perhaps we merely need some other combination of special local environments to complete the picture. On the other hand, we may have misled ourselves somewhere along the way, and we might need an altogether different chemical system to bring life into existence.

As we mentioned, the basic scenario of the Miller-Urey experiment has been challenged during the past decade by scientists who feel that the Earth's atmosphere never had as much free hydrogen as is contained in the mixture of gases originally used in the experiment. However, other scientists have shown that complex organic molecules can form even in a weakly reducing atmosphere, one with only a small amount of free hydrogen. For example, when we shine ultraviolet photons on a mixture of carbon dioxide, carbon monoxide, and nitrogen together with just a small amount of hydrogen molecules, we find that hydrogen cyanide and water appear. Hydrogen cyanide, combining with itself in alkaline solutions (the early oceans), can produce amino acids so long as ultraviolet radiation continues to strike the mixture. This reaction also produces cyanamide. Cyanamide can make amino acids link together, the first step in the formation of proteins, when cyanamide mixes with amino acids in a dilute solution illuminated by ultraviolet. Using the language of chemistry, we write:

3 hydrogen cyanide molecules		
+		1 glycine molecule
2 water molecules	=	+
+		1 cyanamide molecule
ENERGY		

In symbols:

$$3HCN + 2H_2O \xrightarrow{\text{Energy}} C_2H_5O_2N + CN_2H_2$$

We conclude from these experiments that we do not require the extremely hydrogen-rich conditions of the Miller-Urey experiment for the first steps in the production of important compounds for living systems.

Rather, *the essential requirement seems to be the absence of free oxygen in the atmosphere.* In other words, oxygen would have been as surely "poison" to early living systems as it is necessary to highly developed animal life today. But further experiments must be performed, using more realistic versions of the primitive Earth, before we can be certain how organic compounds first arose. We can test the "realism" of these models, to some extent, by calculations concerning the formation of our atmosphere, as well as the chemical reactions that the postulated constituents of the atmosphere would undergo with one another and with likely substances on Earth's surface. We must also follow the escape of small molecules from the Earth's gravitational field. These mathematical models, however, even though we can calculate them at amazing speed, can be only as good as our knowledge of the lists of ingredients and of the chemical pathways they follow.

In other words, calculations can provide a unique answer only if we know all the conditions. But we simply don't know the conditions that existed on the primitive Earth! The same holds true for the experimental approach to the origin of life. This approach has the advantage of allowing the phenomena to speak for themselves, but we cannot design a satisfactory experiment without knowing all the environmental conditions. At best, we can show that some mixtures and some mechanisms produce plausible results. In this limited sense, we have been quite successful.

The External Alternative

To illustrate the great variety of possibilities for environments on the primitive Earth, we can easily postulate a completely different scenario for the origin of life. During the first few hundred million years after the planets formed, each planet must have undergone an intense bombardment by meteoroids and comets of all sizes (see Chapter 11). But life seems to have emerged on Earth during or just after this period! Could the meteoritic bombardment have played some role in the origin of life?

The most intense phase of the bombardment almost certainly had a negative effect on life's ability to emerge. Large impacts would have blown off much of the planet's atmosphere. Somewhat smaller ones would have created temporary atmospheres of vaporized rock, essentially heat-sterilizing the planet's surface. Such considerations suggest that it was not until the main bombardment phase was over that life would have been able to endure once it had begun.

At that time, another aspect of the bombardment could have played a significant role. As soon as the Earth developed an atmosphere that was thick enough to slow a large fraction of the smaller incoming objects, some of the organic compounds they contained, including amino acids, could

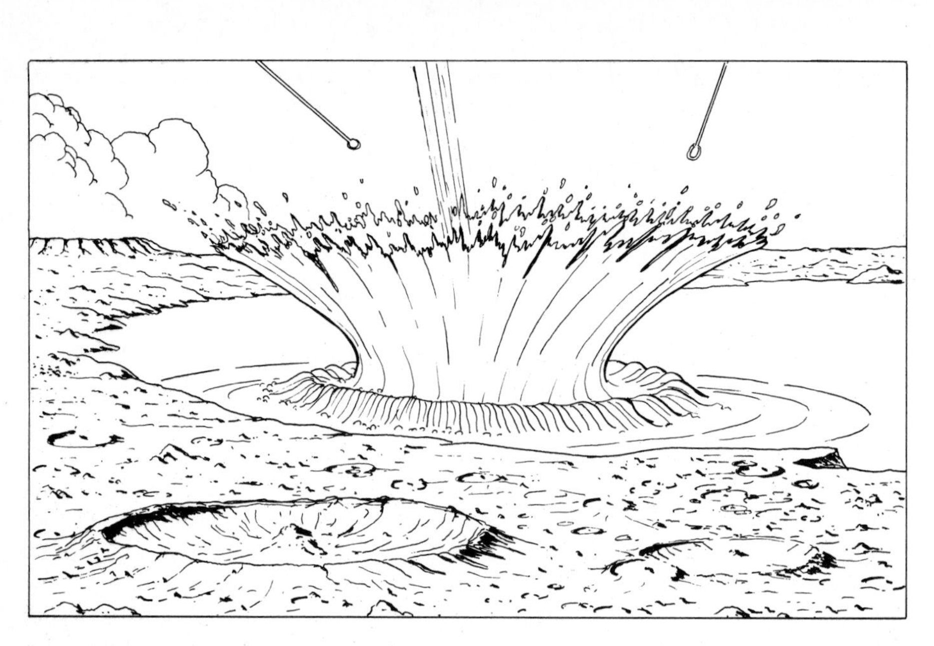

Figure 8.8 In this artist's rendering of a primordial pond, a small comet or volatile-rich meteorite has just splashed into a lake formed by an impact crater on the early Earth. As proposed by Benton Clark, organic materials from the comet are dispersed in the lake, where they are available for chemical synthesis.

have arrived intact on Earth's surface (Fig. 8.8). This would undoubtedly have given a head start to prebiological chemistry on Earth, on Venus, and—to a lesser extent—on Mars. The same head start would have occurred, we may speculate, on any rocky planet in any solar system, provided that the planet had a sufficiently dense primitive atmosphere.

We can easily imagine the build-up of organic molecules in ponds, again as the result of water flowing over a pockmarked landscape. The craters formed by the larger impacts could themselves have become lakes, offering sites for further reactions among these compounds. A continuing source of hydrogen would have come from the breakdown of the organic materials brought by the comets and meteorites, partially compensating for the hydrogen that constantly escapes from the Earth's upper atmosphere. In this way, a hydrogen-rich environment could persist for a longer time than we would otherwise predict.

Polymerization

This cometary or meteoritic scenario does not radically change our approach to the question of how life began. We must still envision spontaneous chemical reactions among likely compounds. But remember where our discussion rests. We have considered only the production of the components from which proteins, lipids, and nucleic acids are made. How do we assemble these components into much larger molecules? DNA and protein molecules are polymers, long chains that repeat a basic pattern over and over. Our problem consists of how to get molecules to join together in a repetitive fashion: the process of polymerization. In the case of DNA, this process must include the development of the marvelous double helix (see page 163). How did all this occur? We don't know. No one has yet achieved the polymerization of these ingredients under natural conditions. We can place DNA in a suitable "broth," one containing enzymes and DNA monomers, and it will be replicated, but we have not yet been able to make the initial DNA occur spontaneously, even though we know its composition and structure.

The formation of DNA and the other polymers found in living cells remains one of the greatest puzzles to be solved by experiments seeking to duplicate the chemical reactions that occurred on the primitive Earth. Present theories suggest that *clays* may have played a critical role in promoting the polymerization that must have preceded the development of cells. The current, sketchy model that some biologists use suggests that molecules began to organize themselves on the surfaces of grains of clay, at the edges of ponds that periodically froze or dried out (Fig. 8.9). The drying or freezing, or both, tended to concentrate the solutions of organic compounds, and helped to remove the water molecules that are released during polymerization.

Clays are important in this process because they possess just about the largest surface areas of any fine material that we know. Furthermore, the organization of the minerals within the clay grains seems particularly well suited for a wide assortment of organic compounds to stick to the grains' exposed surfaces. Both of these characteristics help to concentrate organic molecules and to absorb water. In addition, clay minerals consist of atoms arranged in organized "lattices." The information contained in these lattice structures—where repeating molecular patterns occur, as they do in organic polymers—might have allowed these structures to serve as the first templates for the organization of organic matter.

Experiments have shown that a clay called "montmorillonite" can indeed serve as a template, a guide that lines up important substances such as adenosine and guanine in ways that would promote their polymerization. In addition, sugars, fatty acids, amino acids, and proteins all interact readily with this mineral. Montmorillonite has a wide distribution on Earth today; a similar clay, "nontronite," provides the best model for the soil of Mars (see Chapter 14). Clay minerals may also be among the dust grains observed in the interstellar medium, where they could help to produce the

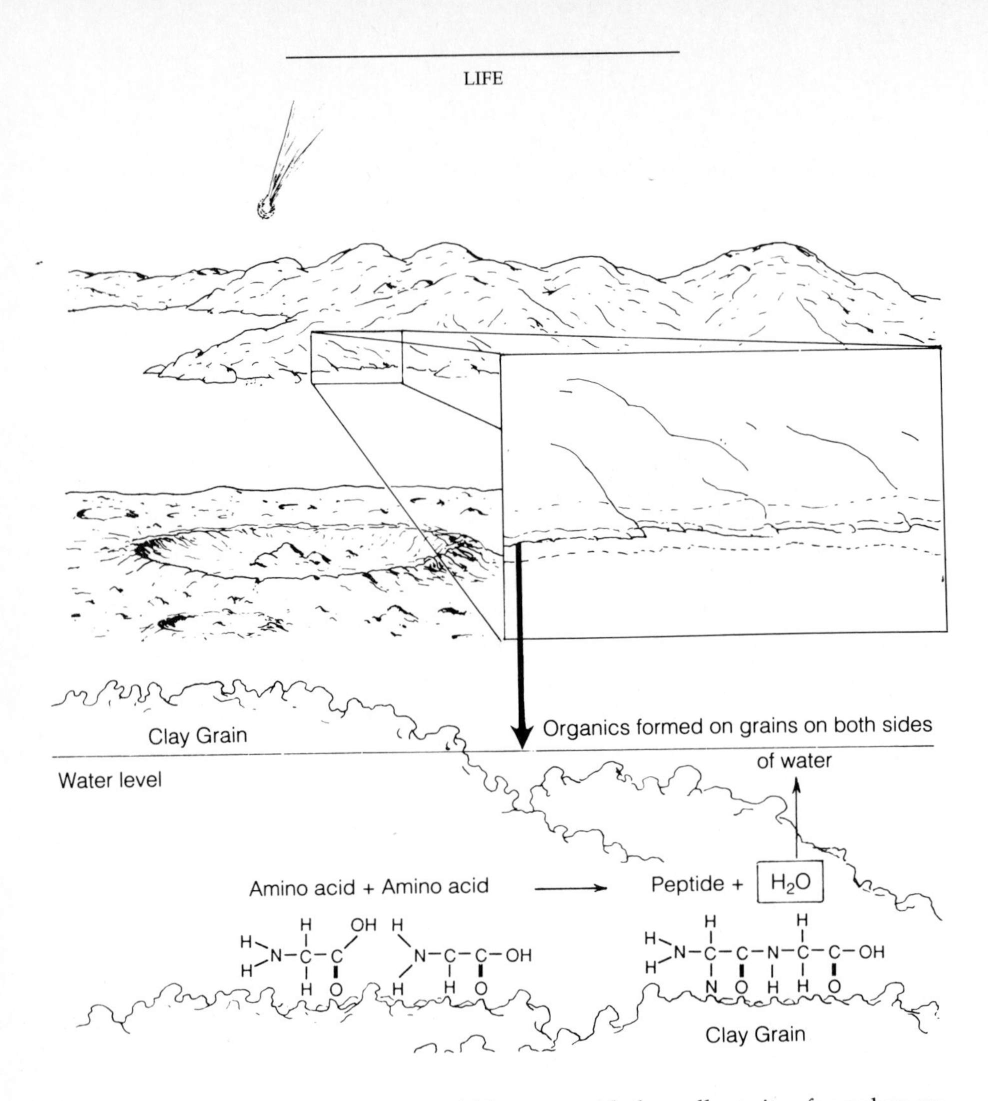

Figure 8.9 Certain types of clays could have provided excellent sites for polymers to form. The grains of clay would help the compounds that collected there to organize themselves into long-chain molecules as the water in which they were dissolved or floated dried up. The chemical equation below shows how two amino acids can join by giving up one molecule of water to form a "peptide." Several peptides can then link together to make a protein, giving up water at each linkage.

organic molecules discovered by radio astronomers (page 94). On Earth, clays appear predominantly at the edges and on the bottoms of bodies of water, just where they could help most with the process of polymerization during the early history of our planet (Fig. 8.9).

Leslie Orgel and his colleagues at the Salk Institute have made considerable progress with an alternative concept, that some specific *organic* molecules can serve as templates that guide the production of other molecules essential to life's origin. The recent discovery that RNA can catalyze reactions on itself (act as its own enzyme) suggests that RNA preceded DNA as the primordial information-carrying molecule

Beyond Polymers

Let's summarize the results of our efforts to understand the origin of life on Earth. We have reconstructed the following history:

1. Four and a half billion years ago, the Earth had a primitive atmosphere consisting of carbon- and nitrogen-containing gases, plus a few percent hydrogen and no free oxygen, above a rocky surface on which there were abundant reservoirs of water. This atmosphere may have been lost and re-constituted several times as a result of comet and asteroid impacts and continued outgassing.
2. During the next few hundred million years, the Earth's surface gradually accumulated a widely dispersed "primordial soup" of organic compounds and phosphates, the result of the energy provided by solar ultraviolet radiation, lightning discharges, and other sources. Some of this material may have arrived in comets and meteorites.
3. Additional compounds formed within the soup as the result of continuing chemical reactions, aided by local sources of heat, water runoff, and evaporation and/or freezing.
4. And this in turn led to the formation of polymers, possibly on the surfaces of clays.

This sequence does not make a unique specification for the origin of life. In fact, we are not even sure of the composition of the Earth's primitive atmosphere. Our model also fails in completeness, for we cannot yet see how all of the chemical units of life have formed, and we do not know how they joined together to make the polymers that we now see as the essential components of life. Nevertheless, we have at least demonstrated that some of the important compounds could form as long as the primitive atmosphere had at least a small amount of free hydrogen, and this gives hope that further efforts will add to our basic picture, and should eventually lead to a solution of the polymerization problem. Our model gains strength from the fact that nature apparently makes organic compounds quite easily, since we have found a whole list of such molecules in the interstellar medium, in comets, and in meteorites.

At the stage where polymers form, we stand at the brink of living systems. But we do not know how nature bridged the gulf between inanimate and animate matter. Quite probably, many different assemblages occurred, and many of them may have been successful at some sort of reproduction. We could generally define "life" at that stage of its development as a system capable of self-replication. But this time was already well advanced. Before this, random protein synthesis might have occurred, in response to the existence of polymers vaguely resembling modern RNA. The production of a protein that could produce more RNA from the original RNA could have started life on its way to success. Presumably, life's first attempts at replication must have been haphazard and extremely crude by the modern standards of contemporary DNA. The early form of the genetic code probably had fewer restrictions and simply discriminated among different types of amino acids having similar structures rather than selecting specific amino acids.

All this may have happened not inside living cells, as life processes do today, but in the ponds and pools where the initial chemical reactions occurred. On the other hand, some biologists think that membrane formation occurs so readily that membranes may have arisen even before polymer replication took place. The key events that led to the evolution of cells—membrane-bounded regions within which organic molecules can interact—remain uncertain. We do know, however, that organic polymers in high concentration are likely to join together and separate from the solution in the form of droplets. Such droplets could serve as the prototypes of cells (Fig. 8.10). If the right kind of long-chain molecules exist, they will tend to form a membrane that encloses the droplets.

Laboratory studies have shown that under appropriate conditions, these droplets can form systems within which simple chemical reactions

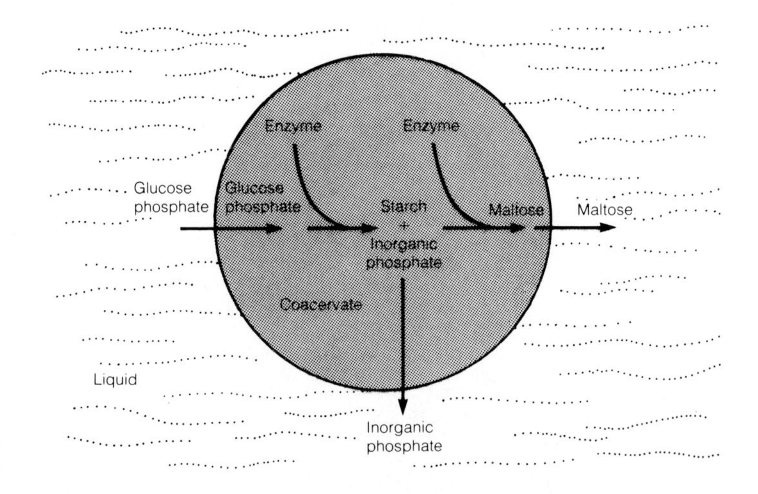

Figure 8.10 Droplets called coacervates can form when a solution of protein polymers (compare Fig. 8.9), nucleic acids, and sugar polymers is shaken. If enzymes are incorporated in the mixture, they can catalyze reactions within the droplets, which then imitate some of the functions of simple cells.

occur. But no one has yet produced a self-maintaining system such as that needed for a valid model of the first cells.

In another approach to the problem, scientists have demonstrated that if proteinlike polymers develop from the amino acids produced in simulated "primitive" environments, these polymers will form microspherules that resemble cells in size and appearance, if the polymers are heated under the proper conditions (Fig. 8.11). Such experiments at least illustrate the opportunities available for the further organization of matter, even if they do not represent the precise path toward the development of cells. The next step toward modern life would be the incorporation of a self-replicating system within the boundaries of these protocells, a system capable of directing the formation, maintenance, and renewal of the entire cell.

This conjecture leads us to a level of organization in matter that we can recognize on the contemporary Earth. The primitive organisms whose development we have suggested resemble a modern virus. This does not mean that an influenza virus is a "living fossil" as a living dinosaur would be, if discovered on some lost island (or in the depths of Loch Ness!). Modern viruses show careful adaptation to their hosts, and the hosts have appeared on Earth quite recently. Indeed, some evidence suggests that modern viruses are all derived from cells, which then would have preceded viruses in the course of life's development. But in their structure and their

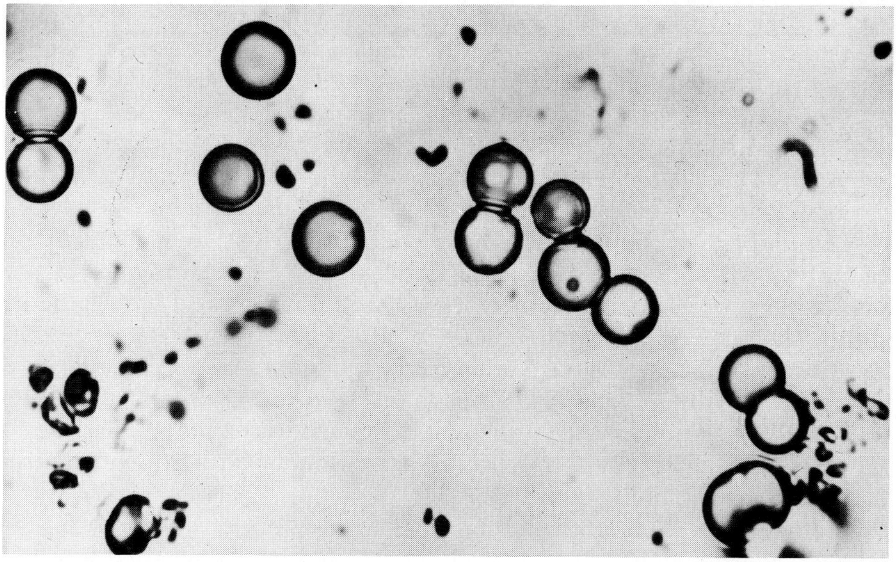

Figure 8.11 If we heat certain polymers that are quite similar to the proteins in animals, some of these long molecular chains will form microspherules that look much like cells.

functioning, modern viruses have great similarity to our hypothetical primitive organism: Each consists of nucleic acid with a protein covering, and each can survive long periods of dormancy. For the primitive virus-like organism, a small, warm pond might represent the host "cell."

SUMMARY

In our attempts to discover how life originated on Earth, we must deal with the fact that life began at least 3.5 billion, perhaps 4 billion, years ago, when conditions on Earth were quite different from conditions now. The geological record of these early eras has vanished, so we must add deductive methods to the actual evidence to estimate the conditions on the primitive Earth.

The chemical evolution that preceded the origin of life requires an environment with no free oxygen. This condition was easily met on the early Earth, since almost all of the oxygen in our atmosphere now has arisen from biological photosynthesis, starting about 2.5 billion years ago. The Earth's primitive atmosphere may have consisted mainly of hydrogen-rich compounds, such as methane and ammonia, or, as recent evidence suggests, it may have been a mixture of carbon dioxide, water vapor, carbon monoxide, and nitrogen. Experiments have shown that if either of these mixtures of gases is placed above a bath of water (to simulate the primitive oceans) and is exposed to ultraviolet radiation (to simulate sunlight unfiltered by the Earth's present ozone layer), simple organic molecules will form. These organic molecules include amino acids, the basic units of much larger protein molecules. Other compounds, also important in forming the small molecules found in living organisms, are produced in these laboratory simulations.

An interesting alternative to this model for a strictly indigenous origin of life on Earth suggests that a "head start" may have been provided by the bombardment of comets and meteorites that the Earth must have suffered during the few hundred million years after its formation. At the very least, this bombardment would have provided large amounts of carbon, nitrogen, oxygen and hydrogen. It is also possible that some of the organic molecules known to be present in meteorites (and expected in comet nuclei) could have reached the Earth's surface, contributing to the warm little ponds that nourished the beginning of life.

Some of the small organic molecules formed on, or delivered to, the primitive Earth would have collected in pools and ponds, where they could have assembled themselves into the long chains typical of proteins and nucleic acids. These long chains, or polymers, repeat a basic pattern over and over again. One way to promote such polymerization would be to collect molecular units on grains of clay, which then undergo periodic freezing and thawing, or evaporation and wetting. Another possibility is the use of other organic compounds as templates.

As various types of molecules passed through these stages, the first self-replicating molecular structures, capable of duplicating themselves when they had split apart, arose as part of the natural trial-and-error process that saw all sorts of molecules appear, if only briefly. Polymers that could reproduce themselves had immense advantages over polymers that formed at random. We are thus not surprised that the basic mechanism for duplication, the one embodied in DNA, appears in every organism that reproduces itself on Earth.

QUESTIONS

1. Why do we think that life on Earth began when the Earth had a much smaller amount of oxygen than is now the case?

2. Why can't we determine what the Earth was like soon after it formed (4.5 billion years ago) by studying the geological record laid down in the rocks as they formed?

3. Where did the oxygen now in the Earth's atmosphere come from?

4. Where did the nitrogen in our atmosphere come from? What processes tend to remove nitrogen from the atmosphere? What processes replenish it?

5. How did life arise on Earth, several billion years ago? What do you think were the essential ingredients needed for this process to occur?

6. What can laboratory simulations of the conditions that may have existed on the primitive Earth show us about the formation of organic molecules? What difficulties stand in the way of making an accurate experiment that duplicates the primitive Earth?

7. Why might clays be important in making complex molecules?

8. Is it possible that organic molecules from outer space played an important role in the formation of life on Earth? How?

9. At what point in the formation of more and more complicated molecules would you say that "life" appeared? Why?

10. What do you think the first ancestors of modern cells were like? What environmental hazards did they have to overcome?

FURTHER READING

Cairns-Smith, Alexander. "The First Organisms," *Scientific American* (June 1985).

Crick, Frances. *Life Itself—Its Origin and Nature*. New York: Simon & Schuster, 1981.

Dickerson, Richard. "Chemical Evolution and the Origin of Life," *Scientific American* (September 1978).

Dyson, Freeman. *Origins of Life*. New York and London: Cambridge University Press, 1987.

Gamlin, Linda, and Gail Vines, eds. *The Evolution of Life*. New York: Oxford University Press, 1987.

Gould, Stephen Jay. *Wonderful Life: The Story of the Burgess Shale*. New York: W. W. Norton, 1989.

Oparin, Andrei. *The Origin of Life*. New York: Macmillan, 1938.

Ponnamperuma, Cyril. *The Origins of Life*. New York: E. P. Dutton, 1972.

9

From Molecules to Minds

Our discussions concerning the origin of life on Earth suggest a process of polymerization: the spontaneous formation of long-chain molecules, perhaps on grains of clay located at the edges of shallow ponds that periodically froze or evaporated. Perhaps some of the polymers, including some that could serve as templates for others, found themselves isolated from the primordial soup in droplets bounded by other compounds. This model for the development of life suggests that the polymer-containing droplets were "protocells" from which the first true cells must have developed.

Prokaryotes

Because viruses cannot function alone, and because they seem to have evolved from living cells rather than the other way around, biologists usually say that the simplest living organisms—those with the least complexity of structural organization—are the simplest *cells:* bacteria.

Bacterial cells are "prokaryotes," cells without special centers or nuclei, whose Greek name means "before the nucleus." Prokaryotes each contain a single long strand of DNA that includes several thousand genes. This is hundreds of times more than the number of genes in a virus. All living cells, and most viruses, use DNA as the master blueprint that transfers genetic information from one generation to the next. Some viruses use the closely related molecule RNA for this purpose, but cells never do. Since a prokaryotic cell contains about a hundred times more DNA or RNA than a virus does, prokaryotes have far more complex reproductive capacities than viruses. The earliest cells were almost certainly prokaryotes. Unlike viruses, prokaryotes are free-living entities: They can make proteins and can reproduce without a plant or animal "host" in which to live. These organisms represent the simplest known biological systems capable of independent life.

The first prokaryotes to evolve probably had to use the nutrients present in their natural environments: organic compounds produced by the interaction of ultraviolet radiation and other energy sources with the chemical mixture in which the prokaryotes found themselves. One can think of the origin of life as the spontaneous transition from "food" to "life" without the intervention of a living organism. The first organisms must have found themselves surrounded by raw materials ("food") that their own chemical systems would have recognized.

As this source of food was extinguished by the changes in our planet's atmosphere (see Chapter 8), some microscopic prokaryotes did an amazing thing. They developed a basic process, one that led to the proliferation of advanced forms of life and to the durability of life on a cosmic time scale: the conversion of sunlight into stored chemical energy for future use. In short, prokaryotes "invented" photosynthesis.

The earliest fossil record of life on Earth was produced by colonies of such organisms (Fig. 9.1). Simple photosynthetic bacteria, formerly called "blue-green algae," still form an efficiently functioning part of the terrestrial biosphere, the integrated system of life on Earth (see again Fig. 8.4 and Color Plate 7). These bacteria may therefore represent both the first and the

Figure 9.1 This is a cross section of a stromatolite found in Australia (see again Figure 8.4) that is 3.5 billion years old. It is also the oldest known example of life on Earth. The layered, domal structure is characteristic of these macroscopic fossils of microscopic life. The layers are produced as silt is trapped by successive colonies of photosynthetic bacteria struggling to receive as much unobstructed sunlight as possible. The rock is 20 centimeters across.

most successful adaptation of life to its environment. Although the earliest form of photosynthesis may have used hydrogen sulfide (H_2S) rather than water (H_2O), and would therefore have liberated sulfur instead of oxygen, even this early photosynthesis represented a tremendous advance for the bacteria that could perform it.

We could easily substitute "starlight" for "sunlight" in our description of bacterial photosynthesis. Blue-green bacteria would be just as happy using the radiation from Alpha Centauri as they are with the light from our own sun, if they were living on a planet orbiting that star. As a result, we can immediately recognize the potential for converting a strange planet located in some distant solar system into an environment that humans could inhabit: We simply seed the planet's atmosphere and oceans with microorganisms capable of releasing oxygen. We might, however, have to adjust the trace-element composition of the other planetary environment before this "directed panspermia" could succeed.

As we attempt to unravel life's early evolution on Earth, we should note that today's prokaryotes still provide evidence of the transition to an oxygen-rich environment. Some prokaryotes live only in the complete absence of oxygen, while others can live with it or without it, and still others require oxygen for survival. This diversity is consistent with the hypothesis that the conversion of the Earth's atmosphere from oxygen-poor to oxygen-rich, through the liberation of oxygen by photosynthesis, probably occurred during the era when prokaryotes dominated the Earth. When we reach the next step in biological complexity—cells with nuclei—we find that almost without exception, these cells require gaseous oxygen in order to live. The prokaryotes had to come first.

The transition to an oxygen-rich atmosphere apparently occurred between 2.5 and two billion years ago, but it might have begun even earlier. At first, oxygen in the atmosphere would have been poisonous to all the prokaryotes, including those that produced it. To them, it was (and still is) simply a waste product, the greatest "pollution" our atmosphere has ever seen. But the availability of oxygen as an easy source of energy for life's metabolism must have created a strong selectional pressure in favor of organisms that could use this new gas.

It is intriguing to find that chlorophyll molecules, which are essential to the release of oxygen by plants, have the same structural unit (called a porphyrin) found in some important oxygen-carrying molecules, such as hemoglobin (the iron-containing pigment in our red blood cells), hemocyanin (the copper-containing equivalent in many marine animals), and cytochromes, protein complexes found in most advanced cells. This fact illustrates an important generalization about life on Earth: In their responses to environmental changes, evolving life forms have used chemical structures already developed for other purposes. In other words, they make do, so far as possible, with the smallest possible changes in what they already have. We see that selectional processes similar to those that affect entire organisms (natural selection) may also operate at the molecular level. In this view, the ability to use free oxygen may have evolved at approximately the same time as the ability to liberate oxygen.

Eukaryotes

The appearance of free oxygen in the atmosphere of our planet coincided in time with the further evolutionary development of many forms of life. We cannot trace the exact path that led from the existence of many types of prokaryotes to the appearance of the "eukaryotes," cells that contain true nuclei (Fig. 9.2). Intermediate stages of evolution undoubtedly occurred, leading to the appearance of eukaryotes on Earth at a time somewhere between 2 billion and 1 billion years ago. (The difficulty in establishing this time arises from the relative lack of fossils that can be positively identified as early eukaryotes.) Eukaryotes represent the highest form of cellular com-

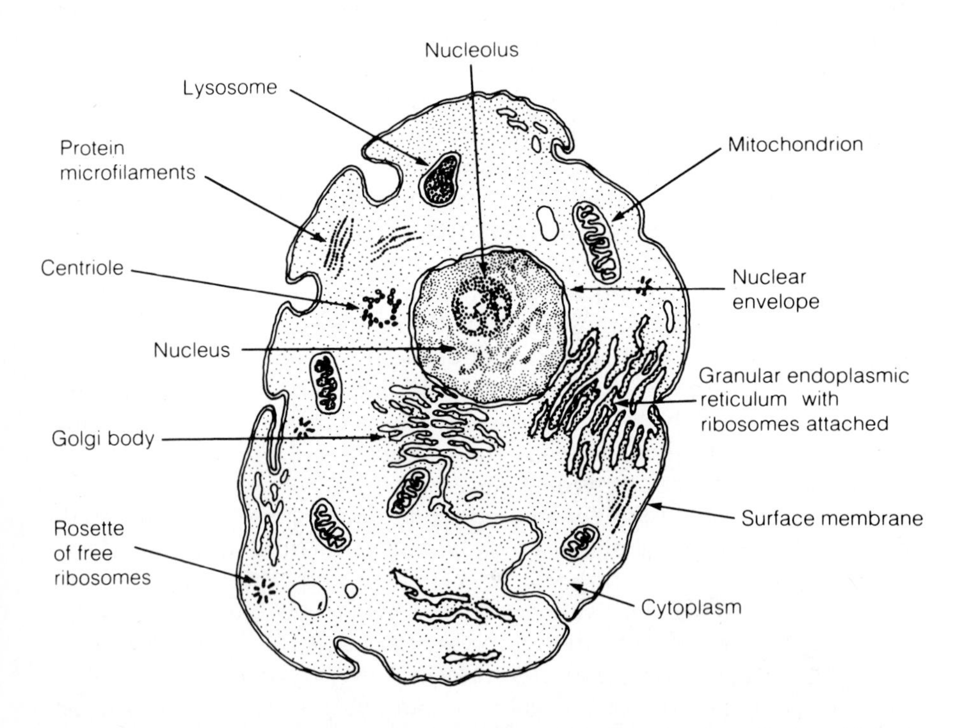

Figure 9.2 Eukaryotes consist of cells with true nuclei—that is, cells in which the DNA-containing chromosomes are enclosed by a membrane.

plexity that has so far evolved on Earth. Eukaryotic cells keep their chromosomes—the DNA-containing structures of the cell—within the protection of the membrane-enclosed nuclei that give eukaryotes their name, which means "true nucleus" in Greek.

In eukaryotic cells, the chromosomes typically contain tens of thousands of genes, and the amount of DNA in a cell is 10 to 1000 times more than the amount found in bacterial cells. Eukaryotes can store much more information in their genetic material than prokaryotes can; their increased complexity of cell structure and cell function reflects this fact. Eukaryotic cells have important internal structures: They contain specialized subunits, called "organelles," that have their own genes and protein-synthesizing machinery, and carry out specific functions (see again Fig. 9.2). Animal, fungal, and plant cells, for example, contain organelles for respiration called mitochondria, and plant cells contain chloroplasts for photosynthesis.

How did eukaryotes evolve from prokaryotes? The traditional view suggested a gradual development, by which prokaryotes slowly acquired a progressively more complex substructure. In fact, it is now clear, principally from the research of Lynn Margulis, that the internal specialization within eukaryotic cells arose from a "symbiotic" (mutually beneficial) relationship between two or more prokaryotic cells. We can imagine a prokaryote capable of benefitting from the ability to perform photosynthesis but unable to do so. If this prokaryote incorporated a simple, chlorophyll-laden relative, it would suddenly achieve photosynthetic capability. This symbiotic chain of ancestry of the eukaryotes implies that mitochondria and chloroplasts may be derived from bacteria. More specific evidence for this evolutionary path exists in the fact that chloroplasts and mitochondria have gene systems that resemble the systems found in bacteria.

From the stage of the first eukaryotes, evolution has led to the increasing specialization of cells in their function and structure, and to the development of larger, far more complex units, such as organs and the large individuals that incorporate them. Each human being or other large mammal consists of about 10 trillion (10^{13}) cells, which appear in about a thousand different varieties. Remarkable as it may seem, the amazing complexity of structure in a human being has arisen through the process of natural selection, nothing more or less than differential success in survival and reproduction among competitors. Our development of self-consciousness, so strikingly important, has come from the same competition.

As organisms continued to reproduce themselves, some of them eventually developed the technique of "sexual reproduction." All eukaryotic cells can reproduce asexually, but the advantages of sex had such great impact that the most complex eukaryotic organisms rely on it. Sexual reproduction allows a new offspring to obtain half its total number of genes from each parent, because each parent contributes half of its genetic makeup

(Fig. 9.3). The reshuffling of gene combinations from generation to generation allows natural selection to proceed much more rapidly than it can for asexual reproduction, in which each offspring is genetically identical to its single parent.

Until now, we have outlined the history of life on Earth in a stately march of billions or hundreds of millions of years at a time. The Earth formed 4.5 billion years ago. The first life of which we have any record (so far) lived 3.5 billion years ago. Geological evidence suggests that the atmosphere became oxygen-rich at a time between 2.5 and 2 billion years ago, consistent with the widespread appearance of stromatolites—evidence of blue-green and other photosynthetic bacteria—in the fossil record. The first eukaryotes appeared about a billion years later, somewhere near 1.6 billion years ago (Fig. 1.9). Thus although life has been in continuous existence on

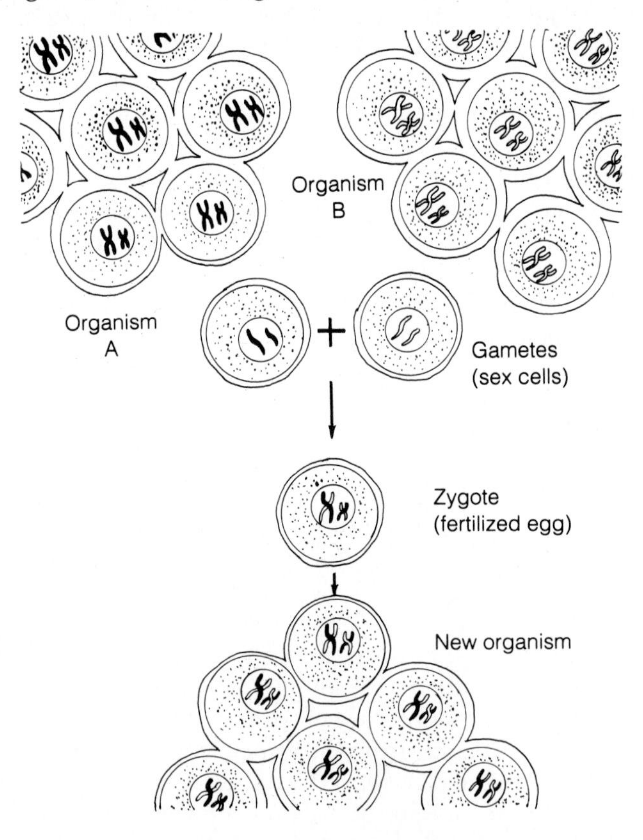

Figure 9.3 In sexual reproduction, each fertilization creates a cell that receives half its chromosomes (the carriers of DNA) from each parent. All of this DNA is faithfully reproduced through subsequent divisions, so an exact copy of each chromosome appears in every cell of the animal.

Earth for most of our planet's 4.5 billion years, during the bulk of that time life has consisted entirely of microorganisms, each too small to be seen as an individual by human eyes.

When we think about the evolution of life, we usually picture monkeys and humans arising from more primitive mammals, or birds and reptiles developing from earlier vertebrates. Such events, important though they are to us, represent the most recent stages of evolution, spanning less than 10 percent of the history of life on Earth. The long, slow changes that produced the first prokaryotes, and eventually the first eukaryotes from prokaryotes, include most of life's developmental history. These microorganisms "invented" almost all of the biochemistry essential to the existence of so-called higher forms of life. We should therefore not overlook their importance to us, and we should recognize that on other planets as well, life will probably spend much of its history in microscopic form only.

The Great Leap Forward

From our emphasis on the billions of years when life consisted exclusively of microorganisms, we should not conclude that an increase in biological complexity always requires such eons of time. Instead, the history of life on Earth suggests that the early stages of life take far longer than the later ones, and that the initial steps toward complexity may be more difficult than the later, extraordinary increases in the complexity of living creatures.

Just about 600 million years ago, some combination of circumstances, presumably involving the large amounts of photosynthetically generated oxygen and the availability of newly habitable environments, led to a sharp increase in the distribution, number, and variety of organisms. The fossils of these organisms mark the start of what geologists call the "Cambrian era." This era must have followed a long period of the gradual evolution of soft-bodied organisms from single-celled eukaryotes. This evolution produced creatures much like today's jellyfish, which flourished just 50 or 100 million years before the Cambrian era began (Fig. 9.4). There is evidence that this transition from single cells to multi-cellular organisms occurred independently several times; modern organisms did not arise from a single multi-cellular ancestor. Multi-cellularity thus appears to be a probable—as opposed to an unlikely—development. We know far less about life before the Cambrian era than we do about later life, for a good reason. Only at the start of the Cambrian era did organisms evolve "hard parts"—shells, carapaces, and exoskeletons—that could be well preserved in sedimentary rocks to form the fossils that we find today.

The Cambrian era brought a remarkable acceleration in the rate of evolutionary diversification. Four billion years elapsed from the origin of the

Figure 9.4 These 700-million-year-old fossils of feather corals from the Ediacaran period appeared shortly before the explosive development of species at the beginning of the Cambrian era. These are among the oldest known multicellular organisms in the Earth's history of life.

Earth to the appearance of trilobites, whose abundant fossils mark the transition to the Cambrian era (Fig. 9.5). With two compound eyes that may have provided binocular vision, these animals mark a significant evolutionary achievement. Yet trilobites remain extremely primitive by human standards; they would hardly be able to construct and to use radio telescopes to communicate with other trilobites on some distant planet in their Cambrian era. But a mere 600 million years after the appearance of these distant relatives of the scorpion, some peculiar, apelike hominids moved from the jungle forests of ancient Africa into the open plains. A scant three million years after this transition, we humans wonder about contact with other inhabitants of the universe.

We can best appreciate the acceleration involved in the different steps of evolution with a diagram (Fig. 9.6), or with an analogy to a calendar year. Think of the entire history of Earth as a single year, starting on January 1. At the beginning of summer, blue-green bacteria made the atmosphere oxygen-rich, but the Cambrian era began only on November 13. Dinosaurs ruled the Earth from December 13 until December 26, when they suddenly became extinct. Our first human ancestors, the hominids,

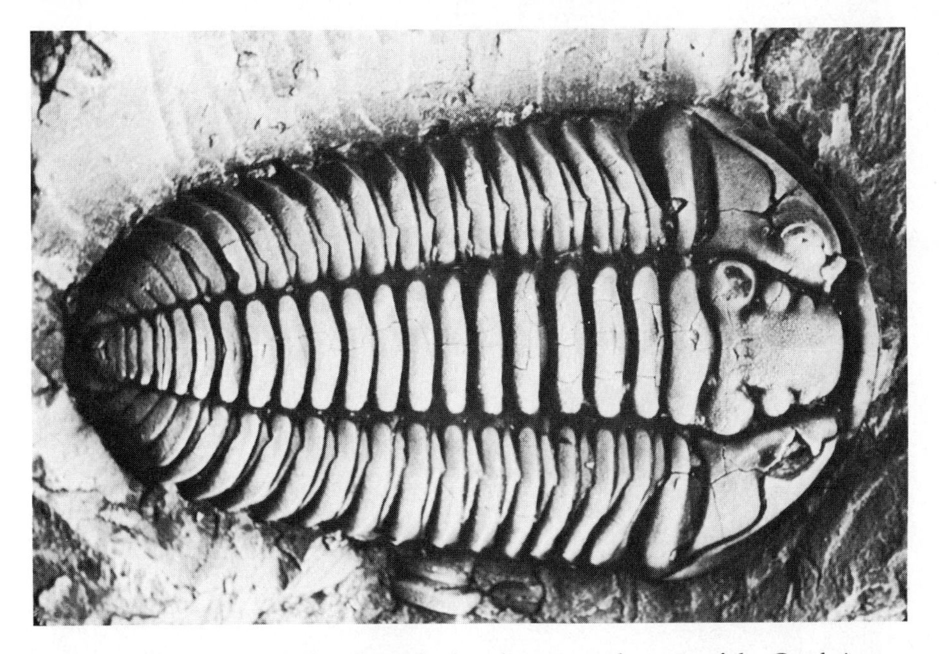

Figure 9.5 The trilobites that flourished in the seas at the start of the Cambrian era had two eyes and a rather complex body structure. This photograph shows a trilobite fossil about 2 centimeters long.

appeared at dinnertime on December 31, and human ability to send New Year's greetings to the other side of the galaxy occurred with less than a tenth of a second to go, at 11:59:59.9155 P.M. on December 31—less than one hundred-millionth of the Earth's lifetime!

This time scale tells us something useful in our quest for life elsewhere in the galaxy. We assume that the development of life on Earth may be fairly typical of life's development, wherever it occurs. This is the *assumption of mediocrity,* as proposed by Josef Shklovskii and Carl Sagan. Thus, even without knowing the details of how life began and how it developed on our planet, we might assume that wherever life occurs, it should usually take about 4.5 billion years—say, 3 to 6 billion years—for intelligent life to develop, once a planet has formed. But intelligent life, in the sense of a civilization capable of galactic communication, has existed on Earth for only the last tenth of a second out of a year's total. Under the most favorable circumstances, the evolution of life on another planet might differ only a tiny bit from that of life on Earth, so that life might require 4.7 billion, or even 5 billion years to reach the same capability. (Always remember that when we discuss human history, we are talking about the last few minutes of a highly eventful year!) In this case, our attempts to make contact with other life

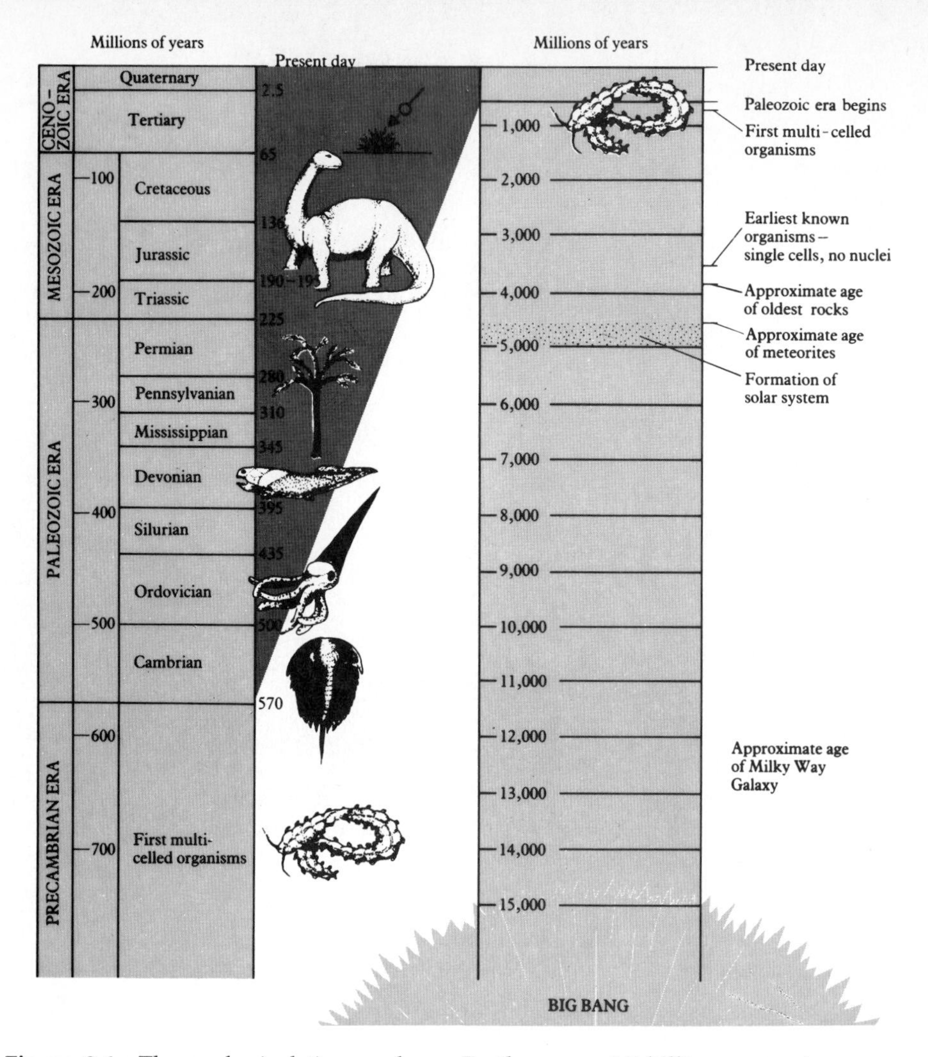

Figure 9.6 The geological time scale on Earth covers 4.5 billion years since our planet formed. The rate of increase in biological complexity has itself incredased markedly during the last 700 million years.

would be doomed to failure if the other planet formed at the same time as Earth. On the other hand, small changes in the rate of evolution might have brought another planet, formed when Earth formed, to our present state of self-consciousness at a time several hundred million years in the past. Finally, since other planets themselves could easily be billions of years older or younger than the sun and its planets, we have another factor that could put us at a very different period of development compared with another kind of life. Until we find another planet with life and compare that planet's evolutionary time scale with our own, we shall have to be content with the rough estimate of 3 to 6 billion years from a planet's formation to the emergence of a civilization capable of interstellar communication.

Suitable Stars for Life

This apparently trivial statement has immediate consequences in directing our decisions about where to search for life in our galaxy. We must obviously limit our search to planetary systems whose central stars have total main-sequence lifetimes that exceed a few billion years. The reason for this requirement is that the luminosity of the star must remain approximately uniform during the period that life evolves. Once the star leaves the main sequence, its increase in luminosity will alter the planetary environments dramatically and irreversibly.

If we impose a lifetime of at least 4 or 5 billion years on main-sequence stars, we restrict our search to stars with spectral types of F5 or cooler; that is, to stars of types F5 through F9, G, K, and M. This might seem a difficult constraint, since the brightest stars in the sky are mostly O, B, and A stars, but in fact the great majority of stars do satisfy the criterion we have just set. For example, all but two of the 25 stars closest to the sun are cooler and last longer than stars of spectral type F5.

The other time requirement implicit from our terrestrial perspective must be of finding a star *at least* 4 or 5 billion years old. In other words, if we pick a G2 star, whose spectral type is identical with the sun's, but a star that has been on the main sequence for only 2 or 3 billion years, we would have to worry that the "best" we could hope to find on any Earthlike planet in orbit around it (if there *is* an Earthlike planet!) might well be a lively population of bacteria. Given our present ignorance and cosmic isolation, this would be an absolutely thrilling discovery. Furthermore, as we shall see in Chapter 14, we need not visit a planet in order to discover life upon it but can instead study the composition of its atmosphere. This allows a search for life from enormous distances, simply by analyzing the light reflected by a planet and its atmosphere. Still, no one denies that the detection of a message from an advanced civilization, or a physical encounter with its emissaries, would be still more thrilling. That goal leads us to ask for a star that has been on the main sequence at least for a time comparable to the present age of the sun, 4.5 billion years.

Life on Other Planets

Even if we have the right kind of star, one that has been on the main sequence for at least 4 or 5 billion years and is shining steadily on an Earthlike planet, what assurance do we have that we would find intelligent life? What does it mean for a planet to be Earthlike? And what are the fundamental conditions we need for intelligent life to develop? Until we find another example of intelligent life, we shall remain unable to answer these questions with any certainty. For our purposes, we can make a general outline of this complex problem, seeking to define those aspects of life and its development on Earth that we can try to generalize to other planets.

From our survey of life on Earth, the first and strongest impression that we derive is *the importance of liquid water.* Terrestrial life originated in water, and water remains vital to life's continued existence. Large organisms did not evolve the adaptations that would permit them to live on land until 350 million years ago, at least 3 billion and perhaps 4 billion years after life appeared on this planet. We can imagine other liquid environments, such as seas of ammonia (although there are drawbacks here, as we shall see in Chapter 10). Nevertheless, life does seem to require a liquid medium that can transport and concentrate important molecules. The liquid state does not arise easily in nature, since it requires that the local temperature stay within a narrow range. And to form the complex polymers needed for life, we seem to require not just liquid in any form but liquid that collects on a solid surface that itself exhibits repetitive molecular structures, in an environment in which evaporation or freezing can occur. The two requirements of solid surfaces and a narrow temperature range are easily satisfied by a planet in a roughly circular orbit at the right distance from its star. In our own solar system these conditions have been satisfied on Earth, Mars, and possibly on some smaller objects. It is surely not a coincidence that our planet currently can boast of both the only oceans and also the most abundant, and probably the only, biological activity in the solar system.

Once life has begun, does it require an exposed land mass for its further development? We cannot answer this question with confidence. On Earth, the most advanced life forms have evolved on land. The greater variety inherent in land environments and the greater availability of oxygen seems to increase the rate at which natural selection occurs: About 80 percent of all known species of plants and animals are found on land. Furthermore, the demands on form and function in a marine environment may not favor the natural selection of intelligent organisms so much as the conditions on land do. For example, the need for streamlining tends to prevent the development of complex limbs with the ability to manipulate objects and ultimately to make tools. Indeed, the most intelligent marine creatures, whales and dolphins, are mammals whose ancestors were land animals similar to otters. Other things being equal, we might look for planets on which water remains liquid but does not cover the planet completely: We should have land *and* oceans!

But would other things be equal? Are we making an error by assuming life elsewhere must be like the life on our planet? After all, octopi, starfish, and lobsters have appendages capable of manipulation and are not especially streamlined. Perhaps an ocean-covered planet could develop intelligent marine life from a line similar to our mollusks rather than to our

Figure 9.7 Sea creatures on an undiscovered planet watch a rare event: the double eclipse of their (fainter, closer) sun by two of their planet's satellites.

mammals. The warmth of the seas themselves might provide the equivalent of the mammalian bloodstream. Would such organisms learn astronomy and consider the possibility of communicating with others like themselves across the depths of space? We must simply leave this possibility open, with no way to study it until we find an ocean-covered planet (Fig. 9.7).

Evolution and the Development of Intelligence

Suppose that we have a planet in orbit around a star of the proper age, that the planet's surface has both seas and land, and that life on this planet has undergone the same type of "early" evolution as Earth. How likely would the appearance of an intelligent form of life then be? In other words, how improbable is intelligent life? To try to answer this question, we must first take a closer look at the process of evolution.

At the molecular level of life, we have described the process that causes an evolutionary change, one that could ultimately lead to new adaptive traits in species, in terms of extremely rare random changes in DNA as it is replicated, leading to new DNA molecules that are just slightly different from the old ones (see page 169). DNA molecules show a remarkable stability against these changes. Mutations occur in advanced animals only about once per gene for every 100,000 times that a cell divides. If such a mistake occurs in the "germ plasm" of an individual, it can change the characteristics of that individual's offspring. As the physician Lewis Thomas has written, "The capacity to blunder slightly is the real marvel of DNA. Without this special attribute, we would still be anaerobic bacteria and there would be no music."

When we examine the record of evolution revealed by fossils on Earth and as we fill in the missing record with our imagination, we notice two important facts. First, *species do not repeat in time.* After the trilobites disappeared, other arthropods came and went, but trilobites have never re-emerged from the ever-branching tree of biological diversity. Second, during the 3.5-billion-year period during which life moved from the first eukaryotes to human beings, *some species have remained essentially unchanged.* We still have blue-green bacteria, along with many other prokaryotes (anaerobic bacteria in particular) that have survived, with presumably only small changes, since their first appearance on Earth.

Although some species do not change, most do. Natural selection does have an orientation, one that leads toward the development of more complex species of life. While the rest of the universe steadily increases its disorder, life heads in the opposite direction at a pace that seems to quicken as complexity increases. No blue-green bacteria exist on the moon, but humans have walked there.[1] The price of this capability comes high, for humans have developed the ability to exterminate themselves voluntarily, a talent that the blue-greens lack.

Is intelligence inevitable? To estimate whether intelligence is likely to be a widespread or an extremely unlikely phenomenon in the universe, we must consider how intelligence arose on Earth, and we must also try to determine whether intelligence, in general, proves beneficial or harmful to a species. Does intelligence always carry with it the seeds of catastrophe, leading intelligent species to their destruction? As yet we have no way to answer this question. So far, we can only say that our example on Earth seems to show that natural selection does favor the development of intelligent, self-conscious life. What happens next remains hidden from us.

Let us first define what we mean by intelligence. For our restricted purposes in searching for life, we call an intelligent species one that has devel-

[1] But we might gain a useful perspective from the realization that over a billion bacteria "walked" with each astronaut!

oped the ability to communicate—either passively or actively—over interstellar distances. In adopting such a narrow definition, we may seem to be excluding Plato, Michelangelo, Shakespeare, and Newton from membership in the "intelligence club." Not at all! As should be evident from the example of crowding 4.5 billion years into one calendar year, the distinction between 2500 B.C. (when the great pyramids of Egypt were built) and today is negligible. We adopt this strictly technological definition of intelligence since it is only with this capability that contact between civilizations in the galaxy is possible, and the quest for such contact is the ultimate objective of our inquiry. By using this definition, we give our imaginations free rein in thinking about the other characteristics an intelligent species might have—for example, in their structure, size, musical ability, and belligerence.

Is Intelligence Inevitable?

Recent scientific thinking about the evolution of life on Earth is divided on the inevitability of intelligence. On the one hand, intelligence may have such great survival value that it must always occur in the development of complex life. (Here biologists are considering a more general definition of intelligence than our technological one.) Aside from our (potential) ability to leave the Earth, we humans already have the means to preserve ourselves against extinction on our own planet, for we can modify our environment to maintain the conditions best suited to our existence. We can heal our sick, feed our hungry, move to any part of the globe, and pass from generation to generation the knowledge of how to achieve these goals. This ability to collect and to transfer information makes us intelligent in the traditional sense and has a tremendous survival value.

On the other hand, the continued existence of anaerobic blue-green bacteria throughout the past 3.5 billion years suggests that intelligence is not the only key to survival! It may be more correct to say that once organisms achieve a certain level of complexity, the development of intelligence will prove useful in their survival. But even this statement will need some testing before we humans prove that we can last for a hundred million years, as the dinosaurs did without being noted for their intellectual brilliance.

The speed with which intelligence has developed on Earth increases our hope that this extraordinary property of matter has also appeared many times over in our galaxy. Let us imagine, for a moment, that all the humans on Earth suddenly disappeared. How much time would pass before the emergence of a new species with intelligence? One hundred million years ago, Earth's only mammals were small creatures that resembled shrews and mice (Fig. 9.8). About 3 million years ago, the first hominids appeared. These two times roughly define our terrestrial example of the interval needed for

Figure 9.8 The first mammals to appear on Earth 200 million years ago resembled modern shrews and mice. They probably hunted insects while keeping an eye out for predatory dinosaurs. See Fig. 8.2 for the configuration of the Earth's continents during this era.

intelligent creatures to evolve, once an appropriate ancestor appears. One hundred million years is a long time, but if we destroyed ourselves, this span of time might bring Earth another intelligent species, perhaps wiser than humans. Then the length of time needed for the emergence of intelligent life on Earth would have increased from 4.5 to 4.6 billion years. Would the stars notice this trivial difference?

Just in case we humans do destroy ourselves in some way that does not annihilate our fellow creatures, we can attempt to identify our cousins most likely to succeed as intelligent masters of Earth. Some might favor a descendant of the dolphins, who could again venture onto the land that the ancestors of these mammals abandoned millions of years ago. Or perhaps another kind of simian could swing out of the trees and begin the hominid pattern of cave dwelling and primitive agriculture. But we can also imagine far different pathways of evolution toward intelligence, such as the appropriately mutated descendants of a tribe of playful otters or clever raccoons, or advanced societies of tool-making octopi. The many examples we find of parallel evolution—for example, the independently evolved eyes of an octopus, different in structure but quite similar in use to the binocular vision of mammals—suggest that if an environmental opening exists, some species will evolve to fill it.

We can thus have some confidence that if humans disappeared, the process of natural selection that led to our development would again produce intelligent creatures from the vast gene pool on our planet. The new lord of Earth might be a species of mammal with binocular vision and tool-making ability. But beyond that, resemblances to ourselves might not be close. The vanished trilobites have left a message: Once a species disappears, it never returns. This tells us, too, that we should not expect to meet humans on other worlds. To quote Loren Eiseley:

> Life, even cellular life, may exist out yonder in the dark. But high or low in nature, it will not wear the shape of man. That shape is the evolutionary product of strange, long wandering through the attics of the forest roof, and so great are the chances of failure, that nothing precisely and identically human is likely ever to come that way again.

Future Evolution on Earth

Has evolution stopped with the emergence of human beings? We have no reason to think so. But humans have introduced a new influence on evolution: attempts at conscious control. We see striking evidence of this influence in the tremendous growth of the human population. One of the factors that originally led Charles Darwin to his idea of natural selection was his observation that succeeding generations of a given species are usually no more numerous than preceding generations. Since most organisms have many offspring, this fact implies that most of the offspring never survive to reproduce. Such differential success at survival and reproduction causes an increase in the number of "successful" individuals, at the expense of the less successful ones.

Human beings seek to modify this process. We work together to overcome the effects of disease, famine, genetic defects, and natural disasters. As a result of our efforts, the number of human individuals does increase from one generation to the next. One big jump in the human population came with the introduction of agriculture; another came when we learned how to avoid the most common fatal diseases. During the past few generations, the human population has doubled every 35 years. This tremendous rate of increase creates enormous problems that will continue to challenge our political, social, and technological abilities. But if we do manage to bring population growth under control, we can anticipate a long future for human beings on this planet. What can we expect to occur in an evolutionary sense, if we manage to avoid crowding ourselves out of existence?

It is easy to guess where we are headed. As human understanding of natural selection, and the urge to control it, have become more fully developed, we have reached the stage of experiments with genetic engineering, cloning, and the prolongation of human lifetimes. Interest in these efforts seems likely to increase, spurred by the desire to achieve the same goal that humans have pursued since first they became aware of the rhythm of their lives: immortality.

Scientists do not see any fundamental reason, as opposed to a present lack of understanding in detail, why we should expect to remain unable to structure forms of life as we choose, and to extend life for very long periods of time. If our distant descendants achieve this capability, will they choose to reach for immortality? Or will they deliberately shun this path and decide to live and die as other forms of life do? Some people might find eternal life a dreadful bore, or wish to yield their place to new humans. We can only guess at the consequences of having the ability to prolong life indefinitely, but one thing seems clear: If advanced civilizations exist elsewhere in the universe, this question must have been asked and answered many times. As the astronomer Frank Drake has pointed out, the results of this inquiry could have significant effects on our search for intelligent life. Somewhere among the stars, a race of immortals may have emerged, spending their time in the rituals and entertainments that please them, perhaps in contact with other races like themselves, and even (unlikely though it may seem) eager to introduce less advanced societies to their level of consciousness.

The Web of Life

We should never forget that despite our importance to ourselves and to each other, human beings represent only a tiny fraction of the living beings on this planet. The importance of these other species becomes increasingly evident to us as we learn more about the science of "ecology," the study of the relationships among organisms and between organisms and their environments. We all know examples of the ways in which one organism depends on another: Viruses have their bacteria, ants have their aphids, humans have their cereal grains. But ecologists now emphasize the fact that *our planet has no closed ecological system.* In other words, we cannot surround some environment with an impenetrable wall and expect the organisms within it to survive. We all depend on one another in complex ways that we now perceive only dimly.

This may seem surprising, but we can see the fact more clearly from our discussion of the Earth's atmosphere, which offers a good illustration of this complex interaction. What would Earth look like if life had not developed? Suppose nature's first experiments in the primordial soup had failed. Then hydrogen would still have escaped, since our planet's gravitational force cannot retain hydrogen atoms. Hence organic compounds, broken apart by solar ultraviolet photons, would gradually become oxidized: They would combine with oxygen, liberated by the ultraviolet destruction of water molecules (Fig. 9.9). With no new sources of organic compounds, the carbon on Earth would gradually become carbon dioxide.

We can estimate how much carbon dioxide would appear on a lifeless Earth by adding up all the organic carbon buried as shales, coal, oil, and the carbon in carbonate rocks. Limestones, for instance, are calcium carbonates, primarily made of calcium, carbon, and oxygen, formed through the presence of living creatures. If we calculate how much carbon dioxide would correspond to the carbon in the Earth's organic molecules and in carbonates, the result may surprise us. If all of this carbon dioxide entered our atmosphere, we would have an atmosphere about 70 times more massive than our present one; in other words, the atmospheric pressure would increase by 70 times. To reach this extreme value, we would have to eliminate not only the effects of life but also the formation of carbonate rocks that arises from weathering caused by carbon dioxide dissolved in water.

In other words, if the Earth had no life and no water (at least, no liquid water), then our atmosphere would consist of nearly pure carbon dioxide, with many times the thickness of our present atmosphere. With water absent, some nitrogen would persist in the carbon dioxide atmosphere and some argon would also be present. This description (see Chapter 12) fits perfectly the atmosphere of our sister planet, Venus!

We may conclude that our analysis of what happens to a planet in the inner solar system from life's presence or absence does make sense, for, as

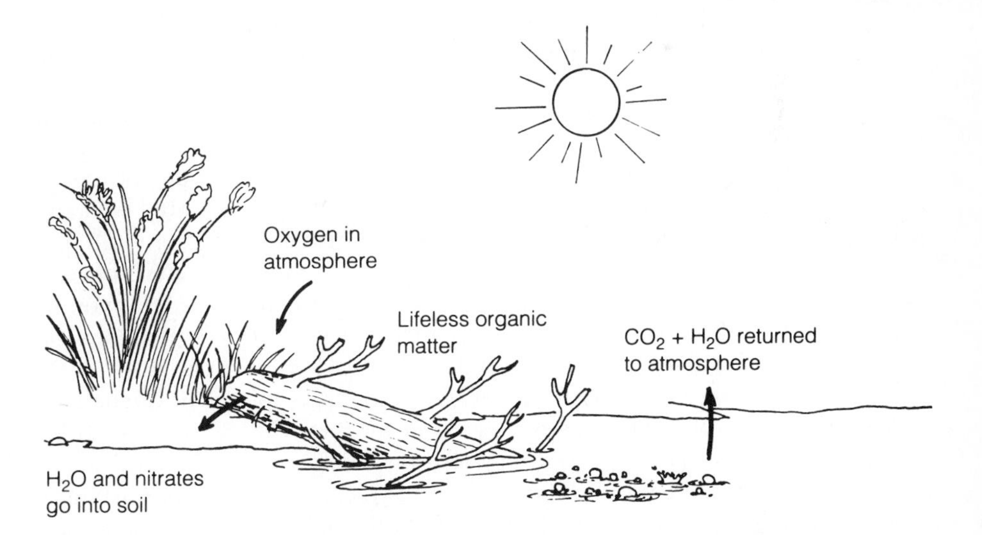

Figure 9.9 Without life to regenerate them, organic compounds exposed to the Earth's strongly oxidizing environment would rapidly combine with oxygen to form carbon dioxide, water, and nitrates.

we shall see, Venus now seems quite incapable of having either life or liquid water. Furthermore, our analysis has enough generality to apply not only to planets in our own solar system but also to other planets in systems far from the sun.

Gaia

A close relationship clearly exists between the composition of our atmosphere and life on our planet. Life began, so we think, because the Earth's early atmosphere contained the compounds essential for the genesis of living organisms. Life persists because the atmosphere provides the medium through which chemical exchange can occur, along with a thermal blanket and a shield against ultraviolet photons and cosmic ray particles.

The intricacy of this chemical interaction between life on Earth and Earth's atmosphere has led two scientists, James Lovelock and Lynn Margulis, to make the radical suggestion that life on Earth actually *regulates* the composition of the lower atmosphere by controlling the amounts of some of the important gases that are present—the gases containing elements that are essential for the continuation of life. In their view, the growth of organisms has a strong influence on some of the chemical reactions taking place in the environment, including those affecting the gases that control the average planetary temperature. The average temperature in its turn certainly affects the growth of organisms, and this potential for a "feedback loop" is found repeatedly as one examines the many ways in which life interacts with the inanimate environment (Fig. 9.10).

The total system of life on Earth—all the organisms as well as the gases, liquids, and solids they produce and consume—is called "Gaia" by Lovelock and Margulis, after the Greek goddess of Earth. Gaia is the product of nearly 4 billion years of evolution, during which life has differentiated and expanded to become the multitude of organisms we find today. This hypothesis suggests that Gaia maintains that particular equilibrium within which life can survive most successfully. Many different mechanisms must be involved in environmental regulation, just as bees use many mechanisms to maintain the optimum temperature and humidity in their hives. For example, if the average temperature on Earth should decrease, perhaps as the result of changes in the eccentricity of the Earth's orbit around the sun, Gaia may respond by promoting the differential growth of those organisms whose life and death would lead to an increase in the carbon dioxide content of our atmosphere. The excess carbon dioxide would trap more of the infrared photons from the Earth and thus would bring the temperature back to its former value.

This hypothesis could be tested (in principle, at least!) by a careful analysis of the geological record. We could seek evidence in the rocks for an

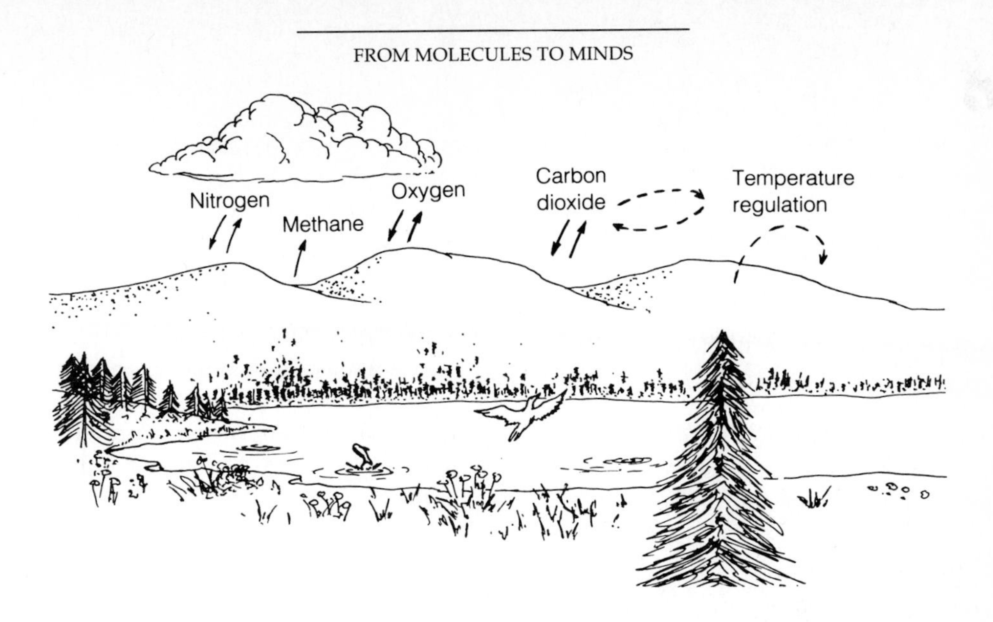

Figure 9.10 According to the Gaia hypothesis, life on Earth controls the amounts of nitrogen, oxygen, carbon dioxide, and other molecules exchanged between the atmosphere and living organisms. While there is no doubt that such exchanges take place, and that even some mineral deposits are biogenic or caused by living creatures (for example, the White Cliffs of Dover, England), most scientists feel that life on Earth is reacting to, rather than regulating, the Earth's climate.

increase in the atmospheric abundance of carbon dioxide after an ice age began, and test whether such an increase shows a correlation in time with a change in the abundances of organisms revealed by the fossil record. No such hard evidence for the Gaia hypothesis has yet been found, leaving most scientists skeptical about this intriguing idea.

The concept of Gaia as a living organism finds some support from the fact that changes in the Earth's average temperature have been remarkably small over the past billions of years, no more than a few degrees Celsius from the value that we find today. Gaia dramatically emphasizes our relative insignificance as a land-based mammalian species, for even if we regard ourselves as Gaia's "central nervous system," we can see that the global ecosystem might benefit if we were replaced one day by a more effective, less destructive species.

In return, the concept of Gaia helps us to see ourselves as part of the web of life. We depend on an intricately intertwined, living system for the food we eat, for the air we breathe, and perhaps for the climate in which we live. The major organisms that we see around us are not the major players in this system of life, despite their sizes. As Lynn Margulis has written, "Kill off all the plants and animals and the planet will recover. But kill off the microbes and in weeks the Earth will be just as sterile as the moon." Tampering with the Earth's ecological balance risks much more than eliminating a few species of organisms or producing esthetically unpleasant effects: We face the potential of seriously disturbing the balance within which all life on Earth operates, and the response may occur in ways that are extremely harmful to us.

We must also keep our sense of the interdependence of life firmly in mind as we examine the other planets in our solar system. Isolated species almost certainly do not inhabit otherwise barren worlds. Instead, we expect to find organisms that live in close harmony with their environment, a harmony in which life itself, if it is anything like life on Earth, has caused environmental changes that we may be able to detect.

SUMMARY

Prokaryotes, the simplest organisms, consist of cells lacking nuclei but possessing both DNA and RNA molecules to govern their functioning and reproduction. Photosynthesis, the conversion of sunlight into stored energy, began in prokaryotes such as the blue-green bacteria that provide much of Earth's oxygen. As this oxygen entered the Earth's atmosphere, organisms that could use this gas as a source of energy developed between 2.5 and two billion years ago.

Eukaryotes, cells with well-defined nuclei in which the cell's genetic material is stored, developed from prokaryotes that formed symbiotic relationships. Eukaryotes have sub-units called organelles that can perform specialized functions, and they contain a much larger quantity of genetic material than prokaryotes. The increasing specialization of functions within eukaryotic cells, and of different cells within larger organisms, has led to creatures with many trillions of cells concentrated in particular organs, and with particular tasks that benefit the entire organism.

About 750 million years ago, multi-cellular organisms first appeared in the fossil record. Some 150 million years later, a tremendous increase in the variety, size, and distribution of plants and animals occurred. From this time, the start of the Cambrian era, we can date the flourishing of animals more complex than soft-bodied sea creatures. Land plants established themselves about 150 million years later. Mammals first appeared only about 100 million years ago, at the height of the age of reptiles, while our hominid ancestors seem to go back about 3 million years.

Thus, five-sixths of the history of life on Earth involves only prokaryotes and single-celled eukaryotes. The planets that might exist in orbits around other stars might well have only such simple forms of life, if any at

COLOR PLATE 1 The Andromeda galaxy, 2 million light years from Earth, resembles our own Milky Way galaxy in size and shape. This galaxy contains about 300 billion stars, along with giant clouds of gas and dust, and has two smaller, elliptical satellite galaxies.

COLOR PLATE 2 A wide-angle photograph looking toward the center of our Milky Way galaxy, in the direction of the constellation Sagittarius, shows stars crowded together by the millions.

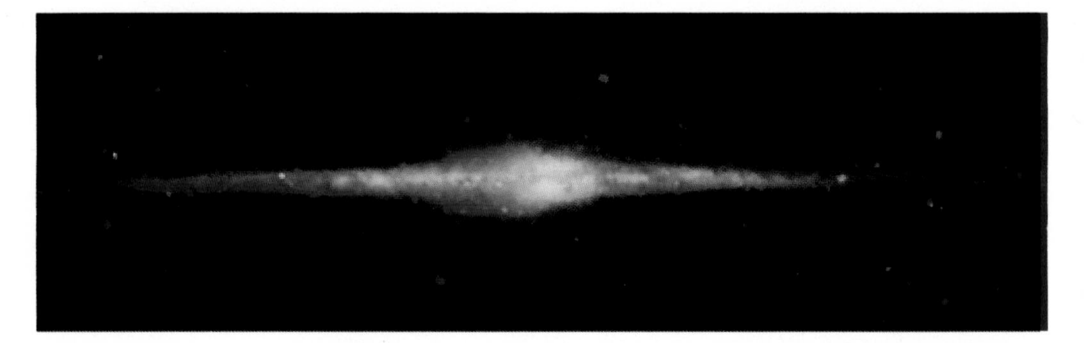

COLOR PLATE 3 In 1990, the Cosmic Background Explorer (COBE) satellite mapped the entire sky at infrared wavelengths. In infrared, we can see the inner structure of our galaxy, including its central bulge and the disk on either side, and can recognize the similarity of the Milky Way to the galaxies shown in Color Plates 1 and 4.

COLOR PLATE 4 The spiral galaxy NGC 4565, which we see edge-on, resembles our Milky Way in its overall flattened structure. This galaxy is 33 million light years away, but has an apparent size more than half as large as the moon.

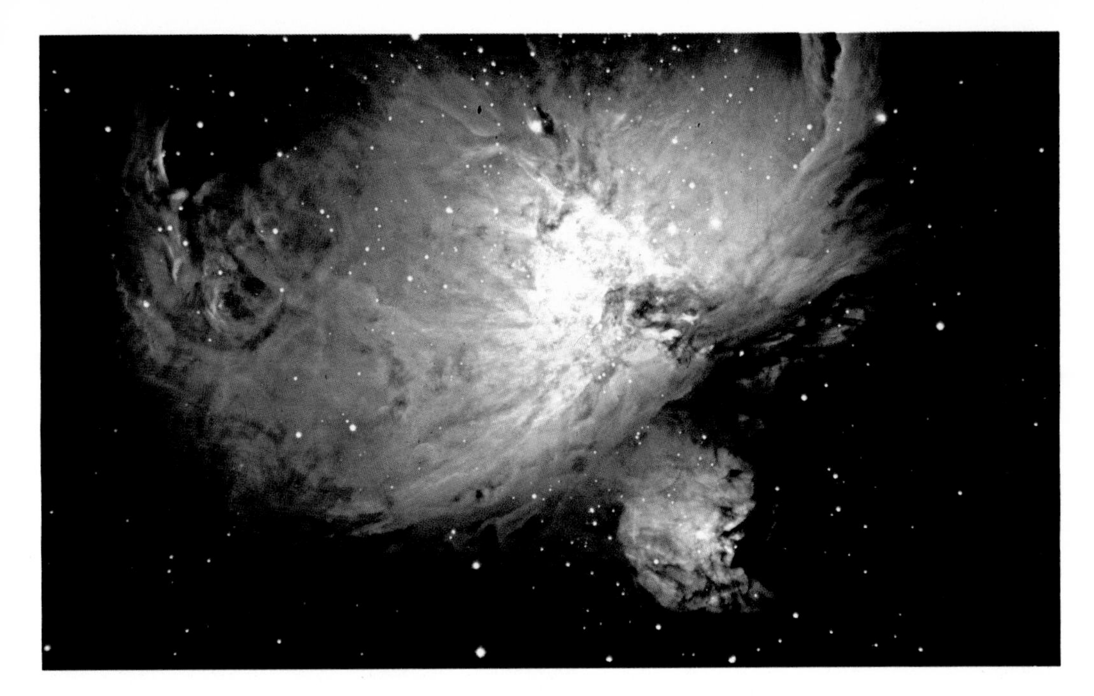

COLOR PLATE 5 The Orion Nebula, one of the closest star-forming regions to Earth, is about 1,500 light years away. It shines with the light produced by hot, luminous stars born within the past million years.

COLOR PLATE 6 The open star cluster called the Pleiades consists of a few hundred stars less than a hundred million years old. If planets orbit any of these stars, life has almost certainly not had time to appear there, or to evolve into complex forms capable of interstellar communication.

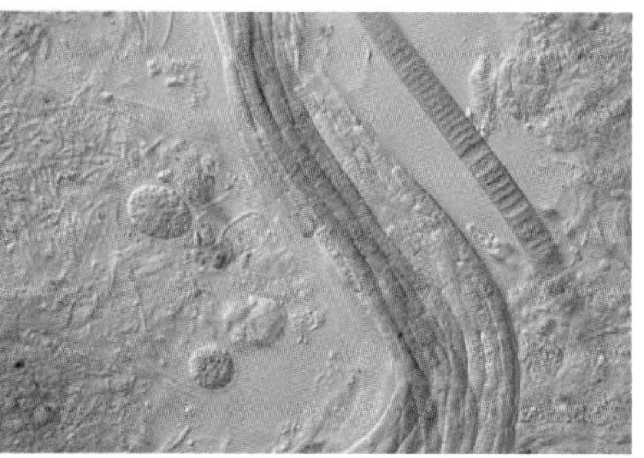

COLOR PLATE 7 Biologists Lynn Margulis and Kenneth Nealson are shown here collecting samples from bacterial mats in Baja California, Mexico. For most of Earth's history, mats like these would have been the only visible signs of life. The adjacent photograph shows a microscopic view of the blue-green bacteria that form the upper part of the mats.

COLOR PLATE 8 A rain forest in Hawaii. The variety of species in some environments like this one is so great that many have not yet been identified. Unfortunately, we are in grave danger of losing these astonishing and beautiful manifestations of life on Earth.

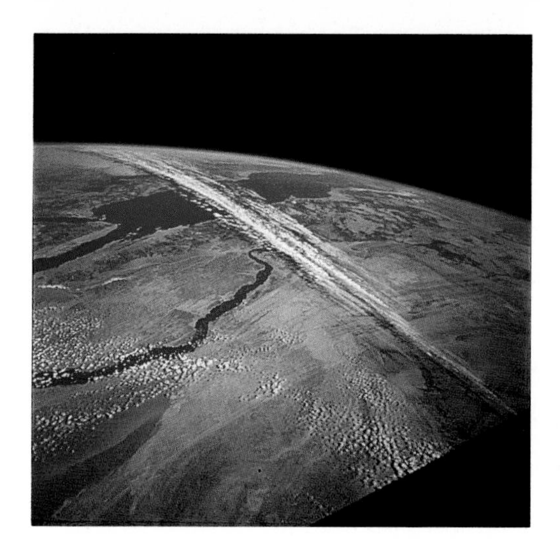

COLOR PLATE 9 Astronauts aboard the Gemini spacecraft took this photograph of the Nile Valley region of Egypt from an altitude of about 200 kilometers.

COLOR PLATE 11 In 1976, the Viking spacecraft obtained spectacular views of Mars and its extinct volcanoes, which include Olympus Mons, by far the largest mountain in the solar system, which rises 25 kilometers above its base.

COLOR PLATE 10 In 1972, the Apollo 17 astronauts investigated the lunar surface—a locale devoid of an atmosphere, running water, and any sign of life.

COLOR PLATE 12 In 1976, the first Viking lander settled onto the plain of Chryse Planitia, to reveal a landscape dominated by rocks and wind-blown dust. Some of the trenches dug to obtain samples of Martian soil are visible at the lower left.

COLOR PLATE 13 Three thousand kilometers from the Viking 1 lander site, the second Viking lander, on the plain of Utopia, found a landscape moderately different in appearance though fundamentally similar. A frost of carbon-dioxide ice ("dry ice") and water ice covers some of the rocks in this view.

COLOR PLATE 14 A dry valley in Antarctica provides the closest representation on Earth to a Martian landscape. Living microorganisms do thrive at some places in these dry valleys, but no larger living creatures exist there.

COLOR PLATE 15 In 1979, *Voyager 1* obtained a global view of the planet Jupiter that shows the swirling cloud patterns, especially those of the Great Red Spot, twice as wide as Earth. The counter-swirling white oval below the Great Red Spot has persisted for more than forty years.

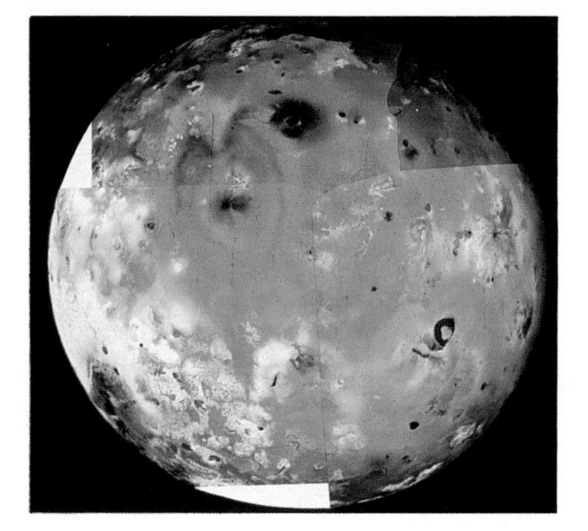

COLOR PLATE 16 The most amazing of Jupiter's moons is multicolored Io, where active volcanoes constantly renew the surface.

COLOR PLATE 17 Jupiter's moon Europa has a cracked, icy crust that may cover a layer of water where life might have developed.

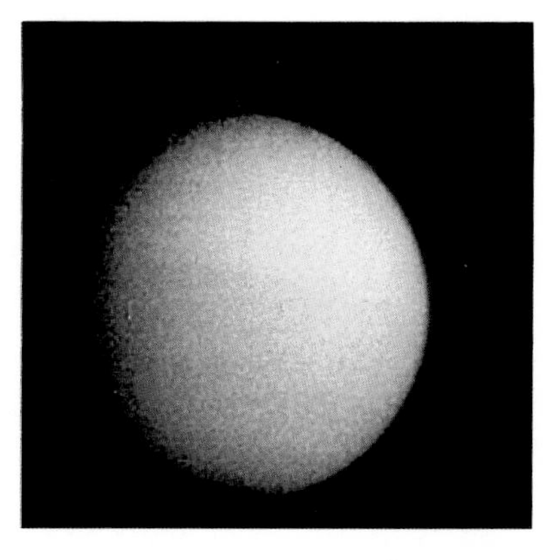

COLOR PLATE 18 In 1980, *Voyager 1* photographed Saturn and its moons, including Titan, half again as large as our moon and the only satellite with a thick, smog-laden atmosphere.

COLOR PLATE 19 The Arecibo radio telescope in Puerto Rico spans a diameter of 300 meters, and can observe radio sources that pass within 20 degrees of the overhead point. This dish will be one of the radio antennas used in NASA's search for extraterrestrial signals.

COLOR PLATE 20 This gathering of scientists at a SETI meeting at the University of California, Santa Cruz, in the summer of 1991 shows several who have obtained personalized license plates to mark their interest in SETI. Frank Drake, who is holding a license plate reading NEQLS L (N = L), is a leading researcher in this field.

all. Or they might be a few billion years ahead of us and have evolved new, perhaps immortal, forms of life. From our use of life on Earth as a typical example, we may conclude that life begins in water but apparently needs land to evolve into complex forms, capable of intelligence, which we define operationally as the ability to communicate with other forms of life on other planets. The final lesson of life seems to be that all forms of life owe their continued existence to their mutual interconnection. A somewhat poetic description of this interrelationship is offered by James Lovelock and Lynn Margulis, who see the entire surface of our planet as a single organism they call Gaia.

QUESTIONS

1. Why are cells important to living organisms? Why can we say that the first cell was a small pond, full of organic molecules and water?

2. What is a virus? Why do we have difficulty in deciding whether or not viruses are alive?

3. What is the key difference between prokaryotes and eukaryotes? Why can some prokaryotes live only if oxygen is absent from their environment?

4. How long ago did eukaryotes develop into multicellular organisms? How long after the start of life on Earth was the subsequent era of great speciation?

5. Suppose that someday we discover a planet similar to Earth that had formed only 3.5 billion years ago, rather than the 4.5 billion years we assign to Earth. If both planets follow roughly similar paths for evolution, should we expect to find many sorts of land animals on the other planet? Should we expect to find birds? Insects? Sponges?

6. Suppose that life took 2 billion years to appear on the hypothetical planet of Question 5. How far would life then have advanced, if its development proceeds at the same speed as it has on Earth?

7. Why has water been so important to life on Earth? Should we expect that life on any planet would require liquid water?

8. What are mutations? Why are they important in the evolution of life?

9. Why does evolution tend to produce more complex organisms from less complex ones, rather than the other way around?

10. Do you think that the evolution of what we call "intelligence" must eventually occur on any planet with life? Why?

11. What fraction of the Earth's total lifetime (4.5 billion years) does humanity's recorded history (about 4500 years) represent? How significantly has the human race changed the face of the Earth within that fraction?

12. What are the major changes that would occur in the composition of the Earth's atmosphere and surface layers if life were to disappear from this planet?

FURTHER READING

Attenborough, David. *The Living Planet.* Boston: Little, Brown, 1984.

Darwin, Charles. *The (Illustrated) Origin of Species,* abridged and introduced by Richard E. Leakey. London: Faber and Faber, 1979.

Gould, Steven J. "The Great Dying," *Natural History* (October 1974).

Kurten, B. "Continental Drift and Evolution," *Scientific American* (March 1969).

Margulis, Lynn. *Early Life.* Boston: Science Books International, 1982.

Margulis, Lynn, and Dorion Sagan. *Microcosmos.* New York: Summit Books, 1986.

Margulis, Lynn, and Karlene Schwartz. *Five Kingdoms.* New York: W. H. Freeman, 1988.

Simpson, George. *This View of Life.* New York: Harcourt, Brace & World, 1964.

10

How Strange Can Life Be?

We have followed the development of life on Earth in order to draw some tentative conclusions to use in assessing the probabilities for the existence of extraterrestrial life. Although we might think at first that *any* kind of life could be possible, so that conclusions based on terrestrial biology would be hopelessly biased, we should remember that the laws of physics and chemistry make some situations far more probable than others. These laws—the summary of our experience in studying the universe around us—appear to be valid as far as we can test them, including the analysis of light from stars and from the most distant galaxies. Unless we choose to abandon the idea that nature follows the same rules in various parts of the universe, our conclusions about the overall probability of life appearing in different environments should have merit, although we must remind ourselves that we are dealing only with probabilities, not certainties. If we do abandon this idea, we can make no scientific predictions at all!

The Chemistry of Alien Life

The history of investigating life on Earth includes repeated, unexpected discoveries of strange creatures, tiny and large, flourishing in environments that seem at first to be unlikely havens for life. For example, we have said that sunlight is necessary for photosynthesis to occur, and our common human bias suggests this must be *visible* sunlight. But a species of bacteria called a "thiococcus" can perform photosynthesis in what we could call complete darkness! These bacteria, thriving in the muddy waters of South American jungles, contain a pigment that absorbs infrared radiation from the sun, radiation whose low energy per photon lies outside the sensitivity range of human eyes.

We can say that life must have water. Indeed life must, but that water can be comparable in acidity to a dilute solution of sulfuric acid, or the water can contain large amounts of dissolved lime and thus be quite basic. Life can assume enormous stationary forms, as in a redwood tree, or be tiny and highly mobile, as in a paramecium. We cannot hope to predict the exact forms of life that we might meet on an alien planet, nor can we predict the kind of adaptation that such living organisms would have made to their environments.

Can we make any definite statement about the *chemistry* of alien life, about the molecules that form living organisms in faraway places? On Earth, we can make generalizations of sweeping breadth. Both the giant redwood and the microscopic paramecium live because DNA and RNA molecules help them make protein. Both these organisms rely on a specific chemistry, which turns out to be the same chemistry for all organisms. This chemistry uses one element, carbon, as the basic structural unit in forming complex molecules, and it uses water as a solvent. Water provides a fluid in which nutrient molecules can float; it preserves chemical equilibrium in living cells, it helps regulate the temperature within them, and it forms a large fraction of each organism's body weight (40 percent in dry plants, 70 percent in humans, 95 percent in jellyfish).

To maintain a solvent in the liquid state requires that the temperature stay within a fairly narrow range, 0 to 100° Celsius in the case of water. Thus the requirements for a common element, carbon, and a common solvent, liquid water, would automatically add a third requirement, the proper temperature range, if life is to arise and persist.

But must life elsewhere use the same chemistry as life on Earth? We hardly expect life on other planets to have molecules exactly like amino acids, proteins, DNA, and RNA, though it might have molecules that perform similar functions. But we must consider this: *Would other forms of life be based on carbon for structure, and on water as a solvent?* Or could we expect a different structural element and a different solvent?

We can attempt to outline the basic requirements for any alien life chemistry. It must be able to form large, complex molecules, for only in this way can it store the amount of information that a living organism needs to function properly. It must use a fluid that carries nutrients and waste products, and provides a bath within which chemical reactions can occur, thus allowing life to persist and to reproduce. Though we can conceive of situations in which this fluid is a gas, a liquid will serve much better because it will keep the chemical reactions within a localized region, as in the cells of living creatures on Earth, and at a constant temperature, thus making it easier for the chemical reactions to continue at a nearly constant rate.

Given these two basic requirements—complex molecules and a liquid solvent—what options are open to life? Has chance alone dictated that we have carbon-based life on Earth, or does something favor this element over all others? And why does terrestrial life depend so much on water? Why not ammonia, or alcohol, or cleaning fluid?

The Superiority of Carbon

Let's consider carbon first. To replace it, we would need a relatively abundant element that can combine with four hydrogen atoms to form a stable molecule (Fig. 10.1). The requirement of high abundance stems from our prejudice in favor of the kinds of life that will be abundant in the universe, not the rare and exotic forms that might exist in some unlikely situation, but rather the life that arose under the conditions that have prevailed throughout most of the universe. The second requirement, the ability to combine with four hydrogen atoms, arises from the need for complexity: Since four atoms are as many hydrogen atoms as any atom can accept, an atom that competes with carbon as the backbone of life must do at least this well.

We can easily illustrate this point. Think of all the different ways that oxygen can combine with hydrogen to form a stable molecule. There are

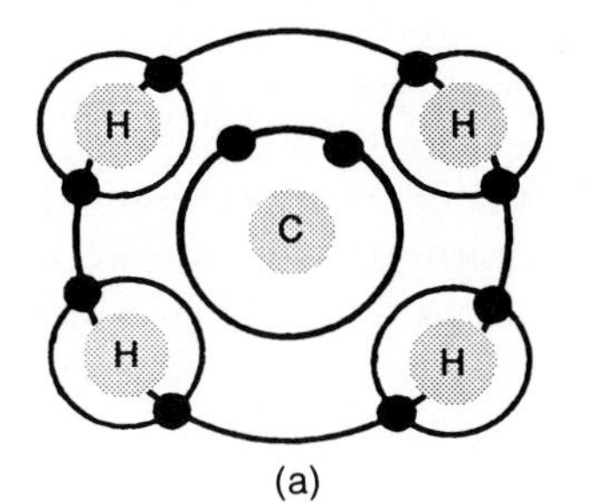

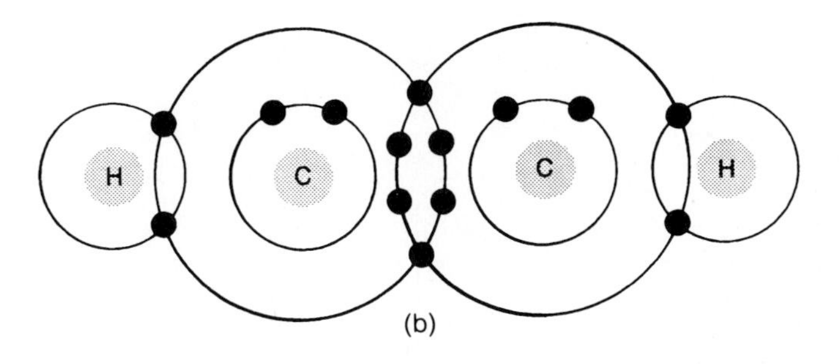

Figure 10.1 (a) A carbon atom can combine with as many as four other atoms, as in methane molecules (CH_4). (b) Carbon atoms can also share electrons with each other to build chains and rings. Acetylene (C_2H_2) is the simplest of these carbon-carbon linked molecules (compare Fig. 7.1).

two: water (H_2O) and hydrogen peroxide (H_2O_2), as shown in Fig. 10.2. How many stable molecules can *nitrogen* form with hydrogen? Two again: ammonia (NH_3) and hydrazine (N_2H_2). Now ask, how many molecules can *carbon* form with hydrogen? If you don't know the answer, you are in good company: Neither does anyone else! But to get an idea, you can consult the *Handbook of Chemistry and Physics.* There you will find the heaviest such molecule listed as enneaphyllin, whose chemical formula is $C_{90}H_{154}$! Remember that these are just the compounds of carbon with hydrogen atoms alone. Colleges teach organic (carbon-based) chemistry as a separate subject because carbon has a truly amazing tendency to form all sorts of compounds with every other element. (See Fig. 10.2).

Does any other element perform as well as carbon? Silicon, the usual candidate to replace carbon, lies directly underneath carbon in the periodic table of the elements. This location means that silicon, like carbon, can combine with four hydrogen atoms. The resulting compound is called silane (SiH_4), the silicon analog of methane (CH_4). Since silicon, like carbon, has a considerable abundance in the universe, we often meet silicon-based life in science fiction. But if we examine silicon more closely, we find that the bond between two silicon atoms has only half the strength of the bond between two carbon atoms. Thus the bonds between silicon atoms can be broken much more easily in chemical reactions than the bonds between carbon atoms. Furthermore, although the strength of carbon-carbon bonds about equals the strength of carbon-hydrogen or carbon-oxygen bonds, silicon-hydrogen and silicon-oxygen bonds have *greater* strength than silicon-

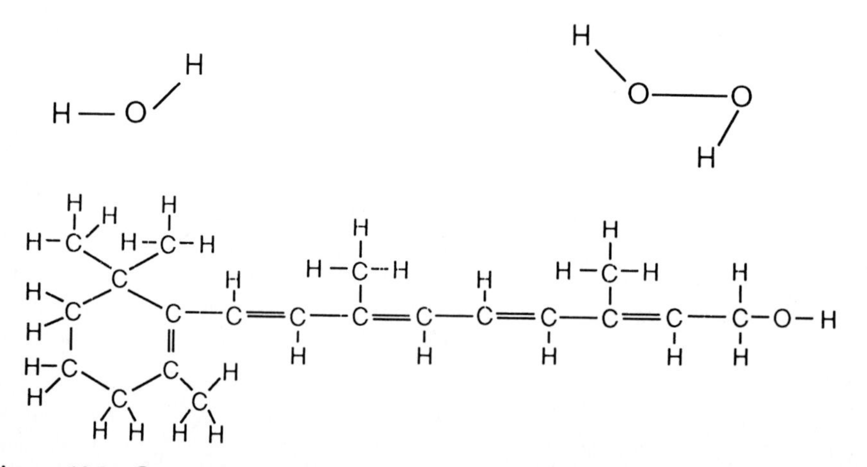

Figure 10.2 Oxygen atoms can combine with hydrogen atoms to form just two stable molecules: water (top left) and hydrogen peroxide (top right). In contrast, carbon and hydrogen atoms can form many complex molecules, such as vitamin A (which includes a single oxygen atom) (bottom).

silicon bonds do. As a result, long chains or rings of atoms based on Si-Si-Si- structure are unlikely, whereas chains based on C-C-C- structure are extremely common. In fact, any straight-chain molecules of hydrogen and silicon containing more than three silicon atoms linked together are highly unstable. In contrast, multiple carbon linkages, forming straight and branched chains, rings, and other shapes, dominate organic chemistry. Some forms of life might skirt this difficulty by making long molecular chains (polymers) with silicon-oxygen bonds instead of silicon-silicon bonds. Such polymers form silicones, which are both stable and chemically versatile, and rocks, which are certainly stable! Silicones, however, tend scarcely to react with other molecules, a fact of importance for their usage as lubricants. Nevertheless, we can certainly imagine that special circumstances, or special catalysts (enzymes), could produce more lively interactions among silicone-like polymers than we find on Earth.

The basic difficulty with silicon as the backbone of life is *silicon's affinity for oxygen*. Even if silicon exists in the most reducing conditions—in an atmosphere loaded with hydrogen—it will form silane (SiH_4) only when the temperatures rise above 1000 K. At lower temperatures, silicon forms silicon dioxide (SiO_2). Proof of this comes from observations of Jupiter's atmosphere, which we know to be predominantly hydrogen, and which contains NH_3, PH_3, AsH_3, CH_4, GeH_4, and H_2O, the fully reduced (hydrogen-laden) forms of the elements nitrogen, phosphorus, arsenic, carbon, germanium, and oxygen. But *no* SiH_4 has been detected on Jupiter. Why not? Because the silicon on Jupiter, instead of forming silane, has combined with oxygen to form silicon dioxide. Since the universe contains 20 oxygen atoms for every silicon atom (see Table 7.1), silicon atoms are extremely likely to end up in silicon-oxygen compounds. And once silicon dioxide or other such rock–forming compounds arise, they are very difficult to decompose.

Consider carbon dioxide (CO_2), the molecule formed when carbon combines with oxygen to the maximum extent possible. Carbon dioxide remains gaseous even at low temperatures (down to –75° C); it is soluble in water; and breaks apart rather easily into its constituent carbon and oxygen atoms. Now compare carbon dioxide with silicon dioxide (SiO_2), which is gaseous only at high temperatures (above 2000° C), is extremely insoluble (in almost everything except hydrofluoric acid!), and requires large amounts of energy to be broken apart into its silicon and oxygen atoms. The former substance has the advantage over the latter in every respect in providing a molecule useful to living organisms. Likewise, methane (CH_4), the fully reduced (hydrogen-laden) compound of carbon, has much greater stability than silane, the fully reduced form of silicon. Methane exists even in our highly oxidizing atmosphere, and although its lifetime there is short (the entire amount must be replaced by methane-generating bacteria on a time scale of about 10 years), it is not nearly so short as that of silane, which bursts spontaneously into flame when exposed to air! This gives us the chance to speculate playfully that if silicon-based creatures did exist and visited the Earth, we would have a natural explanation for the fire-spouting dragons of medieval legends. It seems far more likely, however, that silicon's deficiencies as a structural element rule out the possibility of silicon-based life in almost all situations.

We can test this statement by examining silicon and carbon in the universe at large. The results support our conclusion. Astronomers have found no silicones or silanes in meteorites, in comets, or in the atmospheres of planets. Instead, they find *silicates,* molecules of *oxidized* silicon, in combination with various other elements. We live on an excellent example of silicon chemistry, the planet Earth, which consists largely of silicates. So far, silane (SiH_4) has been found only in the interstellar medium (see Table 4.1) In striking contrast, many complex carbon-based (organic) molecules appear in meteorites, comets, interstellar clouds, planetary atmospheres, and cool stars, along with carbon monoxide and carbon dioxide. Carbon passes easily between its fully oxidized (CO_2) and its fully reduced (CH_4) forms, while silicon does not.

We may reasonably conclude that although life based on silicon may be *possible,* it will be extremely uncommon at best. Life based on carbon seems favored as the dominant kind of life in the universe. This may sound dogmatic, but we should remember that the abundances of the various elements appear to be roughly the same everywhere that we can measure them in the universe. Thus the elements still rarer than silicon have extremely poor prospects as the basis for the chemistry of life elsewhere.

Solvents

Let's now turn to the question of solvents. What makes water so wonderful as a solvent? Why does this substance have such critical importance to life as we know it? To be truly useful, any solvent must remain liquid within a large range of temperatures, so that the variations in conditions on a planet or satellite do not make the solvent freeze or boil at all locations on the object's surface. This temperature range should include temperatures high enough for chemical reactions to proceed at a reasonably rapid pace, yet not so high that collisions destroy important molecules. We would also prefer a solvent that helps the organisms containing it to regulate their temperatures. By definition, the solvent should have the ability to dissolve other chemical compounds, since organisms use it to transport nutrients and to carry off wastes. We can see how well water fulfills these requirements in

TABLE 10.1 Temperature Ranges at Which Solvents Remain Liquid

Solvent	Temperature for Liquid	Range of Temperature
Water	0 to 100° C	100° C
Ammonia	–78 to -33° C	45° C
Methyl Alcohol	–94 to +65° C	159° C

comparison with two other possibilities, ammonia (NH_3) and methyl alcohol (CH_3OH).

We first note that water does indeed remain a liquid over a large range of temperatures (Table 10.1). In this comparison, water outperforms ammonia, but methyl alcohol, by virtue of its greater temperature range as a liquid (despite the lower maximum temperature), seems almost as good as water.

The next comparison among solvents deals with their ability to help regulate temperature. The heat-regulation capability of a solvent depends on both its "heat capacity," the amount of energy required to raise 1 gram of the solvent by 1° Celsius in temperature, and its "heat of vaporization," the amount of energy needed to change the liquid into a vapor once it has reached its boiling point. We want a solvent with as large a heat capacity and heat of vaporization as possible, in order to minimize the effect of sudden temperature changes on the physical conditions within the solvent. A large heat capacity implies that a big change in the external temperature will affect the organism only slightly. A large-scale, nonbiological example of this principle is familiar to most of us: Land near the ocean has much milder climate than does a region at the same latitude far from large bodies of water. The heat capacity of water is 1 calorie per gram-degree, a value intermediate between the heat capacities of ammonia (1.23 calories per gram-degree) and of methyl alcohol (0.6 calorie per gram-degree). In its heat of vaporization, however, water far surpasses both its competitors. To vaporize water requires 595 calories per gram, while ammonia needs only 300 calories and methyl alcohol only 290.

Water's high heat of vaporization means that a relatively large amount of heat is required to evaporate a small amount of water. Thus each drop of water that evaporates from the skin of an organism has a large cooling effect, removing the excess heat released within component cells as life processes occur (Fig. 10.3). Indeed, the fact that we consider the normal temperature for human beings to be 37° C (98.6° F), and not 36 to 38° C (96.8 to 100.4° F) gives an excellent indication of how well our bodies can regulate their temperatures. Although we don't yet understand just what conditions are required for intelligence to exist, we can appreciate the fact that mammals

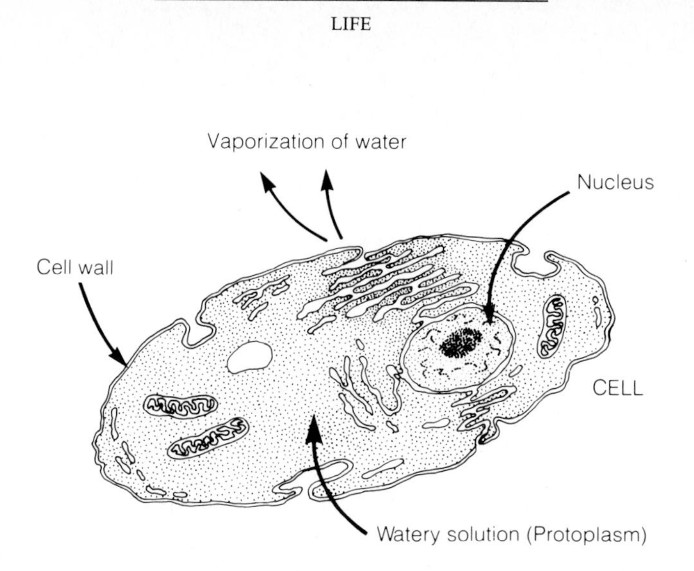

Figure 10.3 Because water has such a high heat of vaporization, a living cell can respond to a temperature increase by vaporizing a relatively small amount of water, allowing it to maintain nearly the same temperature.

need a precise temperature regulation to maintain their rates of chemical reactions at the right level and to keep the structural elements of their bodies in proper condition. Mammals are distinguished from "lower" forms of life by the greater complexity of their brains and by the chemical reactions that go on in these brains; precise temperature regulation allows these reactions to occur as they should. Just think of your own experience; how well do you function with a mere 4° F of fever?

An additional advantage of water as a solvent appears in its "surface tension," which describes the tendency of a liquid to form droplets. The surface tension of water, twice that of ammonia and three times that of alcohol, exceeds the surface tension of any other liquid known. This property undoubtedly played a vital role in making aggregates of organic compounds before cells evolved, for the surface tension would force some compounds together and would preserve the boundaries between water and mixtures of different organic molecules. Surface tension continues to concentrate solutions of solids at the interfaces of different media in an organism—for example, at the boundary of a cell wall. (See again Fig. 10.3).

For its heat capacity, heat of vaporization, and surface tension, water is hard to beat. But in a way these are extra benefits, since life's essential fluid medium must first of all act as a solvent; that is, it must dissolve a wide variety of chemical compounds in order to carry them easily into and out of living systems. Here again, water is outstanding: It has more than twice the ability of either ammonia or methyl alcohol to carry other molecules in solution.

We have compared water to two other fluids and have found that it has several advantages in addition to being the best solvent. Why don't we consider many other possible liquids as well? The answer lies in the cosmic abundances of different types of atoms. Other possibilities, such as hydrogen sulfide (H_2S) and hydrogen chloride (HCl), could remain liquid under reasonable temperature conditions. But these solvents must be far less abundant than our basic three, since they include less common elements (such as sulfur and chlorine) rather than the hydrogen, nitrogen, oxygen, and carbon that form water, ammonia, and methyl alcohol. The alternative possibilities have additional shortcomings that also make them less desirable. As a result, if water wins over ammonia and methyl alcohol, it probably surpasses all other fluids that are likely to exist in other astronomical situations.

We ought to pay a moment's attention to a nearly unique property of water, one it shares with almost no other fluids. *Water expands as it freezes,* which means that solid water (ice) floats on liquid water. If ice sank, ponds that froze in winter would freeze all the way down, killing most of the organisms within them. This sort of freezing would occur in ponds of ammonia (at –78° C) and of methyl alcohol (at –95° C). We may, however, consider our worries about freezing ponds to be an example of "high-latitude chauvinism," since most of the Earth never freezes. Life may in fact have originated in low-latitude areas, surviving and evolving in ponds that never froze even their topmost layers.

What about another planet, with other conditions on its surface? We can imagine several advantages in using a substance, such as ammonia, that does not expand when it freezes. The difficulty in a solvent that expands resides in the stress that freezing produces against the walls that contain the fluid. Frozen water in pipes can burst the steel walls, and the cells of organisms that freeze will rupture, killing the organisms. But if life's chemistry relied on a solvent that did not expand upon freezing, low temperatures might simply produce a dormant state, from which life could easily recover when warm weather reached the place of hibernation.

What an extraordinary advantage for space travel! The immense distances between stars present an insoluble problem for interstellar voyages with conventional rockets, if we require relatively brief journeys. The need for speed arises from the desire to make the trip within the crew's lifetime. Could we extend that lifetime by a large factor simply by putting the crew on ice? Human beings cannot survive such freeze-drying, since they consist mostly of water. But imagine a race of astronauts whose life chemistry depended on ammonia as a solvent: They might use the fact that ammonia does not expand upon freezing as the way to visit most of our galaxy, waking up only briefly at each stop!

Before we grow too excited about ammonia, however, we ought to recall a key advantage water has over ammonia. Liquid water is self-shielding against ultraviolet radiation: Some of the water molecules will be dissociated by ultraviolet photons, releasing oxygen and hydrogen into the atmosphere, and some of the oxygen atoms will link into ozone (O_3) molecules, which absorb ultraviolet (Fig. 10.4). Since most stars produce large amounts of ultraviolet radiation, the self-protection that water provides has great importance, for without such an ozone shield, organic molecules would quickly be destroyed by ultraviolet, and the oceans themselves would slowly be destroyed as their constituent molecules were dissociated. In contrast to water, the dissociation of ammonia produces not oxygen but nitrogen atoms, which can neither form a shield against ultraviolet light nor produce a source of chemical energy similar to that available from oxidation. Hence, a planet with ammonia oceans would require a separate source of shielding—perhaps a permanent cover of hydrocarbon smog—to prevent the destruction of the oceans, and of any life on the planet, by ultraviolet. Water carries this shielding capacity along with the oxygen in every molecule.

Our considerations of the various possible solvents are not exhaustive. Nevertheless, they can serve to make a key point: Life on Earth relies on carbon chemistry, and on water as a solvent, not by accident, but because carbon and water have an inherent superiority. The case for carbon may be stronger than that for water, but we should not feel narrow-minded in anticipating that *most* alien life forms rely on the same element and the same solvent for their basic chemistry of life.

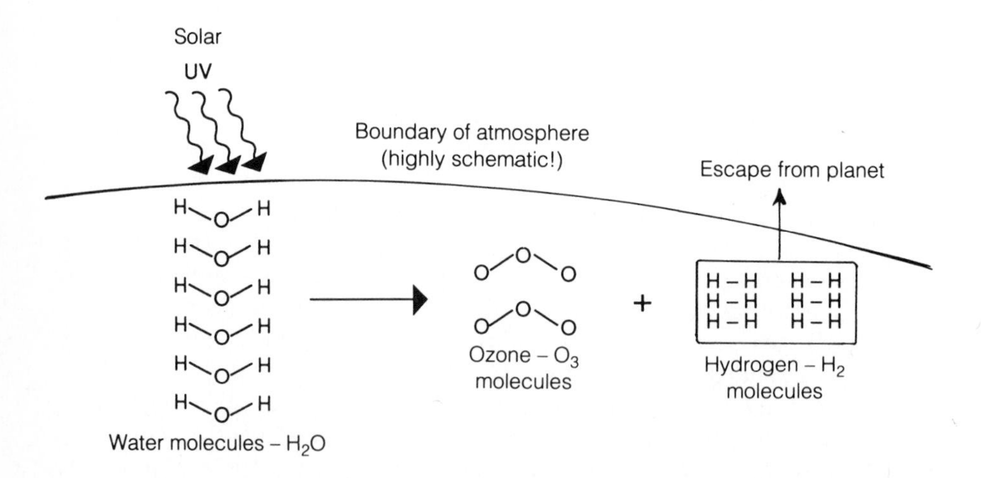

Figure 10.4 When ultraviolet radiation dissociates water molecules, some of the oxygen atoms released will link into ozone molecules, which act as a shield against ultraviolet light.

Nonchemical Life

Until now, we have restricted our discussion of life to the chemistry that we know, taking into account possible variations in the temperature, density, and elemental composition. Confining our considerations to chemistry means discussing the interactions among *atoms,* which can form many sorts of molecules, as simple as water or as complex as DNA. Atoms and the molecules they form interact with one another through electromagnetic forces, which arise from the positive and negative electric charges that the atoms contain. Hence when atoms join into molecules, electromagnetic forces, not gravitational or strong forces, provide the linkage.

But when we try to consider strange forms of life, we must look beyond the chemical interactions that govern life on Earth. Likewise, we must not restrict ourselves to planetary surfaces in our search. Planets do provide the likeliest sites for chemical life to develop, because on planets we expect to find the necessary conditions—a relatively high density of matter, temperatures appropriate to relatively rapid chemical reactions, and climatic conditions maintained within a relatively narrow range by the near circularity of the planets' orbits—that will allow chemical interactions to proceed at a relatively rapid rate. The varieties of environments within these general limits that are provided by a planet's surface—a range of daily and seasonally varying temperatures, along with interfaces between land, water, and atmosphere—are also important factors for the origin and evolution of life.

Thus life based on chemical reactions may well turn out to be widespread on the surfaces of suitable planets. But we must not neglect forms of life much stranger to us. Three important examples, which do not exhaust the entire range of possibilities, are life in dense interstellar clouds, on the surfaces of neutron stars, and of entire galaxies.

These three possibilities envision living entities with characteristic sizes that are much larger or much smaller than the forms of life with which we are familiar. But even here we must bear in mind a key characteristic of life on Earth: the presence of individuals. We humans number in the billions, while most species of mammals, birds, and reptiles have many millions, or at least many hundreds of thousands, of representatives. Only those species close to extinction include just a few thousand, or a few hundred, individual species members. This fact reminds us that life on Earth has evolved through the interactions of enormous numbers of individual animals and plants, precisely because in order for natural selection to make progress—that is, to distinguish those individuals most fit for survival and reproduction—each species must have a large pool of members who could produce offspring with slightly different characteristics. Most evolutionary "experiments" end in failure, so a species of animals or plants with only a few hundred individuals cannot hope to carry its evolutionary development much further in its native habitat. Instead, such a species fights for simple survival.

Should we expect things to be different on another planet, or in interstellar space? Probably not. We can easily imagine conditions in which certain life forms encounter few or no problems of competition for survival and reproduction, and never enter a new environment. Such circumstances offer greatly reduced pressure for evolutionary changes. So long as competition or changes in environment exist, and so long as mutations occur, we expect natural selection to lead to new types of species, and for the new species to evolve into still other ones over many millions of years. The key to this process remains the interplay of a large number of individuals in a species, and the large number of species is the result of this interplay.

We thus tend toward the conclusion that complex forms of life, such as those that have evolved to the point of achieving intelligence, cannot be the product of only a few events that produce just a few individuals. Instead, living organisms by the billions and trillions testify to the natural result of competition for reproductive success on Earth.

Black Clouds

In an intriguing work of fiction, the eminent British astronomer Fred Hoyle has described the arrival near Earth of a giant interstellar cloud, capable of thought, of directed motion, and of an indefinite lifetime, a cloud whose life processes depend on electromagnetic forces (as ours do) and whose thoughts consist of radio messages from one part of the cloud to another (Fig. 10.5). The cloud's excursion into our sun's vicinity occurs because the cloud must periodically replenish its supply of stored energy; it does so by absorbing large amounts of starlight, in this case the light from our own star. During its passage, the cloud discovers intelligent beings (using our restricted, technological definition!) on one of the sun's planets, who manage to establish radio communication with the black cloud, which reveals to them many mind-bending facts about the universe. The cloud's "brain" consists of complex networks of molecules, which can be increased in number and specialization as the cloud chooses. Electromagnetic currents pass among these molecules, and conceptually the cloud's brain works in much the same way that our brains do.

Fred Hoyle's black cloud, a living organism as large as the orbit of Venus and almost as massive as the planet Jupiter, reminds us of the difficulties and advantages that our experiences provide when we speculate about other forms of life. On one hand, we can imagine the existence of a living "black cloud" that uses starlight energy to construct greater and greater

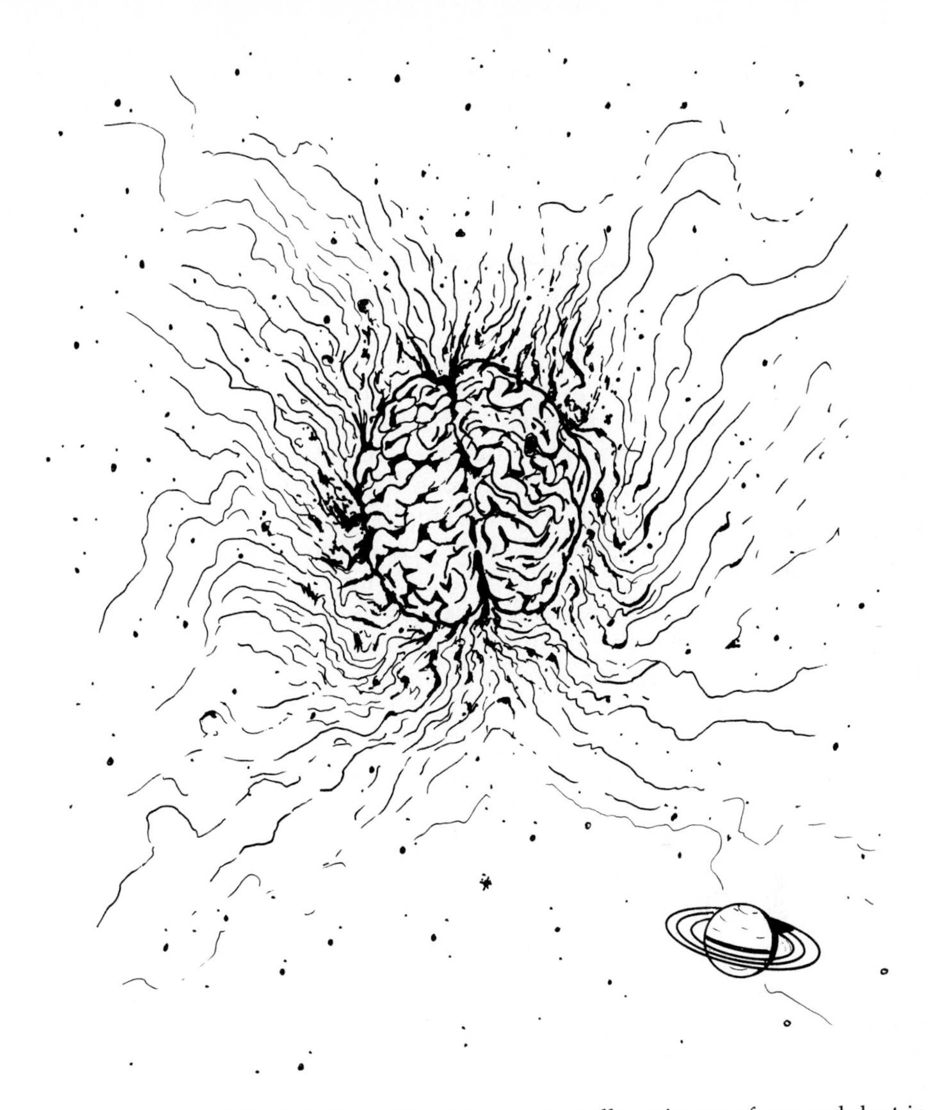

Figure 10.5 Fred Hoyle's Black Cloud is an interstellar mixture of gas and dust in which electrical currents carry thoughts, and starlight provides the basic energy source.

degrees of ordering, that thinks and feels through electromagnetic impulses sent along appropriate pathways. On the other hand, our knowledge of how terrestrial organisms evolved suggests the necessity of many previous steps through which evolution can proceed. A black cloud certainly could *exist*, but how its existence would ever *arise* remains a darker mystery.

If we, like Fred Hoyle, imagine earlier stages in the evolution of black clouds, primitive cloudlets of lesser abilities, then we return to our favoritism for planetary surfaces. Why? Because matter in interstellar space has such low densities (compared with matter on planets) that any interactions among particles occur far more slowly there. As a result, the time for life to arise in interstellar clouds should span not the billion or so years that applies to the Earth but thousands or millions of times longer—far longer than the age of the universe since its expansion began. For the same reason, even if life could originate in matter at the low densities of interstellar clouds, we would expect that all life processes should take longer, because of the greater distances and lower densities of matter that are involved.

Billions of years had to pass on Earth before complex organisms appeared—a period so long, in fact, that the time for "intelligent" life to arise on Earth measures a noticeable fraction of the entire age of the universe (about one-quarter). Although some interstellar clouds might indeed possess the "right" temperatures (about -50 to 80° C) for life to begin in a relatively short time, most are far colder than this (-250 to -200° C). Furthermore, the denser clouds tend to be the colder ones, so they counterbalance a larger number of atoms per cubic centimeter with the slower motions of those atoms. But even those clouds with the proper temperatures and the "right" mixture of elements have a density of matter less than one trillionth of the density of our atmosphere! This presents a serious obstacle to organizing matter sufficiently for life to emerge, given the large distances between one complex molecule and another in one of these clouds.

Life on Neutron Stars

Until now, we have looked at the consequences for life of electromagnetic forces, which are responsible for the laws of chemistry. Following a stimulating suggestion by Frank Drake, let us consider briefly how another type of force, the strong force, could produce an entirely different sort of life on the surfaces of neutron stars.

As we saw in Chapter 6, neutron stars arise from the collapsing cores of highly evolved stars (see again Fig. 6.8). The temperature on the surface of a neutron star must be about 1 million K, more than a thousand times the 300 K that characterizes the Earth's surface. Furthermore, the gravitational force on the surface of a neutron star exceeds that on Earth by about 1 trillion times. The enormous temperatures and immense gravitational forces mean that no molecule or atom could survive for long. High-speed collisions among molecules would break them apart and would knock an atom's electrons loose from the atom. As a result of the enormous temperatures, we can hardly expect to find any sort of life based on electromagnetic forces, which make atoms and molecules, on the surface of a neutron star. Still, we should not lose all hope, for life based on *strong forces* remains conceivable there.

Strong forces hold together the nuclei of all atoms more complex than hydrogen. In nature, we find nuclei with anywhere from one particle (the single proton of hydrogen) to 238 particles (the 92 protons and 146 neutrons of uranium-238). Nuclei with more elementary particles than 238 can be made in physics laboratories, but they fall apart ("decay") into smaller nuclei after an amazingly short lifetime, by human standards. The most massive "artificial" nucleus yet made has 265 particles—105 protons and 160 neutrons—and it decays in less than one-billionth of a second.

We do not, however, live at the size level of protons and neutrons. Suppose that we were, say, a proton some 10^{-13} centimeter in size, traveling at a speed of 1000 kilometers per second, a speed typical of protons at a temperature of 1 million K. We would then move a distance equal to our own size in 10^{-21} (one billion-trillionth) of a second! Compare this with a human being whose size measures 170 or so centimeters, and who walks a distance equal to this size in about 1 second. We can see that 10^{-21} second would have the same meaning to a proton as a second does to a human being out for a stroll: For elementary particles at 1 million K, events tend to occur in times measured in units of 10^{-21} second, just as things happen to humans at 300 K in times measured in seconds. More precisely, the electromagnetic interactions within our body chemistry characteristically occur on time scales of a few thousandths of a second (the time for our eyes to perceive visible-light photons), a few tenths of a second (the time to send a message all the way through our bodies), or a few seconds (the time to reverse position completely).

Now imagine the seething surface of a neutron star, where elementary particles travel and collide at speeds of thousands of kilometers per second, which is equivalent to traveling distances of thousandths of a centimeter in times of trillionths of a second. Such collisions could produce massive nuclei, each made of thousands, or even tens of thousands, of elementary particles. These nuclei would last for perhaps one–million-billionth (10^{-15}) of a second. After this brief time had elapsed, the massive nuclei would decay into smaller nuclei. This seems nearly instantaneous, but from the point of view of an elementary particle, for which 10^{-21} second is a typical time scale, these massive nuclei last millions of times longer than the average time between interactions. In other words, a massive nucleus might have a million different collisions or other interactions before it decayed into other sorts of nuclei.

We can speculate that evolutionary processes on the surface of a neutron star might produce forms of life, individuals that interact with their environment and with other individuals in an organized way. If all this came true, then the development and evolution of such life would happen far more rapidly than our Earthborne experience admits. Since the typical time scale for strong force life, 10^{-21} second, is one billion-billionth of the thousandth of a second that characterizes our electromagnetic force life, we might expect evolution to proceed that much more quickly. Then the origin of life would require not about 1 billion years, but about one-billionth of a year, or one-thirtieth of a second! Short as this time may seem to us, it allows billions upon billions of interactions for each of the complex nuclei that might exist on the surface of a neutron star before that nucleus decays into other types.

If we have the courage to follow this scenario further, we can see that on the surfaces of neutron stars, entire civilizations could rise and fall, and rise and fall a million times over, faster than the human eye can wink. The individual members of these civilizations would have sizes of about 10^{-11} centimeter, would live for 10^{-15} second, and, if they use photons for communication, would favor gamma-ray photons with perhaps 10^{10} times the frequency of visible-light photons. Such gamma-ray photons arise from interactions among elementary particles, just as visible-light photons typically arise from interactions among atoms.

If this sort of neutron-star life does exist, then we probably cannot hope that a civilization that arose would use radio waves, either for their own communication or in an attempt to find other civilizations. Nonetheless, these speculations about neutron star creatures, based simply on the facts we know about the strong forces that attract elementary particles, should not be forgotten when we consider how many civilizations may now exist in our own galaxy. But we do have a difficulty in communication that arises from the great differences in perspective, especially of time, between ourselves and a neutron star civilization. We can hardly hope to establish a meaningful interchange with a civilization that lasts only a billionth of a second. Or would they still have something to tell us?

Gravitational Life

Since we have taken a mental excursion to consider the possibility of fast-moving, strong force life on neutron stars, we may deem it proper to consider the opposite extreme, the idea of life based primarily on *gravitational forces.* Here the typical subunit should be an object large enough for gravity to dominate over electromagnetic and strong forces: a star. If individual stars play the role of individual atoms or molecules in Earth life, or of individual nuclei in neutron star life, what does it mean to notice that stars cluster together in galaxies? Are galaxies alive?

No, we think not. (But consider the speculations about Gaia on page 224; perhaps the Gaias themselves form individual parts of a super-Gaia, the ultimate organism!) Our definition of life does not simply require a great degree of organization. Galaxies are fairly well ordered, especially spiral galaxies, but not as much as, for example, paramecia. Furthermore, galaxies seem to lack any sort of purpose, but here we may suffer from human chauvinism. Things happen in galaxies, but on time scales that range from seconds (for an individual star to collapse) to hundreds of millions of years (for an entire galaxy to rotate once). Stars typically interact with one another on a time scale of many millions of years. If life could originate from the repeated effects of such close interactions—as when molecules interacted in the primordial soup on Earth—then scarcely any time has passed, measured in terms of the basic gravitational unit of time, the time for a single interaction among stars. Hence life based on gravitational forces appears to be at much the same stage in its (supposed) development as life on Earth was just a few years after the Earth had formed. Gravitationally based life could yet appear, but we shouldn't expect this to occur before *billions of billions* of years have passed, not the mere billions of years needed for life based on electromagnetic forces.

The Advantages of Being Average

We have considered the possibilities of life based on strong forces, on electromagnetic forces, and on gravitational forces. These are the basic forces we know, but we may be prisoners of our limited experience and limited imaginations. In the following chapters, we shall deal with the possibility of life based on electromagnetic forces, and usually of life on planetary surfaces. We have seen why this appears to be our best bet, but let us take the broad view and say that other types of life *may* exist. If so, they contribute a bonus to our estimate of the total number of civilizations with whom we can hope to communicate.

If we consider only life on other planets, we may indeed underestimate the number of civilizations by leaving out the possibilities of life in interstellar clouds, on neutron stars, in entire galaxies, or originating on planets but now wandering through interstellar space. Furthermore, we shall be biased not only in favor of life something like our own, but toward civilizations similar to ours. But we elect to take this risk because our current view of the universe suggests that such civilizations will be much more numerous than the other forms of life we have considered.

Even if we do restrict ourselves to looking for life on planets with time scales roughly the same as ours, we still have a great variety of possible environments. In our own solar system alone, these chances range from the ammonia clouds of Jupiter to the Sahara desert, from the Grand Canyon of Mars to the frozen methane of Pluto. Other planets in other systems may present a still greater range of environments where life would have a chance to develop. Thus, even our restriction to planets, which we do not regard as absolute, provides the possibility of finding millions of civilizations in our own galaxy, hence millions of billions of civilizations (at least!) in the universe. These numbers will provide a start in our quest to find other civilizations with whom we may have a quiet talk before we pass on.

SUMMARY

Life that is based on chemical reactions—that is, on the interaction of atoms to form complex molecules—appears to require carbon as its key structural element. Only carbon atoms can form chemical bonds with hydrogen, oxygen, nitrogen (and other less abundant elements) in a way that readily promotes the development of a wide variety of information-bearing polymers. Silicon can also form polymers, but these are too stable under ordinary conditions to serve as the basis of life. The chemical affinity of silicon for oxygen implies that at temperatures low enough for complex molecular structures to exist, silicon will be bound up as silicates, the rocks that supply our footing on this planet. If carbon is the crucial element in all chemical life, we are still not much restricted, since carbon has a high abundance everywhere in the universe.

Life also seems to require a solvent, a fluid medium in which atoms and molecules can encounter one another and undergo chemical reactions. The great ability of water to dissolve other substances makes it one of the most favored solvents. In addition, water's heat capacity, its heat of vaporization, its ability to remain liquid in a temperature range appropriate for many chemical reactions, its cosmic abundance, and its chemical stability all single it out as exceptionally well suited for use by all living organisms on Earth.

Ammonia or methyl alcohol might serve as a solvent instead of water under certain highly specialized conditions, but we expect the majority of living systems in the universe to rely on the same fluid medium that enables us to survive. The requirement for a liquid solvent imposes a restriction on the temperature range that life can tolerate, making planets in nearly circular orbits a highly favored location.

For all types of life, it seems important to have a variety of individuals. Otherwise, the processes of natural selection will not have enough material on which to operate, as it discriminates among various living creatures on the basis of reproductive success.

We can consider at least two other sorts of life besides chemical life: elementary-particle life, made of nuclei that interact through strong forces and last far less than a trillionth of a second; and gravitational life, made of huge objects so far apart that gravitational forces predominate over electromagnetic, strong, or weak forces.

Gravitational life can hardly have arisen in the 15 or 20 billion years since the big bang, because this amount of time does not allow gravitational interactions to produce structures more complex than galaxies, galaxy clusters, and stars. Neutron-star life might indeed exist, since the high temperatures and large densities on neutron-star surfaces allow all sorts of nuclei to form and break apart in tiny fractions of a second. We would, however, have a hard time communicating with any such forms of life, since the natural time scale would be about one-billionth of one-trillionth of one second, and the most favored wavelengths would be those of extremely high-energy gamma rays.

Even if we restrict ourselves to chemical life, we can look beyond planetary surfaces. Interstellar clouds of gas and dust might acquire extremely complicated structures and might be able to reach the state of self-consciousness even without undergoing any reproductive process. Such black clouds could use internal electric currents in much the same way that our own bodies do, to carry messages to and from the central "brain." But here again, the time scale required to achieve the necessary complexity may be prohibitive because of the relatively large distances among individual molecules.

QUESTIONS

1. Why does carbon seem to be the best element for building complex molecules throughout the universe?

2. Would silicon serve as well as carbon in giving structure to large molecules? Why or why not?

3. What does the large abundance of silicate molecules in comets, meteorites, planets, and stars tell us about the willingness of silicon to combine with oxygen? What are the implications of this result for forming large, complex molecules, using silicon atoms as the key structural element?

4. Why does water's high "heat of vaporization" and large heat capacity give water advantages for use as a solvent in living organisms?

5. What are the advantages of a solvent that expands when it freezes? What are the disadvantages?

6. What physical conditions make life in dense interstellar clouds an unlikely event, compared to life on planetary surfaces?

7. Why would neutron star life have fantastically short lifetimes, compared to living organisms on Earth?

8. Why would gravitationally based life be slow to originate and to evolve, compared with life on Earth?

FURTHER READING

Bernal, James. *The World, the Flesh, and the Devil,* 2d ed. Bloomington, Ind.: Indiana University Press, 1969.

Grobstein, Carl. *The Strategy of Life,* 2d ed. San Francisco: W. H. Freeman, 1974.

Feinberg, Gerald, and Robert Shapiro. *Life Beyond the Earth.* New York: Morrow, 1980.

Shklovskii, Josef, and Carl Sagan. *Intelligent Life in the Universe.* San Francisco: Holden-Day, 1965.

Science Fiction

Forward, Robert. *Dragon's Egg.* New York: Ballantine, 1980.

Hoyle, Fred. *The Black Cloud.* New York: Signet Books, 1957.

The Land is a Mother that never dies.
— Maori saying

This painting by Jon Lomberg of "The First Baby on Mars" shows a young human examining the Martian environment.

PART FOUR

The Search for Life in the Solar System

Empty space is like a kingdom, and heaven and earth no more than a single individual person in that kingdom . . . . How unreasonable it would be to suppose that besides the heaven and earth which we can see there are no other heavens and no other earths!

— TENG MU, 13TH-CENTURY PHILOSOPHER

The sun's nine planets, their satellites, and the more primitive material that resides in comets and in meteorites, provide us with numerous sites where life, or at least prebiological complex molecules, might be found. Once we could only speculate about possible sites for life outside the Earth; today we can investigate them in person and with automated spacecraft. Though we have not found any definitive proof of life anywhere except on our own planet, we have acquired vast amounts of useful information during the past two decades that may, in the fullness of time, permit us to understand why life can appear in some environments and not in others. To the extent that our solar system furnishes us with a representative planetary system, our local search for life also brings us information about a multitude of possible habitats for life within our own galaxy and beyond.

11

The Origin and Early History of the Solar System

During our own lifetimes, the search for life on other worlds has begun in earnest. Spacecraft equipped with sensitive and subtle instruments have visited all of the planets except Pluto. We have walked on the surface of the moon, and our robot explorers have landed on Mars and Venus. We have studied our sister planets not simply to discover whether or not they were inhabited but also to learn more about their present environments and what they can tell us about the past history of the Earth. As we review the results of these investigations, we should focus on the central concern of our quest: *How did life begin in our solar system, and in how many places?*

When we move to consider the search for life in the universe, we quickly encounter another question: How many of all the stars that exist—single, double, or multiple—have planets? In other words, how many solar systems like ours exist in our galaxy? We shall return to an explicit discussion of this question in Chapter 16, but we now want to concentrate on the planets we know. Our attempts to unravel their origins and histories should help us to understand how likely it is that other planets have formed around other stars, and that one or more of these planets might harbor some kind of life.

Our attempts to reconstruct the history of how the solar system formed rest on our examination of the most primitive, unaltered parts of this system, those that have changed the least during the 4.5 billion years since the solar system formed. The most easily studied representatives of the early years of our planetary system are the "comets," frozen lumps of dirty ice that orbit the sun at immense distances from us; the "meteoroids" and "asteroids," rocky or metallic bodies in roughly planetlike orbits around the sun; Jupiter, the largest planet; and Titan, the biggest satellite of Saturn. Mercury and the moon, the two members of the inner solar system whose surfaces have not been weathered by chemistry, erosion, and tectonic processes, reveal a record of the last stages of planet formation.

These objects have provided us with most of what we know about the early solar system; not enough, unfortunately, to reconstruct the entire scenario of its formation, but enough to suggest a plausible model, one that can explain most of what we have discovered thus far. We expect to find additional records of conditions in the early solar system during the next decade, when we study the atmosphere of Jupiter with an automated probe, pass close by two large asteroids and rendezvous with a comet, and send a spacecraft into orbit around Saturn that delivers a probe into the chemically evolving atmosphere of Titan.

The Formation of the Solar System

The Earth, the sun, and all the planets and their satellites began their lives when a dark condensation in an interstellar cloud of gas and dust (Fig. 11.1)

Figure 11.1 The Horsehead Nebula in Orion consists of dust-laden material photographed against a background of hot gas. Deep within the clouds of gas and dust that lie behind this obscuring veil, new stars are forming now, some perhaps with planets.

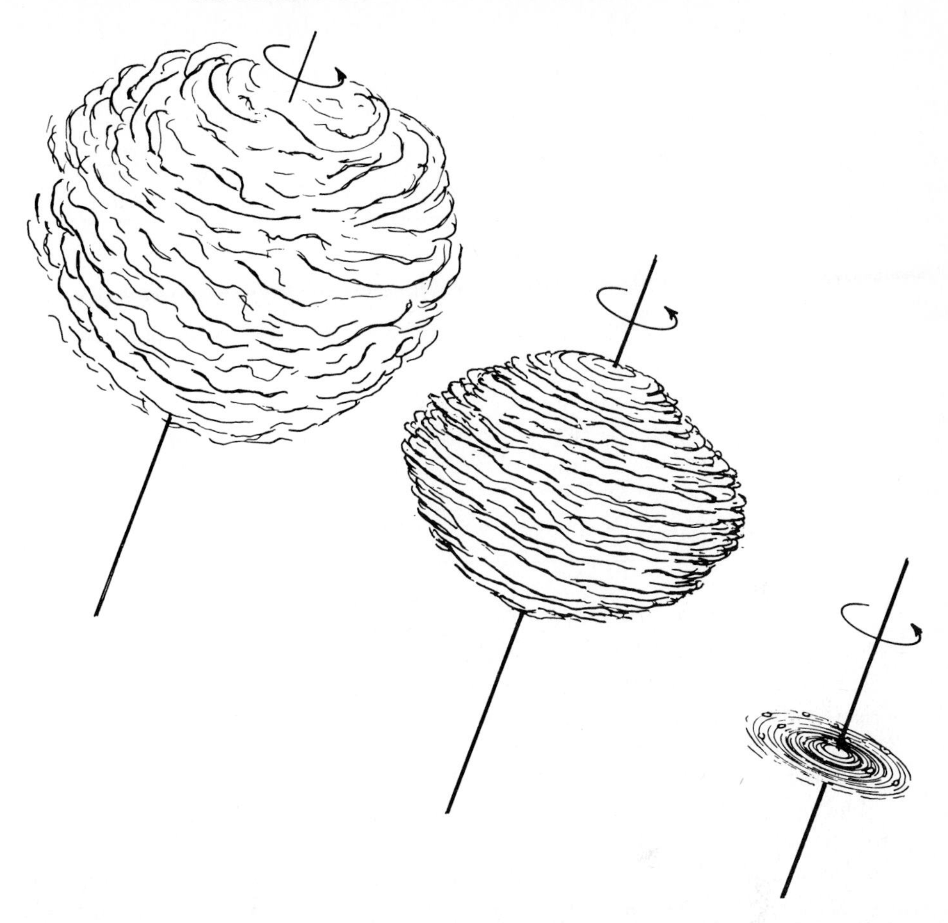

Figure 11.2 As the cloud of gas and dust that formed the solar system began to contract, it must have acquired some rotation, which led to more rapid rotation as the cloud grew smaller. This rotation tended to support the cloud against contraction in directions perpendicular to the axis of rotation, and thus led to a pancake-like shape for the contracted, rotating cloud. Within the disklike configuration, the individual planets accreted from the matter revolving at their present distances from the sun.

somehow grew dense enough to start contracting under the influence of its own gravitational forces (Fig. 11.2). All of the matter we now find organized into discrete objects was once simply gas and dust floating among the stars, only later becoming planets with long and complex histories.

When the cloud that would form the solar system first began to contract, it must have done so as a condensation with some rotation. The rotation was slow at first but grew more rapid as the cloud shrank, for much the same reason that figure skaters who pull in their arms spin more quickly. The

TABLE 11.1 Characteristics of Planets and Their Orbits

Planet	Diameter (Earth = 1)	Mass (Earth = 1)	Average Density (gm per cm³)	Orbital Period (Years)	Distance from the Sun (Earth = 1)
Mercury	0.38	0.06	5.4	0.24	0.39
Venus	0.95	0.81	5.3	0.62	0.72
Earth	1.00	1.00	5.5	1.00	1.00
Mars	0.53	0.11	4.0	1.88	1.52
[Asteroids]					[Avg. = 2.77]
Jupiter	11.2	318	1.3	12	5.2
Saturn	9.5	95	0.7	29	9.5
Uranus	4.0	15	1.3	84	19
Neptune	3.9	17	1.6	165	30
Pluto	0.2	0.002	2.0	249	39

original cloud must have had a size equal to a fair fraction of the average distance between neighboring stars, perhaps a light year, because only then could the cloud include enough matter to make a star and its planets. In contrast, the present size of the sun's planetary system equals 40 times the size of the Earth's orbit—just 1/2000 of a light year.

Only the orbits of comets remind us of the relatively vast size of the primitive cloud that produced the solar system. Most of the comets that orbit the sun reach tremendous distances at the far points of their trajectories, so that the comets as a whole have a spherical distribution around the sun and planets with a total diameter of about three light years. This "cometary halo," often called the "Oort Cloud" after Jan Oort, the astronomer who deduced its existence, represents the true outer boundary of the present solar system. As we shall see, this halo was probably established after the formation of the sun and planets.

As the cloud of gas and dust contracted toward the present size of the solar system, the cloud's rotation tended to support it in the directions perpendicular to its axis of rotation (Fig. 11.2). In other words, gravity could pull matter toward the center of the cloud more easily *along* the rotation axis than *perpendicular to* this axis. This fact caused the cloud to assume a pancake-like shape during the later stages of its contraction, as particles within the cloud began to collide more often and to follow new orbits around the cloud's center as a result of these collisions. At this stage, the cloud became what astronomers call the "solar nebula."

The combined result of the contraction and rotation was a spinning, disk-like solar nebula, within which gas and dust had much greater density than they did before the contraction began. The nebula was densest of all at its center, where the protosun began its final condensation. By the time the sun grew so dense that nuclear fusion reactions began inside it, the pancake-shaped cloud had begun to form agglomerations at various distances from its center. The rather regular spacing of the planets' orbits from the sun (Table 11.1) apparently reflects the way in which matter accumulated within the disklike configuration.

The configuration of the planetary orbits also represents a natural result of the formation process. The nine planets orbit the sun in nearly circular trajectories that all (except for Mercury and Pluto) lie in very nearly the same plane (see again Fig. 2.1). The sun contains 99.9 percent of the mass in the solar system, and the four giant planets, with Jupiter by far the leader, have the bulk of the 0.1 percent residue. The Earth, largest of the four inner planets, has only 1/318 of Jupiter's mass and 1/329,000 of the sun's mass.

The four giant planets differ most strikingly from the four inner planets (Mercury, Venus, Earth, and Mars) in their size and composition. *The giant planets are large, gaseous, rarefied, and hydrogen-rich, while the inner planets are small, rocky, dense, and hydrogen-poor.* Because the giant planets consist mostly of hydrogen and helium, they resemble the universe at large. The inner planets are distinctly different: Though the universe consists mostly of hydrogen, the Earth does not.

A relatively simple explanation exists for the extreme differences between the four giant planets and the four inner planets. As nuclear-fusion reactions began in the sun's deep interior 4.5 billion years ago, the solar nebula close to the sun grew much warmer than the dust and gas at greater distances. This warming had a profound effect on the kinds of material that could condense and accumulate into "planetesimals," the small objects that can collide to form planets.

At distances close to the sun—less than about five times the Earth-sun distance (5 astronomical units, or A.U.)—the sun's heat prevented ice from forming. This fact had significant consequences, since *ice is potentially the most abundant solid in the universe.* We may be surprised by this statement, since we live on a rocky planet, but another look at Table 7.1 will show that it is true. As the table illustrates, *oxygen is 20 times more abundant than silicon, the most abundant rock-forming element.* Any object can easily exhaust all its silicon when this element combines with oxygen to make rocks—on Earth the proportions seem to be about three oxygen atoms to one silicon atom—and still have plenty of oxygen left over to make ice by combining with hydrogen.

The lesson of Table 7.1 is that if you seek to accumulate large masses of solid material from a cosmic mixture of the elements like the solar nebula, you have a much better chance of doing so in a region where you can make ice as well as rock. This rule was apparently obeyed during the formation of the solar system, since we find the giant planets only at distances greater than 5 A.U. from the sun, where ice could condense. We may therefore think of the four inner planets as being made through collisions of rocky planetesimals, producing the relatively small and dense objects that we find today. But in the outer solar system, the planetesimals were made of rock *and ice,* and they built the relatively huge cores of the giant planets. These cores in turn attracted hydrogen and helium gas from the surrounding nebula. The asteroids that we find today probably resemble the planetesimals of the inner solar system, while comets provide a good match to the icy planetesimals, rich in organic compounds and laced with silicate dust, that made the cores of the giant planets.

The resulting difference in the masses of the forming planets was decisive to their further histories. The inner planets never became large enough to attract and to hold hydrogen and helium, the most abundant gases in the solar system and the universe. These inner planets therefore remained small (Fig. 11.3). They developed "secondary atmospheres" from gases released during their formation and their subsequent geological histories. But in the outer solar system, the massive cores built from icy material could capture hydrogen and helium (and everything else!) from the solar nebula around them. They grew steadily larger and became the giants we find today, with "primitive atmospheres" rich in hydrogen and helium.

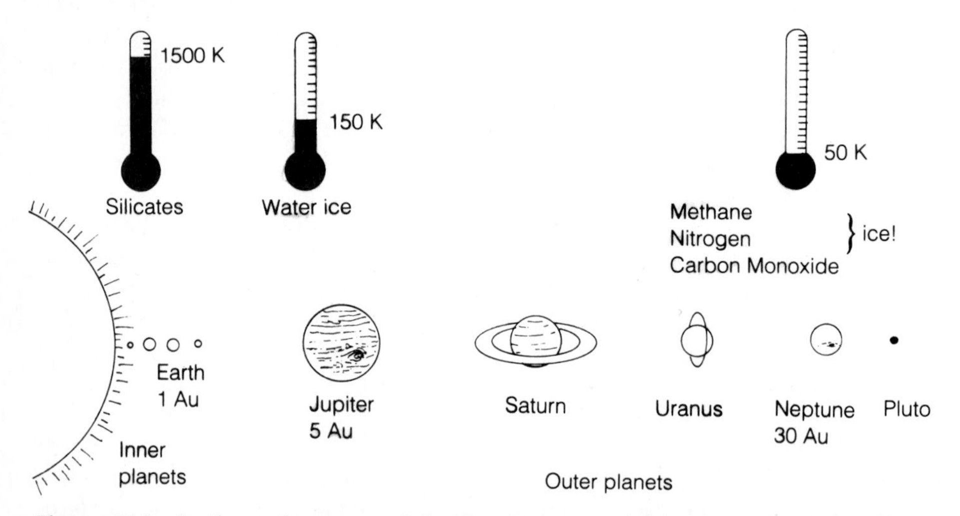

Figure 11.3 In the early stages of the development of the solar nebula, the high temperatures near the sun allowed only rocks to condense, and these accreted to form the small, dense inner planets. In the cooler, outer parts of the nebula, water could condense to form ice, so both rocks and ice were available to build the massive cores of the giant planets. These cores, in turn, could capture hydrogen and helium from the nebula to build the thick atmospheres we find on these objects.

If today we were to move Jupiter to the Earth's distance, only one-fifth as far from the sun as it actually is, Jupiter could still retain its hydrogen and helium, because it has 318 times the Earth's mass and exerts a great deal more gravitational force. But Jupiter could *grow* to this mass only at five times the Earth's distance from the sun. The planet that formed at 1 A.U., our own Earth, was never able to acquire such an enormous mass. If we could add the missing hydrogen and helium to the Earth, so that the ratio of these light gases to the rock-forming element silicon would be the same on Earth as it is in the sun, we would create a planet with the mass of Saturn. This exercise provides a graphic demonstration of the basic difference between the inner and outer solar system. We live on a sort of cosmic cinder, from which most of the cosmically abundant volatile (easily vaporized) elements are missing.

Even the giant planets Uranus and Neptune, with "only" 15 and 17 times the Earth's mass, seem to have lower abundances of hydrogen and helium than do Jupiter and Saturn, with 318 and 95 times the Earth's mass. This implies that Uranus and Neptune were unable to acquire solar abundances of hydrogen and helium from the surrounding nebula as they formed at their immense distances from the sun. We don't yet know why this happened. Perhaps the solar nebula was simply too thin near its outer boundary to supply the amount of gas necessary to make atmospheres as thick as those of Jupiter and Saturn.

The distribution of planets in *any* planetary system into one group of dense, rocky inner planets and another of rarefied, gaseous outer planets seems probable, if our own solar system formed in a standard way. We shall find support for this assertion when we study the satellites of Jupiter (Chapter 15), since the inner moons of this giant planet are denser than the outer ones. Jupiter is so large that it warmed the space around it as it formed, producing a separation of rock and ice similar to that which occurred around the sun. We expect the same effect in other planetary systems, although we do not have even one other example yet that could be used as a test (Chapter 16).

Comets

The small objects most representative of the primitive solar system are the comets, whose highly elongated orbits show no concentration toward the disk of the solar system; instead, cometary orbits have a spherical distribution around the sun. Exactly how the comets achieved this distribution remains unclear. The most popular current notion is that comets formed among the present giant planets in the outer solar system and were later scattered into their present orbits by close encounters with the forming planets. From the observations of Halley's comet made in 1986 by Soviet, European, and Japanese spacecraft, we now know that the nucleus of this comet is an irregular lump of dirty ice, 16 kilometers across. Most comet nuclei are no larger than the island of Manhattan. But some are much bigger, approaching the dimensions of Denmark or Holland. Still larger comets may exist. Pluto and Triton (Neptune's large moon) show striking resemblances to comets (page 372).

The composition of comets was successfully predicted some 40 years ago by Fred Whipple, who developed a "dirty snowball" model that provided an excellent explanation of cometary behavior. The "snow" in a comet consists mostly of ordinary water ice, plus solid carbon dioxide, methyl alcohol, and other as yet unknown frozen gases, probably including more complex compounds such as formaldehyde and cyanoacetylene. The cometary "dirt" consists of grains of organic and rocky material of various sizes that have apparently undergone no melting or other transformations, since they are still imbedded in the cometary ices. Astronomers think that this lack of chemical processing ensures that comets represent pristine samples of the original material from which the solar system formed, and that unaltered grains and molecules from the interstellar medium may be trapped in their snowy interiors.

For most of its life, each cometary snowball orbits slowly around the sun at immense distances, at least a thousand times farther than frosty Pluto (Fig. 11.4). Far from the sun's warmth, the snowball remains in a cosmic deep freeze only a few degrees above absolute zero. Billions, perhaps trillions, of comets orbit the sun, but we know nothing about them until one

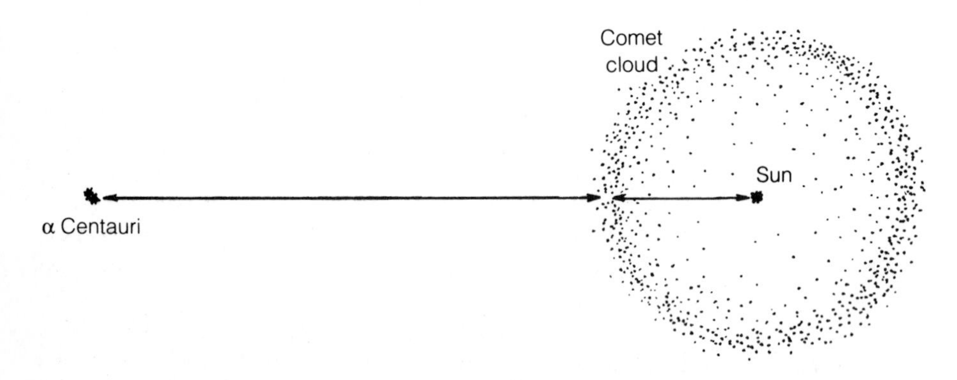

Figure 11.4 Most comets have tremendously elongated orbits that keep them hundreds or thousands of times farther from the sun than any planetary orbits, which are too small on this scale to be visible. The Oort Cloud of comets reaches a sizeable fraction of the distance to Alpha Centauri (Fig 16.1), the nearest star!

of them happens to acquire a new orbit, perhaps through the gravitational attraction from nearby stars, that takes it toward the inner solar system.

In this case, nothing much happens to the comet until it reaches a distance from the sun comparable to Saturn's, when some of the ices start to vaporize, and the gas and dust released from the snowball nucleus spread out to form a fuzzy envelope called the "coma." As the comet approaches still closer to the sun, more and more gas and dust are released, trailing behind the nucleus as the spectacular tail, often many millions of kilometers in length, that gives a comet its beauty (Fig. 11.5). Some of the molecules absorb high energy ultraviolet photons from the sun, losing electrons in the process. Thus the light from the gaseous part of the comet's tail comes mainly from CO^+ ions. Other molecules, such as H_2O, sublime from the nucleus and are split apart by this UV radiation, so we find H and OH in the coma, and H_2O^+ in the tail. At this stage, the hydrogen atoms in the coma have expanded so far into the vacuum of space that they can create a cloud around the nucleus with a diameter larger than that of the sun. Thus

Figure 11.5 Comet Mrkos, named after its discoverer, appeared in 1957. Streaming out behind the comet's nucleus are a straight tail of ionized gas and a curved tail of dust particles.

solar heating transforms the dark, ugly chrysalis of the comet nucleus into a magnificent gauzy butterfly that briefly becomes the largest object in the solar system.

Some comets have two tails, one made of dust and the other of gas, and occasionally they may have several tails of each kind. The gravitational attraction of the sun and of the giant planets can wreak havoc in comets as they traverse the planetary part of the solar system. Astronomers have seen some comets break apart as they pass close by the sun, with two or more distinct pieces following the original comet's orbit. Sometimes the gravitational attraction from one of the giant planets can deflect a comet into an orbit much smaller than its original one. Such "short-period" comets, of which many dozens are known, take only a few years to orbit the sun once, as compared with millions of years for the original "long-period" comets. Halley's comet, most famous of all for its regular returns at 76-year intervals, has an intermediate orbit that carries it past Neptune but nowhere near the immense distances of the long-period comets (Fig. 11.6).

Comets offer us a chance to investigate primitive solar system material, formed directly from the original gas and dust that made the solar system and changed little during the past 4.5 billion years. They even do us the favor of bringing that well-preserved ancient material near the Earth when a

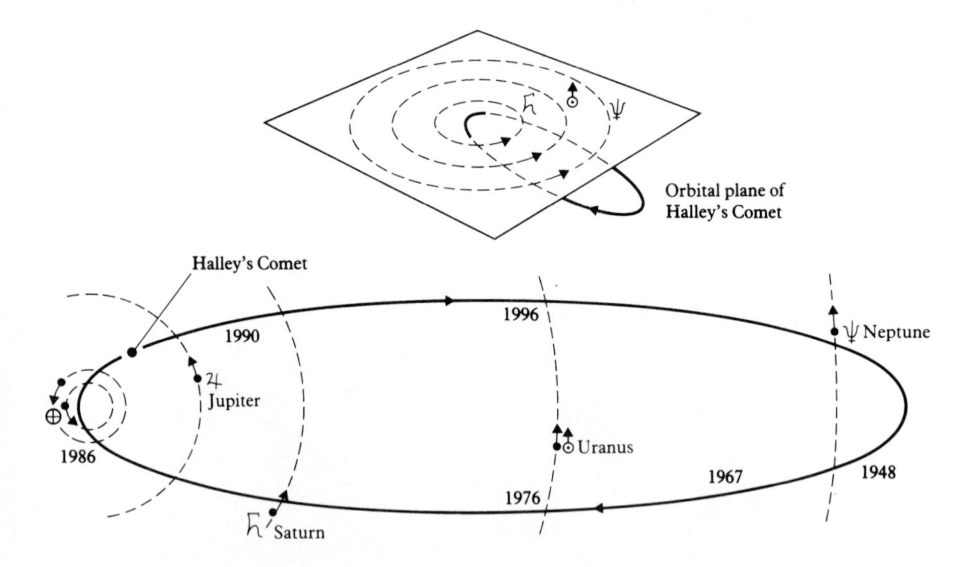

Figure 11.6 Unlike the planets, comets have orbits that are inclined at all possible angles. For example, Halley's comet, depicted here, orbits the sun in the opposite direction to the planets; that is, its orbit is inclined by more than 90° to the planets' orbits.

long-period comet, perhaps on its first trip close to the sun, passes within a few million kilometers of our planet. Even a tough-skinned comet that has made hundreds of orbits should contain material deep inside the nucleus that has been protected from high temperatures, ultraviolet irradiation, and cosmic-ray bombardment, and thus is more primitive than any matter in the inner solar system. For this reason, scientists were eager to send space probes to Halley's comet during its last close encounter with the sun in 1986.

Through spectroscopic analysis of the light that comets reflect, and through the use of radio telescopes to detect the microwave radiation emitted by cometary gases, astronomers have found a number of simple molecules in comets, including fragments of more complex substances. Since comets are believed to have condensed in the earliest stage of solar system formation, we might expect to find within them some of the molecules identified in interstellar clouds (see again Table 4.1). Indeed, the discovery of water, carbon monoxide, hydrogen cyanide (HCN), methyl alcohol (CH_3OH), and methyl cyanide (CH_3CN) in comets strengthens the assumed connection between these icy objects and the interstellar clouds from which they may have condensed (Table 11.2).

The same may be said for the recent discovery of carbon dioxide (CO_2) in the interstellar medium, since this gas has long been known in comets. Ground-based and space-borne spectroscopic observations have also revealed the presence of a cluster of emission lines caused by some substance or blend of substances that must contain C-H bonds. But the exact identification of this organic material remains elusive. It is present in every comet that has been tested for it, including Halley's comet. A similar, but not identical, feature also appears in some spectra of interstellar clouds.

To see whether comets really contain unmodified interstellar compounds, much more work must be done. For example, we would like to know the "parent" molecule of the fragment we observe as C_3. Could it be

TABLE 11.2 Molecules Detected in Comets

Coma			Tail	
H_2O	CN	NH_2	H_2O^+	N_2^+
HCN	CH	C_3	CO_2^+	CO^+
CH_3CN	OH	CO_2		CH^+
NH_3	NH			OH^+
H_2S	C_2			
CH_3OH	CS			

C_4H_2, or HC_5N? And what about the simplest amino acids that are produced in Miller-Urey experiments? Could some of them exist in cometary nuclei? These possibilities could be investigated by a series of spacecraft missions that would first sample the gases close to the comet's icy nucleus, and eventually bring back pieces of comets for study in our Earth-bound laboratories. We would then know what kinds of compounds these icy messengers might have brought to the surface of the early Earth, perhaps to help the formation of the atmosphere, perhaps even to give a head start to some of the chemical reactions that led to the origin of life (see Chapter 8).

Unfortunately, the 1986 missions to Halley's comet did not provide much help in answering these specific questions. Spacecraft observations verified the existence of a solid, icy nucleus at the comet's center, and they revealed that the nucleus is covered with a dark layer of material that is most probably rich in organic compounds (Fig. 11.7). Nothing else would

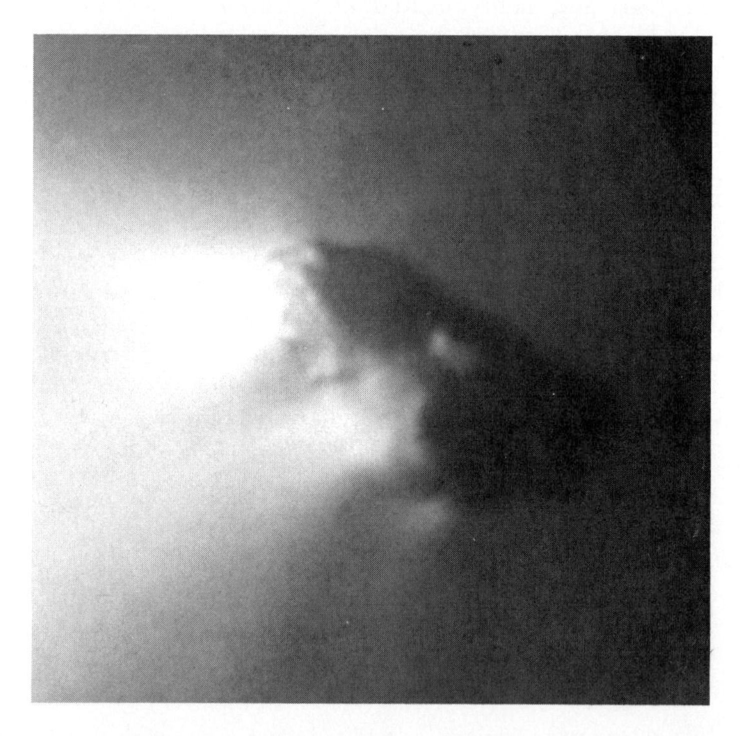

Figure 11.7 The nucleus of Halley's comet is about twice the size of Manhattan. This picture was produced from 60 separate images obtained by the Giotto spacecraft in March 1986. The smallest details are only 60 meters across. Jets of dust and gas that produce the comet's spectacular tails emanate from at least three bright regions along the left (sunward) side of the nucleus.

be as dark as the surface of Halley's comet! When the spacecraft instruments analyzed the dust particles in the coma, some indeed proved to be silicates, while others were purely organic in composition. But the spacecraft sent on these missions were not equipped to carry out the detailed chemical analyses required to answer the questions we have raised. For that we must await the new mission called *Comet Rendezvous Asteroid Flyby* (CRAF), which should reach another short-period comet in the year 2006. As the name of the mission implies, the CRAF spacecraft will rendezvous with the nucleus of the comet, monitoring its increasing activity as the comet approaches the sun. In addition to making remote measurements of the nucleus and studies of the composition of gas and dust in the coma, scientists are exploring ways of using CRAF to carry out direct investigations of the properties of the nucleus itself (Fig. 11.8).

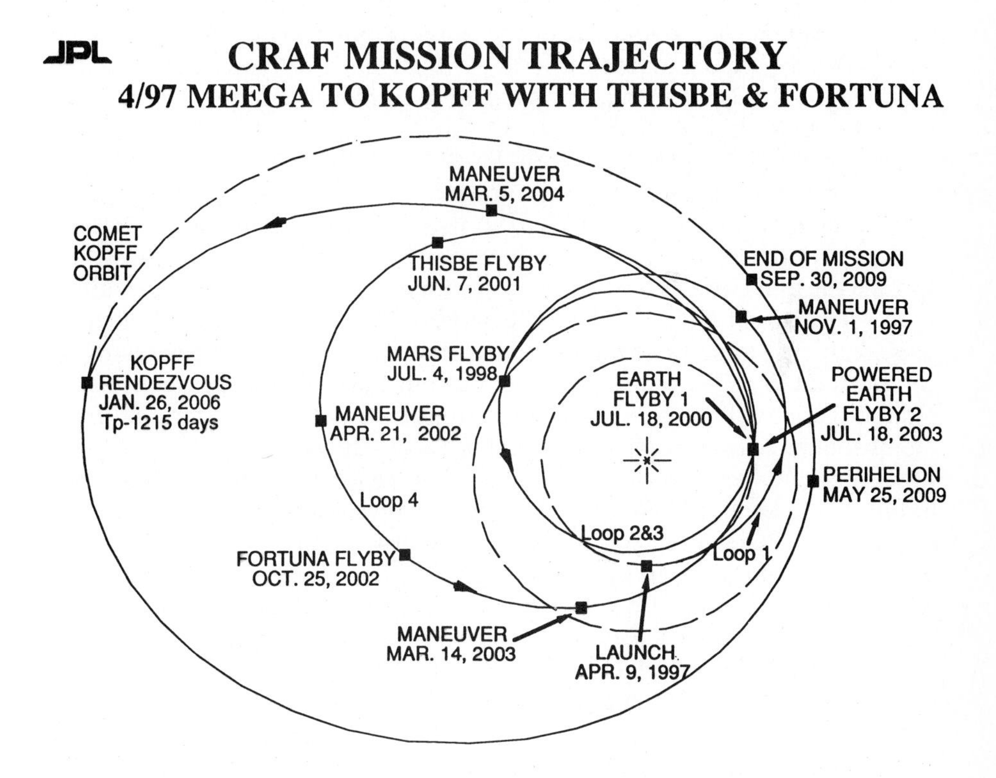

Figure 11.8 The CRAF spacecraft must follow a complicated trajectory in order to rendezvous with a comet and to stay with it along its orbit. Unfortunately, the future of this mission is in grave doubt.

Asteroids, Meteoroids, and Meteorites

We have seen that comets, which spend most of their lives in the frozen depths of space at the boundaries of the solar system, represent the best-preserved primitive material that is relatively accessible to us. Another source of somewhat less primitive matter resides in the lumps of debris called asteroids and meteoroids, which, like short-period comets, orbit the sun in rather elongated trajectories, but which always remain fairly close to the sun (see again Fig. 2.1). The "asteroids," ranging in size from a few objects hundreds of kilometers across down to millions of objects less than 1 kilometer across, orbit the sun, between Mars and Jupiter. Asteroids are thought to represent some of the remaining pieces of a would-be planet that never could form, because Jupiter's gravitational forces kept the material from collecting into a single large object. Most of this material has long since left the scene—crashing into Jupiter, scattered out to great distances or colliding with the inner planets—since the total mass of the remaining asteroids is less than one-tenth of the mass of our moon. These surviving planetesimals could provide us with a valuable set of fossils from the era 4.5 billion years ago, when planets began to form but had not yet grown to their present sizes.

Spectroscopic observations from Earth reveal that asteroids have a wide range of compositions—including some made mainly of iron, some made of rocky material that has been melted in its early history, and some that are dark, carbon-rich bodies that may resemble the satellites of Mars (Fig. 13.16). Our first close look at an asteroid came in late 1991, when the Galileo spacecraft passed Gaspra, one of the rocky objects, on its way to Jupiter (Fig. 11.9). The spacecraft made remote measurements of Gaspra's surface composition and physical properties. But our best knowledge about asteroidal material will probably still come from the meteorites, asteroidal fragments that have been recovered from the Earth's surface.

Meteoroids are basically little asteroids, and astronomers make a distinction between the two only because "meteoroids" are relatively small and have orbits around the sun that cross the Earth's orbit, so that they have the potential to collide with the Earth, if the orbits ever intersect. When a small meteoroid encounters the Earth's atmosphere, its large velocity relative to the Earth, which arises from the difference between its orbit and the Earth's, subjects it to great frictional heating in our planet's upper atmosphere. This heating consumes any small meteoroid, a process that we see as the appearance of a shooting star or meteor. If the meteoroid begins with a sufficiently large mass, it can survive its frictional passage through the atmosphere, and its remnant will reach the Earth's surface as a "meteorite."

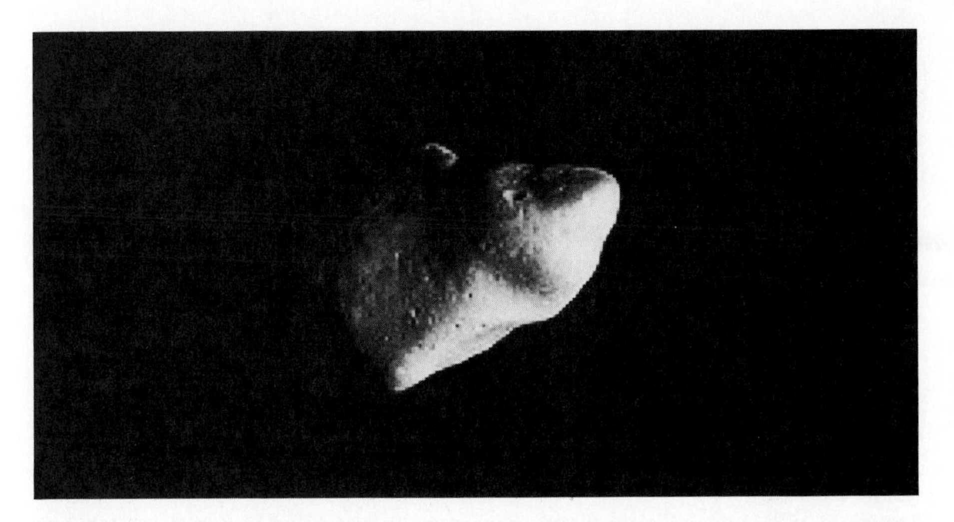

Figure 11.9 On October 29, 1991, the *Galileo* spacecraft passed the asteroid Gaspra at a distance of 16,200 kilometers and sent this photograph back to Earth. The illuminated part of the asteroid is about 16 by 12 kilometers, and the smallest visible craters are about 300 meters across.

Bombardment of the Inner Planets: A Threat to Life?

The surfaces of Mercury, Venus, Mars, and the moon are dotted with many thousands of craters, most of which were made within a few hundred million years after the solar system formed. During this epoch, meteoroids by the millions must have rained down on all the planets of the inner solar system as the last chunks of matter to reach the planets encountered the new planetary surfaces. This early bombardment included comets as well as large objects we would normally call asteroids. A few of these early asteroids may have been as big as the planet Mars. The early bombardment essentially ended some 3.5 billion years ago. Our own planet shows only a few large meteorite craters (Fig. 11.10), representing relatively recent impacts. Although the Earth surely did not escape this rain of terror, the first few hundred million years of our geological record have vanished because of erosion and the movement of the crustal plates.

On August 10, 1972, a reminder of what must have occurred far more often in bygone eras flashed through the skies over Wyoming as a meteoroid burned its way through our atmosphere, missing the surface of the Earth by just 58 kilometers (Fig. 11.11). This object was only four meters across, though it weighed a thousand tons and would have been devastating upon

Figure 11.10 Meteor Crater in Arizona, more than a kilometer in diameter, was formed about 20,000 years ago by a meteorite with a mass of many thousand tons. The impact released as much energy as a 20-megaton hydrogen bomb.

Figure 11.11 The meteoroid that nearly collided with Earth in August 1972 had a speed of 15 kilometers per second, relative to Earth, as it passed through the atmosphere above Wyoming. It is seen here as a streak of light above the Grand Teton mountains.

impact. A related object, apparently a piece of a comet, did hit the Earth in 1908 in Siberia, felling trees for many kilometers around. The impact of the meteoroid of 1972, small as it was, would have released about the same energy as the atomic bomb that destroyed Hiroshima.

An alert reader might be asking at this point why contemporary meteoroids are necessarily small. What prevents an average-size asteroid from hitting the Earth from time to time? The answer is: nothing! The smaller debris is simply more plentiful, but many large projectiles certainly struck the planet in its early history (as we see from the moon's surface, which was subject to a similar bombardment) and we now have good evidence that an asteroid about 10 kilometers across struck the Earth just 65 million years ago.

This startling idea was advanced by Walter Alvarez and his collaborators, who noticed an unusually high concentration of iridium in sedimentary rocks that were deposited 65 million years ago. This epoch marks the boundary between the Cretaceous and Tertiary periods of geological history. Iridium appears in relatively low concentrations at the Earth's surface, because it is a heavy element, and most of it sank toward the center when the Earth was younger and hotter. Hence the high concentration of iridium (still only a tiny fraction of the total) found in rocks formed at a certain epoch wherever they are found on Earth (Fig. 11.12) indicates that some singular event enriched the abundance of iridium at the Earth's surface at this time. The impact of an iridium-rich asteroid would do the trick, and analysis of meteorites indicates that such asteroids must exist.

What makes this especially relevant to our concerns is the fact that *the transition from the Cretaceous to the Tertiary was marked by the wholesale extinction of entire species of plants and animals.* Roughly half of all species existing on Earth at that time vanished in a geologically short period. These "mass extinctions," which included the famous species of dinosaurs, had been known for many years, but their cause was not well understood. While there is still some controversy about the interpretation of the iridium abundances, it now seems likely that the impact of the asteroid wreaked havoc on a global scale, perhaps through a severe temporary change in climate caused by tons of dust lofted into the atmosphere, perhaps through widespread fires.

The "mass extinction" at the end of the Cretaceous period is just one of a dozen or more such mass extinctions found at intervals of about 30 million years through the fossil record of life. Though only the Cretaceous-Tertiary extinctions have been thoroughly linked with an impact signalled by increase in the abundance of iridium it is possible that all, or nearly all, of the mass extinctions arose from large impacts with the Earth.

We thus encounter an important factor to consider in the origin and evolution of life on Earth or any other planet: bombardments from outer space. Evidently we should not expect life to develop until most of the early bombardment is over. Even then, life will be at risk from occasional impacts

Figure 11.12 A layer of clay just a few inches wide marks the boundary between the Cretaceous (K) and Tertiary (T) eras on Earth. Here we see the layer exposed on a cliff wall near Trinidad, Colorado. It was during the time that this clay was deposited that massive extinctions of various types of life occurred on Earth. The clay is strongly enriched in iridium and other metals, suggesting an extraterrestrial origin for part of this material.

by asteroids of 10-kilometer or larger diameter. Furthermore, it *might* turn out that those planets where mass extinctions periodically "clear the undergrowth" for new species to dominate will allow evolution to proceed more rapidly than it can on planets where no mass extinctions occur. A precondition for an inhabited planet may be a planetary system in which the clearing out of debris moving in eccentric orbits—the debris that produces impacts with inner planets—happened early. And a precondition for a planet to evolve intelligent life sooner rather than later could be that at long intervals, large impacts did recur, eliminating most species of life, which were then replaced with new species.

Meteorites

From examination of all the meteorites found on Earth, scientists have developed a general classification scheme based on the chemical and mineralogical composition of these objects. Most of the meteorites that have been found are "stony," basically lumps of rock; a minority are "stony-iron," with metal-rich inclusions; and a few are made mostly of iron, nickel, and other metals. Radioactive dating of meteorites gives maximum ages of 4.5 billion years, which establishes the age of the solar system itself. Most interesting of all is the subclass of stony meteorites called "chondrites." These contain rounded inclusions, noticeably distinct from the rest of the material, which are called chondrules. And of the chondrites, by far the most significant for us are the "carbonaceous chondrites," in which as much as 5 percent of the mass may consist of various types of carbon compounds. Since these objects show the least amount of modification by heating they are the most primitive of the meteorites.

Within the class of carbonaceous chondrites, the most primitive examples contain the highest percentage of carbon, nitrogen, and water of all the meteorites. Some scientists believe that these primitive carbonaceous chondrites may actually be fragments of old comets rather than debris from the asteroid belt. Whether or not this is true, we do not need to wait for samples of a comet to be returned by some future mission to study the products of prebiological organic chemistry in space. Instead, we can begin this study with the compounds found in the various types of carbonaceous chondrites, and such investigations have already given significant results.

Amino Acids in Meteorites

Testing meteorites for organic compounds made outside the Earth has always suffered from the long wait between the time when a meteorite falls to Earth and the time that it is found. Many different paths of contamination can then add Earth-made organic molecules to what may have been primitive organic molecules. Luckily, however, scientists have occasionally been able to recover carbonaceous chondrites soon after they fell to Earth. The first of these, the Murchison meteorite that fell in Australia in 1972, was recovered, at least in part, on the following morning (Fig. 11.13).

Detailed analysis of the Murchison meteorite revealed the presence of 74 amino acids, the basic building blocks of proteins. Eight of these amino acids appear in the proteins of living organisms on Earth, eleven have other roles in terrestrial biology, and the remaining 55 are clearly extraterrestrial. The fact that so few of the amino acids ordinarily appear in terrestrial life suggests that contamination of the meteorite had not occurred. Two further tests verified this conclusion. First, the Murchison amino acids showed equal amounts of left-handed and right-handed molecules, while Earth-made amino acids are almost all left-handed within a living organism (see again Fig. 7.2), and become equal mixtures of left- and right-handed molecules only if undisturbed for hundreds of thousands of years. Second,

Figure 11.13 The pieces of the Murchison meteorite, recovered soon after they fell in 1972, contain 74 different kinds of amino acids, and all five nucleotide bases, as well as many other organic compounds of biochemical interest.

measurement of the ratio of the carbon isotopes in the meteorite, carbon-12 and carbon-13, showed only 88.5 times as much carbon-12 as carbon-13. Living creatures on Earth exhibit an abundance ratio of these two isotopes that varies between 90 and 92, a small deviation from 88.5 but enough to prove the extraterrestrial origin of the carbon in the amino acids of the Murchison meteorite.

The discovery of these amino acids, later verified in the similar Murray meteorite, shows that amino acids could form in a relatively hostile environment—the frigid wastes of interplanetary space—from basic ingredients during the early history of the solar system. This formation of amino acids tends to confirm our basic scenario for the origin of life on Earth, in which these necessary precursors for proteins form naturally from common ingredients on this planet. Further confirmation comes from Cyril Ponnamperuma's discovery of guanine, adenine, uracil, cytosine, and thymine—all five of the cross-linking bases found in DNA and RNA molecules in the Murchison meteorite. Fatty acids and other "life-related" molecules have also been found in carbonaceous chondrites. Therefore the chemical processes that produce important compounds for the origin of life seem to occur in various natural environments, and we can expect these processes to be widespread in the universe.

Comets, too, may contain life-related molecules, and this possibility has nourished a very controversial suggestion that comets may even have seeded the Earth with plagues as they passed by (see page 97). A more reasonable speculation in this vein suggests that meteoritic and cometary material, which was surely falling on the early Earth, may have provided some ready-made organic molecules to add to those being generated in our planet's primordial environment. Did these organic compounds and the temporary reducing conditions produced by their arrival give the origin of life a head start, as we have speculated in Chapter 8? What was happening on the surface of our planet during this intense early bombardment?

Mercury and the Moon

As we have mentioned previously, the first 700 million years of our planet's history have vanished as a result of volcanic activity, erosion, and the slow movements of the crustal plates. Thus to try to reconstruct that history, to see what kind of meteoritic and cometary bombardment our planet must have suffered in the last stages of its formation, we must turn to objects that are small enough to have very little internal geological activity and no atmospheres. In the inner solar system, the best candidates in this category are Mercury and the moon. Our satellite may have additional importance for considerations of the origin and persistence of life on Earth, as we shall see.

We expect the rocks on the moon and Mercury to be more primitive than those on Earth because of the absence of erosion and the comparative absence of disturbances of the surface by volcanoes, mantle plumes, motions of crustal plates, etc. These geological processes are manifestations of the movements of molten rock, requiring a big heat engine which in turn requires a relatively large planet that can release enough heat from radioactive rocks over long enough times to make these processes work. Even Mars, larger than Mercury or the moon, does not have as much geological activity as the Earth or Venus because the heat released by radioactivity is conducted more quickly to the surface from which it rapidly escapes into space. Hence these small planets still preserve large segments of their primitive crusts, showing the effects of the early bombardment that must have affected the Earth as well.

From studying Mercury and the moon, however, we have learned that as far as life goes, they are hopeless. The closeness of Mercury to the sun raises the planets' temperature well above 300° C on the day side, and bathes the

Figure 11.14 The surface of Mercury, seen in a mosaic of photographs taken by *Mariner 10* in 1974, shows a rugged terrain much like that of the moon.

surface more intensely than any other in the sun's ultraviolet radiation. The moon has similar problems, despite its greater distance from the sun.

Mercury's surface resembles the heavily cratered areas of our natural satellite (Fig. 11.14). It also reflects light in much the same way that lunar rocks do, suggesting that Mercury's surface and the moon's consist of much the same sorts of material. The 59-day rotation period of Mercury is locked into its 88-day orbital period, in a delicate sort of resonance in which Mercury rotates exactly *three* times for every *two* orbits around the sun. Therefore, each part of the planet eventually faces the sun's blistering heat before reentering a long night, during which the temperature falls as low as –150° C! Explorers who might land on Mercury to examine its primitive rocks would have to protect themselves against both outlandish heat and sub-Siberian cold, if they planned to remain for several months.

The Early History of the Earth and the Moon

Nearly four centuries ago, the great Italian astronomer Galileo Galilei first discovered that our satellite has a rough surface. The mountains and valleys that he thought he saw on the moon were soon found to consist mainly of craters and their remnants. Astronomers and geologists debated the craters' origin for more than three centuries. Are they the result of lunar volcanos? Are they produced by gas bubbles that rose through molten rock and burst on the lunar surface? Or are they the scars left by the impact of rocky projectiles? Galileo saw that the moon's brighter areas, called the uplands, have more craters than the dark areas, called "maria" (the Latin word for seas) because they reminded him and other early astronomers, incorrectly, of terrestrial oceans (Fig. 11.15).

A clue to resolving the puzzle of the moon's craters comes from the recognition that the difference in the density of craters between the uplands and the maria might be a result of the different ages of these regions. If craters were formed by impact, and if the number of impacting objects decreased with time, then we can explain the difference in cratering *if* the more heavily cratered uplands are older than the maria, which might have been covered with new lava sometime after the moon formed. All of these statements turn out to be true.

Figure 11.15 The bright lunar uplands have far more craters than the dark lunar maria, which we know to be old lava flows that fill enormous impact basins.

Like Mercury, the moon is too small to sustain an atmosphere, or to produce plate motions on its surface of a scale that would remove the early crust. Thus we can again read the record of early events in the solar system by studying the lunar landscape and its rocks. We find the scenery is the same as that on Mercury, countless craters that are the silent relics of bombardment by countless objects, ranging in diameter from a few microns up to tens of kilometers. The energy of motion carried by these asteroids and comets caused them to explode upon impact, thus producing circular craters, regardless of the angle of impact. The heavily cratered surface of the moon bears witness to the fact that its crust must have formed early in the history of the solar system. But did the moon form close to the Earth?

Was the moon captured into its present orbit after it had formed? Or did the moon divide from the Earth? If either of these possibilities did in fact occur, catastrophic effects on the Earth's surface must have resulted, which might have had dramatic consequences for the possible origin of life. Many scientists thought that the arguments about where our moon came from would be answered with certainty after detailed exploration of the moon's surface made from July 1969 through December 1972.

Human Exploration of the Moon

Part of the cultural heritage of the human race seems to be a desire to explore the moon and to speculate about what might be found there. During the 1960s, as instruments circled the unseen side of the moon, landed on its surface, and made tentative chemical analyses, arguments raged not only about the origin of the moon but also about the possibility, admittedly remote, that astronauts returning from the moon might bring back some strange lunar microbes, capable of infecting Earth with unknown diseases.

Most scientists felt it impossible that life could exist on the moon, but a vocal minority, led by Carl Sagan, pointed out that the moon might have shared some of the Earth's early history, which could have included a primitive atmosphere and abundant water, and that nature seems remarkably adept at making at least the first steps toward life throughout the universe. (In view of the later discovery of complex molecules in interstellar clouds, and of amino acids in the Murchison and Murray meteorites, Sagan seems to have had a good argument.) The opposing viewpoint held that the moon's low gravity could not have held volatile compounds such as water long enough for life to develop; even if somehow life had evolved and had survived as spores in the lunar subsoil, such life would not be able to interact with organisms on Earth because of differences in chemical structure between the two types of life. In a memorable public exchange, Edward Anders, an expert on meteorites, offered to eat the first dust brought back from the moon as a demonstration of his confidence in the sterility of the lunar surface!

Still, nobody *knew* whether life existed on the moon, and no one could prove what the effects of bringing such life back to Earth would be, so elaborate precautions were taken to protect the astronauts and life on Earth. This episode has more than passing significance, since we now face precisely the same arguments about samples from Mars and from comets. And in some distant time when we first make physical contact with an alien civilization, we must wonder whether their air and soil contain diseases or poisons for us, or ours for them.

During the 15 years between 1957, when the first artificial satellite was launched, and 1972, when the last manned spacecraft left the moon, more than 50 spacecraft approached the moon or landed on its surface. Twelve astronauts from our planet walked there, gathering over 800 pounds of lunar rocks and dust, and set up experimental equipment that continued to gather data long after our explorers left (Fig. 11.16). One of the first discoveries to come from inspection of rock samples from the moon proved to be one of the most important: The rocks in the dark lunar maria are quite similar to terrestrial basaltic rocks (Color Plate 10). This fact shows that the moon has not always been cold; rather, it must once have been hot enough to produce magma, molten rock that later crystallized into basalts. Studies

Figure 11.16 Manned exploration of the moon during the early 1970s allowed the examination of our satellite to a degree undreamt of only a generation before, and again beyond our reach today.

of lunar rocks in terrestrial laboratories showed differences in composition between lunar rocks from the various maria and from the lunar uplands, and revealed that *all* of these rocks are distinctly different from terrestrial rocks. In particular, the lunar rocks show abundances of volatile elements (those that boil at low temperatures) that are hundreds of times smaller than those in terrestrial basalts.

This difference has great significance. The Earth already shows a notable deficiency in volatile elements, as compared to the abundances of elements in the giant planets and in stars, and the deficient elements on Earth include hydrogen, carbon, and nitrogen, which are vital to life as we know it. A much lower abundance of volatile elements on the moon tends to rule out the possibility of life based on these elements. Furthermore, the evidence suggests that the moon's extreme deficiency in volatile elements has existed ever since the moon formed, so we have no possibility of an early environment in which prelife chemistry could have flourished. Evidently it *is* safe to eat moon dust, at least in small quantities!

Another key conclusion that emerges from the comparison of lunar rocks with terrestrial rocks is that the moon could never have been an unmodified part of the Earth; the differences in composition rule this out. Even though the hypothesis of the moon as ejected from Earth no longer holds water, an emerging consensus of lunar specialists favors the idea that the moon formed from pieces of the Earth, the debris of a giant collision between our planet and a Mars-sized object during the first 100 million years after the solar system formed. This apparently audacious hypothesis is supported by calculations showing that such large planetesimals could indeed form from the solar nebula and crash into the growing planets. Such an impact would efficiently drive off volatiles from the debris ejected into space, thereby neatly accounting for the chemical differences between Earth and moon. Further elaboration and tests of this idea are now a major topic of research among lunar geologists and planetary scientists.

Thus the moon has failed to provide us with primitive solar system material and stands aloof in our skies, lifeless, devoid of any atmosphere, baking to 125° C in the two-week lunar daytime and freezing to –125° C during the two weeks of night. What use, then, is the moon to life, aside from inspiring some of Earth's living creatures to amorous activity during times of full moon?

The answer, surprisingly enough, is that the moon may have played a critical role in the development of life on Earth. The moon has made two key contributions. First, its presence in orbit with the Earth helps to stabilize the orientation of the Earth's rotation axis in space. Were it not for the moon, this axis might wander far more than it does from century to century and over millions of years, producing long-term climate changes that could kill any life that had developed under different conditions. Mars, which has no large satellite, has changed the inclination of its rotation axis in ways that have dramatically affected surface conditions on the planet.

Second, the moon produces large "tides" on Earth. The moon exerts more gravitational force on the near side of the Earth than it does on the Earth's center, and more on the center than it does on the far side (Fig. 11.17). The Earth's seas can respond more readily than the land to these differences in gravitational attraction, and so the seas flow, relative to the land, in a twice-daily tidal cycle. In the early years of the Earth, tide pools may have proven to be crucial to the formation and interaction of complex molecules. As we have seen, the tide pools may have provided in effect the first primitive biological cells. Four billion years ago, the moon orbited the Earth at a smaller distance than it does today, and raised higher tides than it does now, so the impact of tidal motion would have been proportionately greater. Tide pools that filled, emptied, and refilled twice a day could have been far more widespread than they are now, if the average tidal variation equaled, say, two meters instead of half a meter.

If the moon formed a necessary part of the conditions for life to begin on Earth, would a similar satellite be necessary for life on another planet? We cannot answer this question until we find another planet with life and check to see whether its inhabitants sing the praises of a lifeless, yet perhaps life-giving, fellow traveler through space.

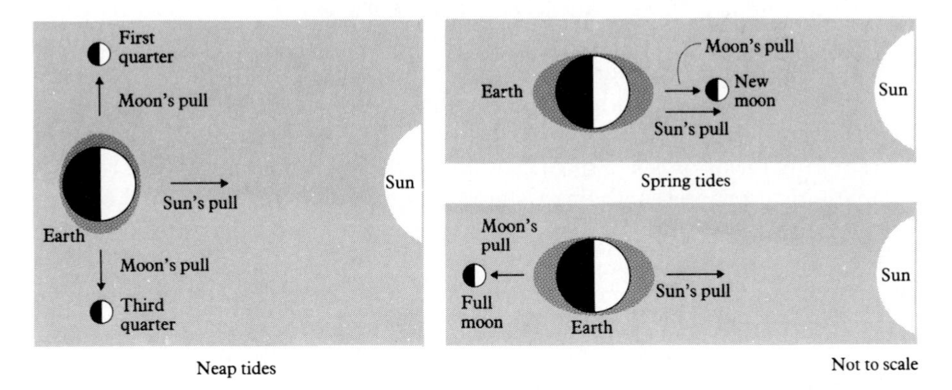

Figure 11.17 The tides arise from the *differences* in the gravitational attraction of the moon (and sun) on opposite parts of the Earth. As a result of these differences, the Earth tends to bulge in directions toward the moon and away from the moon. The sun's tide-raising ability, about 40% of the moon's, can either increase or oppose the tides that the moon raises.

SUMMARY

Comets, meteoroids, and asteroids represent lumps of solar-system matter that have changed very little since the time when the solar system formed 4.5 billion years ago. In particular, comets may contain samples of the most primitive solar-system material, because most comets orbit the sun in highly elongated trajectories that carry them much farther from the sun than any of the planets. At these huge distances from their central star, no heating has disturbed the original state of the condensed material.

Individual comet nuclei are generally only a few kilometers in diameter and consist of frozen "snowballs" of water, carbon dioxide and other gases, silicate dust, plus more complex molecules including organic compounds. Upon approaching the sun, some of the cometary ices vaporize, producing a gauzy coma around the nucleus and a long, rarefied tail that streams away from the sun for millions of kilometers. Meteoroids, which are basically small asteroids, orbit the sun in trajectories that intersect the Earth's orbit. Frictional heating in our atmosphere then vaporizes most of the object to produce a meteor or shooting star. Larger meteoroids can survive this heating and fall to Earth as meteorites. An impact by a 10-kilometer-sized object may well have caused the massive extinctions of many species of life that are demonstrated by the fossil record above and below the Cretaceous-Tertiary boundary (some 65 million years ago). This possibility adds another agent to the list of natural phenomena that have affected the evolution of life on Earth.

An important though rare class of meteorites, the carbonaceous chondrites, have part of their mass in the form of carbon compounds, quite unlike most meteorites that are iron, iron-nickel, rocky material, or a mixture of these. Some carbonaceous chondrites have revealed the presence of 74 different amino acids, the basic units of proteins, and all five of the crosslinking bases of RNA and DNA. Since these monomers have formed in nonbiological processes, their existence suggests that such essential components of life as we know it tend to form rather easily as part of the process of solar-system formation.

The planet Mercury, not subject to any atmospheric weathering, shows a heavily cratered surface, the result of great meteoroid bombardment during its first few hundred million years. The surface of the moon is similarly pock-marked, showing the local effects of this early bombardment. Such craters are missing from the Earth because crustal plate movements and erosion destroyed this early history long ago. However, the moon itself may constitute a record of that time, if it was formed from the debris of a giant impact as many scientists suspect.

Because the moon stabilized the wanderings of the Earth's rotation axis, and because the moon causes tides, we may owe the existence of life on our planet at least in part to our relatively massive satellite. The moon has saved us from death-dealing changes of climate and has provided the twice-daily tides that filled and emptied nutrient-rich tide pools, thereby concentrating the compounds within them and promoting the development of further complexity through polymerization.

QUESTIONS

1. Why do we think that comets are the most primitive members of the solar system?

2. What causes shooting stars? Why do most shooting stars never reach the Earth's surface?

3. Why does a comet produce a spectacular tail when it nears the sun? How big is the comet's nucleus, which has most of the comet's mass?

4. What do comets consist of? What does this imply about the primitive solar system's composition?

5. Why does the discovery of amino acids in the carbonaceous chondrite meteorites seem important for theories about the origin of life on Earth?

6. In what ways do Mercury and our moon resemble one another?

7. Why do the surfaces of Mercury and the moon show so many craters? How has this affected the geological record of the moon's first few hundred million years?

8. Why did some scientists speculate that moon dust might contain microorganisms? How could the airless and waterless moon ever have acquired life?

9. Is the composition of moon rocks identical to that of similar rocks on Earth? What does this tell us about the origin of the moon and the Earth?

10. In what ways has the presence of the moon favored the origin and development of life on Earth?

11. Why were the tides important in producing life on Earth, according to most theories of how life began?

FURTHER READING

Beatty, J. Kelly, and Andrew Chaikin, eds. *The New Solar System* (3rd ed.). Cambridge, MA.: Sky Publishing, 1990.

Chapman, Clark. *Planets of Rock and Ice.* New York: Scribner, 1982.

Chapman, Clark, and David Morrison. *Cosmic Catastrophes.* New York: Plenum, 1989.

Frazier, Kenneth. *Solar System.* Alexandria, Va.: Time-Life Books, 1985.

Goldsmith, Donald. *Nemesis: The Death-Star and Other Theories of Mass Extinction.* New York: Walker Books, 1986.

Hartmann, William. "The Moon's Early History," *Astronomy* (September 1976).

———"The Early History of the Planet Earth," *Astronomy* (August 1978).

Lewis, Richard. *The Voyages of Apollo: The Exploration of the Moon.* New York: Quadrangle Books, 1975.

Morrison, David, and Tobias Owen. *The Planetary System.* Reading, Mass.: Addison-Wesley, 1988.

Sagan, Carl, and Ann Druyan. *Comet.* New York: Random House, 1985.

Schramm, David, and Robert Clayton. "Did a Supernova Trigger the Formation of the Solar System?" *Scientific American* (April 1978).

12

Venus

Venus is a near-twin of the Earth in size and mass, as well as the planet that comes closest to the Earth as the two objects trace their respective paths around the sun. Venus was the Roman goddess of love and beauty, named Aphrodite by the Greeks. These ancient cultures were captivated by the brilliance of this planet in their morning and evening skies. Because Venus hides its surface beneath a blanket of perpetual clouds, early speculation about life on other worlds suggested mysterious creatures on this planet, forever hidden from our view. Modern astronomy has laid this fancy to rest, however, and we now know that Venus stifles under a dense smothering blanket of atmosphere, with its surface more than 350° C hotter than boiling water! As a result of this discovery, we face a serious challenge in our search for life on other worlds: Why did Venus become so different from Earth? How did two planets, nearly alike to begin with, diverge so widely in their later development?

To obtain some perspective on this problem, we must analyze Venus as best we can by sending radar waves and spacecraft through its atmosphere and by making spectroscopic analyses of the gases that surround it. These modern approaches to Venus have allowed us to reach a rather complete understanding of our suffocating sister planet.

The Temperature of Venus

The clouds of Venus, far thicker than those on Earth, continuously obscure the planet's surface (Figs. 12.1 and 12.2). Thus, when we measure the temperature of the parts of Venus that reflect sunlight, as we can by spectroscopic techniques, we obtain the temperature at the level of the cloud tops, some 55 kilometers above the surface. This temperature, a relatively Earthlike –33° C, does not seem particularly threatening. Radio waves, however, unlike visible light, can penetrate all the way through the clouds. Any object that is not at a temperature of absolute zero will radiate some thermal energy at these long wavelengths. Thus astronomers were

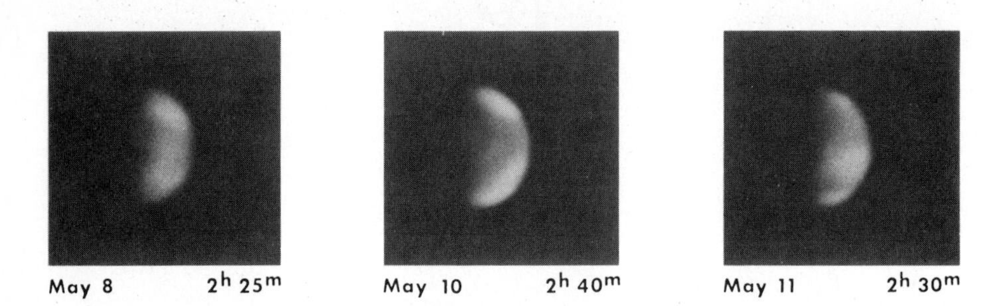

Figure 12.1 Even the best telescopic observations of Venus reveal a nearly featureless expanse of unbroken clouds, though in ultraviolet radiation we can see some patchy shadings that can be used to measure wind speeds.

keen to use the powerful radio telescopes developed in the 1950s to measure the surface temperature of our nearest planetary neighbor. They were astonished to find that the average temperature, all over the planet, is 475° C—hot enough to melt lead!

We can also beam radio waves at Venus in such a way that they will pass through the clouds, reflect from the surface, and return to antennas on Earth, carrying information about the surface that reflected them. To study

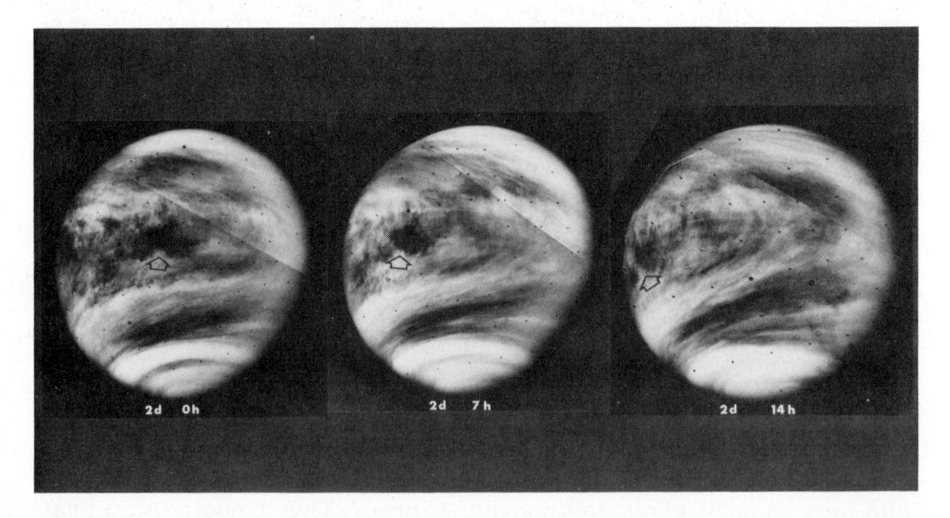

Figure 12.2 These photographs of Venus were taken in early 1979 when *Pioneer Venus* was in orbit around the planet, at an altitude of about 65,000 kilometers above the planet's surface. The spacecraft's camera also used an ultraviolet filter (compare Fig. 12.1) to increase the contrast of cloud features. From this distance, the circulation pattern on Venus is clearly visible.

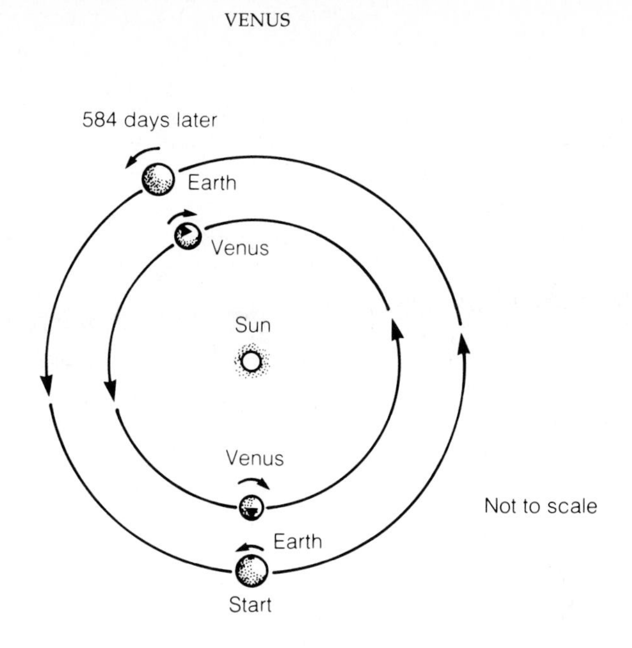

Figure 12.3 Venus rotates once every 243 days in the opposite direction to its motion around the sun. Each time that Venus and Earth are closest to each other, at intervals of 584 days, the same side of Venus faces the Earth.

Venus in this way employs a sophisticated radar system, similar in principle to the devices that police use to detect speeding motorists (but with enormously powerful transmitters for Venus reflections!). These techniques showed that Venus rotates more slowly than any other planet: One day on Venus lasts 243 Earth days. Furthermore, this rotation occurs in a direction opposite to the motion of the planet in its orbit around the sun (Fig. 12.3). This is also opposite to the direction of the Earth's rotation and of the rotation of all the other planets except Uranus and Pluto. As the power and sophistication of Earth-based radars improved, it became possible to map features on the surface of Venus as well. The landscape revealed in this way is strangely different from Earth's, as we shall see below. To begin with, the radar reflections revealed large impact craters on some parts of the surface of Venus. This indicates that these regions are much older than any of comparable size on Earth.

The Atmosphere of Venus

How does Venus maintain such a high temperature at its surface? The answer must surely lie with the planet's atmosphere, since we know from spacecraft landings that the surface contains no particularly large amounts of radioactive rocks to give off heat. The key qualities of the atmosphere that lead to the planet's enormous surface temperatures are its composition and its total amount, expressed in terms of the pressure the atmosphere produces at the planet's surface.

As verified by a series of instrumented spacecraft sent into the atmosphere of Venus by the Soviet Union and the United States, this atmosphere consists largely of carbon dioxide, which provides 96 percent of the total. Most of the remainder of Venus's atmosphere is nitrogen molecules (about 3.5 percent), with a small amount of argon and traces of sulfur dioxide, water vapor, hydrochloric acid (HCl), hydrofluoric acid (HF), carbonyl sulfide (COS), and carbon monoxide. This composition differs drastically from that of the Earth's atmosphere, which is mostly nitrogen, oxygen, water vapor, and argon, with only traces of carbon dioxide. But still more important, the total atmospheric pressure at Venus's surface turned out to be *90 times that on Earth.* In other words, not only is Venus's atmosphere mostly carbon dioxide, but there is about 100 times more carbon dioxide in the atmosphere of Venus than there is nitrogen (78 percent of our total) in the Earth's atmosphere.

With an atmosphere almost 100 times thicker than Earth's and a surface temperature high enough to melt some metals, Venus possesses the hottest surface of any planet, and provides an excellent setting for a movie about the delights of hell. The planet seems as hostile to life as any we could imagine. Pictures radioed back by Soviet spacecraft that landed on Venus in the 1970s and 1980s show that some sunlight does reach the surface, though the intensity of the lighting on the surface of Venus roughly equals that on Earth when the sky has a complete cover of thick clouds (Fig. 12.4). In view of the fact that the surface temperature is high enough to melt lead and zinc, it is reassuring to find rather ordinary rocks on the surface and not some molten goo. To balance this reassurance (such as it is), we have learned that the clouds of Venus that make this planet so brilliant in our skies are composed of droplets of concentrated sulfuric acid (H_2SO_4), not of water or of ice crystals. In brief, our sister planet seems no place for a summer vacation, or even a tourist stop.

The Greenhouse Effect

How did Venus become such a hostile inferno? Since our sister planet orbits the sun at only 72 percent of Earth's distance, we would expect Venus to be hotter than Earth, but not 450 degrees hotter! Mars, which orbits at one and a half times the Earth's distance from the sun, has an average surface temperature just 50° C below the average on Earth. But if we

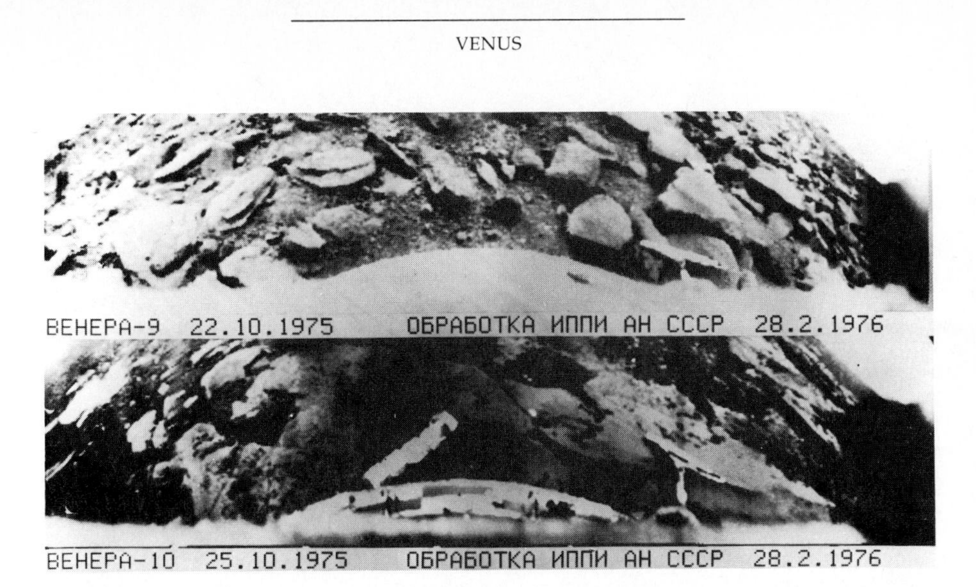

Figure 12.4 The Soviet spacecraft *Venera 10* sent these photographs back from the surface of Venus. The rather familiar-looking boulders show that the surface temperature does not melt rocks.

consider what the thick atmosphere of Venus does to the light that arrives from the sun, we can see how this planet can achieve a surface temperature of nearly 500° C.

The visible light that manages to penetrate the clouds of Venus reaches the planet's surface and tends to warm it up. As a result of this heating, the surface radiates infrared photons, as any object at a temperature above absolute zero will. (Objects at thousands of degrees, such as the surfaces of stars, radiate mostly visible-light photons, but even these high-temperature radiators also emit infrared photons.) Infrared radiation has more difficulty penetrating the atmosphere than visible light does, however (Fig. 12.5). The carbon dioxide molecules in the atmosphere can absorb great amounts of infrared. As they do, the molecules themselves become warmer and also radiate in all directions. The infrared radiated downward contributes to the warming of the planet's surface, while the upward radiation interacts with the atmosphere above in the same way as the infrared photons from the ground, heating still higher layers of the atmosphere. Eventually, the radiation from the uppermost layers of the atmosphere escapes into space. This reradiated energy must just balance the energy absorbed by the planet. For this equilibrium to be achieved, the surface temperature must be extremely high, since direct radiation from the ground is so thoroughly blocked.

These processes of absorption of visible light, radiation of infrared, and the partial trapping of the infrared radiation by Venus's atmosphere produce

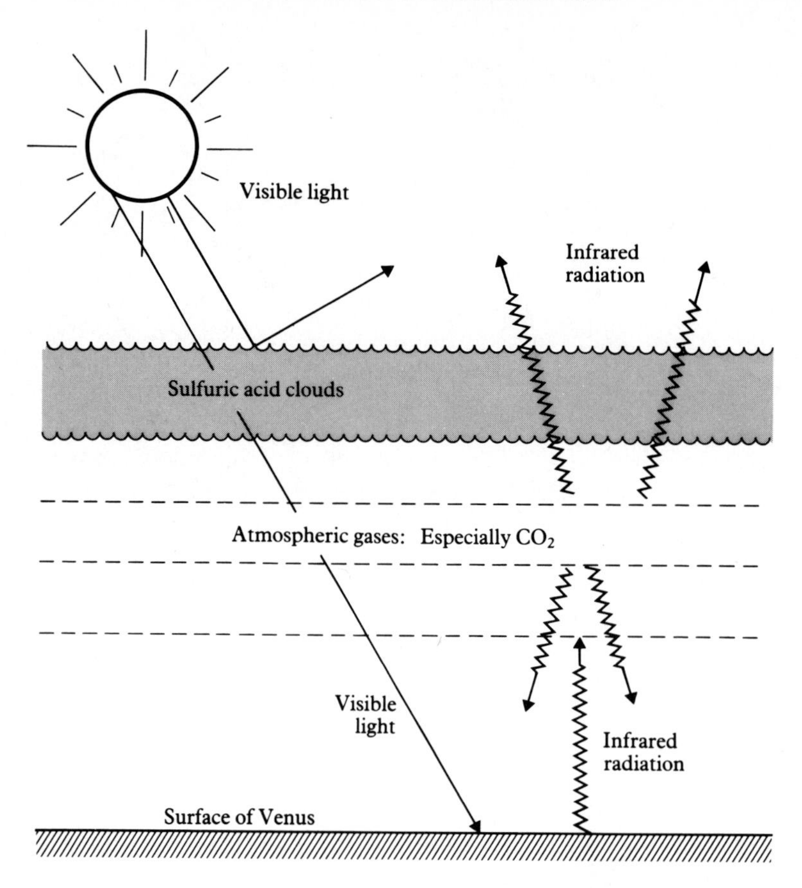

Figure 12.5 Sunlight that filters through the clouds of Venus will heat the surface, which then radiates infrared. Since the carbon dioxide molecules in the atmosphere of Venus absorb infrared extremely well, the energy in this infrared radiation is transferred to the molecules and the atmosphere becomes warmer. The surface is then heated further by energy radiated downward by the atmosphere.

a tremendous heating of the planet's surface and lower atmosphere. Astronomers call the trapping of infrared radiation by a planet's atmosphere the "greenhouse effect," because a gardener's greenhouse works in a similar manner: Visible light streams through the windows, heating the plants and soil, which then radiate infrared that is blocked by the glass panes. Hence the entire greenhouse becomes warmer than it would be if the glass, which plays the same role as a planet's atmosphere, were removed. A still more familiar example of the greenhouse effect occurs inside an automobile on a hot day, when visible sunlight warms the interior, which radiates infrared that cannot escape easily through the glass windows; again, the blocking of infrared radiation produces a noticeable warming.[1]

On Earth, the greenhouse effect plays a crucial role in allowing our kind of life to continue. Our own atmosphere traps some of the infrared radiation from the ground and keeps the Earth's surface and lower atmosphere warm. Denuded of an atmosphere, the Earth's surface temperature would average about -20° C. In fact, the average temperature is about 35 degrees warmer than this, because water vapor and carbon dioxide molecules in our atmosphere trap some of the infrared radiation. These two molecular types are particularly efficient at absorbing infrared, while nitrogen and oxygen, the major constituents of our atmosphere, are not. Human beings have already significantly increased the amount of carbon dioxide (released in fuel combustion) in the atmosphere, and have thus tended to make our planet even warmer. This may prove a serious problem during the next few decades, if the present rate of rainforest destruction and fossil fuel consumption continues.

The greenhouse effect operates full blast on Venus, whose atmosphere, almost 100 times thicker than our own, consists mainly of carbon dioxide. Even though only a small fraction of the arriving sunlight filters down through the clouds, the planet's massive atmosphere, made mostly of an efficient absorber of infrared radiation, raises the surface temperature by 400 degrees, in comparison with the temperature the planet would have if Venus had no atmosphere.

Why Is Venus So Different from Earth?

Now that we understand *how* the atmosphere of Venus keeps its surface so hot, we need to know *why* Venus has such a massive, CO_2-rich atmosphere, while the Earth does not. As we saw in Chapter 8, the differences between the atmospheres of Venus and Earth seem to be the result of life on Earth and the absence of life and liquid water on Venus.

The Earth's atmosphere does not contain much carbon dioxide, because limestone rocks, composed primarily of calcium carbonate ($CaCO_3$), have locked up most of this gas. A typical calcium carbonate is an agglomeration of millions of tiny sea shells, formed by living sea creatures from the carbon dioxide dissolved in sea water. If we ground up and heated all the carbonate rocks in the Earth's crust and released the products into the atmosphere, our Earth would find itself with an atmosphere about 70 times thicker than

[1]Strictly speaking, these examples are not correct, because both the greenhouse and the automobile achieve most of their heating by limiting *convection*, which a planetary atmosphere does not do. Nevertheless, the name "greenhouse effect" is firmly entrenched in the scientific lexicon.

the canopy of air we now enjoy, made mostly of carbon dioxide, just like the atmosphere of Venus! It is worth emphasizing that carbonates would form even in the absence of life, if liquid water were present. In outline, carbon dioxide from the atmosphere dissolves in water to form a weak acid that can combine with silicate rocks to make carbonates. The erosion by water continually exposes fresh layers of rock to this process. Since liquid water is also essential for life's origin and sustenance, we can see that the *key difference between these two planets is the absence of liquid water on Venus.*

Suppose that the Earth's atmosphere grew much richer in carbon dioxide—say, 10 times as rich as it is now. The CO_2 concentration would then be 0.3 percent. Even this relatively small change in the atmospheric composition would produce great changes on our planet, because carbon dioxide gas absorbs infrared radiation so well. We can take this speculation a step further by calculating what would happen if we suddenly moved Earth as close to the sun as Venus is. The higher temperature caused by this greater proximity to the sun would make the oceans warmer, speeding up the rate of evaporation and thereby increasing the amount of water vapor in the atmosphere. Water vapor is simply a gas made of water molecules, which are also good infrared absorbers. The enhanced absorption of infrared radiation would cause an additional increase in surface temperature, leading to more evaporation, more infrared absorption, and more evaporation until finally the oceans would be entirely in the atmosphere. This process that feeds so successfully on itself has been called a *runaway* greenhouse effect, since it proceeds until there is no water left on the planet's surface. At this point, the Earth would be extremely hot, and the water vapor molecules in the atmosphere would rise high enough to be broken apart by high-energy ultraviolet radiation from the sun, as shown schematically by the following equation:

$$H_2O + \text{Ultraviolet Sunlight} \longrightarrow H + H + O$$

The hydrogen atoms produced in this way would escape from the planet's gravitational field, since they have so little mass, while the heavier oxygen atoms would remain behind to combine with other elements (Fig. 12.6).

This theoretical prediction of what would happen to water on Venus was dramatically confirmed when an unusually large amount of deuterium, the isotope of hydrogen in which each nucleus has one neutron along with the proton, was found in the planet's atmosphere by observations made from the orbiter and the main probe of the *Pioneer Venus* spacecraft in 1981. As hydrogen leaves the planet, deuterium, with twice the mass of ordinary hydrogen, has more difficulty escaping. Thus its concentration will increase with time. The *Pioneer Venus* instruments found that the ratio

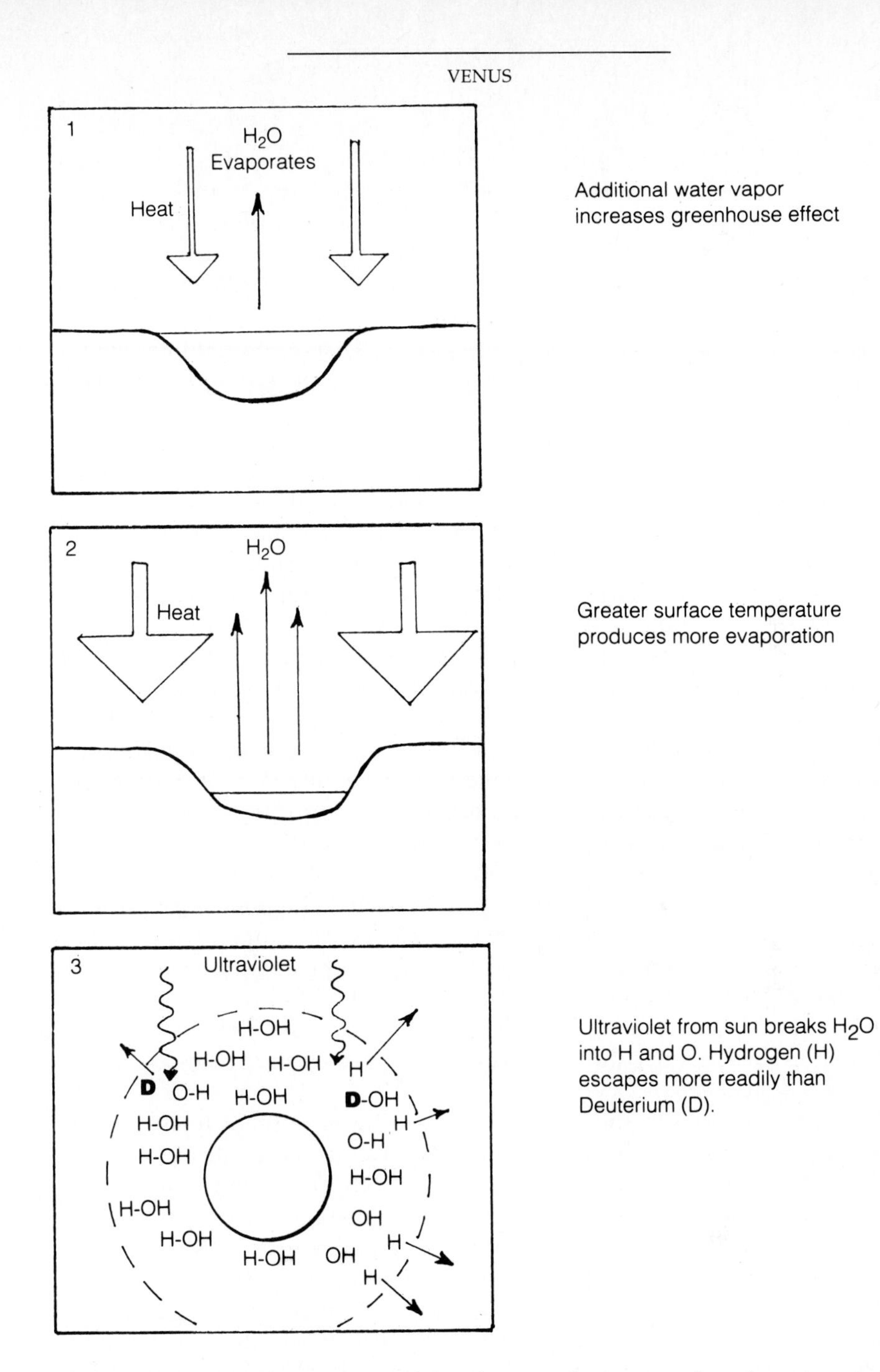

Figure 12.6 At Venus's distance from the sun, the increased surface temperature (compared with Earth) will prevent water from existing on the planet's surface. Instead, a runaway greenhouse will develop that ultimately destroys the water through irradiation by ultraviolet from the sun.

of deuterium to ordinary hydrogen on Venus is 100 times the value found on Earth! This result was so surprising that it initially met with some skepticism. But the high deuterium abundance on Venus has been unequivocally confirmed by measurements of the spectrum of radiation from Venus's lower atmosphere made with an Earth-based telescope in 1989. Absorption lines from both H_2O and HDO were detected in these observations, and the corresponding ratio of deuterium to ordinary hydrogen again showed a hundredfold enrichment compared with the ratio on Earth. Evidently a huge amount of hydrogen has indeed escaped from our sister planet.

Thus by the simple thought-experiment of asking ourselves, What would happen to the Earth if we moved it as close to the sun as Venus?, we have found a good explanation of why our sister planet seems so strange to us. This mental exercise led to a prediction—the enormous escape of hydrogen gas, with the result that the heavier isotope deuterium stays preferentially behind—that could be tested by a set of observations. The tests supported the prediction.

While this all seems neat and reassuring, the problem of the history of water on Venus is one of those that is not yet completely solved. We must bear in mind that other explanations exist for the deuterium enrichment on Venus. Furthermore, the detailed radar mapping of the planet's surface by the *Magellan* spacecraft in 1991 has failed to reveal any sign of early aqueous erosion. No dry river beds have been found, no indications of ancient beaches where once the seas of Venus may have lapped volcanic shores. However, these negative results do not contradict the hypothesis of early oceans on Venus. Old as this planet's surface is in some places, the density of impact craters is nowhere as great as it is on Mercury or the moon, whose surfaces are therefore much older. It is in those very earliest days that Venus must have lost its water, and no record of that time has yet been found, nor is it likely to be discovered in the future. Like the Earth, Venus has lost its earliest history through the action of geological processes driven by its internal heat.

We can use another, indirect but compelling, argument to show that Venus once had water. The comets and carbonaceous meteorites that struck the Earth must have had companions that hit both Venus and Mars, bringing water and other volatiles to all three planets. Strong evidence supporting this conclusion can be seen in the ancient erosional features on Mars (Figure 13.4). Thus the best explanation for the enriched deuterium and the dry atmosphere on Venus at the present time seems to be that Venus began its existence with a large endowment of water, perhaps as much as that now found in our oceans, and that this water has been lost through dissociation caused by sunlight and the escape of hydrogen.

We have thus discovered a relationship that has far-reaching consequences, for what we have learned about Venus will apply equally well to planets in other solar systems. We can summarize this finding as follows: *Any planet that is as close to its light-giving star as Venus is to our sun (closer to the star if the star is less luminous than the sun, farther from the star if the star is more luminous) will not be able to maintain liquid water on its surface.* This seems a devastating conclusion for planets subject to this rule, since liquid water is essential to the continued existence of life as we know it (see Chapter 10).

Let us review this relationship. Liquid water removes carbon dioxide from the atmosphere by dissolving the gas and forming carbonate rocks from it. Life enormously accelerates this process by producing carbonate structures such as shells. Thus without liquid water and without life, the carbon dioxide released into the atmosphere of our hypothetical planet will remain there to produce precisely the hellish conditions that we observe today on Venus. Life on Earth has made our planet quite different from Venus; it would be ironic indeed if human life made Earth more like Venus by releasing ever larger amounts of carbon dioxide into the atmosphere as we burn fossil fuels for our civilization's energy.

When scientists first realized that the surface and lower atmosphere of Venus are extremely hot, they wondered if they could not find some place on Venus where life might exist. What about the polar regions? Couldn't they be cool enough, perhaps, for liquid water to be present? Unfortunately, the same massive atmosphere that raises the surface temperature prevents this hypothesis from being true. Like a gigantic oven, the atmosphere circulates heat all over the planet, keeping the entire surface—dayside, nightside, pole to pole—roasting at 475° C, with temperature differences probably no greater than 10 or 20 degrees anywhere on the planet. This conclusion, predicted by calculations of how the atmosphere behaves, has been verified by detailed observations of the thermal radiation emitted by the planet.

Life on Venus?

Thus the surface of Venus, sweltering under an atmosphere as thick as half a mile of ocean, offers no safe haven for life. But what about life in the atmosphere? Could this be an ecological niche we have overlooked? As an example, we may consider an atmospheric layer 53 kilometers above the surface, where the pressure equals 750 millibars, about the same as that in Aspen, Colorado (Fig. 12.7). The temperature there equals 37° C, the human body temperature. Sunlight filtered through the haze will be more intense than at the planet's surface, since we are now located above some of the cloud deck. Since Venus has no ozone layer, this sunlight may include deadly (to us!) ultraviolet radiation, but we can imagine organisms with

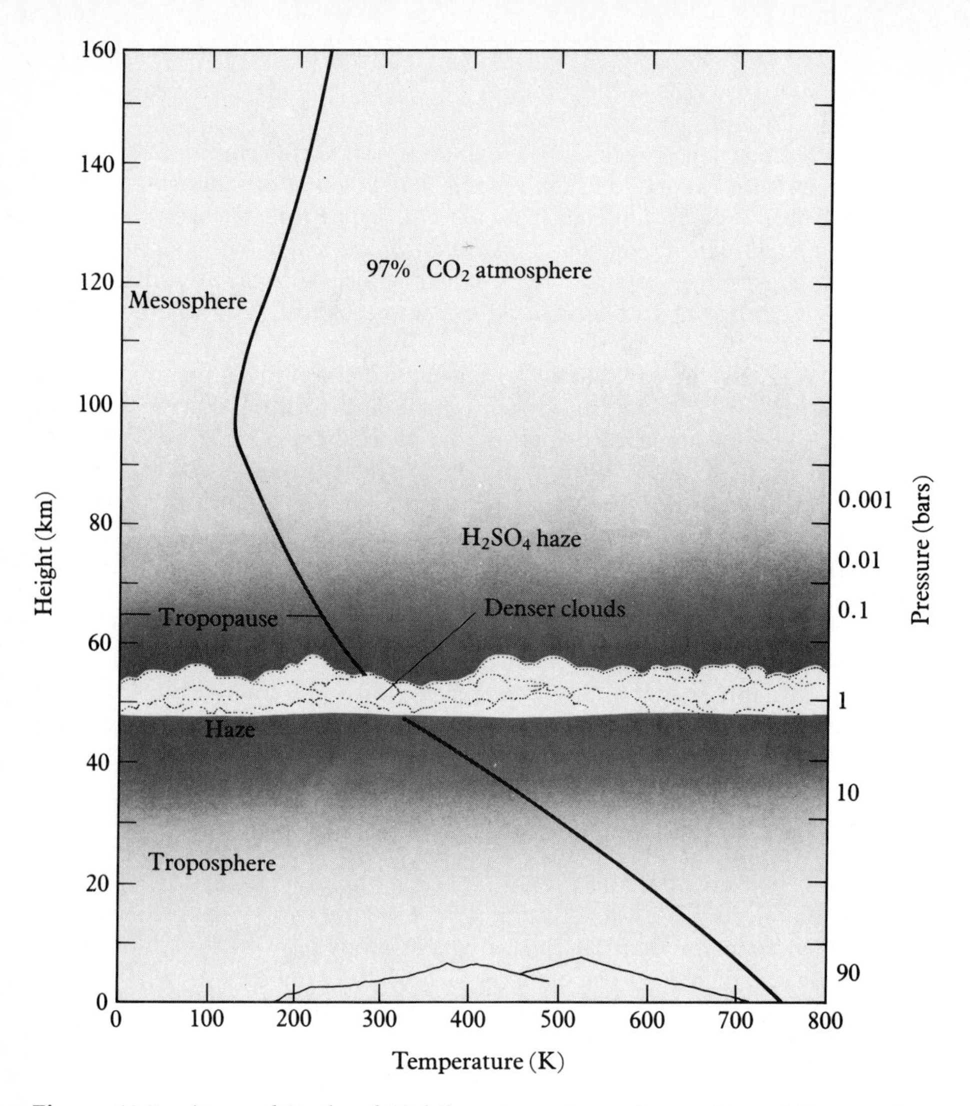

Figure 12.7 At an altitude of 53 kilometers above the surface of Venus, the temperature has fallen by 440° C to a warmish 37° C, and the pressure has decreased to less than 1 percent of its surface value.

simple outer shells that protect them against ultraviolet. Thus, at first glance, the upper atmosphere of Venus seems a pleasant environment; a crew of astronauts could float beneath a balloon, with simple oxygen masks and goggles, searching for the local equivalents of birds and butterflies.

But this survey overlooks two important facts: The atmosphere of Venus is exceedingly dry (the relative humidity never exceeds 0.01 percent) and the haze in which our astronauts are floating consists of sulfuric acid droplets! Some types of life on Earth could, in fact, survive in such environments, but only for a limited period of time. Life as we know it needs water to grow and to reproduce, and the atmosphere of Venus does not seem to provide this basic requirement for our kind of life.

Could a *different* kind of life have begun and evolved on Venus? Could human beings alter the present conditions on Venus, perhaps by bringing water-laden asteroids to the planet, thus making it habitable to an over-supply of space-starved Earth dwellers? The first question cannot now be answered, but all the evidence at our disposal points to a lifeless Venus, smothered by its carbon dioxide atmosphere. As for altering the planet, we have just seen that liquid water is not stable on Venus because it produces a runaway greenhouse effect. Bringing water there won't change this basic condition. The runaway greenhouse would just start up again.

What about conditions in the past? We must always remember that when we examine the planets today, we are effectively seeing snapshots in time. With 4.5 billion years to play with, we can certainly imagine that the climate on Venus was different in the past than it is today. In fact, there may be a slender ray of hope here. Just after the sun reached the main sequence (see Chapter 5), it was about 25 percent less luminous than it is today. This means that Venus would have been significantly cooler. (Whether it would have been cool enough for liquid water to persist on its surface is not altogether clear.) In any case, this would not have been a long period, since the sun's luminosity continued to increase until it reached its present value. Nevertheless, this possibility makes the search for ancient evidence of erosion by liquid water in the landscape of Venus a worthwhile enterprise, one that has always been in the minds of investigators studying radar and lander images of the planet's surface. Unfortunately, the *Magellan* observations imply an average age for the surface of only 1 billion years, far too young to reveal traces of ancient water erosion.

Exploration by Spacecraft

Meanwhile, human beings continue to send spacecraft to investigate conditions on this brilliant and enigmatic planet. Both the Soviet *Venera* and the U.S. *Pioneer Venus* missions that reached Venus in December 1978 included probes that landed on the planet's surface, and the United States also put a spacecraft in orbit around Venus, where it has continued to study the planet for more than a decade. Radar studies by the Pioneer Venus spacecraft in 1979 revealed the presence on Venus of a rift valley 5 kilometers deep, 300 kilometers wide, and at least 1500 kilometers long, the largest canyon in the solar system. In addition, the radar also found a mountain range on Venus called Maxwell, that is higher than Mount Ever-

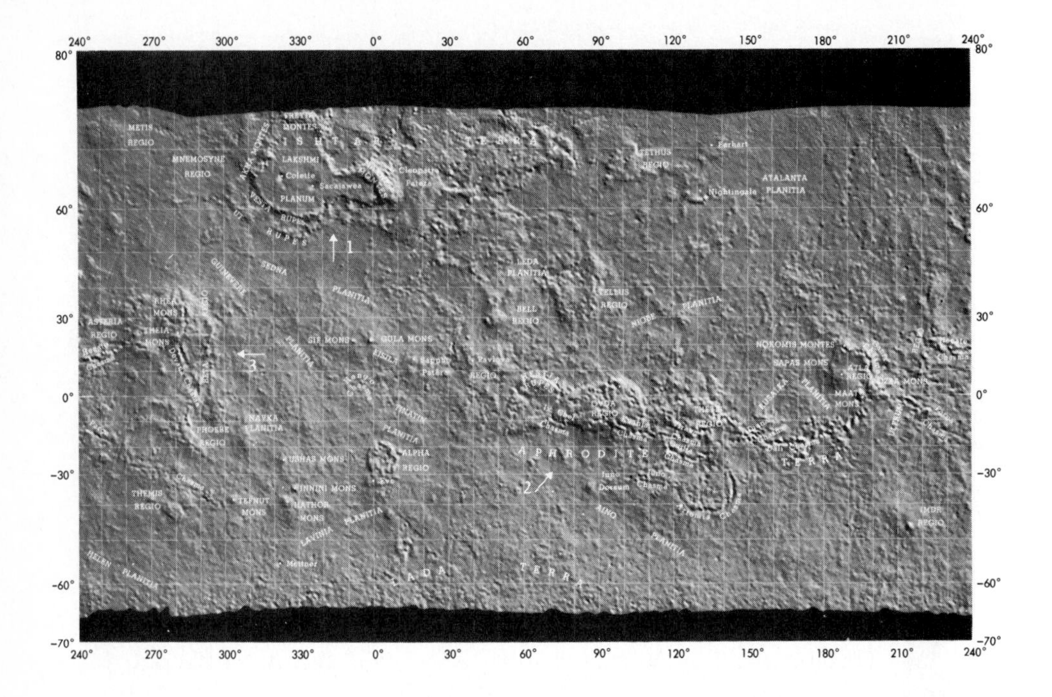

Figure 12.8 This map shows the surface of Venus as revealed by the *Pioneer Venus* radar mapping experiment. This global map will soon be replaced by data from the *Magellan* mission, which produced images of the surface with 100 times better resolution.

est, and a vast plateau region called Ishtar that is larger than the Tibetan Plateau on Earth. These features are shown on the map reproduced in Fig. 12.8. Note the relatively smaller area of raised features (continents) on Venus compared with Earth.

These findings were confirmed and extended by two Soviet radar-equipped orbiters in 1983, that had a resolution of surface detail equal to 1 to 2 kilometers. The Soviet Union followed this success in 1985 with additional probes that deployed balloons to study the planet's meteorology. The *Magellan* radar mission, launched by the United States in 1989, reached the vicinity of Venus on August 10, 1990. *Magellan* has produced images of nearly the entire surface with a resolution of 120 meters, some 10 times better than the earlier Soviet missions. At this scale, it is possible to search not only for signs of ancient water erosion but also for contemporary volcanism. While no signs of water's presence have been found on the surface, there is plenty of evidence for vulcanism (Fig. 12.9). The question is simply whether or not any of the observed volcanoes are still in an active state.

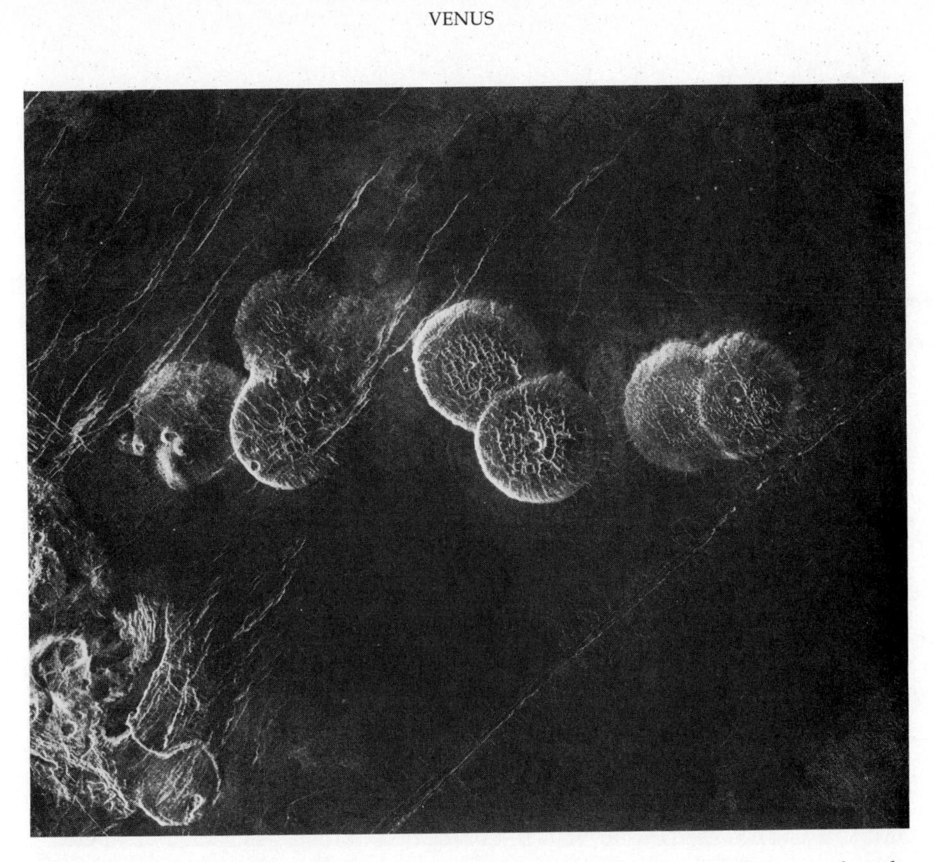

Figure 12.9 The seven circular dome-like hills shown here were discovered at the eastern edge of Alpha Regio on the surface of Venus by the *Magellan* spacecraft in 1991. These features are interpreted as very thick flows that emanated from a central opening. They are about 25 km (15 miles) in diameter with maximum heights of 750 meters. The smallest visible details are 120 meters across.

Active volcanoes have been invoked to explain large changes in the amount of SO_2 in the atmosphere of Venus, but we still lack definitive proof of their existence. The SO_2 variations might be an atmospheric effect involving decomposition of the H_2SO_4 clouds. The presence of impact craters (Fig. 12.10) and the absence of any evidence for plate tectonics—the process that continually modifies the crust of the Earth (see again Figs. 1.2 and 8.2)—have led many geologists to conclude that Venus represents an Earth-like planet in an arrested state of development. Perhaps 3 billion years ago the surface of Earth looked the way the surface of Venus appears today (except for our oceans, which were certainly present on Earth then). The arrested development of the surface may also be an indication of the absence of water from the upper layers of Venus, since the mantle convection that

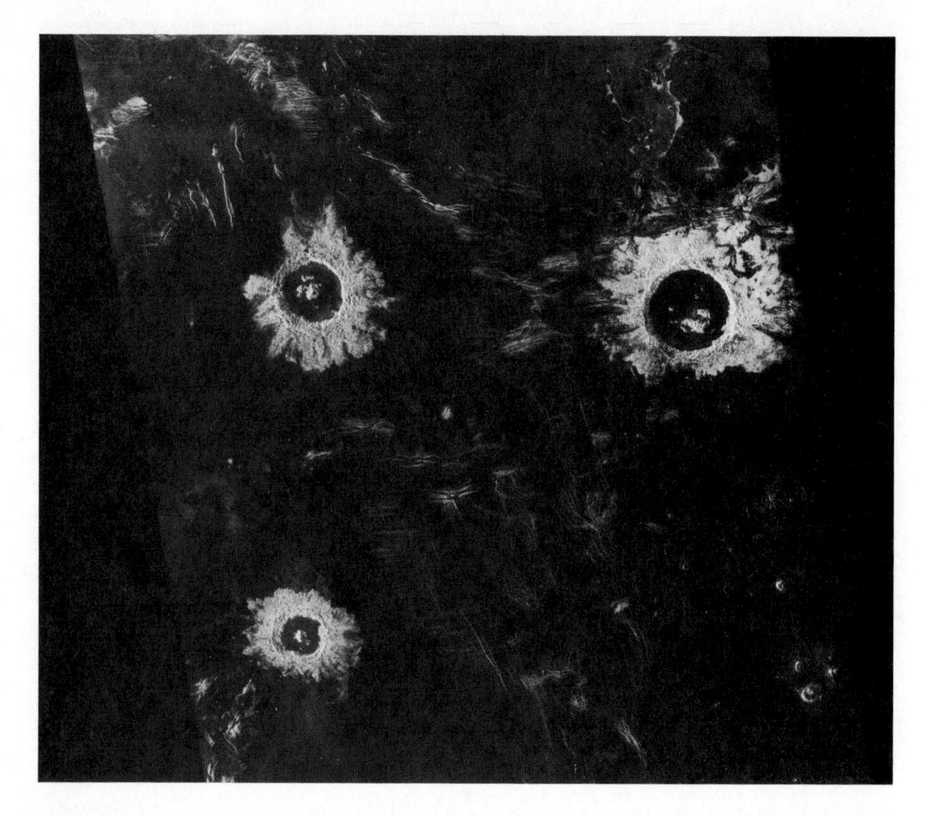

Figure 12.10 A cluster of 3 impact craters on the surface of Venus revealed by the *Magellan* spacecraft. The crater diameters range from 37 to 50 km (23 to 31 miles). While this clumping of large craters is certainly impressive (we have nothing like it on Earth), there are large areas on Venus with no impact craters whatsoever.

drives the movement of crustal plates on Earth relies on water as a lubricating fluid.

These new discoveries simply confirm our present picture of the strangely divergent evolution that our sister planet has undergone. The great lesson that Venus teaches us in the search for life in the cosmos is *the importance of a planet's being at the right distance from its central star.*

SUMMARY

Venus, the planet closest to the Earth's size and mass, has an entirely different atmosphere and surface than our own planet. Wrapped in a continuous blanket of sulfuric acid clouds, Venus never reveals its surface to outside observers in visible light, a surface that bakes at a fantastic 475° C. This high temperature arises from the enormous greenhouse effect that the thick atmosphere, mainly carbon dioxide, produces: The short-wavelength, visible-light rays from the sun can penetrate the atmosphere to reach the surface of Venus more easily than the long-wavelength, infrared photons that the surface radiates can penetrate outward. A similar greenhouse effect on Earth keeps our planet's surface about 35 degrees warmer than it would be in the absence of an atmosphere. But the atmosphere of our sister planet, 90 times thicker than Earth's and composed primarily of carbon dioxide, an efficient absorber of infrared radiation, has a far greater effect.

Radar waves penetrating the clouds of Venus have revealed the planet's slow rotation period (243 days) and the fact that the surface of Venus is geologically diverse. An early runaway greenhouse effect—the absorption of infrared photons by water vapor, raising the temperature and thereby producing more water vapor and still higher temperatures, all of which prevents the formation of limestone rocks that could use up carbon dioxide—has made Venus an entirely different kind of planet than Earth. We can conclude that apparently small differences between one planet and another, such as the Earth's greater distance from the sun, can lead to great variations between planets during the billions of years since they formed.

QUESTIONS

1. How can we tell that the surface of Venus is covered with craters if the planet's surface is not visible to us?

2. In the Earth's oceans, the pressure increases by 1 "atmosphere" for every 10 meters of depth. How deep into the oceans would we have to descend for the total pressure to equal that at Venus's surface, about 90 atmospheres?

3. Why does the atmosphere of Venus produce a much greater greenhouse effect than the Earth's atmosphere does?

4. Why is carbon dioxide particularly important in producing the greenhouse effect on Venus and on Earth? How does the release of more carbon dioxide into the Earth's atmosphere change the temperature at the Earth's surface?

5. Why does the temperature on the dark side of Venus almost equal the temperature on the day side, even though the dark side remains dark for months on end?

6. What has happened to most of the carbon dioxide that was once in the Earth's atmosphere to prevent this planet from having a greenhouse effect similar to that on Venus?

7. The atmospheric pressure on Venus decreases by a factor of 2 for every 5 kilometers we rise above the surface of Venus. Starting from a surface pressure of 90 atmospheres, how high must we ascend into Venus's atmosphere, until the atmospheric pressure falls to 1 atmosphere, equal to the pressure at sea level on Earth?

8. In certain parts of Venus's atmosphere, the pressure equals 3/4 atmosphere (75% of the sea-level pressure on Earth) and the temperature is about 37° C. Why does this fail to present a good environment in which life could develop?

9. How can we test the hypothesis that Venus formed with a large amount of water and then lost it as a result of photodissociation? Why hasn't this happened on the Earth?

FURTHER READING

Beatty, J. Kelly, and Andrew Chaikin, eds. *The New Solar System* (3d ed.). Cambridge, MA: Sky Publishing, 1990.

"Radar Views of Venus," *Sky & Telescope,* February 1984.

Burgess, Eric. *Venus: An Errant Twin.* New York: Columbia University Press, 1985.

Fimmel, Richard, et al. *Pioneer Venus.* Washington: NASA SP-461, 1983.

Morrison, David, and Tobias Owen. *The Planetary System.* Reading, MA: Addison-Wesley, 1987.

Pettengill, Gordon, et al. "The Surface of Venus," *Scientific American,* August 1980.

Sagan, Carl. *Cosmos.* New York: Random House, 1980.

Saunders, Steve. "The Exploration of Venus: A *Magellan* Progress Report," *Mercury,* September/October 1991.

13

Mars

Because of its red color, our ancestors in many cultures came to associate Mars with spilled blood, and named this planet after their gods of war. Many centuries later, human fascination with the red planet continues and has led to the discovery of clouds, dust storms, and seasonal changes on the surface of Mars, and to the controversies over the Martian "canals." The strong feeling among many scientists that we might find life on the next planet out from Earth culminated in the two *Viking* landings on Mars in 1976. Additional missions that will include this same goal are being planned for the 1990s.

Table 13.1 summarizes the physical characteristics of Mars and Earth. Despite the planet's small size and its greater distance from the sun, we can see a striking similarity between Earth and Mars in length of day, in apparent seasonal changes (the result of similar inclinations of the two planets' rotation axes), and above all in the visibility of a solid surface on which these changes occur (Color Plates 9 and 11). Because 71 percent of Earth's surface is covered by water, the total land areas of Mars and Earth are almost equal (Fig. 13.1).

The best observational suggestion that life might exist on Mars was first noticed during the early part of this century. It consisted of seasonal changes in the contrast of markings on the planet's surface. As part of these changes, a yearly wave of darkening was observed to begin in late spring at

TABLE 13.1 Comparison of the Physical Characteristics of the Earth and Mars

Planet	Radius (km)	Mass (gm)	Surface Gravity (Earth = 1)	Length of Day (Hours)	Length of Year (Years)	Distance from the Sun (Earth = 1)
Earth	6378	5.98×10^{27}	1.00	24	1.00	1.00
Mars	3395	6.42×10^{26}	0.38	24.5	1.88	1.52

Figure 13.1 Mars has a diameter just 53 percent as large as Earth's. The most striking apparent difference between Earth and Mars is the huge amount of water on Earth. Because three-quarters of the Earth's surface is covered by oceans, Mars and Earth have almost the same land area. Note the presence of large impact craters on Mars, resembling those on the moon.

the polar cap and then to spread toward the equator as summer advanced (Fig. 13.2). This is the reverse of the direction that spring travels on Earth, but astronomers conjectured that the wave of darkening on Mars might also be the result of growing vegetation. If the availability of water on Mars had more importance than the local surface temperature, and if water came only from the melting of the polar cap, then plants might flourish as windborne water vapor, or running water on the ground, spread toward the equator from the pole. The growing plants would reflect less light than bare ground, and thus the areas with more vegetation would be darker.

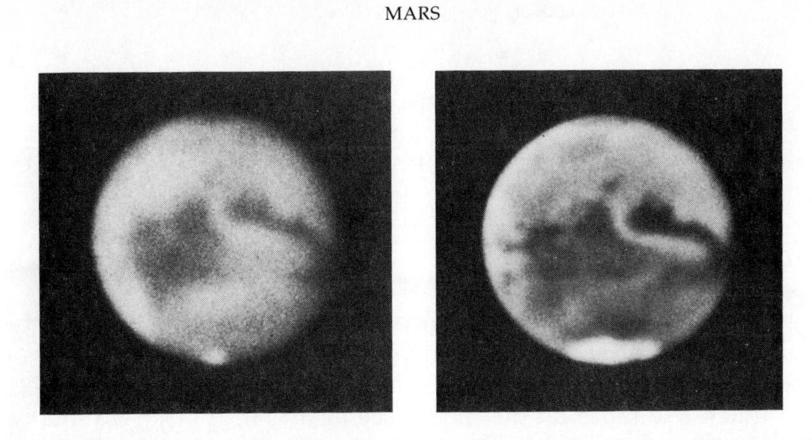

Figure 13.2 Telescopic observations of Mars show darker features that stand out against the brighter and lighter-colored background. In each hemisphere, spring brings an increase in contrast between the dark and light areas. As spring turns to summer, the polar cap shrinks to its minimum size (see left panel).

Modern Observations of Mars

As telescopes improved and the equipment used with them grew more sophisticated, this image of Mars began to change. The first challenge came from a series of temperature measurements. Mars is very cold. The temperature of the warmest spot on Mars, at the warmest moment of the Martian day during the time when Mars comes closest to the sun, rises to 20° C, equivalent to a cool summer day in Los Angeles. But the average temperature on Mars falls far below the freezing point of water. Even the spot where the noontime temperature reaches 20° C has a nighttime temperature near –90° C! Under these conditions, any channels of liquid water, natural or Martian-made, appear highly unlikely.

Spectroscopic observations of Mars confirmed the suspicion that no running water was to be expected on the red planet. By the early 1960s, astronomers knew that Martian "air" consists mostly of carbon dioxide, with only a trace of water vapor. The total amount of gas in the Martian atmosphere, and the pressure that the atmosphere exerts at the Martian surface, must be less than 1 percent of the values for our own atmosphere. The surface pressure plays an important role in speculation about the possible existence of life on Mars, because if the Martian atmosphere does not exert at least 0.6 percent of our atmosphere's pressure, then *water cannot exist as a liquid,* no matter what the surface temperature may be. If the atmospheric pressure falls below this threshold value (that is, below 6 millibars of pressure), then ice will not melt; it will simply form water vapor when it warms up—a process known as "sublimation"—just as dry ice (frozen carbon dioxide) sublimes at the Earth's surface. At these pressures, water cannot exist in the liquid state, only as a solid and a gas. Without liquid water, life as we know it cannot persist, so the total atmospheric pressure at the surface of Mars seems to be a key to the possibility of life there.

Results from Space Probes

On July 15, 1965, a spacecraft called *Mariner 4* sailed past Mars, carrying a small complement of instruments that provided our first close-up look at another planet. *Mariner 4* showed us that at least some parts of Mars have craters much like those on the moon (see again Fig. 13.1). Two follow-up spacecraft, *Mariners 6* and *7,* went past Mars in 1969, and finally humans sent *Mariner 9* into orbit around the planet in 1971. The instruments aboard these vehicles had become increasingly advanced, so that the ultraviolet spectrometers on *Mariners 6* and *7* could discover the presence of ozone on Mars in a proportion of less than one part per million, while the infrared spectrometer on *Mariner 9* found no methane on Mars, setting an upper limit of less than 25 parts per billion of methane in the Martian atmosphere. (For comparison, note that the Earth's atmosphere contains 50 parts per billion of ozone and 2000 parts per billion of methane.)

The Mariner spacecraft showed that the average surface pressure on Mars falls close to the critical level of 6 millibars, which would preclude the existence of liquid water. Careful study of the thousands of pictures sent to Earth by *Mariner 9* failed to show any evidence of the notorious "canals" promoted by Percival Lowell (Chapter 1, Fig. 1.4). Evidently these linear features were only an optical illusion resulting from chance alignments of dark spots on the surface, as seen in Fig. 13.3. Spectrometers on *Mariners 6* and *7* revealed that the Martian polar caps consist primarily of frozen carbon dioxide, not frozen water. This is a dramatic illustration of the extreme cold on Mars, since in order for carbon dioxide to freeze, the ground temperature must fall to –125° C. Measurements of infrared radiation from the planet confirmed the earlier deduction that even at its equator, the surface temperature of Mars falls to –90° C just before dawn on an average day. The temperature changes by more than 100° C each day, far more than even the most extreme temperature changes on the deserts of Earth. These impressive variations testify to the thinness of the Martian atmosphere, which has a density only 1/150 of Earth's. Despite the fact that the Martian atmosphere is mostly CO_2, like that of Venus, it is so thin that it provides a greenhouse effect that warms the planet by only 5 degrees.

The small amount of ozone in the Martian atmosphere is located close to the ground and is distributed in patches over the coldest regions of the planet. It therefore does not provide an effective shield against ultraviolet light from the sun. When some scientists put this fact together with the

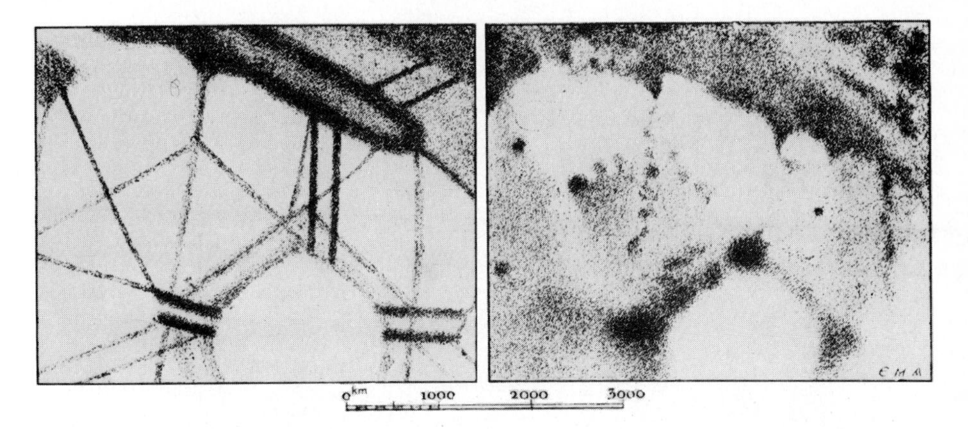

Figure 13.3 When the astronomers Percival Lowell and Giovanni Schiaparelli drew maps of Mars a century ago, they saw numerous lines or "canals" (left panel, showing a drawing by Schiaparelli). Eugenio Antoniadi drew the same region on Mars, but he saw no such canals (right panel). We now know these canals were only optical illusions, despite Lowell's claim that they showed the technological handiwork of intelligent Martians.

probable absence of liquid water and the extreme cold, they concluded that the prospects for life on Mars seemed to be vanishingly small. Other scientists, however, stressed the fact that some terrestrial microorganisms could exist under the harsh Martian conditions, if they obtained suitable shielding from ultraviolet radiation, perhaps by living under rocks, and if they had access to some liquid water, just enough to coat grains of soil for some small fraction of a day. These scientists pointed to the existence of sinuous dry valleys with branching tributaries as evidence that liquid water once flowed on the surface of Mars (Fig. 13.4). The immense Martian volcanoes (Color Plate 11)—among them the mighty Olympus Mons, whose summit towers 25 kilometers above the surrounding plain—show that the planet's crust has undergone activity within relatively recent eras (Figs. 13.5 and 13.6). Could the cosmic luck of the moment have brought us to the red planet during an extreme ice age? Might conditions have been more favorable to life on Mars in the past? Could they become favorable again?

We do know that the size of the angle by which Mars' rotation axis tilts with respect to the planet's orbit around the sun changes with time. The tilt of the axis varies periodically through greater and lesser values over millions of years (Fig. 13.7). These changes of tilt arise from the combined gravitational forces of Jupiter and the sun on the planet Mars, and they have an important effect on the Martian climate.

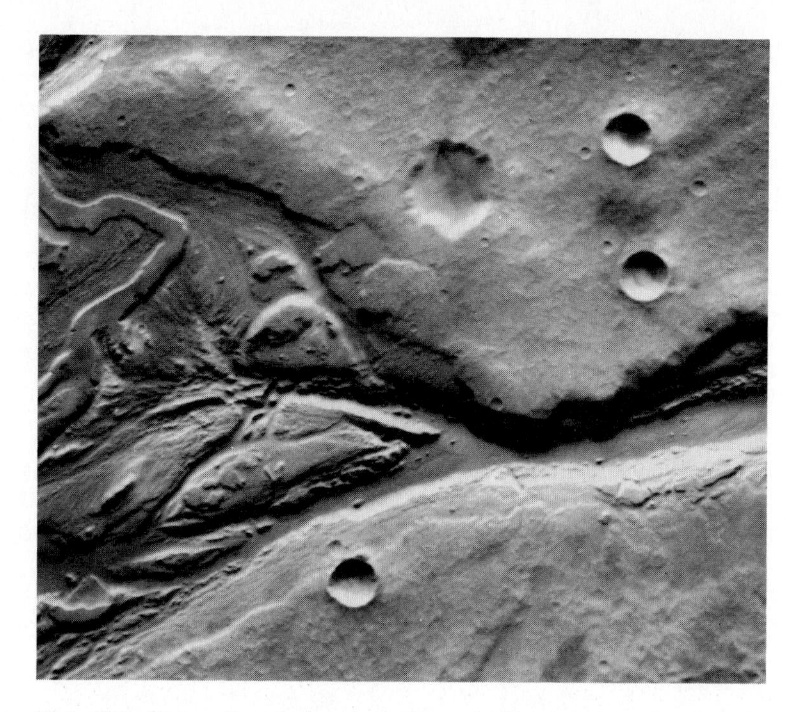

Figure 13.4 Winding valleys, often complete with branching tributaries, resemble dry river beds on Earth and strongly suggest that liquid water once flowed on the Martian surface.

Consider what happened when the angle of inclination of the rotation axis exceeded its present value. In those years, the change between summer and winter on Mars must have been more pronounced than the change now, and we can imagine that the polar caps might have disappeared completely in summer for a given hemisphere. When this effect was first discovered, astronomers thought that this sublimation could release sufficient gas (mostly carbon dioxide) into the Martian atmosphere to raise the atmospheric pressure enough for liquid water to exist. This would provide an explanation for the braided channels, closely resembling dried-up river beds, that *Mariner 9* had photographed (Fig. 13.4).

With such a cyclic appearance and disappearance of liquid water, life on Mars might have developed the ability to survive the hundreds of thousands of years when no liquid water exists by means of some sort of superhibernation that would await the years when the polar caps would sublime once again, allowing liquid water to revive the desiccated organisms. Unfortunately, most of the water-carved channels appear to be not a million years old, but a billion or so years. This means that these imagined cycles of liquid water

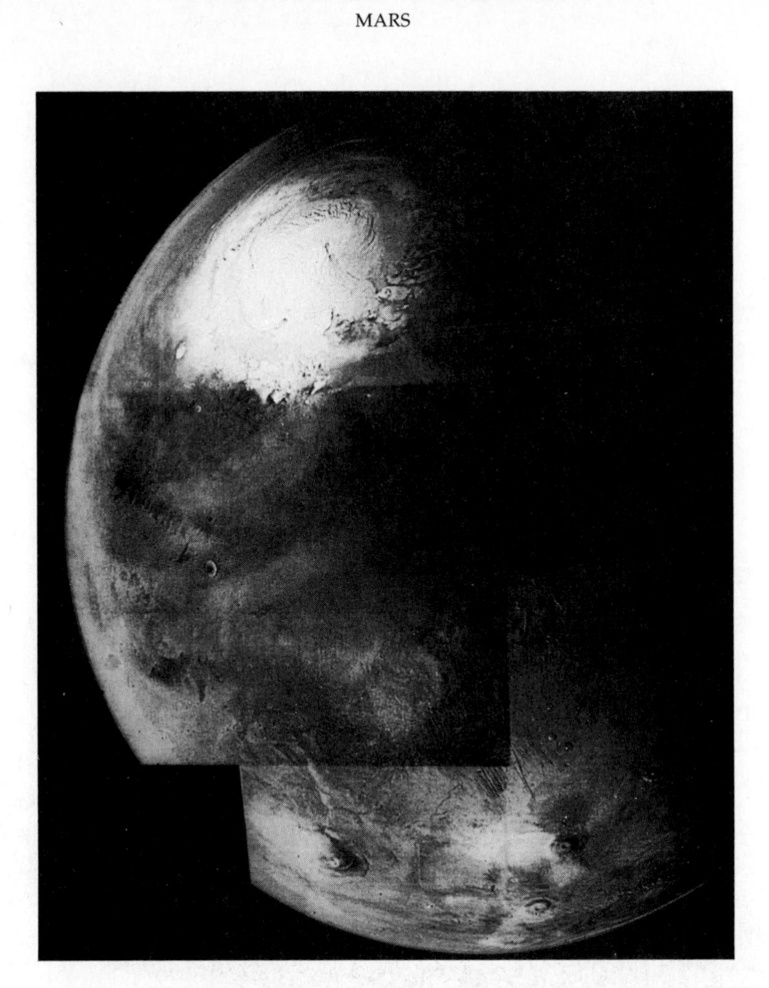

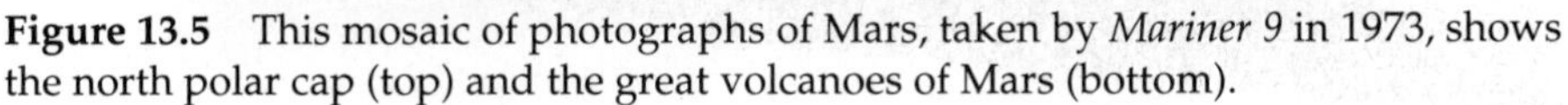

Figure 13.5 This mosaic of photographs of Mars, taken by *Mariner 9* in 1973, shows the north polar cap (top) and the great volcanoes of Mars (bottom).

may have occurred only during much earlier eras of Martian history. Nevertheless, the evidence for liquid water on the surface is plain. How and when were these channels carved? What else was happening on Mars at that time?

The Viking Project

After 100 years of controversial observations made from Earth and eight years of equally controversial studies from spacecraft, the preparations for the first landing on Mars unfolded in a highly charged atmosphere. The *Viking* mission plan called for two spacecraft, each with an

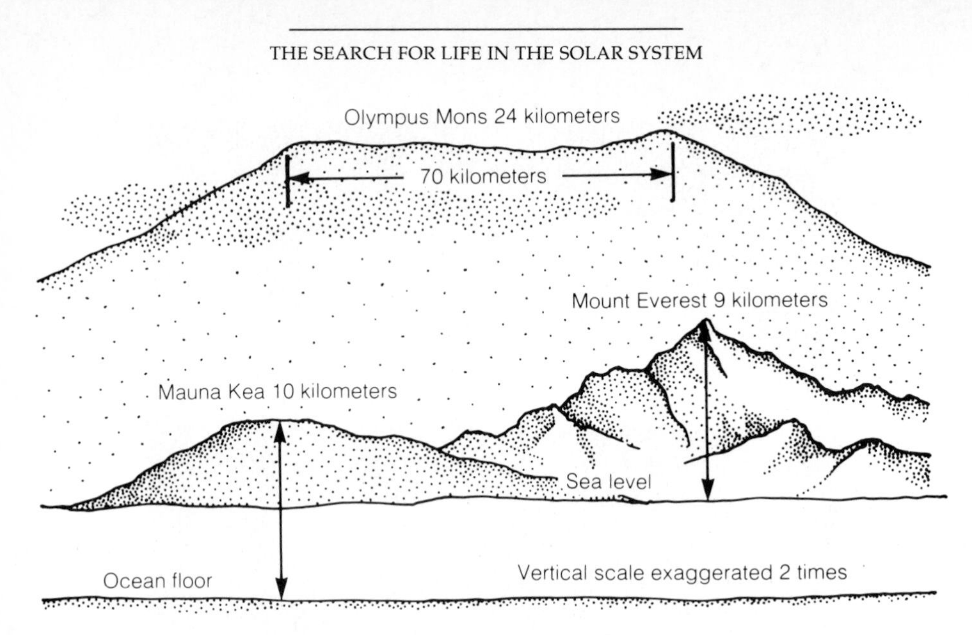

Figure 13.6 Olympus Mons, the largest known mountain in the solar system (so far!), rears its enormous bulk 25 kilometers above the plains of Mars. This volcano has a diameter of 300 kilometers, more than five times the width of Mauna Loa, the largest volcano on Earth.

orbiter and a lander, to arrive at Mars in the summer of 1976. The orbiters could study the suitability of the landing site before the landers made the actual descent and would serve as relay stations to send back the data collected on the surface as well as platforms for remote investigations of the entire planet.

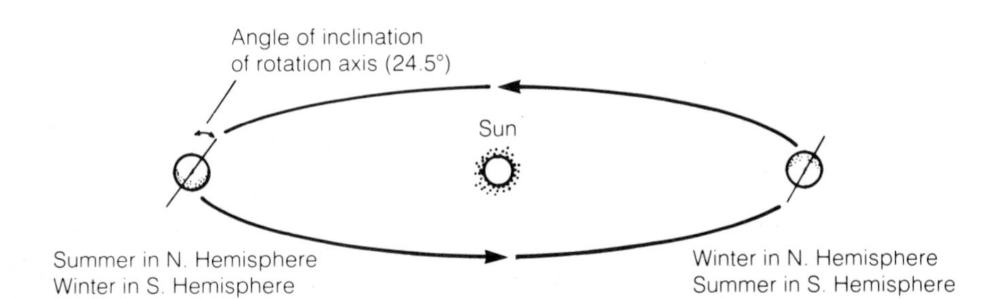

Figure 13.7 The angle of inclination between the rotation axis of Mars and the perpendicular to the planet's orbit around the sun changes periodically by several degrees. When this angle has a greater value than it does now, the difference between Martian summers and winters must be more pronounced than it is at present.

Table 13.2 shows the scientific payload of the *Viking* spacecraft, the experimental equipment designed to answer the most pressing questions about Mars. What is the composition of the Martian soil? The inorganic analysis would tell us. Is the interior still active enough to produce "Marsquakes"? A seismometer would give the answer. What does the Martian atmosphere consist of? Data would come from mass spectrometers on the aeroshell that protected the lander during its descent, and on the lander itself, as well as from a spectrometer on the orbiter. What is the temperature on Mars? The orbiter's infrared radiometer could measure both the ground and the atmospheric temperatures, while the lander could sense local variations precisely. What is the weather like? The Viking lander could measure

TABLE 13.2 Scientific Payloads for the Viking Missions to Mars

Investigation	Instruments
	Viking Orbiter
Visual imaging	Two television cameras
Water vapor mapping	Infrared spectrometer
Thermal mapping	Infrared radiometer
	Viking Lander
Imaging	Two facsimile cameras
Biology	Three analyses for metabolism, growth, or photosynthesis
Molecular analysis	Gas chromatograph–mass spectrometer (GCMS)
Inorganic analysis	X-ray fluoresence spectrometer
Meteorology	Pressure, temperature, and wind velocity gauges
Seismology	Three-axis seismometer
Magnetic properties	Magnet on sampler observed by cameras
Physical properties	Various engineering sensors
	Radio Propagation
Orbiter/lander location	Orbiter and lander radio and radar systems
Atmospheric and planetary data	
Interplanetary medium	
General relativity	

wind speed, wind direction, and atmospheric pressure. What does Mars look like? Images from the orbiter could show surface features slightly larger than the Rose Bowl; later, once the orbit had been adjusted, we could distinguish the grandstand from the playing field. Meanwhile, the landers would take pictures of their surroundings with resolutions that varied from a few millimeters near the lander to a few hundred meters at the horizon. Two cameras on each lander gave the ability to take stereoscopic pictures, and filters enabled the acquisition of color photographs from Mars. What does Mars smell and taste like? The molecular analysis instrument would tell us, by examining the composition of the atmosphere and the soil. And does life exist on Mars? A good question! Many instruments could help answer it, but three were designed specifically for this purpose alone. We shall discuss the performance of these three experiments in the next chapter.

With the choice of experiments and of the scientists to build and operate them, the *Viking* project began in earnest in 1969, seven years before the actual landing. As the pictures of Mars came back from *Mariner 9* during 1972, the planet assumed an entirely new identity: In addition to the drab, moonlike landscape that previous spacecraft had glimpsed, there emerged huge volcanos, deep canyons, and those intriguing, sinuous channels. Where were the two best places on Mars to look for life? The search for suitable landing sites had to satisfy three constraints that had nothing to do with the scientific objectives of the mission. First, the local surface pressure had to be at least 4 millibars, or the atmosphere would not provide sufficient friction to slow the lander safely. This eliminated all the higher-elevation parts of Mars. Second, the landing had to occur relatively close to the Martian equator, and not near the poles, so that the lander could establish good radio links with Earth. Third and most important, the site had to be "safe," meaning that the lander could not survive if it settled down on a rock larger than a basketball. Yet the smallest surface features visible from the orbiter were the size of a football stadium!

Fortunately, radar studies made from Earth could tell us the small-scale roughness of various areas on Mars, and were used to eliminate some otherwise attractive landing sites from final consideration. In view of the difficulties that the scientists and engineers faced in finding good landing sites for spacecraft 400 million kilometers from Earth, they deserved their celebrations for landing safely not once but twice at this distance. Furthermore, the landing sites turned out to be soft enough for the Viking sampler to dig trenches in the soil, rather than hard slabs of rock from which no samples could be taken aboard the spacecraft.

The scientific criteria for a good landing site leaned heavily on the search for life on Mars. Simply put, the Viking team wanted to land in places that were warm, wet, and inhabited. In view of what was already known about Mars, quite liberal definitions were needed if any sites were to make the grade. The final decision before the spacecraft reached Mars was to land once in a low region where several of the sinuous channels observed by *Mariner 9* had their courses and a second time at a higher latitude, in the hopes of finding more water near the surface, since the landing would occur in summer for that hemisphere.

Once the first *Viking* spacecraft went into orbit around Mars in June 1976, however, the scientists saw that the high-resolution photographs of the prime landing area showed impact craters right in the sinuous channels (Fig. 13.8). These craters implied a span of billions of years since the time

Figure 13.8 The original site chosen as the prime landing area for *Viking 1* turned out to have impact craters in the dried-up stream beds, evidence that billions of years had passed since water carved these channels.

that water last flowed there. Worse yet, the site seemed to present too many hazards to risk a landing, so the prime landing area was shifted, after many discussions, to the western edge of the Chryse basin, downstream from what appeared to be well-defined water erosion (Fig. 13.9).

On July 20, 1976, seven years after the first human walked on the moon, the first Viking lander settled down comfortably on the surface of Mars. The initial experiment made there was to take and transmit a picture of one of the lander's footpads, and subsequent images filled in the landscape (Fig. 13.10). The choice of a landing site had been good, for only one rock in the immediate vicinity was large enough to have overturned or punctured the lander, if

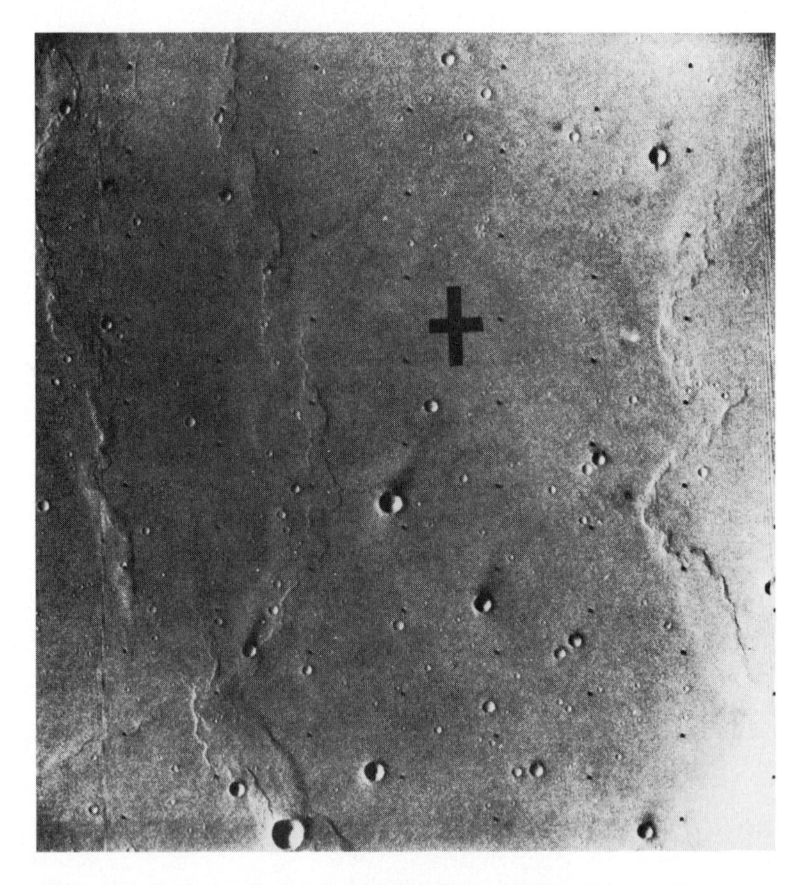

Figure 13.9 The final site chosen for the landing of *Viking 1* lay at the western edge of the Chryse basin, and was favored for its overall smoothness. The cross marks the projected landing site, as photographed from the Viking 1 orbiter. The landing in fact occurred about 15 kilometers to the left of the center of the cross, which itself spans 6 by 8 kilometers.

Figure 13.10 Rocks a few centimeters or even a few dozen centimeters in diameter dominate this view of the immediate surroundings of the Viking 1 lander. The housing at lower left center contains the sampler to scoop up soil.

the landing had occurred upon it. Color photographs confirmed the fact that the Martian surface is red, the result of iron oxides in the soil (Color Plate 12). Any intelligent Babylonian could have predicted that! What surprised some people, however, was the reddish cast of the Martian sky, very different from the customary blue skies of Earth. The reason for this red tint is visible in the dirt around the spacecraft: Some of the particles are small enough to remain suspended in the atmosphere, where they give the Martian sky its unusual color. A similar effect occurs during violent sandstorms on Earth, when somewhat larger particles seem to turn the sky yellow.

Windblown dust plays a large role on Mars, forming fields of dunes (Fig. 13.11) and a changing pattern of light and dark streaks over the planet's surface. Scientists are now convinced that the "wave of darkening" once thought to be evidence for life on Mars (see again Fig. 13.2) is nothing more than the effect of seasonal wind patterns transporting dust from place to place on the surface while periodically reducing the transparency of the atmosphere by suspending greater quantities of the finest dust grains. Some of this dust gets transported to the polar regions over periods that are still not well defined, building up layers of sediment that could be interesting places to study Martian history (Fig. 14.6).

The second landing on Mars, on September 3, 1976, put a spacecraft at 40 degrees north latitude but revealed a landscape much like that at the first site (Fig. 13.12 and Color Plate 13). Even though two sites hardly provide much of a sample of the planet, comparisons of their detailed similarities and differences revealed a huge amount of information.

Figure 13.11 This view from the Viking 1 lander shows several nearby dunes. The boom carrying the meteorology experiments extends to the right. This frame shows the landscape just to the left of the view shown in Figure 13.10, where the shadow of the meteorology boom is visible.

Figure 13.12 The panorama seen from the Viking 2 landing site resembles the landscape seen from the Viking 1 lander, despite the distance of 2500 kilometers between the two sites. Rocks ranging in size from a few meters to a few centimeters in diameter dominate both views; no signs of life are visible.

The New Mars

To the eyes of the Viking scientists, the Martian landscape seemed hauntingly familiar. The sharpness and variety of the rocks on Mars, together with the undulations of the surface and the dunes visible near the horizon, all resemble scenes on Earth, which also has regions with reddish soils. Both Earth and Mars contrast sharply with the moon, where the lack of an atmosphere has exposed the soil to relentless bombardment by micrometeorites and to the bleaching effects of the sun's full glare of ultraviolet radiation (see Color Plates 10 and 12). The examination of Martian soil revealed other similarities to the Earth, and showed that the chemical composition of the soil at the two landing sites is virtually identical.

Table 13.3 shows the composition of the soil at the two Viking landing sites in terms of the chemical elements; the inorganic soil analyzer could not distinguish among various compounds and minerals formed from individual elements. The abundance of sulfur on Mars seems unusually high, but otherwise the Martian ratio of elements is similar to that in nontronite, an iron-rich clay found on Earth. Fine particles of this material may form the dunes in Fig. 13.10 and 13.11 and may also cause the distinctive reddish tint of the Martian sky. But the key conclusion drawn from Table 13.3 is that the surface of Mars consists—at least at the two landing sites sampled—of elements in the abundance ratios familiar to us on Earth.

TABLE 13.3 Elemental Composition of Soil at the Two Viking Landing Sites

Element	Percentage of Total Composition of Soil	
	Site 1	**Site 2**
Silicon	20.9 ± 2.5	20.0 ± 2.5
Iron	12.7 ± 2.0	14.2 ± 2.0
Magnesium	5.0 ± 2.5	
Calcium	4.0 ± 0.8	3.6 ± 0.8
Sulfur	3.1 ± 0.5	2.6 ± 0.5
Aluminum	3.0 ± 0.9	
Chlorine	0.7 ± 0.3	0.6 ± 0.3
Titanium	0.5 ± 0.2	0.6 ± 0.2
All Others (thought to be mostly oxygen)	50.1 ± 4.3	

The same holds true for the Martian atmosphere. The predominance of carbon dioxide, first discovered from Earth, was confirmed by the Viking mass spectrometers, which sorted atoms and molecules by mass and thus allowed scientists to deduce which compounds must form the atmosphere. Ninety-five percent of the Martian atmosphere is carbon dioxide, while nitrogen molecules provide 2.7 percent, argon 1.6 percent, oxygen molecules about 0.1 percent, and water vapor and carbon monoxide provide still smaller traces of the Martian "air." The overwhelming predominance of carbon dioxide reminds us immediately of Venus. Once again, we meet a planet whose atmosphere seems to have evolved in the absence of abundant life and liquid water.

For Venus, water is almost totally absent, because the planet orbits so close to the sun that the resulting high surface temperature forced the water to escape soon after Venus formed. On Mars, we seem to meet the opposite problem: Much of the original supply of Martian water is probably still on the planet, but most of it is frozen in the subsurface soil as permafrost, similar to the wintertime freezing of water in Arctic soils on Earth. We seem to have the classic situation of "Goldilocks and the three planets": Venus is too hot, Mars is too cold, and Earth is just right!

From measurements of the abundances of the various gases in the Martian atmosphere, and from our knowledge of how gases escape from a planet such as Mars, we can try to reconstruct the composition and density of the planet's early atmosphere. As on Venus, we again find that deuterium is enriched on Mars, but this time by only a factor of 5 instead of 100. Evidently a large fraction of the last water available at the Martian surface was destroyed by ultraviolet radiation with the escape of hydrogen. The heavy isotope of nitrogen is also enriched, indicating that a large fraction of the nitrogen originally present on Mars has also escaped. Putting these results together with other information such as the abundances of the noble gases—neon, argon, krypton, and xenon—we conclude that it is virtually certain that Mars once had a much denser atmosphere than it does today. Despite the fact that carbon dioxide forms 95 percent of the present Martian atmosphere, most of the carbon dioxide near the planet's surface seems to be locked up in the polar caps and in carbonate rocks. We can calculate that if all the carbon dioxide that now seems to be missing were present in gaseous form, then the atmospheric pressure could rise to a value similar to that presently encountered at the Earth's surface.

This interesting fact allows us to explain the mysterious channels, apparently carved by running water. With a surface pressure of 1000 millibars, liquid water could easily exist, and this atmosphere would have been able to keep the planet warm enough to maintain a fully developed rain-river-pond-rain cycle. In fact, there is even possible evidence for rainfall in the form of multibranched channels that look like drainage systems. Signs of ancient standing water have been found in apparent shorelines and in craters whose walls were breached by flowing streams (Fig. 13.13). But the large-scale modifications of the surface by running water, such as the region shown in Fig. 13.8, look like the scars left by huge quantities of water surging across the Martian plains in giant flash floods. It is still not clear exactly what the conditions were under which these various manifestations of water erosion took place. Some of these channels, especially the larger ones, could have been formed under a thin atmosphere, with the running

Figure 13.13 This Viking photograph shows a flood channel that intersects Martian craters. Apparently water in the channel breached the wall of the large crater at the left. These craters may have provided the equivalent of ponds or tide pools on the early Earth.

water protected by a layer of ice. But if it really rained on Mars, if slow-moving small streams existed, a thick atmosphere must have been present.

Most of this water activity apparently occurred billions of years ago, and one of the major questions still to be answered is how long this phase lasted on Mars. The presence of impact craters in the floors of the large channels suggests that the last massive floods on Mars took place between 3.0 and 3.5 billion years ago. Some scientists have argued that the presence of runoff channels on the slopes of geologically recent volcanoes indicates that *all* water activity did not stop at that point. In any case, it is clear that the early history of the planet included conditions allowing the existence of liquid water, so it is impossible to rule out the *origin* of life on Mars because the planet has no water today. Certainly all the major elements that we have identified as essential to life exist on Mars. The famous polar caps, made mostly of dry ice, also include a significant amount of water ice underneath the solid carbon dioxide. We can be sure of this from the fact that the summer sunshine raises the temperature of the north polar cap to the point where all the dry ice sublimes, leaving a small cap of water ice (Fig. 13.14). Before the water ice can sublime completely, the next fall's carbon dioxide frost covers it over, preserving some ice season after season. There is also plenty of water frozen beneath the surface in the form of permafrost. We know this both from the presence of the water vapor in the atmosphere, which must be coming from some source, and from various features in the visible surface (Fig. 13.15). We saw that the best terrestrial analogue for the Martian soil analyzed at the two landing sites is an iron-rich clay. Remember that clays may well have played a key role in getting life started on Earth by serving as templates for the formation of complex molecules (page 198).

It appears that all the necessary ingredients and conditions required for the origin of life were probably present on Mars during the first billion years of the planet's history. The remaining questions have to do with whether they were all present at the right place in the right circumstances. Thanks to the advent of the space age, we can stop speculating and actually go to Mars to get the answers, as we shall see in Chapter 14.

Meanwhile, the present Martian prospect certainly seems grim by terrestrial standards. Consider the challenge of boiling an egg. Hardy scouts on Mars could never build a fire, since no fuel exists, and the atmosphere has too little oxygen for anything to burn. If the scouts brought an electric stove, chipped ice from the polar cap, and tried to produce water, they would merely sublime the ice into vapor. The ice wouldn't melt into water. Only with a pressure cooker could anything as simple as boiling an egg take place, and even then, a simple doubling of the pressure, such as occurs in such cookers on Earth, would result in boiling water at 5° C! To make liquid water, and keep it liquid at, say, 60° C, the scouts would need a pressure cooker that

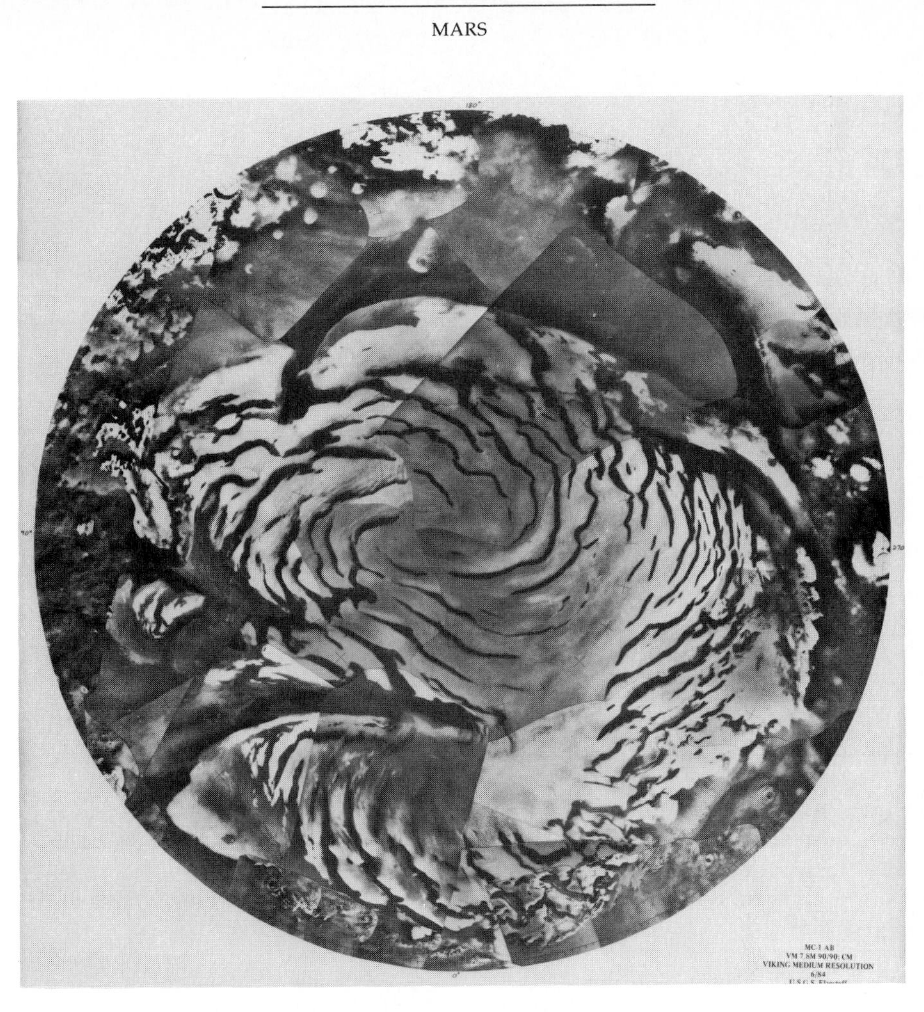

Figure 13.14 During summer in the northern hemisphere of Mars, the carbon dioxide in the polar cap sublimes into vapor completely, leaving a small cap of water ice, which requires a higher temperature to sublime.

raised the pressure 100 times above the atmosphere's. It is thus evident that scouting on Mars would be no picnic, yet a fair amount of scouting seems called for if we are ever to find life on the fourth planet of the sun.

Phobos and Deimos

Mars has two tiny moons, named after the chariot horses of the god of war: Phobos, meaning fear, and Deimos, meaning panic. They were discovered in 1877 by Asaph Hall, though Johannes Kepler, Voltaire, and Jonathan

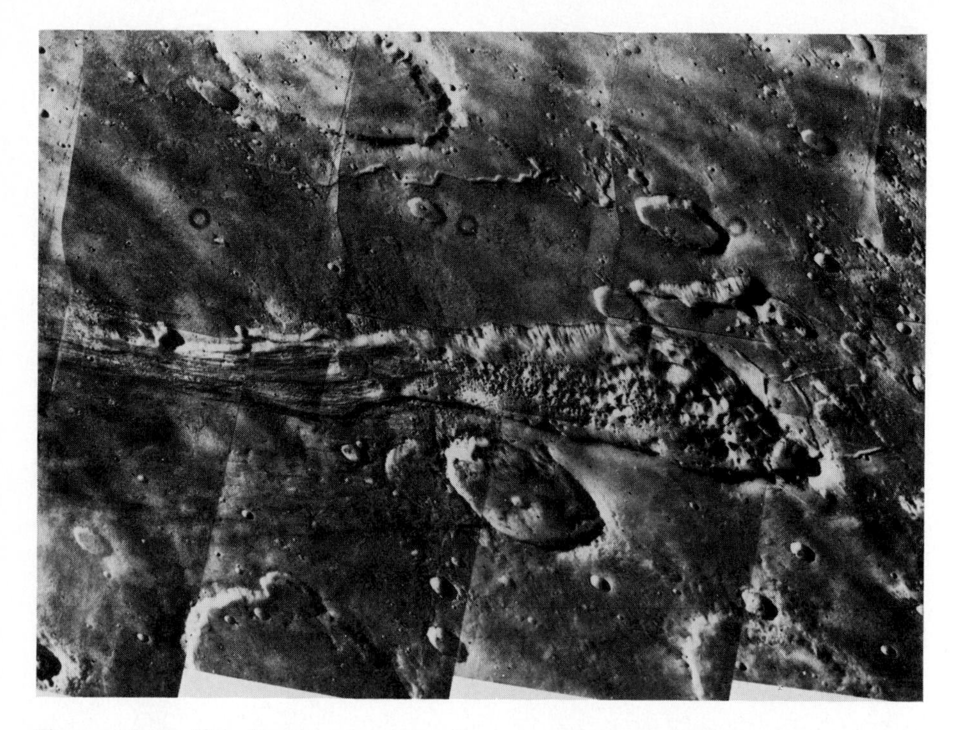

Figure 13.15 This large triangular valley, several dozen kilometers across, may have arisen from subsidence (collapse) of the Martian surface, perhaps a result of melting of permafrost or lenses of ice below the surface. This subsidence apparently caused a gigantic flash flood that carved the existing channel to the left in the photograph.

Swift had all speculated that Mars ought to have two moons.[1] In contrast to the Earth's own moon and to the four large satellites of Jupiter, the two Martian moons have diameters of less than 20 kilometers, making them far smaller than the larger asteroids, small enough that a tourist on either of them could launch rocks into space with a good hard toss.

The Viking orbiters took some marvelous photographs of Phobos and Deimos, laying to rest the supposition that these small planetary satellites might be the artificial creations of an advanced civilization. As Fig. 13.16 shows, these moons are too small for their own gravitational forces to

[1] At the time when Kepler, Voltaire, and Swift wrote, the Earth was known to have one satellite and Jupiter four, so to make a proper geometrical progression, Mars was imagined to have two moons. This simple explanation has not prevented some fantastic speculation that Kepler, Voltaire, and Swift must have been in touch with advanced civilizations who told them of the moons of Mars.

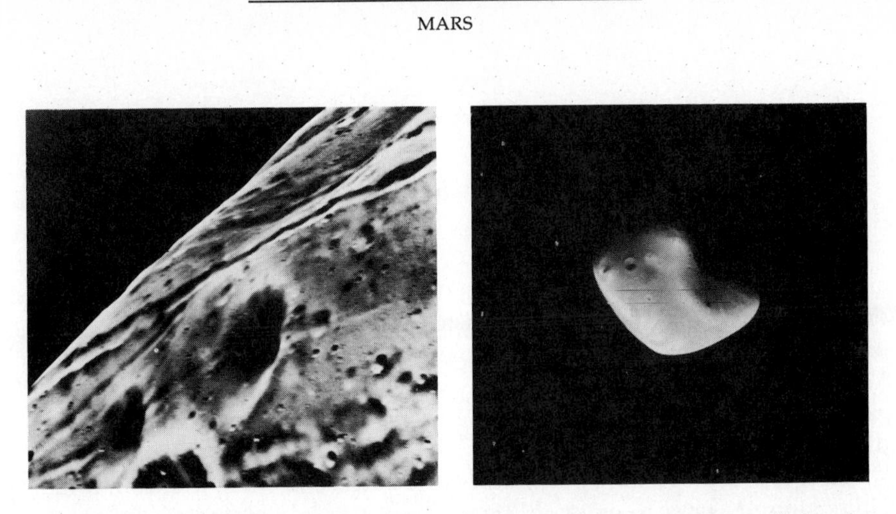

Figure 13.16 Mars's two small satellites, Deimos (right) and Phobos (left), show the effect of cratering. The impacts of Phobos have also apparently produced the series of parallel grooves visible on the surface of the satellite. Deimos is seen in full phase in this view, with no shadows to enhance the contrast.

deform them into nearly spherical shapes, as occurs for all large objects, such as the Earth and the moon. Battered by impacts from countless meteoroids, Phobos and Deimos resemble asteroids (see Fig. 11.9). In fact, these satellites may *be* captured asteroids, rather than objects that accreted in their present orbits when Mars formed. Spectroscopic studies suggest that the composition of Phobos and Deimos closely resembles that of the carbonaceous chondrite meteorites (see page 271).

Hence some future investigation of these two bodies may reveal the presence of organic compounds like those discovered in the Murchison meteorite. Determining the relationship of these satellites to the asteroids and comet nuclei promises to be an interesting problem. There is tantalizing evidence from the Soviet *Phobos* mission that visited Mars in 1989 suggesting that water vapor may be escaping from Phobos, presumably subliming from ice buried deep in its interior. But to find life itself, our best prospect still seems to be Mars. The Viking biology experiments represent the best direct search for alien life that we have made so far, and the story of their accomplishments requires a separate chapter.

SUMMARY

Mars, with just one-half Earth's diameter and one-tenth Earth's mass, does resemble Earth in its rotation period, in its seasonal changes in appearance, and in its possession of an atmosphere. Its greater distance from the sun, however, and—most importantly—its atmospheric pressure of less than 1 percent of that on Earth combine to preclude the existence of liquid water on Mars. The bulk of the water has either frozen into the subsurface soil as permafrost or into Mars's polar caps, which are mainly frozen carbon dioxide; a tiny fraction exists as water vapor in the Martian atmosphere, which is also mostly carbon dioxide.

The low atmospheric pressure on the surface of Mars makes it impossible for liquid water to exist on Mars except under extremely special conditions. Like carbon dioxide on Earth, which passes from dry ice to vapor directly, water on Mars must be either solid or gaseous. Nevertheless, photographs of Mars taken by the Mariner and Viking spacecraft show evidence for liquid water on Mars in the distant past. This evidence consists of sinuous, branching channels that must have been carved by flowing water.

The surface of Mars reveals a heavily cratered terrain, proof that little erosion and weathering has occurred during the 4 billion years since most of the craters formed. Gigantic Martian volcanoes, some three times as high as any mountain on Earth, are symbols of geological activity that may even extend to the present day.

The Martian "canals," once highly touted by some astronomers, clearly do not exist at all but are instead optical illusions of a particularly exciting variety. Seasonal changes in color over wide areas of Mars's surface apparently arise from windblown dust, and certainly not from the growth or decay of vegetation. The two moons of Mars, Deimos and Phobos, are so small that they maintain irregular shapes. Their battered surfaces reveal a history of early bombardment. These satellites probably resemble small carbonaceous asteroids in appearance and even in composition, leaving open the possibility that they contain organic compounds of biological interest.

QUESTIONS

1. What sort of observations made before spacecraft visited Mars encouraged people to believe that Mars might have vegetation, and even intelligent beings, living on its surface?

2. What are the "canals" of Mars?

3. Why is it impossible for any liquid water to exist on the surface of Mars?

4. On Earth, the atmospheric pressure decreases by a factor of 2 every time we gain 6 kilometers of altitude. About how high would we have to rise before the atmospheric pressure would fall to the value at Mars's surface, about 1/150 of the pressure at sea level on Earth?

5. Why does the Martian atmosphere fail to shield the planet's surface against ultraviolet radiation from the sun? What does this fact imply for the possibility of life on Mars?

6. What evidence suggests that Mars once had liquid water on its surface? If this were so, what conditions must have been different for liquid water to exist? What other evidence supports the possibility of such a massive climate change?

7. Discuss the difficulties that scientists faced in choosing a landing site for the *Viking* expedition to Mars.

8. What do the Martian polar caps consist of?

9. Where do scientists think that most of the carbon dioxide that was formerly in the Martian atmosphere may now be found?

10. What technique would you use to boil an egg on Mars?

FURTHER READING

Baker, Victor. *The Channels of Mars.* Austin: University of Texas Press: 1982.

Beatty, J. Kelly, and Andrew Chaikin. *The New Solar System* (3d ed.). Cambridge, MA: Sky Publishing, 1990.

Carr, Michael. *The Surface of Mars.* New Haven and London: Yale University Press, 1981.

Haberle, R. M. "The Climate of Mars," *Scientific American* (May 1986), 54.

Littman, Mark. *Planets Beyond: The Outer Solar System.* New York: Wiley, 1988.

Morrison, David and Tobias Owen. *The Planetary System.* Reading, MA: Addison-Wesley, 1987.

Veverka, Joseph, et al. "The Puzzling Moons of Mars," *Sky & Telescope* (September 1978).

Washburn, Mark. *Mars at Last!* New York: G. P. Putnam, 1977.

Science Fiction

Bradbury, Ray. *The Martian Chronicles.* New York: Doubleday, 1950.

14

Is There Life on Mars?

How can we determine whether life exists on Mars? The analysis provoked by scientists' interest in this question during the last century has led to the suggestion of four chief ways to detect life on another planet. In order of increasing difficulty (historically) for us on Earth, they run as follows: (1) We can look from Earth for any major changes on Mars that we might ascribe to the efforts of another civilization. (2) We can listen at various radio wavelengths for signals produced by another civilization. (3) We can analyze the composition of the planet's atmosphere, using advanced spectroscopic techniques, to spot subtle changes caused by the presence of life. (4) We can go to another planet, to perform complicated experiments that test for the presence of both large and microscopic living organisms.

Note that the first three of these techniques can be applied directly from Earth, provided that we have sensitive telescopes, radio antennas, and spectroscopes. Note also that the fourth and most direct means of search costs considerably more than any of the first three, though not so much that humans have refused to pay it. Before we look at the results from this fourth test, let us pause to consider what an intelligent Martian might find out about life on Earth, using the first three techniques.

Until quite recently, the first two techniques—looking and listening—would have failed to signal the presence of life on Earth to a Martian with the same technology that we possess now. The distorting effects of the Earth's atmosphere would obscure our planet's surface so much that changes in the landscape caused by humans could not be seen, and the nighttime glow of our cities would not have reached the level of easy detection. Similarly, no radio transmissions would have been detectable, since humans began to generate large amounts of radio power only in the 1920s. The third test, however, would have yielded positive results as early as two billion years ago, even if no civilizations had arisen on Earth.

Spectroscopic observations made from Mars would reveal the presence of a large amount of molecular *oxygen* in the Earth's atmosphere. This would puzzle an intelligent Martian, since oxygen is a highly reactive gas that rapidly combines with rocks that are continually being brought up from the interior to the Earth's crust. This process removes oxygen from our planet's atmosphere, so the continued existence of this gas requires a source, one that can use energy to liberate this reactive gas into the atmosphere. That source, of course, is life, the green plants that release oxygen through photosynthesis.

If all that oxygen wasn't enough of a clue, Martian scientists would find more evidence for life when they discovered that a small amount of methane is also present in our atmosphere. Methane is a major component of the fuel we commonly call "natural gas." It burns in the presence of oxygen, and burning is simply a rapid form of oxidation. Ultraviolet radiation from the sun provides enough energy to oxidize all the methane in the Earth's atmosphere in just 10 years, converting it to carbon dioxide and water. The continued existence of methane on Earth requires the presence of a source that replaces the methane as it oxidizes. Once again, life is that source, in this case bacteria that live in the stomachs of grass-eating animals and in swamps. (Natural gas is associated with deposits of coal and oil, both the products of dead organisms.)

Finally, a Martian who examined the Earth's spectrum in 1972 and again in 1992 would notice a striking change. The modern atmosphere contains tiny amounts of chlorofluorocarbons, which are known in the United States as "freons," and are used to pressurize spray cans and to serve as the working fluid in refrigerators and air conditioners. These substances do not occur in nature, hence their existence indicates the presence of an advanced (but not yet so intelligent!) civilization.

Thus the third test would show Martians that life exists on Earth long before they landed any spacecraft to sample the immediate environment. Likewise, if we had discovered large amounts of oxygen in the Martian atmosphere, we would have considered life to be likely there. But we now know that oxygen forms only 0.13 percent of the thin Martian atmosphere, which has a total pressure only 0.7 percent of our own. This tiny amount of oxygen can be easily explained as the result of photochemical reactions that occur when sunlight strikes the small amount of water vapor and carbon dioxide in the Martian atmosphere. Methane and other hydrocarbon gases (to say nothing of freons!) remain undetected on Mars, so the planet shows no signs of the unexpected conditions that mark Earth as biologically active.

In short, the first three tests have failed to show any sign of life on Mars: no "canals," no cities, no radio or television broadcasts, no unexpected atmospheric gases. As a result, humans put nearly a billion dollars worth of effort into the fourth test, and built the *Viking* spacecraft, capable of landing on Mars and performing delicate experiments to test for the presence of Martian microbes.

How to Find Martian Microorganisms

Although the greatest excitement we might expect from our search for life would be the discovery of large, advanced creatures capable of communicating with us, the history of life on Earth reminds us that microscopic organisms far outnumber large ones and are far hardier. The fossil record shows that microbial life was the only kind of life on Earth for billions of years, far longer than larger creatures have existed. Nor have microorganisms decreased in numbers or in adaptive ability. The dirt in your backyard contains more organisms than the number of stars in our galaxy. We have no reason to expect another planet with life to differ from Earth in the overall development of living systems, so we greatly increase our chances of finding life when we include a search for microorganisms. This realization underlies the design of experiments to search for life on Mars.

But how do we find these "cavorting beasties," as their discoverer, Anton Leeuwenhoek, called them? We cannot use an ordinary camera to see tiny organisms, and it is difficult (as design studies showed) to send a powerful microscope to Mars and to provide it with samples in a suitable form to be examined. Instead, Viking had to rely on experiments that would detect microorganisms in slightly more complex ways. Furthermore, the Viking experiments had to deal with the possibility that the carefully planned and marvelously miniaturized laboratory that landed on the surface of Mars might simply detect Earth microbes that had been carried millions of kilometers to Mars aboard the lander itself!

The Viking scientists managed to overcome this last problem through immensely careful heat sterilization of the entire spacecraft. They also wrestled with the best way to obtain samples of Martian soil that might contain microbes unaffected by the descent of the spacecraft. A simple extendable boom with a scoop at the end was finally judged best, because tests showed that the landing would disturb the soil only slightly, and then only directly beneath the spacecraft (Fig. 14.1). Suppose an uncontaminated soil sample could in fact be tested for microorganisms. What would they like to eat and drink? Here the Viking scientists fell back on the basic principles of biology and chemistry, principles that must be at least somewhat biased from the fact that we derive them from one single example of biology, life on Earth.

With this bias firmly in mind, we can offer the following guidelines: Life on Mars, if it exists, ought to be based on carbon chemistry and should involve a fluid solvent, some common substance that can exist as a liquid under Martian conditions. For Mars, which has no ammonia or alcohol, this means a carbon chemistry with water; that is, a system of life essentially identical in its chemical outlines with life on Earth. This may seem a restrictive conclusion, but the imposition of a chemical similarity at such a basic

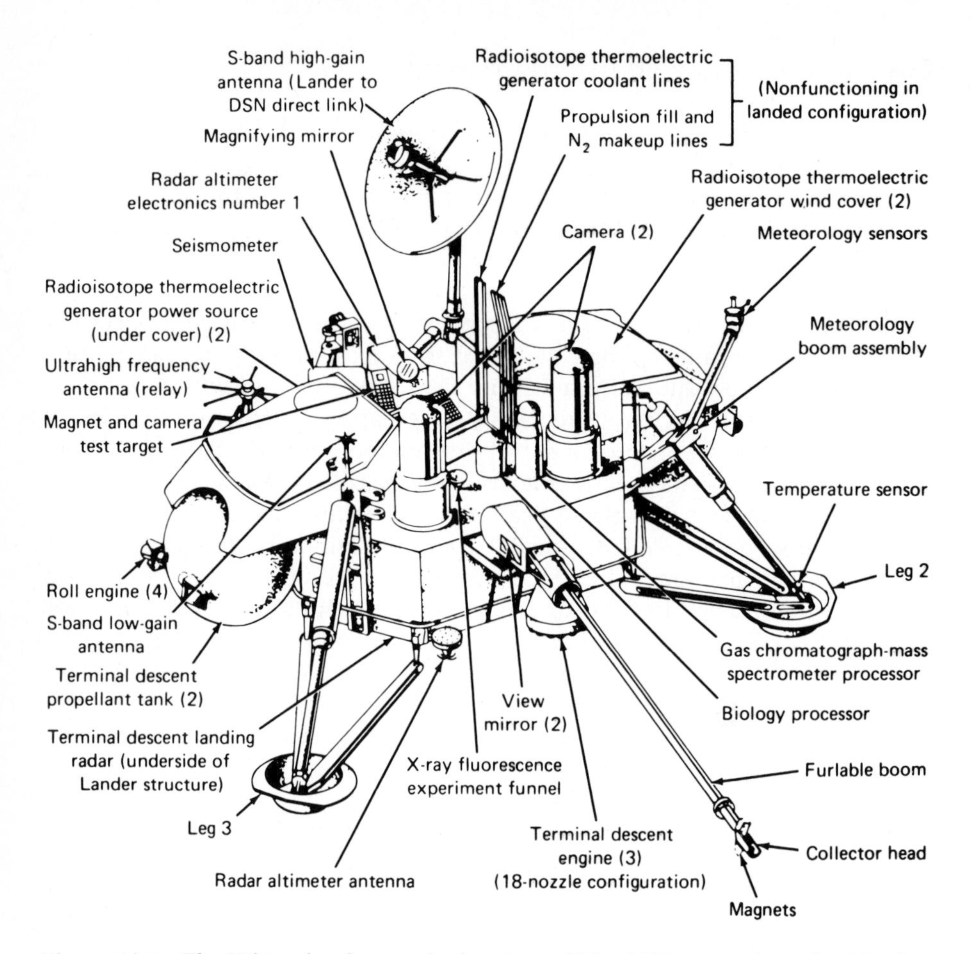

Figure 14.1 The Viking landers packed an incredible ability to analyze the Martian atmosphere and surface into a small volume, including an extendable boom with a scoop that could bring soil samples into the chambers of various experiments on board.

level permits enormous flexibility in the type of life that might actually exist (see Chapter 10). Then how can we detect tiny Martian microbes of unknown form and behavior?

A logical first step is to study the Martian atmosphere and soil in detail. All organisms, even microbes, develop an intimate connection with the places they inhabit. If microorganisms are sufficiently abundant, they will modify the chemistry of their surroundings, thus producing evidence of life that should be detectable through analysis of either the atmosphere or the soil of Mars.

The Viking Results: Atmospheric Analysis

We have already seen that observations of the Martian atmosphere from Earth found no evidence for any nonequilibrium gases that would imply the existence of life. The Viking measurements simply supported this analysis. The mass spectrometers did find that nitrogen forms between 2 and 3 percent of the atmosphere of Mars. This was an important discovery, because nitrogen gas cannot be detected by Earth-based spectroscopic observations.

This may seem surprising, since we know that nitrogen constitutes 78 percent of our own atmosphere. The reason is that this gas is almost completely transparent to visible and infrared radiation, only absorbing ultraviolet photons at wavelengths we cannot detect at the Earth's surface because of the ozone layer. Hence it is much easier to study nitrogen with a mass spectrometer that is immersed in the atmosphere. This instrument ionizes any gas that enters it, and then uses a magnetic field to separate one ionized gas from another according to their masses and their degrees of ionization.

Since nitrogen exchange between organisms and the atmosphere of Earth forms an essential process for life on this planet, the discovery of nitrogen in the Martian atmosphere seemed a hopeful sign to those who searched for life on Mars. But no evidence of any gases unexpected without the presence of life appeared; the upper limit on the methane abundance at the two landing sites fell below a few parts per million. The infrared spectrometers on the *Mariner 9* spacecraft had already set a planetwide upper limit of 25 parts per billion of methane in the entire atmosphere. Thus there seems little chance that extensive herds of cattle exist on Mars! The spectrometers also found no traces of silane, the silicon analogue of methane. (Silane gas would in fact be unstable in the Martian environment, so silicon-based life would bring us back to the possibility of fire-breathing dragons (see page 234.)

The Viking Results: Soil Analysis

Once the first Viking lander descended on the surface of Mars and the first soil sample rode the scoop into the test chambers, the results came quickly back to Earth. No evidence of organic compounds appeared in the Martian soil. The Viking lander made its soil analyses with a highly sophisticated instrument called a gas chromatograph–mass spectrometer, or GCMS. The GCMS baked soil in an oven to drive off volatile gases. These gases adhere to the gas-chromatograph part of the instrument with different degrees of stickiness, and heating them makes the volatiles leave in sequence. The mass spectrometer analyzes the departing stream of gases to determine which compounds are present.

Figure 14.2 shows an analysis of Antarctic soil on Earth with a laboratory version of the Viking GCMS. Although this soil contains barely enough living organisms to give weakly positive results in the three life-detection experiments, the GCMS shows that many organic compounds are present. In other words, the organic compounds associated with life can be seen more easily than the living creatures themselves. For instance, the Antarctic soil sampled in Fig. 14.2 contains 10,000 times more carbon in organic molecules than in microorganisms (Color Plate 14).

We must realize, however, that not all organic compounds require living creatures to produce them. Figure 14.2 also shows a GCMS analysis of a meteorite that contains amino acids, the building blocks of proteins. Again we find carbon-rich molecules, but these organic molecules were not made by life. Thus the detection of organic molecules in general would not show that life exists on Mars, since the simpler kinds of organic compounds could have been brought to Mars by meteorites, or could have even arisen from photochemical reactions in the Martian atmosphere.

In practice, these ambiguities did not plague the GCMS experiment, because it found *no* organic compounds in the soil of Mars. On the scale of Fig. 14.2, the Martian soil results would be indistinguishable from zero, as defined by the horizontal axis of the graph. These results apply to both landing sites, and to two different samples at each site, including one from underneath a rock that might (so it was thought) have sheltered organisms from deadly ultraviolet light. The only compounds found in the soil were water and carbon dioxide, whose presence was hardly surprising since these gases exist in the planet's atmosphere. The upper limits on all likely organic compounds in the soil, such as hydrocarbons, acetone, furan, and acetonitrile, fall at a few parts per billion.

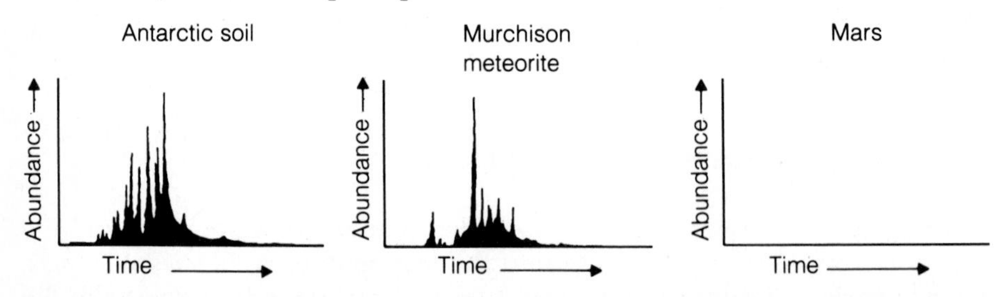

Figure 14.2 A test model of the Viking GCMS analyzed Antarctic soil (left) and a piece of the Murchison meteorite (compare Fig. 11.12), which contains amino acids (center). The graphs indicate that a rich variety of organic substances is contained in each of the two samples since each peak in the graph represents one or more compounds. On this scale, the GCMS analysis of Martian soil is a straight line at zero with no peaks at all (right).

These negative results set powerful constraints on any models of Martian biology: How could life exist on Mars without leaving any trace of its presence? Could Martian organisms be such efficient scavengers that all traces of their wastes, their food, or their corpses could not be found, even with the great degree of sensitivity that the Viking landers brought to Mars? The biologists involved in the Viking project considered this an extremely unlikely possibility.

The Viking Biology Experiments

In view of the negative findings of the soil analysis and the equally negative returns from the atmospheric analysis, the prospects for finding life on Mars with the experiments designed for precisely that purpose seemed poor. So the Viking scientists received quite a shock when all three of the biology experiments showed positive results!

After considering many possibilities, the Viking project had selected three experiments to search for life on Mars: (1) the gas exchange (GEX); (2) the labeled release (LR); and (3) the pyrolitic release (PR). These experiments all reflect scientific experience with life on Earth. Thus, for example, all the organisms that we know derive their energy from two basic processes: oxidation (removal of hydrogen, combination with oxygen) and reduction (removal of oxygen, combination with hydrogen). Both processes deserve investigation on Mars. But how? What would the Martians "eat" for energy? Under what conditions would they eat?

We have seen that it was reasonable to assume that any life that existed in the environment provided by the surface and atmosphere of Mars would be carbon-based and would rely on water as a solvent. Thus scientists felt justified in offering the Martian microbes some suitable mixture of organic nutrients dissolved in water, to see what—if any—reactions would occur in this diet. The GEX and LR experiments followed just this procedure.

The GEX experiment put a soil sample in contact with several dozen "likely" nutrients (Fig. 14.3). These nutrients have such broad appeal for Earth-based life that the GEX quickly became known as the "chicken soup" experiment. Once the soil sample came in contact with the chicken soup, the question was: Would the gas chromatograph analyzing gases in the chamber detect any changes in the composition of the gas above the soil? Such changes should arise from the life processes of the Martian microbes. On Earth, the chicken-soup approach would reveal the presence of life through changes in the amount of oxygen, carbon dioxide, or hydrogen in the air above the soil caused by the metabolic activity of organisms in the soil.

The labeled release (LR) experiment aimed at checking on biological activity more directly. This experiment used a set of compounds that were

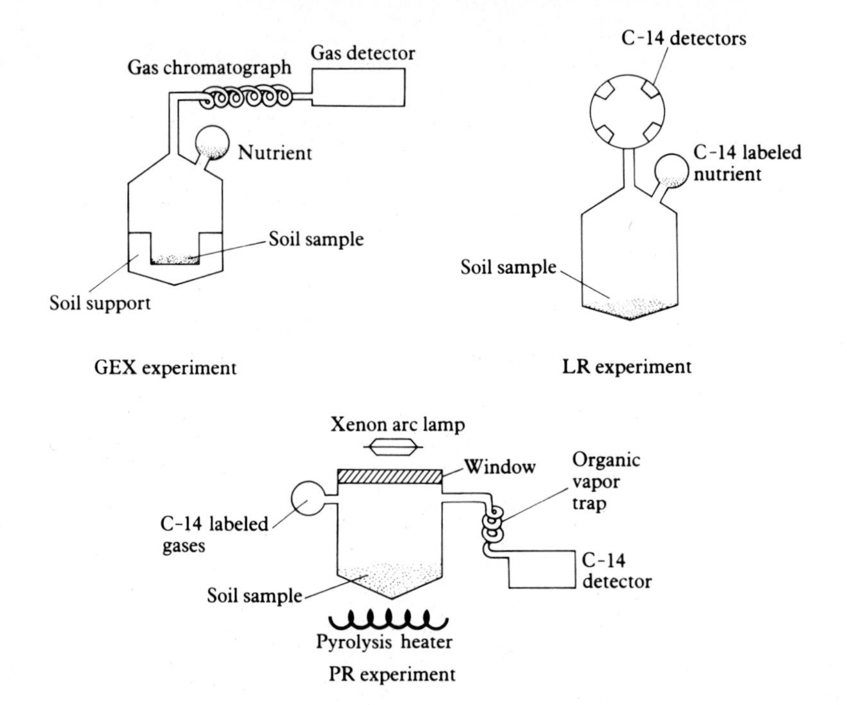

Figure 14.3 A schematic representation of the GEX, LR, and PR experiments shows that each of them analyzed the Martian soil in a different way to test for the presence of life. The GEX experiment exposed a broth of nutrients to a few grams of soil and then looked for changes in the gas above the soil-and-nutrient mixture. The LR experiment tagged carbon-rich compounds with radioactive carbon-14 atoms in place of some of the usual carbon-12 atoms. These labeled compounds then dripped over the soil sample. Any biological processes should have caused some tagged compounds to appear in the gas above the sample. The PR experiment replaced the normal Martian atmosphere with an equivalent set of gases labeled with radioactive carbon atoms. Any organism that ingested some of these labeled molecules would produce a radioactive signal when the soil in which they lived was roasted.

labeled by substituting radioactive carbon atoms for some of the ordinary carbon atoms in the compounds (see again Fig. 14.3). This labeled mixture dripped onto the soil sample, and the gases above the sample were tested to see whether any radioactive carbon compounds, such as carbon dioxide or methane, had been released by the life processes of Martian organisms. On Earth, we could call the LR a respiration experiment, to see whether organisms are releasing gases into the atmosphere.

The GEX and LR experiments have two basic drawbacks. First, liquid water cannot exist on Mars now, so Martian microbes may be completely unused to a watery medium. Second, organic nutrients found delicious by terrestrial organisms may be poisons to Martian microbes. The third biology experiment, the pyrolitic release (PR), dealt with these drawbacks (see again Fig. 14.3). In this experiment, transported Martian soil enjoyed an environment virtually identical to that on the planet's surface, but the atmospheric gases in the test chamber were labeled by adding carbon dioxide and carbon monoxide that had been tagged with radioactive carbon. After allowing any organisms to live for a while, the soil was heated to 750° C, and the volatile gases released by this heating passed into a vapor trap and then into a counting device that could measure the radioactivity of the gases.

Only carbon dioxide and carbon monoxide could pass all the way through the vapor trap. Once these two gases had left the system, the vapor trap was heated to drive off any organic vapors that might have been produced. These vapors could be recognized because they would contain some of the radioactive atoms from the added carbon dioxide or carbon monoxide. In other words, the PR experiment aimed at roasting the remains of Martian microbes to release carbon atoms that the microbes had incorporated through biological activity. The most important terrestrial activity of this kind consists of plant photosynthesis, in which carbon dioxide in the Earth's atmosphere is converted into organic compounds by green plants. Baking the plants would vaporize these organic compounds. In the PR experiment these vapors would be radioactive.

Tests of the PR experiment on Earth showed, however, just how difficult it might be to separate biological activity from chemical reactions. When biologists exposed a sterilized sample of simulated Martian soil to short-wavelength ultraviolet radiation in the presence of a mixture of gases that copied the composition of the Martian atmosphere, they found that some simple organic molecules, chiefly glycol and formaldehyde, were formed. If the same chemical reactions occurred in the atmosphere of the real Mars, they would exactly mimic the behavior of a simple photosynthetic organism that was turning carbon dioxide into more complex carbon compounds. Hence, the experimenters decided to filter out the short-wavelength ultraviolet radiation from the light source that simulated the sun inside the experimental chambers, in order to exclude the kinds of photochemical reactions that they had discovered in their laboratory tests on Earth.

Results of the Viking Biology Experiments

On the eighth day after the first landing on Mars, the lander scoop dug a trench in the soil and distributed samples to the various experiments (see Color Plate 12). The GEX placed about a gram of soil into a tiny, porous container positioned above the nutrient medium. Two days later, the first analysis of the gas in the container showed an exciting result: A large quantity of oxygen had appeared in the chamber, 15 times the proportion in the planet's atmosphere! The simple exposure of Martian soil to the humidity in the test chamber (caused by the nutrient-laden fluid) had apparently been sufficient to liberate oxygen from the soil. Was this an indication of life on Mars? After months of testing, which included the discovery that a previously heated sample of soil produced the same release of oxygen, the biologists concluded that they were observing not biological activity but merely the chemical interaction of Martian soil with a higher pressure of water vapor than had been present for millions of years. In other words, not the "chicken soup" itself, but the humidity it produced, had led to oxygen-releasing *chemical reactions* in the Martian soil.

The day after the first GEX data, the LR experiment reported back: The result was again positive! After checking to be sure that the background radioactivity level was low, the LR added about two drops of the radioactive nutrient material to the soil. A sudden rise in the radioactivity of the gases above the soil sample appeared, a more dramatic reaction than biologists find with many life-bearing soils on Earth. Unfortunately for those who hoped for proof of life, the Viking scientists soon realized that the radioactive gas, almost certainly carbon dioxide, could be produced by simple chemical reactions that involve peroxides. If, for example, hydrogen peroxide (H_2O_2) exists in Martian soil, it could easily react with an organic compound in the nutrient medium, such as formic acid (HCOOH), to form water and carbon dioxide. A second wetting of the soil showed no increase in the amount of radioactivity in the gas in the test chamber. In fact, the additional nutrient apparently absorbed some of the radioactive carbon dioxide that was originally released. Hence, the scientists concluded not that life exists on Mars, but rather that the Martian soil may contain chemicals such as peroxides that release carbon dioxide when exposed to simple organic compounds.

Since the first two experiments yielded information that seemed ambiguous, the Viking team eagerly awaited the results from the PR experiment, which did not use a water-based nutrient and thus avoided one of the primary agents (water) that the biologists suspected of causing purely chemical reactions in the Martian soil. The PR experiment required an incubation time of five days, during which radioactive carbon monoxide and carbon dioxide stayed in the sample chamber. Analysis of the initial experiment revealed that radioactive carbon had indeed become part of some compounds in the soil (Fig. 14.4). Weak as this signal was, it seemed clearly positive: To the PR experiment, Martian soil behaved much like an Antarctic soil on Earth, nearly sterile but not entirely so. Thus this experiment, apparently the most difficult to fool by nonbiological reactions, yielded an

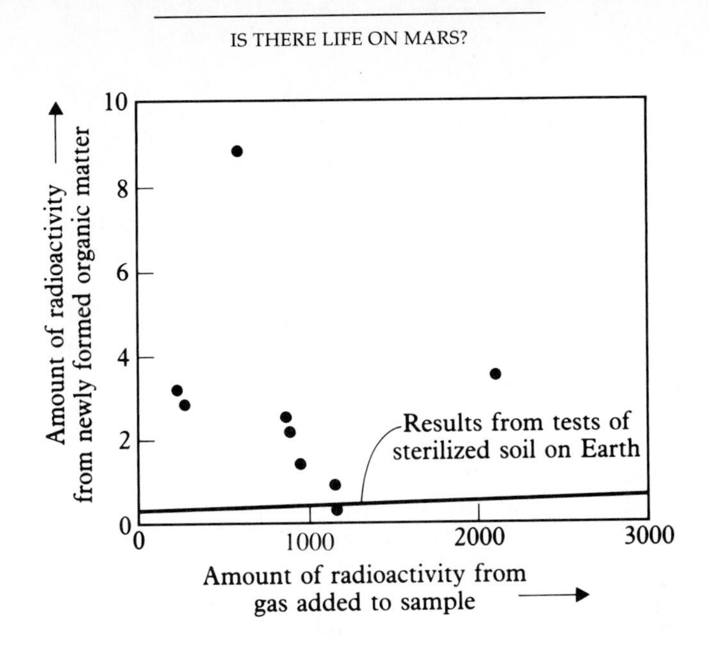

Figure 14.4 In the PR experiment, Martian soil was exposed to CO and CO_2 tagged with radioactive carbon under simulated Martian conditions. Gases produced by subsequently heating the soil contained some of that radioactive carbon, showing that a new organic compound had formed in the soil.

undeniably positive result. Yet even in this case the Viking scientists were skeptical that they had found life on Mars.

The reason for this skepticism comes from the fact that when the scientists arranged to heat the Martian soil to 175° C and to 90° C for several hours before the radioactive gases were injected, they still found positive results from the PR experiment. The higher temperature reduced the reaction by 90 percent, but a positive signature still emerged; the lower temperature had no effect. Since Mars's surface temperature never reaches even 40°sC, the Viking scientists did not believe that any Martian life could have adapted to survive 3 hours at 175° C, which no terrestrial organisms can do. Furthermore, because the GCMS showed that the soil lacks any organic material down to a level of a few parts per billion, the likelihood that a tiny amount of the (supposed) heat-resistant microorganisms could provide the positive results of the PR experiment seemed minuscule. In fact, further testing has suggested that a small amount of ammonia in the first PR soil sample—contributed by leakage from the descent engines on the Viking lander—may have been responsible for the soil chemistry that produced the first, weakly positive, signal.

Nonetheless, all three biology experiments yielded positive results. The fact that these results arise from chemical reactions that mimic the effects of microorganisms, as the Viking scientists have concluded, simply shows the difficulties we face in trying to distinguish an alien form of life from a vast array of possible chemical reactions. It is a problem remarkably similar to that faced by the scientists on board *H.M.S. Challenger* a century earlier (see page 5). The consensus among the Viking biologists is that Martian soil contains loosely bound oxygen in various compounds, such as peroxides. Because the temperature on Mars is low, and because water is entirely absent from the planet's surface, these compounds can remain far from chemical equilibrium for long periods of time, but if we add even a small amount of water, we produce chemical reactions that imitate the effects of biological activity.

The Viking missions have revealed that Martian soil is chemically active but probably contains no carbon-based living organisms, at least at the two landing sites. But could we have killed the existing Martian microbes by landing close to them or by the act of bringing them inside the spacecraft? This seems unlikely, since any such microbes must have adapted to daily temperature variations of almost 100° C, and the temperature inside the spacecraft (20° C) did not exceed the maximum temperature on the planet's surface. The PR experiment, in fact, was designed to duplicate the actual Martian environment as closely as possible. Tests carried out on Earth to simulate the landing on Mars showed that at the distances reached by the sample arm, the Martian surface would not have been heated sufficiently to kill an indigenous population of microbes. Could life on Mars be based on silicon? The GEX experiment, which showed that oxygen appears upon exposing Martian soil to moisture, might be a characteristic reaction of silicon compounds and water. The fact, however, that the reaction was found to die out rapidly following the initial exposure to moisture speaks against a life-based interpretation rather than a chemical one. (See Chapter 10 for other arguments against silicon-based life.)

We do have one device to detect life on Mars that remains completely independent of chemistry—namely, the cameras on the two Viking landers. The pictures from Mars have been searched with extreme care for any evidence of movement, any burrow, footprint, trail, or artifact of any kind, and nothing has been seen, except the effect of the Martian winds on the dust in the empty landscape (Fig. 14.5). Since the atmospheric composition of Mars, as well as the Viking soil analysis, argues against the presence of life—especially of life in large quantities—the prospects for any type of life at the

Figure 14.5 Two photographs of the 2-meter boulder near the *Viking 1* lander (known familiarly to the experimenters as "Big Joe") were taken 6 months apart, in September 1976 (left) and March 1977 (right). During the six-month interval, a small amount of sediment shifted from a rock in front of "Big Joe's" right flank (arrow).

two landing sites must be judged poor, despite the tantalizing positive results from the three biology experiments.

Did the Vikings Land in the Wrong Places?

But what about life elsewhere on Mars? The similarity between the soil analyses at the two landing sites, separated by more than 2000 kilometers, argues against great variations from one place to another. This argument gains strength from current models of the planet's surface, which suggest that the soil is "gardened" (turned over) by meteorite impacts and by wind down to a depth of several meters, on time scales of the order of tens of thousands of years, and that the soil particles are blown over the whole surface of the planet. We can see this process in action during the intense Martian dust storms, which can obscure the surface of the entire planet. Thus every part of Mars's surface seems likely to come in contact with every other part, so that sampling the dusty material at one place should be equivalent to sampling at all locations, as was indeed the case for the soil analyses made at the two Viking landing sites (see page 315).

A possible exception to this rule arises in the polar regions (Fig. 13.14). The north pole of Mars has a permanent cap of water ice, and we can imagine specialized environments at the edges of this polar cap where conditions could be more favorable to life during the peak of summer. A

Figure 14.6 Layered terrain near the Martian poles testifies to the build-up of material through cycles of dust deposition, erosion, and fresh transport of dust particles. Such layered sediments contain a historical record of climate cycles on Mars and store materials formed at earlier epochs on the Martian surface.

relatively abundant source of water, one that we know enters the atmosphere and then freezes out again, lies in the polar cap. The ground temperatures there, as measured by the Viking orbiters, reach −33° C, about as warm as the highest temperatures measured near the second landing site but not subject to extreme changes as Mars rotates. In fact, simply because the polar regions stay cold, organic material might concentrate there, since it would condense at these temperatures, and once on the ground it could never heat up and blow away. The south polar region of Mars, not as well observed as the north polar cap, seems to be still colder than the north pole at this epoch, and keeps a small permanent cap of frozen carbon dioxide that probably covers a layer of water ice. Thus the north pole of Mars seems to offer the best high-latitude environment at the present time.

But "best" may not be good enough. Color photographs of Mars show that even near the pole, the bare ground reveals the same reddish color as the rest of the planet. Here, too, the soil may be gardened, blown about by the winds, and certainly irradiated by ultraviolet photons from the sun. Thus this soil should be as thoroughly oxidized as the soil tested by the biology experiments at lower latitudes. Models for the wind circulation on Mars show that the dust that becomes airborne in middle-latitude regions will be carried to the poles, where we can see layers of material that have built up as the result of repeated cycles of erosion, airborne transport, and deposition of fine dust particles (Fig. 14.6). Therefore, even here no living organisms should exist in the soil. Perhaps under the ice, at the frigid poles, we might one day expect to find some protected carbon compounds, but here the continual low temperatures seem to rule out the existence of active life.

We might imagine isolated "oases" that could harbor life on Mars, but we must remember that every oasis requires a well-populated external community for its continued existence. As the sands shift to cover an oasis or dry spells exhaust it, seeds and microorganisms must arrive from somewhere else to repopulate it; otherwise, without this outside reservoir of life, any oasis must soon become lifeless.

Dogmatic statements have fared poorly in the history of science, and we must be reluctant to state categorically that no life exists on Mars. The evidence we have accumulated so far points toward the absence of life, though some new discovery may someday appear that convinces biologists that Mars does harbor some living organisms. We *can* say that no known terrestrial organisms, including the toughest of all microbes, could grow in the present Martian environment.

An Ancient Eden? Goals for Future Exploration

As we saw in Chapter 13, we have good reason to believe that primitive conditions on Mars resembled conditions on the primitive Earth. Those dry river beds and the evidence that exists for standing water (see Fig. 13.4) are strong indications that the planet went through an early period that included a CO_2-rich atmosphere with a surface pressure about equal to the present sea-level pressure on Earth. Such an atmosphere could produce a greenhouse effect that would be sufficient to warm a large fraction of the planet above the freezing point of water. This scenario is supported by detailed studies of the inert gases and the isotope ratios of abundant elements in the Martian atmosphere. True, the last major water activity on the Martian surface apparently died out about 3.0 to 3.5 billion years ago. This is a highly significant date, however. In Chapter 9, we looked at a picture of the most ancient evidence of life on Earth, fossilized colonies of blue-green bacteria known as stromatolites (Fig. 9.1). These fossils are 3.5 billion years old. If life could have arisen on Earth and evolved to this level of complexity during the first billion years of our planet's history, then why not on Mars? With the evidence we presently have available, we cannot exclude this tantalizing possibility.

So life may have begun on Mars, only to die out as the planet's climate changed and the atmosphere approached its present low density. We should search for evidence of these early beginnings in the oldest sedimentary rocks on Mars, just as we look for evidence of microbial fossils in the oldest rocks on Earth. Finding the equivalent of Martian stromatolites may not sound as exciting as encountering exotic alien beings, but think what such a discovery would mean. At the present time, we know of only one example of life in the universe: our own. Is this astonishing property of matter a rare or common phenomenon? We simply do not know. Finding evidence that life originated on Mars, even if it all died out, would at last offer us an answer to this question. If life indeed began on both of the two planets in our solar system on which we would expect it, then we can reasonably hope to find life occurring with some frequency on the planets of other stars.

Searching for fossils calls for a much more elaborate set of missions than we have sent to Mars thus far. We need "rovers" that can move through the Martian canyons and into craters flooded in ancient times, beaming pictures back to Earth and taking samples as directed (Fig. 14.7). We would hope to find fossil evidence of colonies of microorganisms, similar to the ancient stromatolites found on Earth. In the best case, we should bring back samples to our laboratories, where we can study them in detail. Missions that would involve this sort of sample return from Mars are now under consideration by NASA, Russia, and the European Space Agency for possible development during the late 1990s.

Before this ambitious goal is achieved, there will be two more spacecraft put into orbits around Mars: *Mars Observer,* a NASA mission to be launched in 1992, and *Mars 94,* a Russian mission planned for 1994. These spacecraft will provide more detailed pictures of selected areas of the surface than we have at present, while making a number of compositional measurements of the surface and atmosphere. *Mars 94* may also include balloons and a package of instruments deployed on the planet's surface, possibly in the form of a small rover.

Figure 14.7 Future missions to Mars might use automated rovers that could roam through dried-up stream beds, taking samples, analyzing them, and sending pictures and other data back to Earth by radio. The greatest advance will come when the samples themselves are returned to our terrestrial laboratories.

What Went Wrong on Mars?

Meanwhile, we can try to understand why the Mars we see today differs so much from our own planet. The current thinness of the Martian atmosphere is surprising, given the rich store of volatiles on both Venus and Earth. Many scientists think that Mars probably lost a large fraction of its original endowment of gases early in its history, as a result of catastrophic bombardment. All of the inner planets suffered from this rain of material, but Mars, because of its smaller size and its proximity to the asteroid belt, was particularly vulnerable.

If Mars had been a larger, more massive planet, it would have accumulated and maintained a store of volatiles comparable to the ones we find on Earth and Venus today. These would include large amounts of carbon and nitrogen that are essential for the existence of a dense, stable atmosphere within the inner solar system. A larger planet would have "outgassed" more volatiles through a greater amount of tectonic and volcanic activity, driven by its greater internal heat. This hypothetical Mars would no longer show a surface pockmarked by craters, for the primitive crust would have disappeared, just as it has on Earth. But would the thicker atmosphere have managed to maintain itself over billions of years? This would depend on how large a greenhouse effect Mars would have established to counteract the lower temperatures that arise at its distance from the sun.

It is this greenhouse effect that would alter the current set of inner planet environments, which we described as a case of "Goldilocks and the Three Planets" in the previous chapter. The larger Mars we have postulated could be warm enough to maintain liquid water on its surface, at least near the equator. Thus we do not reach a conclusion so sweeping as the one we found for Venus. While it seems to be impossible to have liquid water on a planet at the analogous distance from its central star that Venus has from the sun, it is certainly possible in principle to have liquid water on more distant planets, if they have a sufficient atmospheric greenhouse effect. Venus taught us the importance of being the right distance from the central star. *From Mars we learn the importance of having sufficient mass to produce and sustain a thick atmosphere.*

On the real, smaller Mars that we have in our solar system, the early atmosphere that was dense enough to allow liquid water on the surface may have been about as thick as our own atmosphere is today. Made primarily of carbon dioxide, this early atmosphere began to disappear as the carbon dioxide was converted into carbonate rocks through the action of the running water, while some molecules of nitrogen broke apart into atoms after absorbing ultraviolet photons and escaped from the planet. Thus the activity of water on Mars proved self-limiting, for the atmosphere eventually became too thin for liquid water to exist, precisely because liquid water *had* existed. Without extensive tectonic activity, there was no means to feed the carbon dioxide back into the Martian atmosphere once it was locked up in limestone. Any organic material that was brought to the surface layers by meteoritic and cometary impact or by atmospheric photochemistry has long since been oxidized into carbon dioxide by the action of ultraviolet light, unhindered by atmospheric ozone. And so we find Mars as it is today, with a thin atmosphere, with evidence for subsurface water in the form of permafrost, with water and carbon dioxide freezing at the poles, with nitrogen and hydrogen slowly escaping into space—and with no sign of life.

Epilogue: What About That Face on Mars?

This chapter has concentrated on the search for *microbial* life on Mars, following the argument (page 327) that such forms of life are far more likely than larger ones. Despite the fact that the discovery of such life would be absolutely extraordinary, the headline "MICROBES FOUND ON MARS!" cannot appeal to human nature as deeply as the headlines conventionally used to sell tabloid newspapers. As past history amply demonstrates, speculation about alien forms of life focuses on projections of ourselves, from H. G. Wells's Martian invaders (see Chapter 1) to Bigfoot and the Yeti (abominable snowman). The latest example of this tendency, and of the enduring fascination that life on Mars holds for us, appears in stories about the "Face on Mars" (Fig. 14.8). The "Face on Mars" provides an almost perfect tabloid story: a faraway place, a

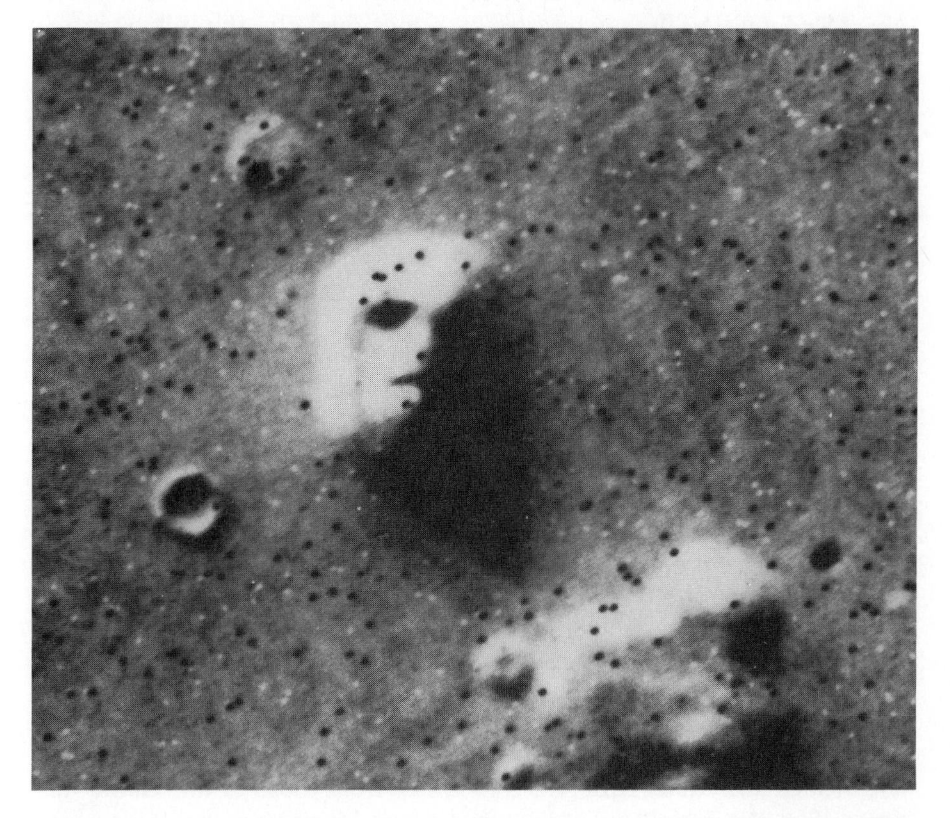

Figure 14.8 The "Face on Mars" could be a monument left by an Earthlike civilization, or a natural rock formation. Distinguishing between these two possibilities lies within the domain of science rather than the provinces of wishful thinking and Earth-centered belief.

strange phenomenon, a group of scientists shocked by what they have discovered, a denial of anything extraordinary by the "establishment."

As it happens, one of the authors (T.O.) discovered the "Face on Mars" while studying pictures of the planet as a member of the Viking project's landing site selection committee. To most people, this natural rock formation is a simple caprice of nature, illuminated from an angle that—with help from the imagination—suggests a human face, as the "man in the moon" has for untold millennia. Rock formations on Earth provide numerous examples, such as "Camel Rock" near Santa Fe, New Mexico and the "Great Stone Face" near Franconia Notch, New Hampshire (Fig. 14.9). Human minds love to find a face—the first object we learned to identify as babies—in an apparently unfeeling, inanimate landscape, testimony to our urge to identify with nature and to identify a human component within it.

Does this really explain the "Face on Mars"? How can we be sure that this "face," half a kilometer across, was not made by the inhabitants of Mars, perhaps as an appeal to creatures from a neighboring planet to identify it? The scientific answer is that we cannot entirely rule out such a possibility, but that we must assign the same probability to Camel Rock and the Great Stone Face being the result of the work of ancient visitors—even intelligent dromedaries!—who left Camel Rock in their own image. Of course, the interpretation of any particular formation remains a human affair. If you rotate Fig. 14.8 by 90 degrees, you may see a human female torso rather than a face—testimony to the difficulty of judging forms across distances of many millions of kilometers.

The deepest evidence against the "Face on Mars" being the artifact of another civilization rests with what we now know about the possibilities of life on Mars. The Viking results are compelling: Not only did the spacecraft search for microbial life and find none, but the cameras on the two landers looked for larger forms of life for more than a year. Even the two seismometers on the landers could have detected life if a herd of rhinos (or camels) ran past while the cameras looked in another direction. The GCMS on each lander, and the spectrometers studying Mars from Earth and from orbit around Mars, could have found traces of life—the departure of atmospheric gases from the equilibrium established when no life exists—produced by any Martian organisms, large or small, anywhere on the planet. The limits established by these experiments are one ten-thousandth of the amount of such disequilibrium gases produced by life on Earth.

During the late 1990s, new missions to Mars will survey the planet with new means, and under different conditions, from those of the Viking mission in 1976. We eagerly await the new knowledge of Mars that these spacecraft will provide. The cameras' better resolution of detail will surely furnish new "faces" and "animals" on Mars, in addition to the wealth of scientific results.

Figure 14.9 Nearly all who see it agree that "The Great Stone Face" or "Old Man of the Mountain" in New Hampshire is a natural rock formation. Does this look less like a human face than the "Face on Mars"—or does the greater distance to Mars encourage more active speculation?

And what of the possibility that the "Face on Mars" was made not by Martians, but by another civilization from a faraway planetary system? The proposition that superior beings visit the solar system has such appeal that we shall examine it in detail in Chapter 20. Meanwhile, the best explanation for the "sculptures" shown in both Figs. 14.8 and 14.9 seems to be the action of three agents well known for their effects on terrestrial landscapes: wind,water, and time. The only forms of intelligent life that the sands of Mars are ever likely to see live in comfort on a nearby planet called Earth (page 250).

SUMMARY

The two *Viking* spacecraft that landed on Mars in 1976 conducted thorough chemical and biological analyses of the soil and atmosphere at the two widely separated landing sites. The results of these studies include evidence that suggests the presence of life on Mars, but a closer examination of this evidence reveals that nonbiological processes are more likely than life to be the mechanisms that give the observed responses.

First of all, we should recognize the fact that nothing that resembles life has been seen by the Viking cameras, so that our hopes for life on Mars rest with microscopic organisms. Second, the analysis of the Martian atmosphere and soil shows nothing we would judge "typical" of life; instead, the soil and atmosphere resemble environments drier and colder than the driest deserts on Earth. No traces of life-indicating methane gas have been found, down to a planetwide limit of 25 parts per billion in the Martian atmosphere. The 0.13 percent of oxygen is produced photochemically.

The Viking landers each performed three experiments to test for living organisms directly. The first of these, the gas exchange (GEX) or "chicken soup" experiment, exposed a sample of Martian soil to a bath of nutrients thought to be favorable to life. The labeled release (LR) experiment dripped compounds labeled with radioactive carbon atoms onto the soil sample, to see whether the soil would produce any (radioactive) compounds typical of life. The pyrolitic release (PR) experiment also labeled carbon atoms, but this time within the Martian atmosphere; after giving any microbes in the soil a chance to interact with this labeled atmosphere, the soil was roasted to see if it now contained any of the labeled carbon monoxide and carbon dioxide.

Impressively enough, all three of these experiments gave results that might signal the presence of life: The GEX showed the release of a large quantity of oxygen; the LR revealed an increase in radioactive compounds above the Martian soil; and the PR showed a positive reaction, similar to that in nearly sterile Antarctic soil. However, since the analysis of Martian soil showed the total absence of any organic material down to a level of a few parts per billion and less, the Viking scientists took a closer look at the three life-detecting experiments. They concluded that inorganic, nonbiological chemical reactions, such as those we would expect if Martian soil contains peroxides, could produce the results found by the GEX, LR, and PR experiments. This does not *prove* that no life exists on Mars—not even that no life exists at the two landing sites. However, in light of the negative results from the biochemical analyses of Martian soil and atmosphere, and the fact that simple chemical reactions could mimic the effects of organic processes, we would have to be immensely optimistic to conclude that the Viking experiments have found life on Mars.

The remaining best hope for finding evidence of life on the red planet seems to be a search for fossils of colonies of microorganisms that had evolved to the point of being able to carry out photosynthesis 3.5 billion years ago. Such fossils might even be similar to the stromatolites found on Earth.

QUESTIONS

1. How could an intelligent Martian detect life on Earth? Describe four different methods for such detection.

2. Why do we think that microbes are more likely than large plants and animals to exist on Mars?

3. The GCMS experiment on the Viking landers failed to find any organic compounds in Martian soil. Does this prove that life on Mars does not exist at the landing sites? Why?

4. What does the absence of methane from the atmosphere of Mars down to a detectable level of 25 parts per billion tell us about the possibility of animal life on Mars?

5. Why was the GEX (gas exchange) experiment on the Viking landers called the "chicken soup" experiment? What did this experiment reveal about the Martian soil?

6. Why did the LR (labeled release) experiment use radioactive carbon atoms? Did any of these atoms appear in the atmosphere immediately above the soil sample? What does this prove?

7. The PR (pyrolitic release) experiment showed that some carbon atoms in carbon monoxide and carbon dioxide gas became "fixed" in the Martian soil. Does this suggest that life exists on Mars?

8. How would you estimate the chance that the two Viking landing sites happen to lie on parts of the planet that are particularly hostile to life?

9. Why might the regions at the edge of the Martian polar caps be more favorable to life than regions closer to the Martian equator?

10. Why do we think that Mars would have a thicker atmosphere if it were a larger planet? What effects would a thicker atmosphere have for the chances of life on Mars?

FURTHER READING

Beatty, J. Kelly, and Andrew Chaikin. *The New Solar System* (3d ed.). Cambridge, MA: Sky Publishing, 1990.

Cooper, Henry. *The Search for Life on Mars.* New York: Holt, Rinehart & Winston, 1980.

Gore, R. "Sifting for Life in the Sands of Mars," *National Geographic* (January 1977).

Horowitz, Norman. *To Utopia and Back.* New York: W. H. Freeman, 1986.

Klein, Harold. "Where Are We in the Search for Life on Mars?" *Mercury* (March/April 1986).

Washburn, Mark. *Mars at Last!* New York: G. P. Putnam, 1977.

15

The Giant Planets and Their Satellites

The four giant planets in our solar system—Jupiter, Saturn, Uranus, and Neptune—differ greatly from the rocky inner planets with which we are more familiar. Because of their enormous masses, the giant planets can retain atmospheres rich in hydrogen and the other light elements, and thus are similar in composition to the primordial solar nebula from which the entire solar system formed. This is especially true of Jupiter, the largest and nearest of these objects. Jupiter thus provides an enormous natural laboratory in which we can test our ideas about the kinds of chemistry that we think occurred in the early solar system. Titan, the largest satellite of Saturn, possesses an atmosphere in which chemical reactions that resemble some of those postulated for the prebiological Earth are taking place today. Investigations of this part of the solar system thus seem highly relevant to our central theme.

The large masses of the giant planets help them to maintain rings, extensive satellite systems, and powerful magnetic fields that dominate the space around them. If we could penetrate the atmospheres of these enormous star-like worlds with probes that would take us toward their centers, we would find that these planets do not have solid surfaces. Instead, the deep, hydrogen-rich atmospheres become denser and denser until they gradually merge into liquid, then (perhaps!) solid matter.

We omit Pluto from this discussion since it is not a giant planet. Pluto is a small icy object, about two-thirds the size of our moon, with a still smaller satellite of its own called Charon. It has a thin, methane-containing atmosphere and patches of dark material on its surface. In several respects, Pluto resembles Triton, the largest satellite of Neptune, so some of the phenomena discovered during the *Voyager 2* flyby of Triton in 1989 may well exist on Pluto. Both of these small, icy worlds provide evidence that organic chemistry can take place under highly unusual circumstances, although after our discussion of the interstellar medium (Chapter 4) this should not be too surprising.

Spacecraft to the Outer Solar System

During the early 1970s, the United States sent two spacecraft, *Pioneer 10* and *11*, on journeys of several years to Jupiter and (in the case of *Pioneer 11*) on to Saturn, which was reached in September 1979. These two robot emissaries from Earth are continuing to travel away from our planet on trajectories that will take them completely out of the solar system. While these spacecraft were indeed "pioneers," giving us our first close-up information about Jupiter and Saturn, most of what we know about the outer solar system was learned during the course of an epic 12-year journey of discovery by the two *Voyager* spacecraft (Figs. 15.1 and 15.2). Launched in August and

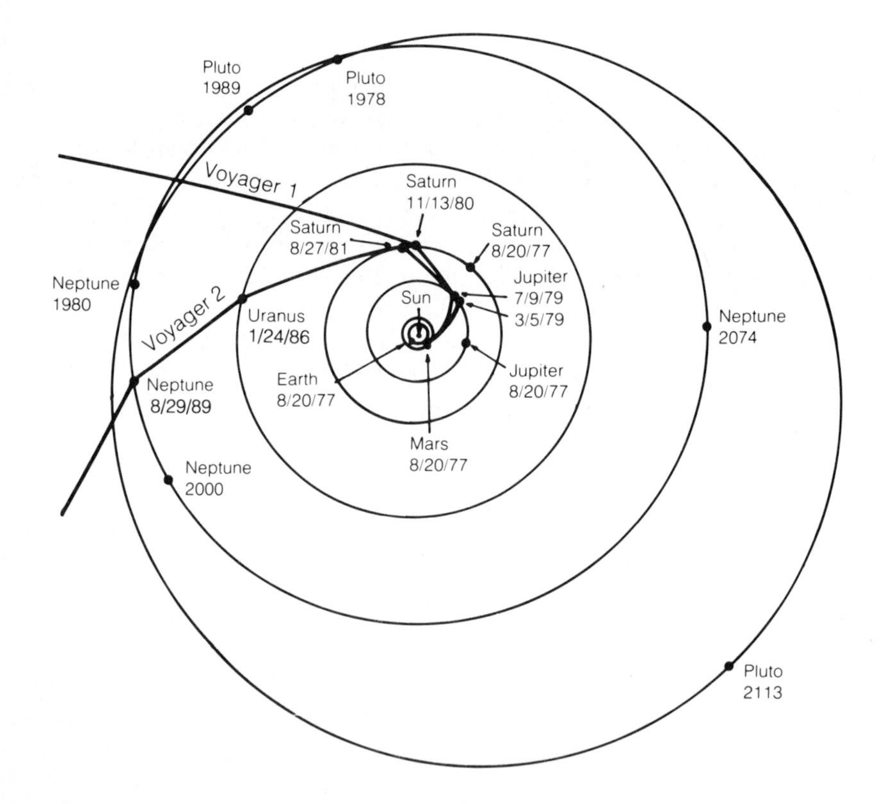

Figure 15.1 Because of a favorable line-up of the giant planets that occurs only once in 176 years, the *Voyager* spacecraft launched in 1977 could fly from Earth to Jupiter, Saturn, Uranus, and Neptune in less than 12 years. The decision to send *Voyager 1* close to Titan kept it from going on to Uranus and Neptune, since *Voyager 1* passed Saturn at too large a distance to receive a sufficient "gravitational boost" to turn its course toward Uranus.

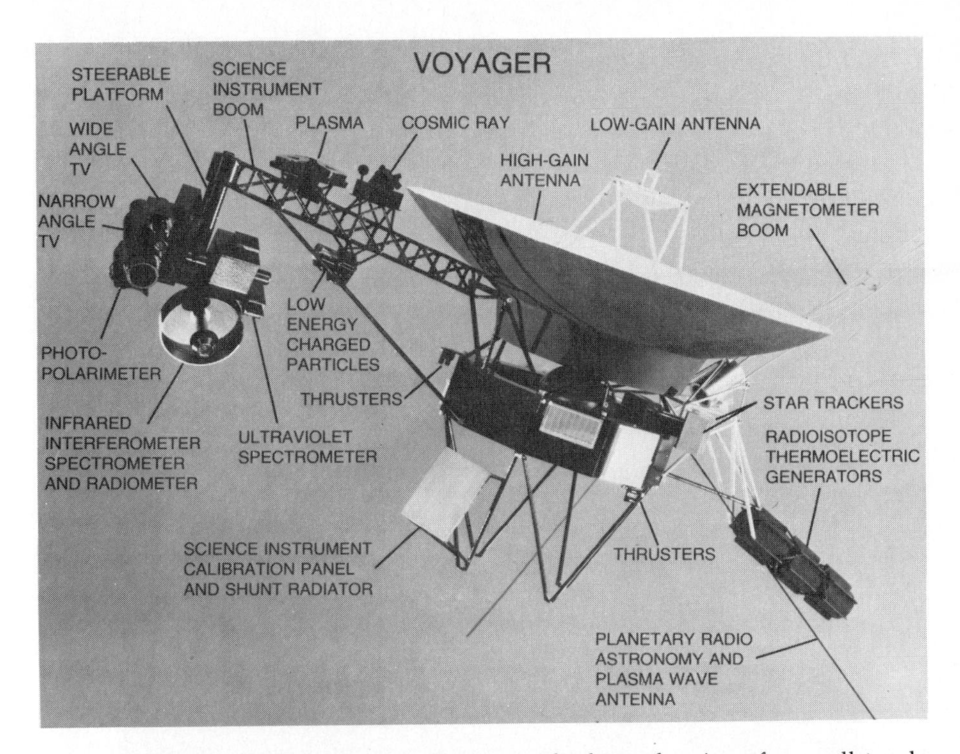

Figure 15.2 The *Voyager* spacecraft were each about the size of a small truck. Packed into this volume were a host of scientific experiments and the computer and nuclear-electric generators to run them. The large, dish-shaped antenna for communicating with Earth is approximately 4 meters across.

September of 1977, the two *Voyagers* reached Jupiter in 1979. Highlights of these encounters included the discovery of active volcanoes on one of Jupiter's moons, the existence of a ring around the planet and an immensely variegated meteorology, revealed by the motions of Jupiter's pastel-colored clouds (Plate 15). The most famous of these is the Great Red Spot, well-known from ground-based observations, but the Voyager pictures revealed many other smaller storm systems, rotating about their centers as they were swept along by atmospheric wind currents (Fig. 15.3).

Two years after the *Voyagers* sailed past Jupiter, the spacecraft arrived in the vicinity of Saturn (Fig. 15.4). Far out from our home, almost 10 times Earth's distance from the sun, the *Voyagers* surveyed the planet's satellites and its rings, as well as the great ball of Saturn itself. *Voyager 1* made a close approach to Titan, a satellite with an atmosphere denser than our own, and thus could not go on to the other planets. Instead, like the *Pioneer* spacecraft, it headed out of the solar system. But *Voyager 2* passed close enough

Figure 15.3 Jupiter, with 11 times the diameter of Earth, shows banded atmospheric patterns parallel to the planet's direction of rotation. The Great Red Spot, 25,000 kilometers long, has persisted for at least the past few centuries (see Color Plate 15 and Fig. 15.8).

to Saturn to use that planet's gravity to boost it on to Uranus, where it arrived in January 1986. This distant, cold world proved to be virtually featureless, but one of its moons, Miranda, was found to have some spectacular cliffs and other features suggesting an unusually violent history for such a small (radius of 250 kilometers) object.

Voyager 2 passed through the Neptune system in August 1989. Neptune was certainly more photogenic than Uranus, having a large, dark, oval cloud system reminiscent of Jupiter's Great Red Spot (Fig. 15.5). As on Uranus, there was evidence of atmospheric chemistry on Neptune; even at these immense distances from the sun, methane in the atmospheres of these planets is being converted into more complex molecules, forming

Figure 15.4 This *Voyager 2* image of Saturn shows that the upper cloud deck on this planet is much more homogeneous than is the case for Jupiter (see Fig. 15.3).

ultraviolet-absorbing hazes. Once again an unusual satellite captured our attention: Triton, an icy body slightly larger than Pluto that was found to have vertical plumes of material rising from a surface—covered with frozen nitrogen and methane—whose temperature is only 37 K!

After passing Triton, *Voyager 2* is continuing on a path that will take it out of the solar system. But this is not the end of our exploration of the outer planets. Launched in October 1989, the *Galileo* spacecraft is scheduled to reach Jupiter in December 1995, after a voyage that has already included encounters with Venus and an asteroid (Fig. 15.6 and Fig. 11.9). *Galileo* consists of an orbiter to study the magnetosphere, satellites, rings, and atmosphere of the largest planet, and a probe that will be deployed to descend deep into the atmosphere, measuring composition, pressure, temperature, and detecting lightning discharges. The probe mission will be over in less than an hour, but the orbiter will tour the system along a family of orbits for at least two years.

A similar mission called *Cassini-Huygens* has been approved by the European and U. S. space agencies for a launch to the Saturn system in the late 1990s. Once again there will be an orbiter and a probe, but this time the probe will be sent into the atmosphere of Titan, Saturn's largest satellite. The current schedule calls for arrival in the year 2004, followed by a four-year tour of the satellites and the magnetosphere. To understand why all

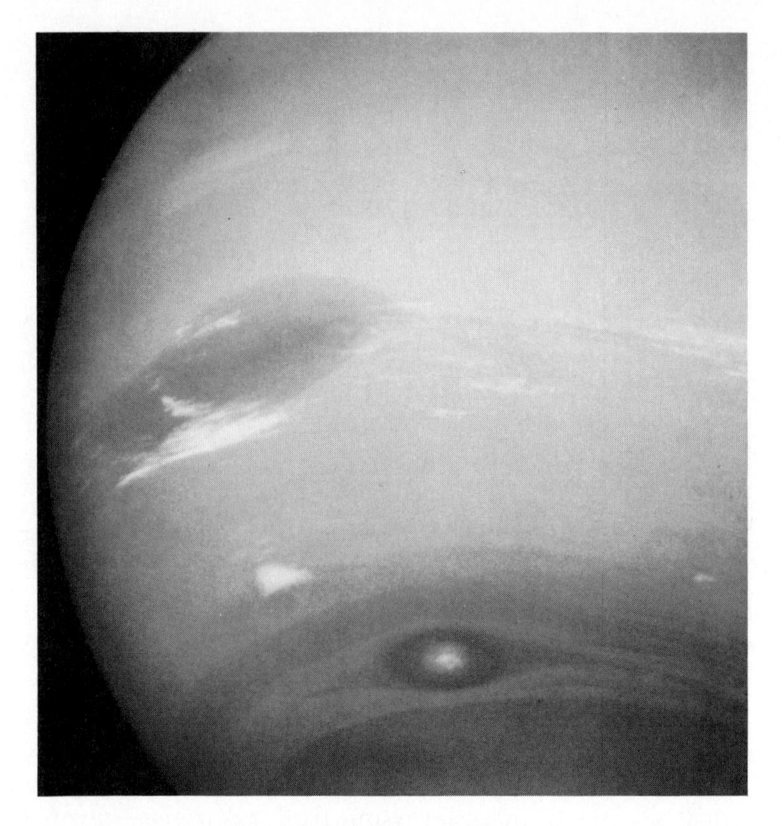

Figure 15.5 On Neptune, *Voyager 2* again found discrete cloud systems reminiscent of Jupiter, but without the giant planet's pastel colors.

this activity is underway, let us examine what we have learned so far about these strange, distant worlds.

The Composition of the Giant Planets

We can begin by performing a simple computer experiment: Consider a mixture of the chemical elements in the same proportions that exist in the sun and in other stars, and let these elements combine with one another to form molecules in every possible way. If we specify an approximate pressure and temperature, we can model the conditions in the atmospheres of the giant planets as they formed from the original cloud of gas and dust that made the solar system. The chief compounds that we predict from this exercise match the ones that dominate the atmospheres of Jupiter and its

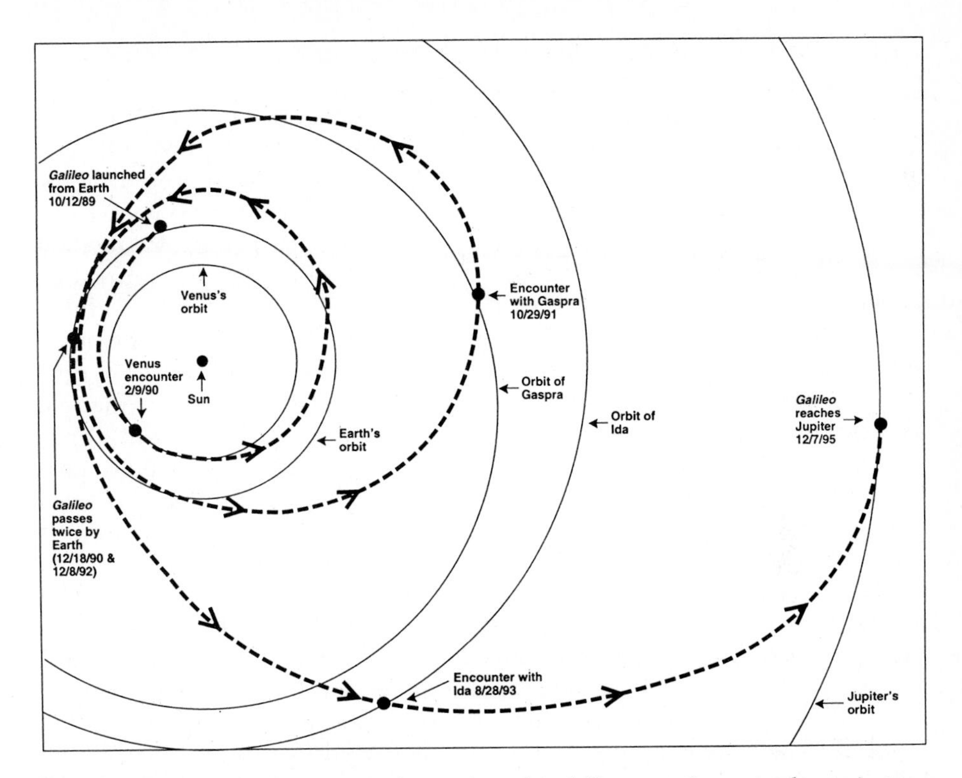

Figure 15.6 The *Galileo* spacecraft was forced to follow a rather complex trajectory en route to Jupiter, since it did not have the benefit of the powerful launch vehicles that sent out the *Voyagers.* Nevertheless, this excursion is yielding additional scientific benefits in the form of encounters with Venus and the asteroid Gaspra (see Fig. 11.9).

neighbors: methane, ammonia, water vapor, and everywhere an excess of hydrogen. (Helium is also highly abundant, and neon should be present in about the same proportion as ammonia, but because helium and neon do not participate in chemical reactions, we do not need to pay much attention to them in this discussion.)

Now compare these gases with those found in the secondary, outgassed atmospheres of the inner planets. Any original hydrogen and helium have long since escaped from these small objects, so we find carbon dioxide rather than methane (unless life is present) and molecular nitrogen instead of ammonia. Hydrogen still escapes, as H_2O is slowly but continuously broken apart by solar ultraviolet radiation. There is no chance of recreating a hydrogen-rich chemistry here, except, as we have seen, in the form of life itself. Jupiter, Saturn, Uranus, and Neptune appear to have primordial atmospheres, made from the primitive matter in the solar system, which they have retained since the time of their formation 4.5 billion years ago.

Actually, the atmospheres of these giant planets must be a mixture of gases captured from the solar nebula and gases that were trapped in the icy planetesimals that formed the planetary cores. These cores grew first, as ices accumulated from the solar nebula, until they were large enough to attract an envelope of gases from the surrounding solar nebula. While they grew, they produced atmospheres of their own from the gases trapped or frozen in their ices. Hydrogen, helium, and neon cannot be trapped in this way, but carbon monoxide, nitrogen, methane and ammonia can. The original CO and N_2 from the icy cores were converted to CH_4 and NH_3 as hydrogen was attracted from the surrounding solar nebula. The result is that we find methane and ammonia to be more abundant in the present atmospheres than a strictly solar composition would predict. This becomes particularly marked on Uranus and Neptune, where the atmospheres are relatively thin and the proportion of gases produced by the cores is much higher than on Jupiter and Saturn.

Many scientists have considered this hydrogen-rich environment to be representative of conditions on the primitive Earth. But we must emphasize that the analogy between Jupiter and the early Earth suffers from several important differences between these planets. First, Jupiter's enormous mass, 318 times Earth's mass, keeps hydrogen from escaping; this means that Jupiter will always have a huge number of hydrogen atoms to combine with any elements or molecular fragments. In contrast, Earth rapidly lost its hydrogen, either the original hydrogen from its formation (if any existed) or the hydrogen produced by the dissociation of molecules, a loss which actually led to an environment more suitable for the evolution of a complex biochemistry. Second, Jupiter does not have a solid surface, and thus has no likely microenvironments, such as Earth's tidal pools or transient ponds, in which the products from chemical reactions in the atmosphere could become concentrated and undergo additional reactions. Nor does Jupiter offer the opportunity for chemical reactions to be catalyzed by soil surfaces, such as the clay minerals that apparently played this role on Earth.

In addition to these key differences between Jupiter and Earth, another important problem for the origin of life arises from the vertical convection in Jupiter's atmosphere (Fig. 15.7). This convection creates a circulation pattern between the relatively cool upper regions and the lower atmospheric levels, where temperatures stay hot enough (above 700° C) to destroy complex molecules. The circulation most likely occurs in a time much shorter than a year, so any large molecules formed in the upper atmosphere would be broken apart by collisions at relatively high temperatures within this span of time. An intermediate level exists where conditions are better. Scientists estimate that water vapor will condense to form clouds in a region

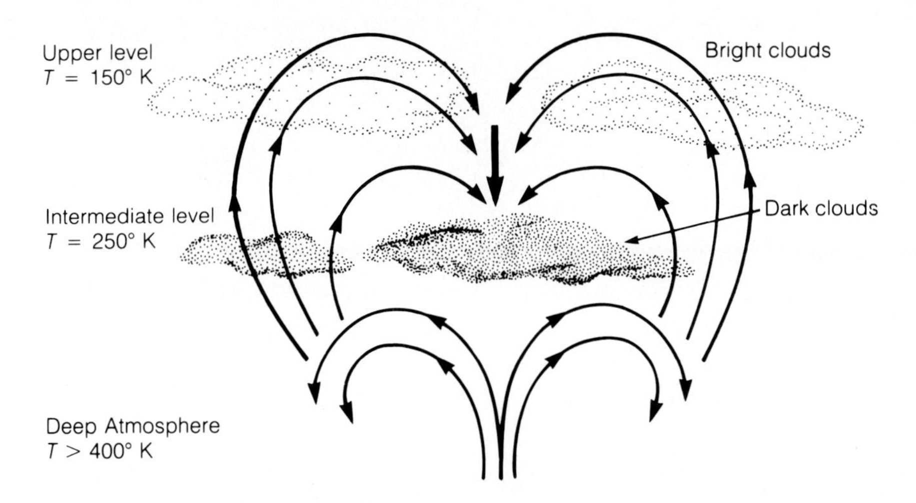

Figure 15.7 Vertical convection currents carry material upward from the warmer layers of Jupiter's atmosphere to form bright clouds. Where the currents descend, we can see darker clouds at lower levels in the atmosphere.

with a temperature of about 27° C and a pressure just a few times greater than the surface pressure on Earth. But the gas in this region will be in constant motion, circulating between the upper and lower regions; so at this level, water vapor continually condenses and evaporates. Higher in the atmosphere, ammonia and ammonia-sulfur compounds take the place of water vapor, forming the clouds we see from Earth. Wherever liquid droplet clouds are present, thunderstorm activity is likely because of the strong vertical convection that can lead to charge separation. Indeed, lightning discharges within the Jovian clouds were detected by the Voyager spacecraft in 1979.

The Great Red Spot (Fig. 15.8 and Color Plate 15) offers an interesting possible exception to the rule of rapid destruction of complex molecules that might form in Jupiter's upper atmosphere. This enormous feature is a giant vortex—a Jovian analogue of a terrestrial hurricane—and it may be able to keep particles of the proper size suspended for long periods of time. If so, molecules could avoid destruction by the heat of the lower atmospheric levels. This long-lived, high altitude suspension could lead in turn to a different kind of photochemistry from that taking place elsewhere on the planet. Such a difference seems to be required to explain this feature's unusual reddish color, whose cause still has not been determined. It is surprising to find a discrete feature with a lifetime that may be hundreds of years in such a dense, turbulent atmosphere, but the sheer size of it—the surface area is greater than the area of our entire planet!—must be part of the explanation.

Figure 15.8 The Great Red Spot of Jupiter was photographed by *Voyager 1* in March 1979. The smallest details in the photograph are about 50 km across.

Investigation of the chemistry taking place here promises to be a rewarding enterprise.

If we descended into Jupiter's ever-thickening atmosphere in our imaginary probe, we would find that the temperature steadily increased, even though we would soon lose sight of the sun. The reason for this is surprising: Instead of a stupendous greenhouse effect, Jupiter has an internal source of heat, which generates more energy each second than the planet receives from the sun. This heat is left over from the formation of the planet. The potential energy of the hydrogen and helium gas in the gravitational field of Jupiter's growing core (see page 258) was converted to energy of motion as these gases collapsed onto the core to form the bulk of the planet's mass. The resulting rapid movements of the molecules generated thermal radiation which is still escaping Jupiter's interior. The same conversion of potential to thermal energy on a much larger scale during the formation of the sun made the interior of our star hot enough for nuclear fusion reactions to begin. But Jupiter, with a thousandth of the sun's mass, never became sufficiently hot inside to turn into a star. Similar internal heat sources are found on Saturn and on Neptune, but Uranus does not radiate more energy than it gets from the sun.

Chemistry on the Giant Planets

Despite the extreme differences between Jupiter and Earth, we remain interested in this giant of our solar system as we consider the ways in which life can begin. Jupiter gives us a chance to study conditions similar to those that existed 4.5 billion years ago as the solar system formed. This opportunity is similar to encountering a primitive tribe of humans just as they are developing language. What sounds would be identified with what objects? How would such decisions be made? How and when would concepts such as time, space, and love be expressed? If we could observe without being noticed, we could expect to learn about our own history by watching the experiments of these people. In much the same way, we hope to learn something about possible pathways for prebiological chemistry in the primordial solar nebula by studying chemical reactions on Jupiter and other objects in the outer solar system.

We know that such reactions are occurring on the largest of the planets because we see some associated effects. First, astronomers have discovered that in addition to the gases that we have already listed, Jupiter's atmosphere contains traces of substances that our computer model does not predict, including carbon monoxide (CO), hydrogen cyanide (HCN), acetylene (C_2H_2), and ethane (C_2H_6). These gases cannot continue to exist in Jupiter's upper atmosphere unless they are continually produced by reactions among the other atmospheric constituents, since they rapidly interact with hydrogen to form methane (CH_4). Ultraviolet solar photons provide the energy to form C_2H_2 and C_2H_6 from CH_4 in Jupiter's upper atmosphere, while thermal energy from the planet's interior forms CO from CH_4 and H_2O at deeper levels, far below the visible clouds, where temperatures reach 1200° K. We also find traces of phosphine (PH_3), arsine (AsH_3) and germane (GeH_4) that must be brought up from these great depths. Other compounds of elements with hydrogen must also be present. The ones discussed so far happen to have very strong absorption lines that are not overlapped by absorptions from methane or ammonia and/or fall in regions of the spectrum where we can "see" to great depths in Jupiter's atmosphere. Thus their detection is favored. Some of the C_2H_2 and the HCN may be produced by lightning discharges at intermediate levels (Fig. 15.9). Are there more complex molecules being formed by these various processes as well?

The second indication that chemical reactions are taking place suggests an affirmative answer: Jupiter's clouds do not have just the white or gray colors that we expect to see when sunlight is reflected from frozen water or

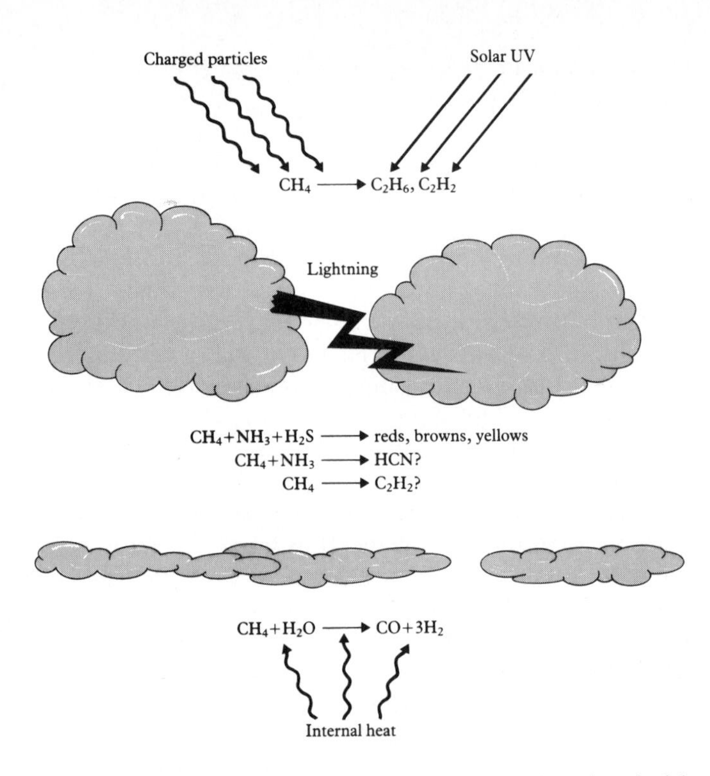

Figure 15.9 Gases in the atmosphere of Jupiter are bombarded by charged particles from the sun and affected by lightning discharges. In addition, heat from within the planet and solar ultraviolet radiation can drive chemical reactions that form and re-form molecules.

ammonia. Instead, the clouds show various subtle shades of color, including the salmon-colored tint of the Great Red Spot (see again Color Plate 15). What substances produce these colors? How were they formed? Do these chemical reactions resemble the prebiological reactions that must have occurred on the primitive Earth?

These key questions remain unanswered. At this time, two competing theories strive to provide an explanation of Jupiter's colors. On one side, we find those who believe that all of Jupiter's coloration can be explained by inorganic compounds that arise from atmospheric chemistry. The yellows and browns could be caused by sulfur compounds formed by lightning discharges in the lower clouds. The Great Red Spot might owe its redness to the presence of red phosphorus that could be produced by chemical reactions from the gas phosphine (PH_3), which is known to be present on Jupiter. The opposing side of the color controversy points to laboratory simulations of Jupiter's atmosphere in which compounds with the colors of Jupiter were produced by shining ultraviolet radiation on an appropriate mixture of methane and ammonia, or by subjecting these gases to some other source of energy such as an electric spark, simulating Jovian lightning. These experiments invariably produce a wide range of organic compounds, some of which have the colors of Jupiter's clouds. It seems quite possible that both types of chemistry are actually occurring on this planet, and each may contribute compounds that cause the colors we observe. What other compounds are being formed we can only guess at present, but it is already clear that we shall find at least the preliminary stages of prebiological organic chemistry on Jupiter as we explore this planet more thoroughly.

We have concentrated on Jupiter in this discussion because it is the giant planet about which we know most. But real differences seem to exist among these giants. Beautiful as it is, Saturn does not have colorful clouds like those of Jupiter (see again Fig. 15.4). Instead, we seem to confront a thick layer of ammonia cirrus, which completely covers the lower, warmer atmosphere that might in fact have characteristics like those of Jupiter. These dense clouds arise from the lower temperature of Saturn's upper atmosphere, the result of the planet's greater distance from the sun, and its lower gravity, which allows clouds to form over a greater altitude range. The visible levels of the atmosphere of Uranus and Neptune are so cold that we can not even find evidence of ammonia, which is probably frozen out at low altitudes. The white clouds we do see are probably frozen methane. Both planets reveal thin hazes of ultraviolet-absorbing material in their upper atmospheres (Fig. 15.10), and Neptune has a large dark vortex in its main cloud deck (see again Fig. 15.5), whose composition we do not know. At still deeper levels, we have even less information, although thermal emission at radio wavelengths from these regions shows that they are warmer than 0° C. Astronomers found hydrogen cyanide (HCN) in Neptune's stratosphere using Earth-based radio telescopes in 1991. Hence chemistry is occurring even here, as N_2 and CH_4 are broken apart and the fragments combine in new ways.

Could Life Exist on the Giant Planets?

We have seen that the outer planets offer interesting environments for the study of prebiological chemistry. Can we be sure that life itself does not exist on these giants? No! So say Carl Sagan and Edwin Salpeter, two well-known astronomers from Cornell University. Sagan and Salpeter have argued that since we really do not know how life began on Earth, we cannot specify the necessary conditions for life to appear on a planet as different from ours as Jupiter. Once life begins, living organisms themselves can regulate and adapt to their environment. Quite possibly, therefore, living organisms on Jupiter or Saturn would be able to overcome some of the obstacles to life that we have described. Giant gas-bag creatures that use warm hydrogen to maintain their buoyancy might be drifting or flying in Jupiter's upper atmosphere as you read this, participants (perhaps) in a Jovian ecology within which some species prey on others. Jupiter furnishes the most likely home for such life forms, because Jupiter appears to have the most chemical activity of the four giant planets. At the simplest level of speculation, however, we can imagine similar creatures floating in the atmospheres of any of the giant planets.

Unfortunately, this speculation does not advance our knowledge. Our increased understanding of life on Earth, and our inability to find life on Mars, should leave us a bit more conservative in speculating on the likelihood of finding life on every celestial object, life that would have adapted uniquely to each particular environment. In close-up studies of the sun's planets, we are so far one for three in finding life. With this record, we cannot argue that every planet should have life upon it, or within it. And yet, in the face of our basic ignorance concerning both Jupiter's environment and life's universality, we ought not discard these intriguing possibilities for life on Jupiter too quickly.

Rings and Satellites

While we are in a speculative mood, let us not restrict ourselves to the planets alone. Each of the giant planets has an extensive satellite system, ranging in number from Neptune's eight moons to Saturn's 18 (or more!). All four have systems of rings; Saturn's are easily visible with a small telescope, while the rings of Uranus remained undiscovered until 1977, when they blocked the light of a star just before and after the occultation by the planet itself. They are much darker than the rings of Saturn, and they share the unusual orientation of Uranus itself, which keeps the planet's axis of rotation nearly in the plane of the orbit (Fig. 15.10). The rings of Jupiter were discovered by the *Voyager* spacecraft in 1979. They are also much harder to see than the rings of Saturn. Neptune joined this club in 1985, when indications of rings were first discovered around the planet, again by stellar occultations. They were clearly seen by *Voyager 2* when it reached the planet in 1989.

Each of these ring systems lies close enough to its planet that the difference in the gravitational attraction exerted by the planet on two adjacent particles in the ring is greater than the gravitational attraction of the particles for each other. The boundary within which this condition exists is known as the "Roche limit," named after its discoverer, Edward Roche. The rings thus consist of material that could not accrete to form a satellite

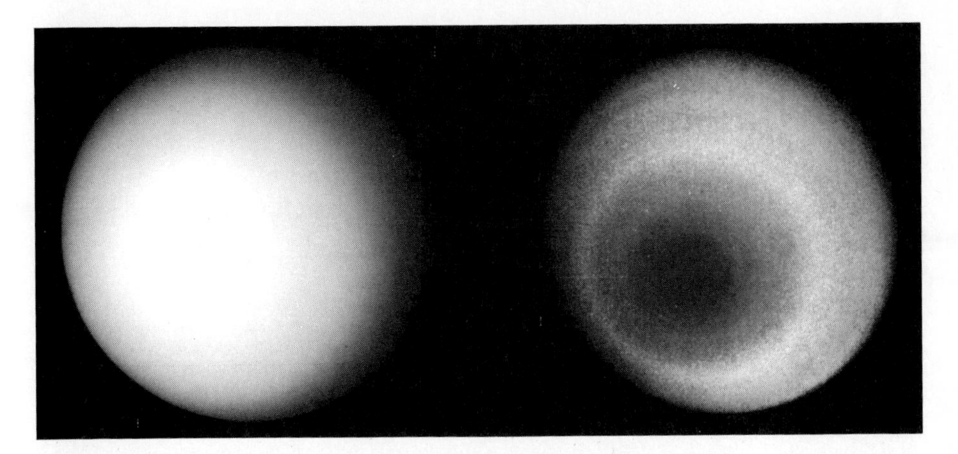

Figure 15.10 *Voyager 2*'s images of Uranus revealed few patterns in the planet's atmosphere, except when specially processed to show the polar haze (right).

because it was too close to the planet, whose disruptive force kept any agglomeration of matter from growing. The millions of tiny satellites—rocks and ice balls—that constitute the rings we see today are kept in stable orbits by the interaction of the gravitational forces from the planet and its satellite system. These boulders do not disintegrate into dust for the same reason that our artificial satellites can survive in their orbits within Earth's Roche limit: The electrical forces that hold together a silicate or ice lattice (or stainless steel!) are far stronger than the disruptive effects of gravity.

Our knowledge of the number of objects that circle the outer planets, not to mention their structure and composition, remains incomplete. Jupiter and Saturn surely have more satellites than we know now, since one or two have been glimpsed and lost again during recent years of observation. The same may be true of Uranus and Neptune, where the Voyager flybys added a total of 16 moons to those we already knew. A curious object called Chiron, with an orbit that lies mostly between that of Saturn and Uranus, was discovered by Charles Kowal in 1977, while James Christy found in 1978 that Pluto has a satellite (Charon) whose orbital period equals the planet's period of rotation. Chiron is now known to have a coma of particles and gas, indicating that it is a giant (radius of 100 kilometers) comet nucleus. Charon, and probably Pluto as well, must have similar compositions to that of Chiron, but they are too far from the sun to exhibit cometary activity.

Let us consider the satellites that we do know. The four largest moons in Jupiter's family were first recorded by Galileo, whose contemporary Simon Marius named them after four of Jupiter's illicit loves: Io, Europa, Ganymede, and Callisto. These four satellites are large objects with solid surfaces (Ganymede and Callisto are about the size of Mercury), and they imitate the gross characteristics of the solar system rather well: The innermost moons, Io and Europa, have densities of matter similar to those of the inner planets, while Ganymede and Callisto represent the outer planets, with low densities implying (in this case) a high abundance of ice (Table 15.1).

All four of these objects received intensive scrutiny with the *Voyager* spacecraft in 1979 (Fig. 15.11). Surprises abounded, including the lack of large impact craters on Callisto, evidence for plate motions of the crust of Ganymede, a complex network of lines on Europa, and above all, the active volcanoes of Io (Fig. 15.12 and Color Plates 16 and 17). Callisto and Ganymede have densities consistent with a structure consisting of half ice, half rock, with the ice on the outside. The lines on Europa appear to be fissures in an icy crust that may (or may not!) cover a layer of liquid water. The liquid water is expected to come from the same cause as the volcanoes of Io—an interior heated by the dissipation of tidal forces from giant Jupiter. On Io this tidal force, the result of Io's proximity to Jupiter and its orbital resonance with Europa and Ganymede, is so extreme that the crust of the satellite may be just a few tens of kilometers thick, covering a molten interior. Dark spots on the surface have been interpreted as lakes of liquid sulfur. The orbit of this moon is surrounded by a torus of sulfur, sodium, and potassium atoms that are driven off Io's surface by charged-particle bombardment and expelled by volcanic eruptions.

These satellites are fascinating little worlds that we would be sure to visit on a trip to Jupiter, but we must admit that they do not help us much in our search for life and life's precursors. Despite its volcanic activity, Io is too small to possess a thick atmosphere; it must have lost all its water, and the only volatile now found to emerge from its volcanoes is sulfur dioxide (Color Plate 16). The other moons of Jupiter are so much smaller than the big four that they arouse little interest for our specific purposes.

TABLE 15.1 The Four Large ("Galilean") Satellites of Jupiter

Satellite	**Distance from Jupiter (km)**	**Diameter (km)**	**Average Density (gm/cm^3)**	**Inclination of Satellite Orbit to Jupiter's Orbit**
Io	422,000	3650	3.6	0°.0
Europa	671,000	3138	3.0	0°.5
Ganymede	1,070,000	5262	1.9	0°.2
Callisto	1,883,000	4800	1.9	0°.2

Figure 15.11 The four Galilean satellites of Jupiter each showed a different appearance to the Voyager cameras in 1979. Io (upper left) reveals a surface mottled by volcanic activity. Europa (upper right), is extremely smooth, Ganymede (bottom left), vaguely resembles our own moon, and Callisto (bottom right), looks like fractured glass.

Europa, however, is worth a second glance (Color Plate 17). If the 100-kilometer thick layer of ice on its surface is in fact a thinner layer of ice overlying tens of kilometers of water, there may be an environment here of possible biological interest. Some scientists have called attention to the fact that the giant clams, the tube worms, and the host of microorganisms found near the hot, sulfurous vents on some of our ocean floors might have analogues on Europa. The trouble with this speculative analogy is that most of these terrestrial submarine creatures rely on dissolved oxygen for their metabolism. Only the bacteria are anaerobic. Nevertheless, finding Europan bacteria would be

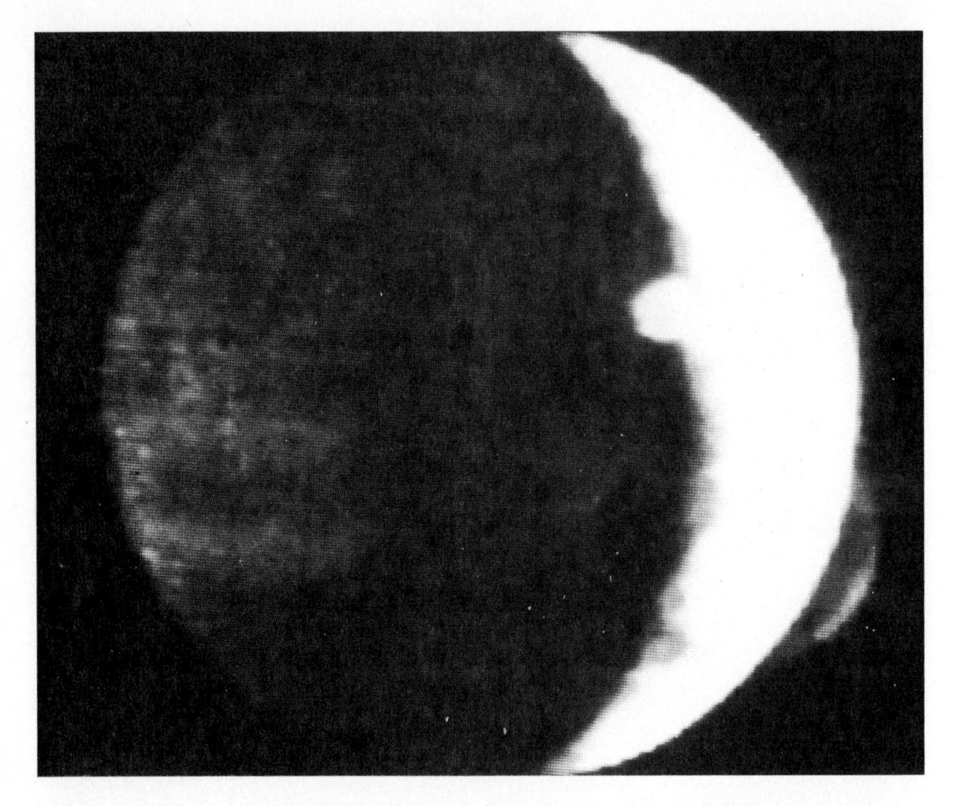

Figure 15.12 This photograph taken by the *Voyager 1* spacecraft shows two active volcanoes on Io. On the lower right, just at the satellite's limb, clouds of SO_2 are rising more than 260 kilometers above Io's surface. The second volcano is revealed by an irregular extension of the line between light and dark on Io, where a volcanic cloud is catching the rays of the rising sun. We can see the dark side of Io dimly because Jupiter reflects sunlight onto it, just as the Earth sends sunlight to our crescent moon (see Fig. 20.5).

an exciting discovery, so we should at least determine whether an ocean is indeed hidden beneath Europa's bland exterior. If such an environment exists and is potentially habitable, even if only by microorganisms, it would greatly extend the distance from a central star where we might expect life to exist.

Titan

Of all the planets and satellites in the outer solar system, the object that is most interesting in our search for life is Titan, Saturn's largest satellite. Like Ganymede and Callisto, this moon is more massive than Mercury. Unlike those oversize orbiters of Jupiter, however, Titan has an atmosphere, and it is the presence of that atmosphere with the range of chemical reactions taking place in it that makes this Satúrnian moon so intriguing (Color Plate 18).

How can a satellite have an atmosphere? It is simply a question of mass and temperature. The more massive the object, the stronger its gravitational field and thus the more energy of motion required to escape into space. The kinetic theory of gases tells us that the average speed of a gas molecule depends on its mass and the ambient temperature. Thus a massive satellite at a low temperature has a better chance of retaining an atmosphere than an object of the same mass located closer to its central star, with a correspondingly higher temperature. This explains why Titan (T = 85 K) has an atmosphere while Mercury (T = 540 K) does not.

This argument is not yet complete. We can understand why Titan has an atmosphere while Mercury doesn't, but what about Ganymede and Callisto, which have approximately the same mass as Titan? The answer here seems to be the difference in the temperature at which these different satellites *formed*. Evidently it was too warm in the region around Jupiter to allow the ices that condensed there to keep large amounts of gases such as nitrogen, methane, ammonia, etc. In the colder environment around Saturn, the ices making up Titan were able to retain these atmosphere-forming compounds. At the same time, it was not *so* cold that such compounds were largely frozen out on Titan's surface, as seems to be the case on Triton and Pluto.

Note that any planet or satellite will be able to retain heavy gases such as methane, which has a molecular weight of 16—that is, each methane molecule has a mass 16 times the mass of a hydrogen atom—or argon (whose atomic weight equals 40) far better than it can retain hydrogen molecules (atomic weight 2) or helium atoms (atomic weight 4). Indeed it is the heavier gases that we find in Titan's atmosphere.

And such an interesting atmosphere it is! The surface pressure on Titan is 1.5 times the sea-level pressure on Earth. Like air, this atmosphere is mostly nitrogen, but instead of oxygen or water vapor, the other major constituent is methane. (As much as 15 percent of argon may also be present, but this gas is not directly detectable with presently available techniques.) Like the atmosphere of Venus, Earth, and Mars, Titan's atmosphere is secondary in origin, produced from the solid materials making up the satellite. Hydrogen is continually escaping from Titan, just as it does from the inner planets. It is the low temperature of Titan's surface that maintains methane in the atmosphere instead of carbon dioxide. At 85 K, water ice is so cold that very few water molecules have the energy to sublime, that is, to leave the solid state to join the methane and nitrogen in Titan's atmosphere. Once in the atmosphere, the water vapor could be broken apart by solar ultraviolet photons and the constant rain of bombarding electrons from Saturn's magnetosphere that drive the other chemical reactions we observe on Titan.

With water trapped as low temperature ice on the satellite's surface, there is no large source of oxygen available to convert the methane to carbon dioxide, an inexorable transformation on the much warmer inner planets.

The result is a direct benefit to us, since we find a planet-sized object with a solid surface, small enough that hydrogen can escape into space but cold enough that no oxygen can alter the primitive chemistry. Titan is a kind of time machine, allowing us to study chemical reactions that were once common in the solar system but have stopped operating long ago in environments that became polarized into those dominated by hydrogen—the giant planets—and those where oxygen ruled—the inner planets.

Not surprisingly, we find a rich assortment of organic molecules on Titan, starting with the simple hydrocarbons formed from methane (C_2H_2, C_2H_6, C_3H_8, etc.); compounds formed with nitrogen, such as HCN, and C_2N_2; and even traces of CO and CO_2 (Table 15.2). More complex substances must also be present, given the ubiquitous smog layer that totally hides Titan's surface. This smog is composed partially of droplets of the simple compounds we have just mentioned, but it must also include polymers of acetylene, hydrogen cyanide, and other compounds that have not yet been identified. These must resemble some of the substances synthesized in laboratory experiments that simulate the bombardment of Titan's atmosphere by cosmic rays and by electrons from Saturn's magnetosphere (Fig. 15.13).

The particulates from the aerosol layers rain down on the surface of Titan, covering the low-lying areas to depths that may reach hundreds of meters. Ethane, the most abundant of the hydrocarbons produced in the atmosphere, will be a liquid at the temperature of Titan's surface. Thus there may be lakes or even seas of this substance on the satellite's surface, full of dissolved smog particles and methane, forming rich sources of readily available energy for future explorations by astronauts if they can just find the oxygen to burn this fuel! (Fig. 15.14).

Given 4.5 billion years, we can imagine that further, low probability reactions have taken place in Titan's seas among the species dissolved

TABLE 15.2 Trace Gases Identified in the Atmosphere of Titan

H_2	C_2H_2	HCN	CO
	C_2H_4	C_2N_2	CO_2
	C_2H_6	HC_3N	
	C_3H_4	C_4N_2	
	C_3H_8		
	C_4H_2		

Figure 15.13 Titan was a disappointment to the scientists in charge of the Voyager cameras, for the giant satellite revealed only a smog-laden atmosphere. This image shows Titan in a crescent phase, with a second, high-altitude smog layer visible above the outer edge of the crescent.

there. Hence the prospect of sending a probe to this distant world to study chemical evolution, determining the level of complexity that has been achieved and the (probably) surprising pathways that have led to this level, is an alluring one. The first step in this direction is the *Cassini-Huygens* mission described above. The Huygens probe will be sent into the atmosphere of Titan, gently descending to its surface as it makes measurements of atmospheric and surface properties over a period of three hours. The Cassini orbiter will have spectrometers that will also study Titan, and a radar that can penetrate the Titanian smog and map the surface topography, searching for those inflammable swamps. It promises to be a fascinating journey!

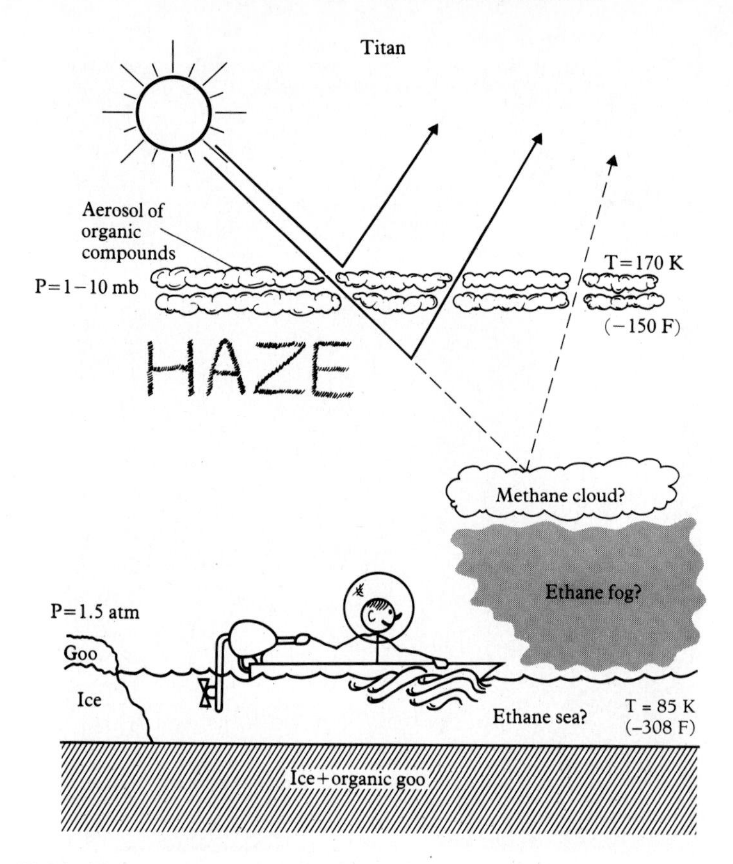

Figure 15.14 Future astronauts on Titan may have the opportunity to explore the satellite by boat. On this world fuel is cheap—it's oxygen that requires considerable effort to obtain.

Iapetus: An Intelligence Test for Earthlings?

Saturn has at least 17 other satellites besides Titan, but only one of these draws our special attention: Iapetus. This unusual moon is the only object in the solar system that we might seriously regard as an alien signpost: a natural object deliberately modified by an advanced civilization to attract our attention, and to leave us no rest until we have deciphered its meaning.[1] What is this apparent modification? Take a look at Fig. 15.15. Iapetus changes its brightness by a factor of seven as it moves around its orbit. In other words, one hemisphere of this satellite appears to be seven times darker than the other. In fact the darkest part of this hemisphere is fully 10 times darker than

[1] Arthur C. Clarke's novel and screenplay *2001: A Space Odyssey* incorporated this idea.

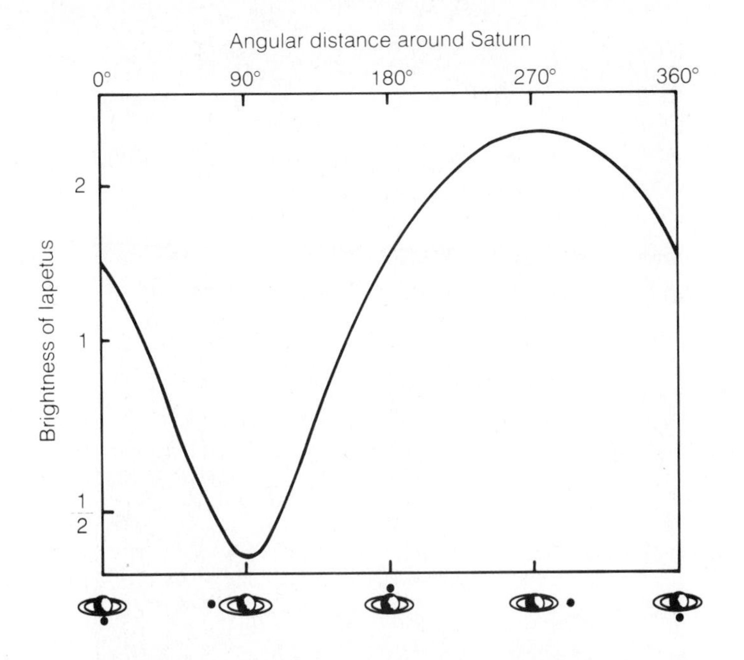

Figure 15.15 Iapetus, the second largest of Saturn's satellites, changes its apparent brightness by a factor of seven during the course of each orbit around the planet.

the bright side (Fig. 15.16). No other natural satellite comes close to this variation in light—not our moon with its familiar mottled surface, nor Io with its volcanoes, nor Titan with its nitrogen-methane atmosphere. One explanation for this brightness variation suggests that frosts are more common on the bright hemisphere of Iapetus, because encounters of the satellite's dark, leading hemisphere with debris, or with charged particles, have worn the frosts away, exposing dark material. Alternatively, the dark material may be debris that the satellite swept up some time in its distant past. Either of these hypotheses would certainly explain the observed variation in brightness but we can't help wondering whether the signpost hypothesis also deserves checking, so that a trip to Iapetus would reveal an obelisk with ...? Speculation is fun, but getting answers can be thrilling, and we should be able to solve this enigma when the same spacecraft that sends a probe into Titan's atmosphere pays Iapetus some close visits in the next century.

Triton: Chemistry at Low Temperatures

Neptune's largest satellite was known to have methane and nitrogen on it from ground-based observations. But no one knew what proportions of

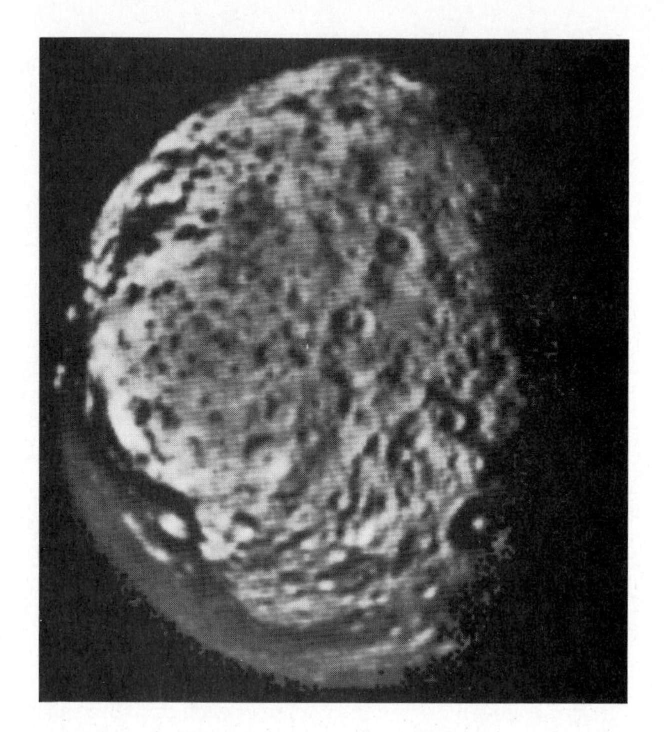

Figure 15.16 This *Voyager 1* image of Iapetus shows the satellite fully illuminated by the sun. The reason we do not see the left-hand side of Iapetus is because this is the beginning of the region covered by the dark material.

these substances were in the atmosphere or on the surface. We didn't even know the size of Triton before *Voyager 2* arrived in August 1989. We discovered that Triton is smaller and brighter than we had thought (Fig. 15.17). It has a tenuous, nitrogen-dominated atmosphere with a surface pressure just 16 millionths of our own and a surface temperature of only 37 K. Triton is thus distinguished by having the coldest sunlit surface in the solar system, since even Pluto is distinctly warmer at the present time. The pinkish color of Triton's surface, the dark, windblown splotches of material observed there, the haze in the atmosphere and above all, the active "geysers"—jets of material shooting some 8 kilometers vertically upward from the surface, then stretching out in windblown plumes for hundreds of kilometers—all demonstrate that this object is highly active despite its low temperature. Is Pluto like this too? Given its similar size and location in the solar system, we might well expect this, but our experience with the *Voyager* journeys has made us extremely cautious about extrapolating from one object to another. Ground-based observations have demonstrated that the surface of Pluto is

Figure 15.17 Frozen nitrogen covers the coldest sunlit surface in the solar system on Neptune's satellite Triton. Despite the low surface pressure—16 millionths of the sea level pressure of our own atmosphere—winds on Triton are capable of moving material around on the satellite's surface, as the dark, aligned, fan-shaped streaks shown here attest.

indeed mottled with dark material and that it has a tenuous, methane-containing atmosphere.

In the interstellar medium, CO and N_2 are respectively the most abundant carbon and nitrogen containing gases. Hence we anticipate that they will be trapped in ices forming at low temperatures in the outer solar nebula, so they should appear in the atmospheres of cold icy objects. The problem is that they are hard to detect by remote observations, especially nitrogen, which has no permitted infrared absorptions. Astronomers found evidence for solid CO and CO_2 on Triton using an Earth-based telescope in

1991. How closely does Pluto resemble Triton? The best observations from Earth suggest some real differences. A mission to Pluto, planned for the next millennium, will eventually tell us just how similar (or different) these tiny frozen worlds really are.

Cosmic Messengers

Having passed by Triton, *Voyager 2* continues to coast on out of the solar system. Like *Pioneers 10* and *11,* both Voyager spacecraft carry messages from the creatures that made them. Instead of the *Pioneer* plaques (see page 143), each Voyager spacecraft carries a long-playing phonograph record, anodized in gold for protection against erosion by interstellar dust particles (see page 378). If another civilization should someday find the spacecraft and realize that the records carry not only sounds but pictures as well, coded into the equivalent of audio signals, they could listen to sounds from Earth and view pictures of our planet and its inhabitants.

The two identical records each present about a hundred scenes from Earth, showing landscapes, human activities and development, the oceans, various animals and plants, and a few astronomical objects for easy recognition. The sounds on the record span an array of ethnic music, as well as greetings in 80 different languages, and selections from Bach, Beethoven, Stravinsky, and Chuck Berry. As the *Pioneer* and *Voyager* messages travel in space, we seem to be repeating ancient cultural patterns by creating artifacts that will preserve our accomplishments beyond the time span of a human life. Despite the tiny probability that such objects will ever be recovered, we send our messengers into space like bottles cast adrift in a celestial ocean. Do we hope that some cosmic beachcomber will find our spacecraft and read their messages? And are we ready for a reply?

SUMMARY

Each of the four giant planets, Jupiter, Saturn, Uranus, and Neptune, far exceeds the mass and size of Earth, the largest of the four inner planets. Unlike the rocky composition of Mercury, Venus, Earth, and Mars, the four giants are spheres of gas, mostly hydrogen and helium, which resemble the stars and the rest of the universe in their composition far more than the inner planets do.

Jupiter, Saturn, and Neptune each radiate significant amounts of heat from their interiors. Uranus does not exhibit this effect, although it otherwise resembes its giant cousins. The giant planets owe their gaseous, predominantly hydrogen-helium composition to the fact that their cores grew rapidly to such large masses that they could capture these light gases from the primordial solar nebula. In contrast, the smaller inner planets have relatively thin secondary atmospheres. The giant planets' initially great masses, together with their greater distances from the sun, have allowed this fundamental difference to persist.

Jupiter, being both the largest and the nearest of these objects, has received the closest scrutiny. In addition to hydrogen and helium, its atmosphere contains methane, ammonia, and water vapor, just the gases we would predict for an object that had a composition determined by the ingredients in the solar nebula. But we also find traces of nonequilibrium gases and pastel-colored clouds, indicating that more complex substances are being formed by ultraviolet radiation, lightning discharges, and thermal energy. Some of these compounds could be similar to those formed on the early Earth or delivered to our planet's surface by comets and meteorites before the origin of life. The most famous example of a colored region is the Great Red Spot, whose chemistry remains unknown. The lower atmosphere of Saturn may exhibit similar phenomena, but they are hidden by a thick ammonia cloud layer. Uranus is even more bland than Saturn, but Neptune again exhibits Jovian-like cloud features, without the pastel colors.

Each of the giant planets has an extensive system of satellites: Jupiter has 16 moons, three of which exceed our own moon in size, Saturn has 18, Uranus 15, and Neptune a mere eight moons. These satellites formed along with the planets, except for the small, outer moons of Jupiter and Saturn, which may be captured asteroids or comet nuclei. The rings of these giant planets represent swarms of minimoons, millions of tiny particles in satellite orbits. These rock- or pebble-sized fragments represent satellites that never managed to form, because the material composing them is too close to the planet.

Saturn's largest satellite, Titan, is especially interesting to us because it has a dense atmosphere in which chemical reactions are continually converting simple molecules into more complex ones. The ubiquitous, chemical smog that fills Titan's atmosphere prevents us from seeing the surface, on which drifts of hydrocarbons and seas of liquid ethane may exist. Iapetus, another of Saturn's moons, piques our curiosity by having one hemisphere that is seven times darker than the other. Is this a marker left by an advanced civilization? It seems unlikely. Uranus, Neptune, and Pluto show us that even at immense distances from the sun, chemical reactions will occur in mixtures of cosmically abundant gases, creating more complex species. Despite its small size and low temperature, Neptune's satellite Triton not only reveals evidence of such chemical reactions but has active, geyser-like plumes rising from its surface. Triton and Pluto show many similarities that suggest they may belong to the same family of icy planetesimals.

Their missions to the outer planets completed, four spacecraft are presently heading out from the solar system, carrying messages from Earth. We can only wonder if one day an advanced civilization will find these cosmic messengers traveling blindly through the vast dark emptiness of interstellar space.

QUESTIONS

1. Compare the amount of sunlight that falls on each square meter of Earth's surface with that which falls on each square meter of Jupiter (5.2 times Earth's distance from the sun) and of Saturn (9.5 times Earth's distance).

2. Why are Jupiter and Saturn warmer than we would expect from the small amount of sunlight that reaches these planets?

3. What does it mean to say that the giant planets have primordial atmospheres, while the inner planets have secondary atmospheres? What has happened to the primordial atmospheres of the inner planets?

4. What would happen to a spacecraft that tried to land on Jupiter?

5. What produces the various colors of Jupiter's clouds? What is the Great Red Spot?

6. Could life exist on the giant planets? Where would be the most likely spots for such life to appear?

7. What is unusual about Titan, Saturn's largest satellite? What do we hope to learn by studying this object?

8. What kinds of markers might be left in the solar system by an intelligent civilization to attract the attention of emerging intelligent species? Why is Iapetus a possible example of such a marker?

FURTHER READING

Beatty, J. Kelly, and Andrew Chaikin. *The New Solar System.* Cambridge, MA: Sky Publishing, 1990.

Cooper, Henry. *Imaging Saturn.* New York: Holt, Rinehart and Winston, 1982.

Ingersoll, Andrew. "Jupiter and Saturn," *Scientific American* (December 1981).

———. "Uranus," *Scientific American* (January 1987).

Johnson, Torrence, and Lawrence Soderblom. "Io," *Scientific American* (December 1983).

Johnson, Torrence, et al. "The Moons of Uranus," *Scientific American* (April 1987).

Littman, Mark. *Planets Beyond: The Outer Solar System.* New York: John Wiley, 1988.

Miller, Ron, and William Hartmann. *The Grand Tour.* New York: Workman, 1981.

Morrison, David, and Tobias Owen. *The Planetary System.* Reading, MA: Addison-Wesley, 1987.

Owen, Tobias. "Titan," *Scientific American* (February 1982).

Pollack, James, and Jeffrey Cuzzi. "Rings in the Solar System," *Scientific American* (November 1981).

Soderblom, Lawrence. "The Galilean Moons of Jupiter," *Scientific American* (January 1980).

Washburn, Mark. *Distant Encounters: The Exploration of Jupiter and Saturn.* New York: Harcourt Brace Jovanovich, 1983.

KONZERT F-dur

Brandenburgisches Konzert Nr. 2

Johann Sebastian Bach's *Brandenburg Concerto No. 2* (top) begins the music section of the information carried by each of the two gold-anodized records (bottom) mounted aboard the *Voyager* spacecraft that are now leaving the solar system after visiting the outer planets. These records are protected by covers that explain how to play them.

PART FIVE

The Search for Extraterrestrial Intelligence

And for the soul
If it is to know itself
It is into a soul
That it must look.
The stranger and the enemy, we have seen
him in the mirror.

— GEORGE SEFERIS

We now stand at the beginning of an era in which we possess the capacity to communicate with other beings at interstellar distances for the first time in human history. To turn the potential dialogue into reality, we should do more than wait for others to find us. Instead, like multitudes of other civilizations that may well exist in the Milky Way, we should analyze the means by which contact could occur, in order to determine the best way to proceed. If this analysis remains the fanciful pastime of a few astronomers, we seem unlikely to meet with success. But if we undertake a careful plan to search for our neighbors, we may yet achieve contact, perhaps not during the first year of our efforts, but within a span of time that human society would judge reasonable.

16

Is Earth Unique?

Our age will appear in terrestrial history as the first great era of solar-system exploration, for at last we are making the acquaintance of the cosmic community in which we live. As yet, however, we have found life nowhere but on Earth, despite our best efforts to study our moon, Mercury, Venus, Mars, Jupiter, Saturn, Uranus, and Neptune with instrumented spacecraft and (on the moon) with human landings. On our own planet, we have made detailed chemical investigations of meteorites that have arrived from interplanetary space but have found no living organisms within them.

Our failure to find life in any of these environments may seem frustrating, but this very failure signals that the search is well underway. Furthermore, our knowledge of the conditions that have produced life on Earth, and of those that may have led to life on other planets, has advanced considerably as a result of the *Viking* landings on Mars and the extensive investigations of Venus. In the outer reaches of the solar system, we have found environments that resemble those some scientists have postulated for the primitive Earth: hydrogen-rich atmospheres, in which thermal energy and ultraviolet radiation convert simple molecules into more complex compounds. In some meteorites, we have found evidence for the products of these reactions, since we have discovered organic material that includes some of the amino acids and all of the nucleotide bases that are used in our kind of life. These discoveries encourage us because they increase our knowledge of the conditions that existed when life began on Earth, and also because they suggest that the fundamental processes that can lead to the origin of life occur commonly throughout the cosmos.

But closer to home, as we have learned more about Mars and Venus, we have come to appreciate the combination of conditions that a planet must achieve if it is to change from an object with interesting chemical reactions into a world that can develop and sustain life. Why, in our solar system, have these circumstances apparently arisen only on Earth? What are the special properties of our planet that have singled it out among the other planets and their satellites as the present home of abundant life? Is chance alone responsible for the fact that this book was printed on Earth rather than on Venus? Is it even possible that the Earth is unique in the galaxy as the only abode of intelligent life?

Distinguishing Characteristics of the Earth

If we confine ourselves to the most basic characteristics of planets, those that are not the result of the others, we can identify three facts that in combination make the Earth unique in the solar system: its distance from the sun, its size, and, perhaps less important, the relatively large mass of its natural satellite. Each of these three characteristics of Earth has an important impact on the possibilities of the origin and evolution of life.

Suppose that the Earth's orbit changed, so that it circled the sun where Venus now does. We can calculate what would happen: The increased intensity of solar radiation would raise the average temperature of our planet, thus producing greater evaporation of water and hence more water vapor in our atmosphere. The water vapor would make it more difficult for the surface of the Earth to radiate its heat into space, and this would raise the temperature still further, which would produce more evaporation of water, and so on and on. The Earth would therefore undergo the same runaway greenhouse effect that has led to the drying out of Venus (Fig. 12.6). The high temperature would also lead ultimately to the production of a thick carbon dioxide atmosphere on Earth similar to that on Venus.

Suppose instead that the Earth exchanged its position with Mars. Then the decrease in incoming solar energy, arising from our greater distance from the sun, would cool the oceans and increase the size of the polar caps of ice; this would leave less water vapor in our atmosphere and would also increase the reflectivity of our planet. Hence the temperature would decline still further, and so on and on until we had a runaway refrigerator. In the extreme case, the Earth would acquire a complete covering of water ice. We can imagine intermediate situations, however, in which a thick carbon-dioxide atmosphere maintained a warmer climate, or in which a temperate environment might exist near the equator. But in an orbit still farther out from the sun, the Earth would lose even this possibility. We are clearly better off at roughly our present distance from the sun, even if we could tolerate a somewhat greater distance if we had a CO_2-rich atmosphere.

Being the right size is also extremely important. Large planets cannot lose their hydrogen, while the smallest planets cannot retain an atmosphere. This constraint actually goes beyond a simple ability to hold an atmosphere once it exists. As we saw in Chapter 14, a planet the size of the Earth has a better chance to develop a temperate environment than a planet as small as Mars or Mercury. The larger object has a greater probability of producing a dense atmosphere, because it had a greater accumulation and outgassing of volatile elements during its formation and later history. It also has a better chance of preserving its atmosphere during the intense bombardment that all inner planets experienced.

Thus *size* and *location* form two key aspects of a planet's suitability for life, along with the *composition of the planet's atmosphere.* What about a moon? The fact that our planet has a large natural satellite may seem incidental to the development and continued existence of life on Earth, despite the familiar evidence that the moon influences such widely varied activities as human romance and fish spawning. But the moon apparently does provide two benefits of great importance to us. First, the moon produces large tides, which may have been a key element in making microenvironments within which life could begin. (On the other hand, the sun also produces tides. Though less than half the size of those raised by the moon, these tides might have been sufficient by themselves.) The moon has a second effect on Earth beneficial to life: It stabilizes the orientation of the Earth's rotation axis. We have seen that changes in the angle by which rotation axis of Mars deviates from being perpendicular to the plane of its orbit can produce large variations in the Martian climate (see page 305), but a large Martian moon would suppress them. Such changes in the "tilt" of Mars arise from the combined gravitational pulls on Mars from the sun and Jupiter. Calculations have suggested that the Earth's ice ages, some of which have made many species of life extinct, may arise from much smaller changes in the inclination of the Earth's rotation axis and in the eccentricity of the Earth's orbit around the sun. These changes could be much more dramatic if we did not have the nearby gravitational pull of a large satellite to suppress them. Variations greater than the ice ages in our planet's climate would probably have been fatal to developing life—at least to life on land.

On the other hand, it is *not essential* to have a large satellite to bring about the desired stability. If the rotational period of the Earth were much faster or much slower, the same effect would occur. The 243-day period of Venus, for example, produces a stable inclination of the planet's rotation axis. So what may be special about the Earth-moon system is that it allows something like a 24-hour day with stability. How would life develop on planets with shorter or longer periods of rotation? We can only guess, but we can suggest that with the wide variety of accommodations to different daylight periods exhibited by the various forms of life on Earth (for example, by penguins and polar bears, which live through long summer days and equally long winter nights), there seems to be no intrinsic drawback to the development of life on a planet with a dramatically different day-night cycle than ours. The absence of the large tides caused by the moon would also not be fatal. Solar tides, the fluctuations of water levels caused by weather, and freeze-thaw cycles might well be sufficient to produce suitable microenvironments.

The Planetary Systems of Other Stars

Nothing in our theories for the origin and evolution of our sun and its planets is unique to the solar system. We may therefore reasonably expect to find planets around the stars that are most like the sun in their lifetimes and luminosities—that is, stars of spectral type F5 to K8. We anticipate that all planetary systems will have a set of rocky inner planets, with atmospheres produced by outgassing, weathering, and escape, for the same reasons that our own rocky inner planets have atmospheres. Judging from our own example, the chances seem good that one of these inner planets will orbit its star at the "right" distance. In our own solar system, the Earth is "right" and Mars and Venus are not far away; all three appear to have had liquid water at some time in the past, though only Earth does today. Our solar system has two planets (Earth and Venus) that are the right size, and one of them is in the right place. This suggests that something like one in every two planetary systems will be similarly favored with a planet of the right size in the right place.

We say one in every two to be conservative; the optimistic guessers would estimate that since we exist, almost every planetary system should have a planet of the right size in the right place! Such optimism requires either that Earth-sized planets will always be present, or that a planet the size of Mars could support life if it were closer to the sun. Either or both of these hypotheses may be true, but, lacking more evidence, we do best to be cautious.

The remaining worry, oddly enough, is the moon! We don't understand why our satellite is so large (relative to ourselves), or even how it formed, so we cannot make accurate predictions for other planetary systems. But as we have seen, the need for a massive satellite is probably not as great as the other two requirements. We can therefore estimate that perhaps one in every four planetary systems should have an Earthlike planet in the right position, with sufficient stability of climate for life to develop. This estimate of one in four rests on the idea that half of all planetary systems will have a planet of the right size in the right place, and half of those will either not need a massive moon to damp their inclination changes, or will have such a moon, as the Earth does.

Don't think that this deduction must be rigorous simply because numbers appear! Other features, as yet unrecognized by us, may make the Earth unique, or nearly so. For example, we have not dealt with the fact that only the Earth among the sun's planets has oceans of water. A large abundance of liquid water clearly helps to nurture life as we know it, and we have assumed that water will follow naturally once a planet has the right size in the right place. But other factors, such as the distribution and abundance of volatile-rich meteorites and comets in the primitive solar system, could be significant in determining whether or not liquid water appears. Finally, we must emphasize that *the existence of a planet identical to the Earth does not guarantee that life will develop on that planet; nor are we in a position to state that only an Earthlike planet can harbor life.* But our experience thus far tells us that Earthlike planets are the most probable habitats, and probabilities are what we must work with when we search for life outside the solar system.

To carry this argument further, we would clearly like to be able to investigate other planetary systems to see whether our own system of planets has some unusual property that is not shared by the others. Unfortunately, we can't carry out such a comparative study now, since we know of only one planetary system, just as we know only one example of life. But our intuition suggests that other such systems exist. The growth of human knowledge about the universe has been accompanied by a steady erosion of our sense of being "special." Early astronomers thought that the Earth was the center of the universe. The discovery that the Earth orbited the sun still left an impression that the sun must be the center of the stellar system that we now call the Milky Way galaxy. Once the sun turned out to be located in a spiral arm far from the center of our galaxy, astronomers still thought that the Milky Way was one of the largest galaxies in the universe. We now know that even this is not true. It is not simply that we are not special, but that the astronomical universe contains extremely few, if any, truly unique objects. There are many quasars and pulsars, although at first only one or two were known; many supernovae, many europium-rich stars, and many dense interstellar clouds. This perspective suggests to us that there are probably many other solar systems, despite the fact that at the present time we have no direct observations of them.

Support for this assumption comes from studies of multiple-star systems. Most stars in our own galaxy, and presumably in other galaxies as well, come in doubles, triplets, quadruplets, and even higher-number combinations (Fig. 16.1). Most double-star systems show the two stars quite close to each other, with a typical separation approximately equal to the distance from the sun to Neptune. This relatively small distance, less than one one-thousandth of the average separation of neighboring star systems, is a clue that the formation of stars from interstellar clouds is not likely to result in a single object. The abundance of stellar systems with dimensions on the same order as our planetary system suggests that in those cases where we find a star that does *not* have a close, visible companion, we might expect that a small, dark star or a series of planets exists instead, as a result of the fragmentation of the original cloud that produced the star itself.

We see evidence for this tendency toward fragmentation during the formation of massive objects in our own solar system. The giant planets are each accompanied by an extensive retinue of satellites, and the innermost

Figure 16.1 The Alpha Centauri star system is the closest to our sun. Each of the two sunlike stars called Alpha Centauri A and B could have one or more planets similar to Earth. The distance between A and B roughly equals the distance from the sun to Neptune. A fainter component of this system, Proxima Centauri, is 2000 times farther from A and B than this distance.

satellites clearly formed along with the planets themselves. Indeed, with the recent discovery of a satellite for Pluto, only two of the nine known planets in our solar system have no satellites at all. Mercury and Venus may simply be too close to the sun: Solar tidal forces may have prevented satellites from forming in or being captured into stable orbits around these planets.

Searching for Other Planetary Systems

These indirect arguments are encouraging, but we would obviously prefer to have some observational evidence of the existence of other planetary systems. Unfortunately, as we discussed in Chapter 5, this evidence has proved very difficult to obtain. Planets shine by reflected light, far more weakly than the stars that illuminate them. Furthermore, the relatively close nestling of planets around stars means that we must look for a weak point of light at almost the same place on the sky as a bright star. A planet of Jupiter's size, and at the same distance from Alpha Centauri A (a member of the nearest star system, shown in Fig. 16.1) as Jupiter is from our sun, would have an apparent brightness in visible light only one-billionth of the apparent brightness of the star! With this tremendous difference in brightness, even the best telescopes we have on Earth could not reveal a planet as large as Jupiter in orbit around Alpha Centauri A or B, assuming that the planet-star distance is roughly the same as the distance from Jupiter to the sun.

But how can we detect planets around other stars if we can't *see* the planets? Four good possibilities exist: First, build a better telescope, and place it outside the Earth's atmosphere to avoid the blurring effects caused by the air. Second, try to find planets not by their weak reflected light, but by the perturbations that their gravitational forces produce on the motion of their parent stars (Fig. 16.2). Third, and directly related to the second possibility, look for Doppler-effect changes in the spectrum of the parent star as the planet orbits around it, pulling it first one way and then the other (Fig. 2.10). Fourth, try to detect evidence of planetary systems in the process of formation, when there is still enough gas and dust around the star to make this process discernible from our distant vantage point (see Chapters 4 and 11).

Attempts to find planets by their effect on the apparent motions of the stars they orbit (methods 2 and 3) have not yet led to an unequivocal detection of another solar system, although there are some intriguing hints. We do not yet have a telescope in orbit dedicated to planet-seeking (method 1); the Hubble Space Telescope was not designed to detect the faint, reflected light from planets circling much brighter stars. However, this powerful device will be able to help with the fourth approach to the problem—the search for solar systems in the first stages of formation.

Astronomers are already achieving considerable success in studying the earliest phases of the formation of planets by using observatories on the Earth. One approach has been to employ large radio telescopes to look for disks around young stars, possible evidence of planetary systems being formed. Several such disks have been identified, and in one case (so far!) there are indications that the gas in the disk is revolving about the star according to Kepler's laws for planetary motion. Other indications of disks have come from observations by the infrared satellite IRAS in 1983 (see again Fig. 4.4), which found some stars radiating more heat than could be expected from the star itself. Evidently grains of dust are orbiting these stars and are warmed by starlight, just as the Earth or Venus is heated by the sun and then radiates this energy at infrared wavelengths.

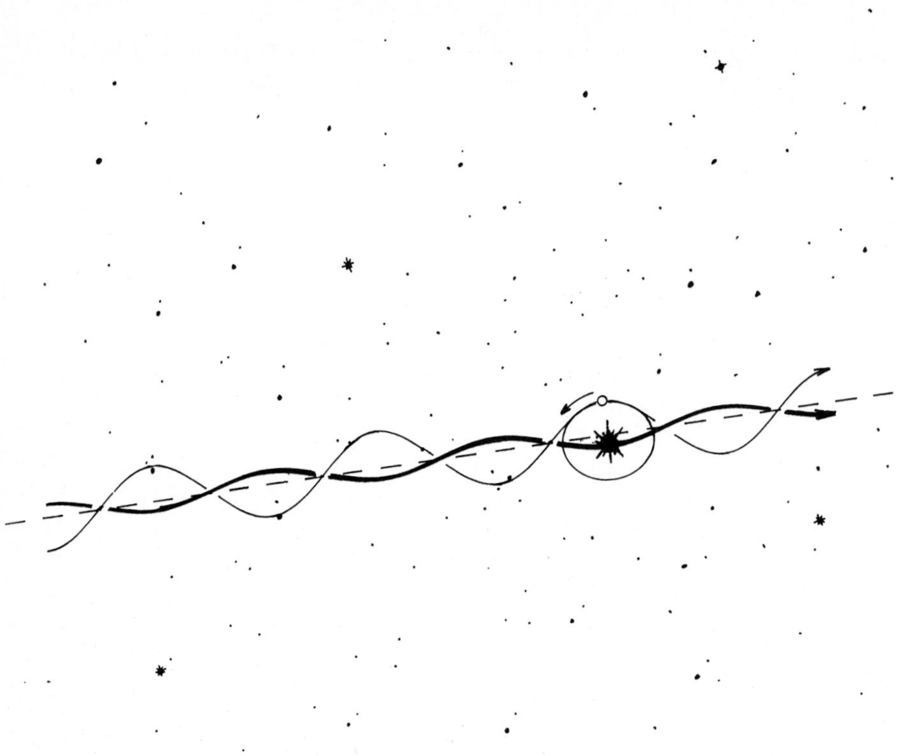

Figure 16.2 If planets orbit a star, then the star will make a small orbit around the center of mass of the star-planet system. Since this system slowly orbits the galactic center, this small orbit adds a tiny wiggle to the star's motion, which would otherwise appear to be a straight line. The closest stars offer a chance for us actually to detect such wiggles.

But are these grains arranged in a cloud or in a disk? In at least one case, a star called Beta Pictoris that is visible from the southern hemisphere, we have direct, optical evidence of a disk. In 1984, using a specially constructed device called a coronagraph that blocked the light from the star itself, Bradford Smith and Richard Terrile were able to record the light from the disk surrounding this star (Fig. 16.3). The Beta Pictoris disk has a diameter of approximately 2000 A.U., far larger than the 80 A.U. diameter of Pluto's orbit in our own solar system. But there are indications that close to the star, there is a relatively empty space in the disk with a diameter of 30 to 40 A.U. Some large objects must be present in orbit about the star to keep this region clear. A giant planet at 5 to 7 A.U. and another one at 30 A.U. would do the job nicely, although there are certainly other possibilities.

Figure 16.3 This picture is a composite of a positive image of Beta Pictoris divided by a negative image of Alpha Pictoris. Both stars were viewed through a coronagraph. The composite cancels out almost all the scattered light, although a bright crescent from Beta Pictoris remains. The disk of material surrounding Beta Pic is visible as a bright diagonal band that can be traced to a distance of nearly 1000 A.U. on either side of the star.

So far, Beta Pictoris is the only star around which such a disk has been detected. But the search continues, and the Beta Pictoris system itself will receive much additional scrutiny in the years ahead. Are there really some forming planets circling this southern star? If so, we would finally have good evidence that our solar system is not unique. Meanwhile, the current "most likely planet" outside the solar system has been found where no one had expected it: orbiting a neutron star, which is the burnt-out remnant of a supernova explosion (Chapter 6)!

In 1992, radio astronomers at the Arecibo Observatory in Puerto Rico announced the results of their search for objects orbiting pulsars, the rapidly rotating remnants of supernovae (see page 144). The search for planets around pulsars relies on extremely accurate timing of the arrival of the radio pulses. In one case, changes in the arrival times of the signals from the pulsar PSR1257+12, located about 1600 light years from Earth, seem to reveal two planets.

The timing measurements show that sometimes the radio pulses arrive slightly earlier than expected, sometimes slightly later; furthermore, these differences repeat in a cyclical manner. The radio astronomers concluded that the neutron star producing the pulsar must itself be moving in a tiny orbit around the center of mass of a pulsar-planet system. As a result, the pulsar is sometimes a bit farther away, sometimes a bit closer, than we would expect if no such orbit existed (see again Fig. 16.2). The pulses were timed with an accuracy sufficient to reveal that the pulsar is being tugged by two objects orbiting around it at distances about equal to the distance of Mercury from the sun. These objects have masses of 2.8 and 3.4 times the Earth's mass—far less than Jupiter's mass, and hence indicative of a planet rather than a star or a burnt-out stellar cinder.

Not bad for observations made from so far away that all our conclusions refer to what was going on 1600 years ago! The most amazing fact about the objects orbiting the pulsar is this: Astronomers can tell how close to *circular* their orbits are. A non-circular orbit would reveal itself through *non-steady changes* in the deviation of the arrival times of the pulses from what we would expect if no orbit existed. Instead, those deviations change in such a regular way that we can conclude that the two objects orbiting the pulsar have orbits not much farther from circularity than the orbit of Earth around the sun.

And that amazes astronomers. How could a star explode (the only way that we know to produce a pulsar) and leave behind not only two small companion objects, but objects moving in nearly circular orbits? Any explosion that ejects a large fraction of the star's mass into space would cause the orbit of any nearby planet to deviate noticeably from circularity, and we know of no way for the object to reacquire a circular orbit in any reasonable time. Some astronomers speculate that in fact the planets must have *formed* within the million or so years since the star exploded! If this is so, it would certainly testify to the ease with which planets can form from debris orbiting stars. Stan Woosley and Robert Lin, two astronomers at the University of California, Santa Cruz, have tentatively suggested that such planets be named either Phoenix (if they formed from the ashes of the exploded star) or Zombie (if they actually survived the supernova's explosion).

Because the new "planets" have such strange characteristics (by our standards), many astronomers maintain serious reservations about the existence of the objects orbiting PSR 1257+12. But the high quality of the data obtained from the pulsar suggest that strange though they may be, the objects might prove to be the first confirmed planets found to orbit a star (or former star) other than our sun.

The Likeliest Stars

While we are waiting for these techniques (and others that will undoubtedly be developed) to find other solar systems, we can speculate about which stars among our neighbors would be good candidates for such searches. Many scientists are sufficiently confident of the abundance of planetary systems to suppose that any apparently single star with a spectral type similar to the sun's must be accompanied by planets. We are most interested in possible planets around *nearby* stars, since they are the ones with which communication—using either rockets or radio—would be easiest. We therefore begin our search with a consideration of the 23 star systems that lie within 13 light years of the sun (Table 16.1).

Note that half of these systems are multiples. Only three of the stars (Alpha Centauri A, Sirius A, and Procyon A) have a greater luminosity than the sun does; that is, only these three stars lie above the sun on the main sequence. The other stars either lie far down the main sequence, typically in spectral classes K and M, or (for the companions of Sirius and Procyon) they have evolved to the white dwarf stage (see Chapter 6). It seems that although the sun does provide us with a generally average star, we can count ourselves lucky in having a source of energy that exceeds 90 percent of all stars in mass and thus in true brightness. Furthermore, a much brighter star such as Sirius A (which has 23 times the sun's luminosity) cannot last as long as the sun; in fact, Sirius A must be less than a billion years old, or it would have evolved away from the main sequence, as its white-dwarf companion clearly has done.

We have already stated our reasons for wanting stars to be on the main sequence for at least 5 billion years to improve our chances of finding a planet inhabited by an intelligent civilization. This requirement eliminates Sirius A and makes Procyon A a bad risk. If a star has planets, we also want at least one of the planets to be warm enough for life to exist on it. If we assume that either water or ammonia will provide the necessary solvent for life (see page 235), we require a temperature that remains within the range of 0 to 100° C (if water is the solvent) or –108 to –33° C (if ammonia does the job). So we are looking for temperatures in the range of –108 to +100° C, with a strong preference for values somewhat above 0° C.

In our own solar system, we know that only the planets Earth and Mars have surfaces that include temperatures in this range. Venus, which would have a temperature of about 45° C if it had no atmosphere, has instead a carbon dioxide atmosphere that keeps its surface temperature at 475° C. If we exclude Venus from the proper temperature zone, we find that the sun, or another star like it, can produce the proper temperatures on planets that orbit between about 0.85 and 2.0 times the Earth's distance from it

TABLE 16.1 Stars Less Than Thirteen Light Years from the Sun*

Star Name	Distance (light years)	Spectral Type	Luminosity (Sun = 1)	Mass If Known (Sun = 1)
Proxima Centauri	4.2	M5	.00006	.1
Alpha Centauri A	4.3	G2	1.53	1.1
B	4.3	K1	.44	.88
Barnard's Star	6.0	M4	.0004	
Wolf 359	7.7	M6	.00002	
Lalande 211385	8.2	M2	.005	
Luyten 726-8 A	8.4	M5	.00006	.044
B	8.4	M6	.00004	.035
Sirius A	8.6	A1	23.0	2.31
B	8.6	wh.dw.	.0020	.98
Ross 154	9.4	M4	.0004	
Ross 248	10.4	M5	.0001	
Epsilon Eridani	10.8	K2	.30	
Ross 128	10.9	M4	.00033	
61 Cygni A	11.1	K3	.082	
B	11.1	K5	.038	
Luyten 789-6 A	11.2	M5	.0001	
B	11.2	M5`	.0001	
Epsilon Indi	11.2	K3	.14	.8
Groombridge 34A	11.2	M1	.006	
B	11.2	M4	.0004	
Procyon A	11.4	F5	7.6	1.77
B	11.4	wh.dw.	.0005	.63
Sigma 2398 A	11.6	M3	.003	.4
B	11.6	M4	.002	.4
CD-36°15693	11.7	M1	.02	
G51-15	11.7	M7	.00003	
Tau Ceti	11.8	G8	.47	.9
BD +5° 1668	12.3	M4	.0015	
Luyten 725-32	12.5	M5	.0003	
Lacaille 8760	12.5	K6	.027	
Kapteyn's star	12.7	M0	.004	
Kruger 60 A	12.9	M3	.0015	.27
B	12.9	M3	.0004	.16

*Boldface type indicates the stars that are most like the sun in this list: spectral types F5 through K3.

(Fig. 16.4). This relatively narrow "habitable zone," within which temperatures are appropriate to life, includes the orbits of Earth and Mars, the two planets that have had liquid water at some time in the past, though Mars no longer does. The limits of the habitable zone will vary somewhat with the composition and thickness of the atmosphere, which affects how much heat is trapped by the greenhouse effect. We must also keep open the idea of Europa-like environments, where tidal heating of a satellite by its planet extends the outer limit of clement temperatures.

If we turn again to the list of stars in Table 16.1, we find that our desire to duplicate the proper temperature conditions means that to be habitable, planets must orbit far closer to the faint K and M stars that comprise the majority of the list than to stars like the sun. These low-luminosity stars do possess tremendously long lifetimes; so if habitable planets orbit around

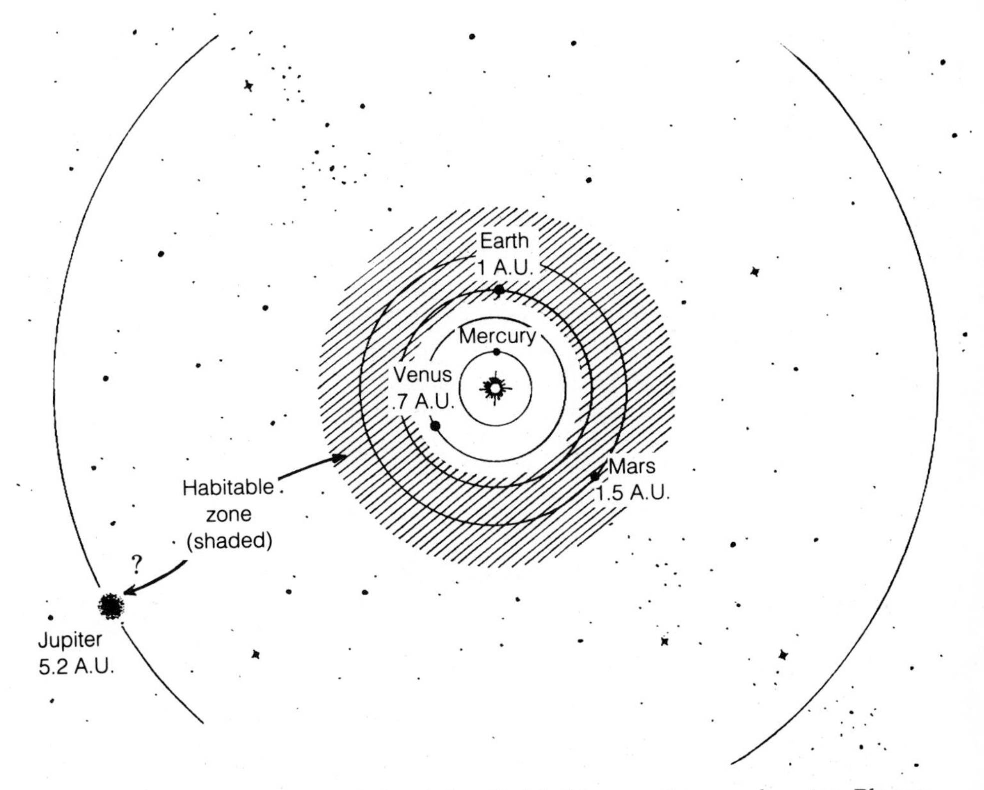

Figure 16.4 Temperature defines the "habitable zone" around a star. Planets orbiting too close to the star will be too hot for molecules to exist, and thus for complex reactions to produce life; planets too far from the star will be too cold for complex molecules ever to form. An exception might exist for planets such as Jupiter, that heat nearby satellites through the dissipation of tidal forces.

them, they can expect many tens of billions of years of steady starshine rather than the mere 10 or 11 billion years that our sun will provide the Earth.

But how close must a planet be to, say, an M4 star such as Barnard's Star? This second-closest star system consists of a single dull red star that emits less than one two-thousandth of the sun's luminosity each second! For a planet to receive as much starlight energy from Barnard's Star as we receive from the sun per square meter per second, the planet's distance from Barnard's Star must be 1/45 the Earth's distance from the sun. In other words, even Mercury's distance from our sun far exceeds the maximum distance a planet could have from Barnard's Star and yet stay warm enough for life to exist.

At such close distances, another worry comes into play: the possibility of a gravitational lock between the orbital and rotational motions of the planet that keeps the same hemisphere always facing the star. This has happened for our own moon and for at least five of the inner satellites of Jupiter; the eccentricity of Mercury's orbit has led to a 3:2 resonance instead (see page 274). How hostile to life would such a rotation lock be? We would like some alternation of conditions—wetting and drying, freezing and thawing—to help the initial chemistry that starts life along its way. But perhaps there are other pathways that lead to the same results. The presence of a sufficiently thick atmosphere on one of these rotationally locked planets might provide adequate modulation of the perpetual day or perpetual night conditions. We simply can't rule these M dwarfs out *a priori* as possible centers for inhabited planets; but they do seem less likely prospects than their less numerous but more luminous sisters.

The habitable zone around stars, the place where planets would have the right temperatures for life to exist, often receives the title of "ecosphere." Because most stars shine less brightly than the sun, their habitable zones are considerably smaller than the sun's (Fig. 16.5). We can therefore shrink the length of the list in Table 16.1 when we look for the stars most likely to have habitable planets. In fact, since Sirius A and Procyon A can be eliminated because of their short lifetimes, the best candidates within four parsecs of the sun with reasonably large ecospheres are Alpha Centauri A and B, Tau Ceti, and Epsilon Eridani. Let us take a brief look at each of these possibilities.

The Alpha Centauri system (Fig. 16.6) consists of two fairly bright stars (A, with 1.5 times the sun's luminosity, and B, with 0.44 times) plus a third, much fainter star, Proxima Centauri, with less than one ten-thousandth of our sun's luminosity. Astronomers once thought that multiple star systems were highly unlikely places for planets, because the gravitational attractions of the various stars would either prevent the formation of planets, or would forbid the existence of stable orbits for any planets that did form. The present

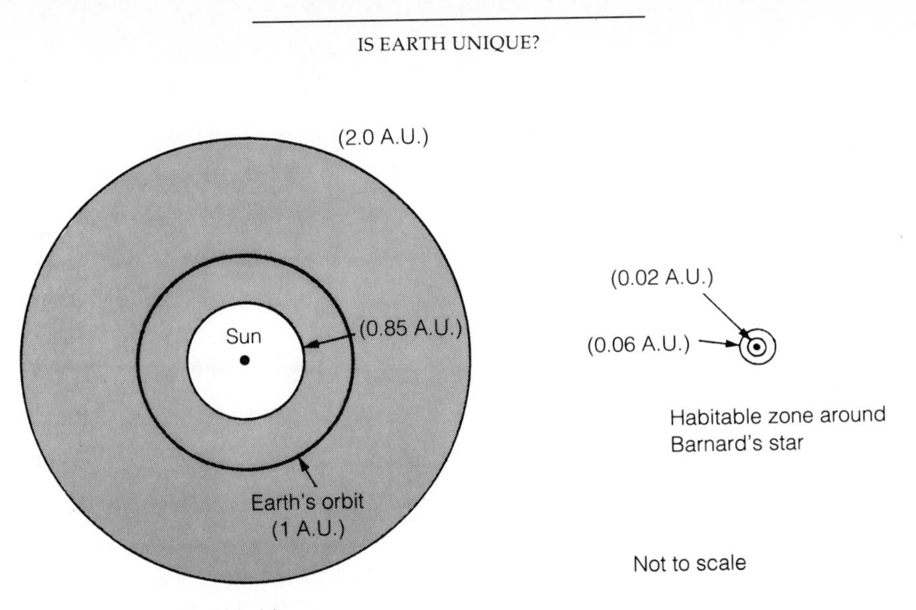

Figure 16.5 A sunlike star has a habitable zone far larger than that around a dim red star such as Barnard's Star. If we seek a temperature range between -108 C and +100 C, the habitable zone around the sun extends from 0.85 to 2.0 times the Earth's distance, whereas the habitable zone around Barnard's Star extends only from 0.02 to 0.06 times the Earth's distance from the sun. Note that the habitable zone around Barnard's Star is even smaller than drawn in the figure.

view is that we do not really know whether planets are less likely to form in a double- or multiple-star system than around a single star. Also, we were too pessimistic about the possibility of stable orbits. In a double-star system, stable orbits are possible if one of two possibilities occurs. Either the planets must move in orbits close to one of the two binary stars as the stars themselves orbit around their common center of mass, or the planets must orbit at a large distance from both members of the double-star system.

Both Alpha Centauri A and B, with spectral types G2 and K5, are quite capable of supporting life on planets in orbits around them, provided that the planets' orbits are sufficiently close to one or the other of these two stars. Since the separation of A and B is about 25 times the distance from the Earth to the sun, we could imagine planets around both stars with orbits similar to those of the inner planets in our system. In this case, inhabitants of a planet moving around one star would see the other star as a sort of a giant moon, a thousand times brighter than our moon though still a thousand times fainter than the star which kept the planet in orbit. This "supermoon" would be out in the daytime for half the year, and then out at night for the other half of the year. The two stars have slightly different colors, yellow and orange, which would add to the beauty of the effect (see Fig. 16.7).

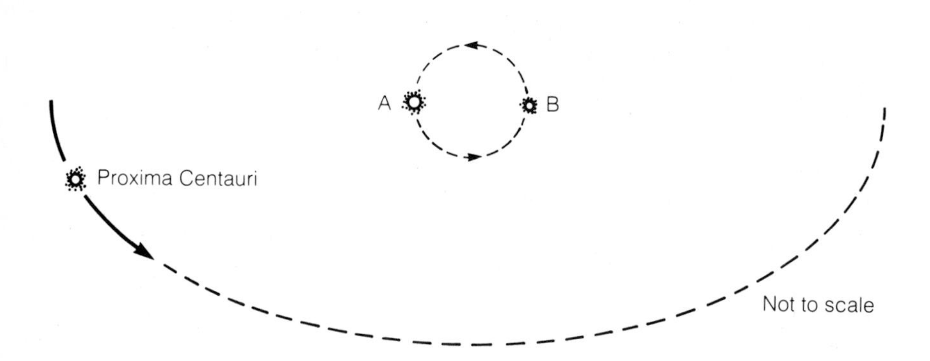

Figure 16.6 The Alpha Centauri system consists of three stars, two of which (Alpha Centauri A and B) closely resemble the sun. All three stars orbit the system's center of mass, but A and B are separated by about 25 times the Earth-sun distance, while Proxima Centauri lies 2000 times farther from the center of mass—50,000 times the Earth-sun distance. Figure 16.1 shows a photograph of Alpha Centauri A and B.

The other case, in which a planet orbits at great distances from both of the double stars, makes the planet follow an orbit similar to that of Proxima Centauri (see again Fig. 16.6). Although such orbits are stable, they do not help us in our search for life, because the great distance of Proxima Centauri from either Alpha Centauri A or B at any time means that both of these stars appear no brighter than our moon. Hoping for life on a planet that occupied such an orbit would therefore be similar to hoping that life would develop by moonlight, which appears unreasonable.

We could, however, imagine that a planet orbits extremely close to Proxima Centauri itself. But this M5 star is so faint and so cool that it presents us with the same conditions we faced in considering Barnard's Star: Habitable planets must have extremely small orbits. If the Earth orbited around Proxima Centauri at our present distance from the sun, we would see our new "sun" as a small red disk 60 times brighter than the full moon (as illuminated by our present sun) though only one-tenth as large on the sky. With this as our parent star, our hopes for survival would be dim indeed, although we would certainly enjoy the celestial spectacle, with nearby A and B shining brightly in our heavens.

Farther out in space than the Alpha Centauri system, the next good candidates we encounter are Epsilon Eridani and Tau Ceti. Epsilon Eridani, a K2 star, has only 30 percent of the sun's luminosity, about the same as Alpha Centauri B. A planet in orbit around this star at the distance of Mercury or Venus from the sun would still fall within the habitable zone. Tau Ceti, a G8 star, resembles the sun a bit more than Epsilon Eridani does, for its luminosity equals 47 percent of the sun's. Again, we can expect a reasonably large

Figure 16.7 A planet orbiting either Alpha Centauri A or B would see two bright "suns," one yellow and one orange, though one would be considerably brighter than the other. Proxima Centauri would appear as a dim red object.

ecosphere around Tau Ceti, larger than Epsilon Eridani's but smaller than the sun's. Epsilon Indi and 61 Cygni A are also likely candidates, since they have spectral types not far below Epsilon Eridani's. But with luminosities that are respectively 53 and 74 percent lower, they do not rank as high on our list.

Both Tau Ceti and Epsilon Eridani drew early attention from scientists who speculated about life around other stars, and from the first radio search to the present day, astronomers have pointed their antennas in the direction of these two stars with the hope, so far unfulfilled, of detecting civilizations there. The Alpha Centauri system has not received comparable attention, primarily because the radio telescopes located on the northern hemisphere of our planet cannot observe it, and these are the principal instruments that have been used to search for indications of artificial signals up to the present time.

TABLE 16.2 Approximate Numbers of Main-Sequence Stars in the Milky Way

Spectral Type	Approximate Mass (Sun = 1)	Luminosity (Sun = 1)	Percent of Total*	Number of Stars
O	30	50,000	0.00002	60,000
B	6	300	0.09	270,000,000
A	2	10	0.6	1,800,000,000
F	1.25	2	2.9	8,700,000,000
G	0.9	0.9	7.3	22,000,000,000
K	0.6	0.2	15	45,000,000,000
M	0.2	0.005	73	209,000,000,000

* We assume that the Milky Way contains 300 billion stars. Stars not on the main sequence, and therefore not included in this list, form 0.8 percent of the total number.

As our discussion of the likelihood of finding another civilization on another planet will show (see Chapter 18), we should not expect life to be so easy that the first two or three stars to be studied would reveal another civilization, especially after only a short period of observation. Nevertheless, we should not abandon the idea that even one of these nearby stars may be circled by a planet carrying a civilization far more advanced than ours, to be discovered when we learn the proper methods for making contact. What a splendid way to prove the existence of other planetary systems!

Meanwhile, our consideration of the 23 star systems within 13 light years of the sun has indicated what an average region of our galaxy is like. We can see that most stars have too low a luminosity to be promising candidates for planet-based life, if we confine ourselves to the kinds of environments we have come to know in our own solar system. But with so many stars in a galaxy such as ours, we need not be discouraged even if we take the conservative view of eliminating the small, cool stars from consideration. As Table 16.2 shows, the Milky Way galaxy contains over 60 billion stars with characteristics comfortably close to those of our sun. We may simply need to extend the boundaries of our search to greater volumes of space before we encounter the technological tendrils of an alien intelligence unfolding in our direction at the speed of light.

SUMMARY

We began this chapter with a question: Does some extremely rare attribute characterize our planet, our solar system, or our star? If so, we may be led to the conclusion that we are the only intelligent civilization in the Milky Way galaxy. This question arises from the realization that at this stage in our exploration of our solar system, our planet provides the only environment where we have found evidence for the existence of life. But there seem to be good reasons for this anomaly: the Earth's size and distance from the sun. If Venus and the Earth changed places, it would be Venus that would be inhabited, provided that Venus had formed at the Earth's present position, so that it would have retained the same mixture of volatile elements. Our moon has played an important role in the origin of life by stabilizing the Earth's inclination and increasing the magnitude of oceanic tides. However, faster or slower rotation might provide similar stability, and more gentle tides could provide adequate environments for the early stages of life.

Other solar systems should have rocky inner planets like ours, and the chances of finding a planet of the right size in the right place with the right rotational period seem to be about one in four, if we assume that our system is typical. Unfortunately, for the time being we have no direct observational evidence of any other solar systems. Indirect evidence abounds, however, and points to the frequent formation of planets as a natural part of star formation. Several techniques for detecting solar systems around nearby stars are becoming available and may provide the evidence we seek within the next decade.

As we look about us in the galaxy, we find that the volume of space within just 13 light years of the sun contains three stellar systems that are good candidates to have planetary systems with habitable zones suitable for life. Two other stars in our list are marginally acceptable, so even this tiny sample of the stars in our galaxy may contain another planet like the one on which we live. This does not mean that such a planet must harbor an intelligent civilization, or even that life must have originated on its surface. But it does suggest that the Earth is almost certainly not unique. To *find* other life, we have to begin a sophisticated search that may extend many more light years away from our solar system.

QUESTIONS

1. What basic characteristics distinguish the Earth from the other planets in our solar system? Do you expect that another planetary system would be likely to contain one or more planets with characteristics similar to Earth's? Why or why not?

2. What would happen to the Earth if it were suddenly moved to the position of Venus?

3. What evidence suggests to astronomers that solar systems are relatively common in our galaxy? What evidence argues against this conclusion?

4. Why is it so difficult to detect planets around nearby stars, if they exist? What methods have been proposed to search for solar systems associated with nearby stars? Can you think of other methods?

5. What is the most common type of star in the galaxy? Why do these stars seem less likely than others to have Earthlike planets?

6. What is an ecosphere? How do its dimensions vary with the spectral type of star it surrounds?

7. Among the nearest stars, which ones are most likely to have Earthlike planets? Why?

8. Describe the Alpha Centauri system. What would it be like to live on a planet orbiting one of the brighter components of this system?

9. Suppose that most stars in the Milky Way galaxy have planets in orbit around them, but only one star in 1000 has planets within its habitable zone. If each such star, on the average, has one planet in the habitable zone, how many planets within the habitable zone are to be found around the 300 billion stars of our galaxy?

10. Consider the nearby star Lacaille 9352, which has a luminosity only 1/100 of the sun's. How many times closer to this star would a planet have to be for each square centimeter of the planet's surface to receive the same amount of starlight energy each second as is the case in our solar system?

FURTHER READING

Billingham, John, ed. *Life in the Universe.* Cambridge, MA: MIT Press, 1982.

Covey, C. "The Earth's Orbit and the Ice Ages," *Scientific American* (February 1984).

Feinberg, Gerald, and Robert Shapiro. *Life Beyond the Earth.* New York: Morrow, 1980.

Goldsmith, Donald. *The Astronomers.* New York: St. Martin's Press, 1991.

Sagan, Carl. *The Cosmic Connection.* New York: Dell, 1973.

Science Fiction

Asimov, Isaac. *The Gods Themselves.* New York: Fawcett, 1972.

Benford, Gregory. *In the Ocean of Night.* New York: Dell, 1977.

Brin, David. "The Crystal Spheres." In *The River of Time.* New York: Bantam, 1987.

Brin, David. *Startide Rising.* New York: Bantam, 1983.

Clement, H. "Uncommon Sense." In *Space Lash.* New York: Dell, 1969.

Niven, Larry. *World of Ptaavs.* New York: Ballantine, 1965.

Pohl, Frederick. *Man Plus.* New York: Bantam, 1976.

Tiptree, J. "Love is the Plan the Plan is Death." In *The Alien Condition,* ed. S. Golden. New York: Ballantine, 1973.

The Very Large Array of radio telescopes near Socorro, New Mexico, includes 27 antennas, each 25 meters in diameter, that can move on railroad tracks forming three arms that are each 20 kilometers long. Computers link these telescopes so that they function as a single telescope with extraordinarily high angular resolution. However, the total collecting area of the 27 antennas remains relatively low. To gain more collecting area, an advanced civilization could build an "antenna farm" similar to the one shown at upper left, a sketch from the proposed SETI "Cyclops" project. Such an instrument would combine high angular resolution with a total collecting area of several square kilometers, making interstellar communication possible at power levels too faint for our present abilities.

17

The Development of Extraterrestrial Civilizations

The prospect of discovering extraterrestrial life has fascinated humans for centuries. When we think of this possibility, we usually have in mind something much more exciting than simply finding *life:* We are looking for extraterrestrials with whom we can share ideas and swap stories. In other words, what fascinates us most is the unprecedented opportunity to communicate with "intelligent civilizations," which we may define as groups of beings with self-consciousness and with complex interactions among themselves—beings who can share their music, poetry, and learning.

Although civilizations could have as many different structures as the beings that create them, our simplifying "principle of mediocrity" suggests that *we are most likely to communicate with societies that are relatively like our own.* In particular, we are far likelier to encounter civilizations with a highly developed curiosity about the universe than those that, for one reason or another, have no such curiosity. As we shall see, this urge to examine the universe influences the ways we expect civilizations to communicate.

The search for other civilizations offers a wide-ranging choice of approaches. As we enter the last years of our millennium, we finally have a good chance to find other civilizations in our galaxy. Four questions seem paramount:

1. How many civilizations exist?
2. How long, on the average, do they persist?
3. How eager are they to meet with us or talk to us?
4. How does communication proceed?

In order to estimate the difficulty of making contact with other civilizations, we must attempt to answer each of these four questions as best we can. The answers—if we can only discover them—should tell us how much effort will be required to open communications with other civilizations. We can then decide whether or not we want to invest this amount of effort for a chance of obtaining a fascinating, though unknown, reward.

For civilizations that do not make contact by accident, this process of evaluation must occur over and over again. Our own civilization, however, has an age of only a few decades to a few thousand years (depending on just how we define "civilization"). This means that human civilization must rank among the youngest in the universe. In contrast, we may guess that many other civilizations developed not thousands, or even millions, but *billions* of years ago. If these civilizations have not yet disappeared, their development must surely have so far surpassed our own present status that we can scarcely imagine their abilities and ideas—until and unless we communicate with them.

How Many Civilizations Exist?

All of our speculation about other planets that may harbor life, and about other civilizations, must rely on our single known example of an inhabited planet, our parent Earth. If we could locate even one other civilization, we would take a giant step toward proving that Earth's conditions represent a fair sample of those that occur in planetary systems. Working together, any two examples of a single phenomenon possess far more impact, and allow far more accurate extrapolation, than a single, possibly unique, example does. For instance, before explorers sailed from Europe to America, they imagined strange and varied races of human-apes in different regions of the Earth, and there was no good argument against these possible variations on the human form. But once explorers found human beings in America as well, the conclusion that our planet has a single race of hominids grew immensely stronger, and was indeed verified by later investigation.

For the time being, we on Earth remain in pre-Columbian ignorance of our neighbors on other planetary systems. We have no convincing evidence that our civilization does not represent an extremely rare event, perhaps a unique one. But when we examine the different factors that appear to influence the development of a civilization, we soon conclude that within our own galaxy, many civilizations should exist. Our own situation seems to possess no unique circumstances that should mark it as the locale of *the* galactic civilization. But if we are not unique, how many civilizations exist? Their numbers—and thus the *distances* to the nearest such extraterrestrial civilizations—depend on the factors that determine whether or not a civilization will develop in a particular place. *The number of extraterrestrial civilizations determines the difficulty of finding another civilization.* We ought therefore to look carefully at the ways in which we can estimate the number of civilizations in the Milky Way, and at how reliable this estimate may be.

Compiling an estimate of the number of civilizations in our galaxy (and in the entire universe) requires consideration of all the factors that influence the result. In order to achieve an intuitive understanding of the difficulties and simplifications involved in estimating this number, it is worth considering a problem closer to our experience: the search for the perfect restaurant.

The Search for the Perfect Restaurant

Suppose we ask ourselves a question: How many restaurants now exist that are just like our favorite restaurant? We all know that our favorite owes its perfection to a combination of many factors. For example, the restaurant must have the right sort of decor: If we like antique mahogany furniture and inlaid mirrors, our restaurant must have these to please us; if we prefer art nouveau, a different interior design is required. Second, the restaurant personnel must be our kind of people. Third, the food must be to our taste, and cooked the way we like it. Fourth, the price of meals must not exceed what we are comfortable paying. Fifth, the restaurant's hours of operation must correspond to the times when we like to eat: We cannot enjoy an early-closing restaurant if we are night owls, or a late-opening one if we want to eat before show time. Finally, the perfect restaurant must be in business *now:* It is no use telling us about the great restaurants that have ceased operation, or imagining the great ones that may open some day, for we seek the perfect place to eat at the present time.

Given these criteria, we can attempt to estimate the number of restaurants now in existence in this country that correspond to our vision of the perfect restaurant. There are many ways to make such an estimate, but here is one that seems reasonable. Because we think that most, if not all, such restaurants will be located in cities, we begin by determining the number of cities in the United States. This number equals about 5000, if we define a "city" as a place with more than 5000 inhabitants. Next, we must multiply this number of cities by the fraction of cities that have restaurants. (Some cities may have laws that prohibit true restaurants—those that serve alcoholic beverages as well as food—and call such establishments "clubs" instead.) Since most cities do have restaurants, we shall set the fraction of cities with restaurants equal to 0.99. Multiplying this figure by the 5000 cities gives us a total of 4950 cities with restaurants.

Next we must consider the number of restaurant locations in each city. Careful sampling by the authors has shown that this number approximates 200 in the average city. Naturally, the larger cities have far more restaurants and the smaller cities have far fewer. If we multiply the number of cities with restaurants by the average number of restaurants per city, we find the total number of restaurants in U.S. cities. This number, although interesting, does not provide us with the number of perfect restaurants. We must first multiply the total number of restaurants by the fraction that have the *decor* that we consider the right one. This number might be one in five, or 0.2. Next, we must multiply our result by the fraction of those restaurants that have the right *personnel.* This number may be one in 10, or 0.1, which implies that only one restaurant in 50 (0.02 of all restaurants) has both the right decor *and* the right personnel.

We must still pass through four more steps in our search. We first estimate the fraction of restaurants that meet our qualifications so far that also have the right sort of *food*—say, one in five, or 0.2. Then we must estimate the fraction of these restaurants that have the right *prices,* which could be one in two, or 0.5. And then we must determine the fraction of these that have the right *hours of opening,* probably as high as 0.8.

If we multiply the numbers that we have estimated for each of our criteria, we shall obtain the number of restaurants in the United States *at some time*—past, present, or future—that satisfy our demands and qualify as "perfect." But we all know that although a restaurant may exist for years, or one may replace another at a particular location, the elusive "perfect" restaurant may last for only a brief fraction of the total time that the restaurant exists. Personnel changes, decor changes, menu changes, price changes—these and a host of other variations may cause a restaurant to lose its classification as "perfect," whether or not a restaurant continues to function at a particular location. What we seek is the number of perfect restaurants *now,* in a world where the lifetime of a restaurant may be measured in months. Hence we must multiply by one final factor: the fraction of our lifetime (more properly, our lifetime as a seeker of restaurants) that a "perfect" restaurant lasts once it has qualified as perfect. If the restaurant goes out of business, or has not yet come into being, it does not belong in the number we seek to estimate—the number of perfect restaurants that now exist.

To find the number of restaurants in the entire United States that match our idea of the perfect restaurant, we must multiply together the factors we have estimated. We have seen that the number of restaurants in the cities of the United States equals the number of cities (5000) times the fraction of cities with restaurants (.99) times the average number of restaurants per city (200). We must multiply this number by the estimated fraction with the right decor (0.1), then by the fraction of these restaurants with the right personnel (0.5), and then by the fraction with the right food (0.2), then by the fraction with the right prices (0.1). We must go on to multiply this product by the fraction with the right hours of opening (0.8), and then by the ratio of the restaurant's lifetime to our lifetime as a restaurant connoisseur.

If we write an equation that expresses these calculations, thus estimating the number of perfect restaurants now in existence in the United States, we obtain:

$$\text{Number of perfect restaurants at this time} = \text{Number of cities in the U.S.} \times \text{Fraction of cities with restaurants} \times \text{Number of restaurants per city}$$

$$\times \text{Fraction with right decor} \times \text{Fraction with right personnel} \times \text{Fraction with right food}$$

$$\times \text{Fraction with right prices} \times \text{Fraction with right hours}$$

$$\times \frac{\text{Average lifetime of a ``perfect'' restaurant}}{\text{Lifetime of restaurant seeker}}$$

After multiplying all these figures, we find that the number of perfect restaurants is equal to:

$$(5000) \times (0.99) \times (200) \times (0.2) \times (0.1) \times (0.2)$$

$$\times (0.5) \times (0.8) \times \frac{\text{Restaurant lifetime}}{\text{Diner's lifetime}}$$

Our calculations yield the result that the number of perfect restaurants is equal to

$$(1584) \times \frac{\text{Restaurant lifetime}}{\text{Diner's lifetime}}$$

This answer highlights a few key facts about estimates such as the one we have just performed. First, the final result may look exact, but it inevitably contains some error. Second, the result depends on the *product of various factors,* some of which are just educated guesses. The estimates of these factors are subject to human error, and, in some cases, to human prejudices. Note that if any one of the factors that enter the multiplication turns out to be far wrong—for example, overestimated by a factor of two—then the final product will be wrong in the same proportion. Third, we cannot make a good estimate without being able to estimate *each* of the factors that enter the equation. Suppose that we resolve the question of what the final factor should be by estimating that, on the average, a "perfect" restaurant lasts for about three years. Then if the diner's lifetime as a restaurant connoisseur equals 60 years, the last term becomes 1/20, and the final answer becomes:

$$\text{Number of perfect restaurants} = 79.2$$

We can summarize our restaurant estimate by saying that the number of perfect restaurants in the United States now seems to be about 80, but we would not be surprised if the number turned out to be 50, or even 2 or 250. If, for example, the fraction of restaurants with the right food were 0.1 instead of 0.2, the number of perfect restaurants would be 40 instead of 80. (Note that in view of our uncertainties, it would be foolish not to round 79.2 at least to 80.) This sort of analysis is very similar to our attempts to estimate the number of civilizations in the Milky Way galaxy.

The Number of Civilizations: The Drake Equation

To estimate the number of advanced civilizations in the Milky Way galaxy, we need to perform a series of multiplications analogous to those that enter our search for the perfect restaurant. In the search for civilizations, we begin with the number of stars in our galaxy rather than with the number of cities in the United States. We must then multiply this number by the fraction of sunlike stars, those that last long enough for life to develop around them. Next, we must multiply by the average number of planets per star, and these three terms together will give us the total number of planets in the galaxy in orbit around stars that last long enough for life to develop. We must multiply this number by the fraction of planets with conditions favorable to life, and then by the fraction of those planets upon which life actually develops. This product gives us the number of planets in the galaxy with life on them at some time, and we must multiply this result by the fraction of planets with life upon which intelligent civilizations arise. The final factor is the length of time during which a civilization has both the ability and the desire to communicate with other civilizations in the galaxy: We must multiply our product by the ratio of the lifetime of a communicative civilization to the total lifetime of the galaxy.

Our equation for the estimated number of civilizations now in the galaxy with which we can communicate thus becomes:

Number of communicating civilizations in our galaxy now N	=	Number of stars in the galaxy N_*	×	Fraction of sunlike stars $\mathbf{f_s}$
	×	Average number of planets per star N_p	×	Fraction of planets suitable for life $\mathbf{f_e}$
	×	Fraction of those planets where life actually develops $\mathbf{f_l}$	×	Fraction of planets with life where intelligent civili- zations arise $\mathbf{f_i}$
	×	$\dfrac{\text{Lifetime of civilization with ability and desire to communicate}}{\text{Lifetime of Milky Way galaxy}}$ L/L_{MW}		

This equation is called the Drake equation by astronomers, who named it after Frank Drake, the astronomer who first devised it. Previous chapters in this book have helped us in producing estimates of the factors that enter the Drake equation (Fig. 17.1). As we read from left to right in the equation, we go from the relatively sure numbers to the less sure, finally arriving at the totally unknown. For example, the number of stars in our galaxy, about 300 billion, is known to within a factor of 1.5, so we think. This is not perfect accuracy, but it is far better than our estimate of, say, the fraction of planets with life where intelligent civilizations develop. Here we have only ourselves to guide us, together with whatever conclusions we can draw from the way we believe we evolved. Finally, when we come to the last term in the equation, we shall encounter great difficulties in estimating the length of time that a civilization survives with communications ability and desire. Nonetheless, this equation has such usefulness that we must proceed to fill in the numbers as best we can and thus estimate the number of civilizations now in existence with whom we could hope to exchange messages.

First of all, for the number of stars in our galaxy we have relatively little doubt of the 300 billion that we discussed in Chapter 3. Chapters 5 and 6, which examine the ways stars evolve, have shown us that the fraction of stars that live for at least 5 billion years, the minimum time that we think is needed for intelligent life to evolve, equals at least half of all stars, and probably a greater number (see Table 16.2). Let us, however, use 0.3 as the fraction that fills the second place of our equation, because stars with the lowest masses may not be able to maintain life on their planets.

The average number of planets per star remains unknown, but if we use our solar system as an example, which we think reasonable because the sun seems a perfectly average star, we come out with a number of about 10 (in our case 9).

The next factor, the fraction of planets with conditions suitable to life, can be estimated by examining our solar system, to the extent that we believe our solar system to be a representative planetary system. We find one planet, Earth, that appears eminently suitable to life, and one other,

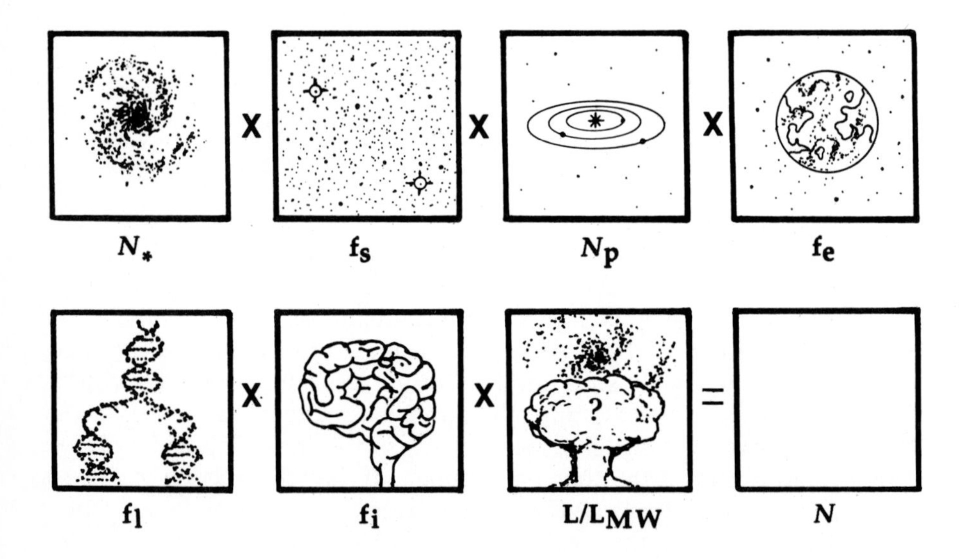

Figure 17.1 The Drake equation attempts to estimate the number of civilizations in the Milky Way capable of communication at any representative time, such as the present. To do this, the equation multiplies terms that express the factors thought to be necessary for the existence of a civilization capable of communication: a planet orbiting a relatively long-lived star, with conditions suitable for life, on which life actually develops and evolves to achieve communication capability. The average lifetime of such "intelligent" civilizations will determine the number in existence at any representative time.

Mars, that appears very close to being suitable, if not exactly so. Our discussion in Chapter 16 suggests that a conservative estimate gives to one planetary system in every four a planet with suitable conditions for life; some astronomers have concluded that nearly every planetary system should have a planet suitable for life. If we use the one-in-four estimate, and if planetary systems average 10 planets per system, this figure implies that one planet in 40 has conditions that favor the origin of life.

We also estimated in Chapters 9 and 16 that the fraction of planets suitable to life on which life actually develops rises close to unity, but in line with our conservative approach, we set this fraction at 0.5, with the risk of underestimating the number of planets with life by a factor of 2.

How Long Do Civilizations Last?

The next number we encounter in our equation is the fraction of planets with life where intelligent civilizations develop. Again, if we were to extrapolate from the sole example we know of a planet with life, our own, we would set this fraction at unity; again, to be conservative, we shall take this fraction to be 0.75. Finally, we must consider the lifetime of a civilization once it has achieved the ability to communicate and desire to do so.

With this last term, we face an almost insurmountable difficulty. Our own civilization has achieved the ability to send messages across interstellar space—or to produce signals that may "leak out" to other planetary systems—only within the last two generations. We shall discuss these processes in Chapter 19. The most important question of all—how long shall we last as a civilization with the ability and the desire to communicate with other civilizations in our galaxy?— continues to be shrouded in mystery, for we cannot look into the future. Since this lifetime plays a crucial role in determining the number of civilizations we can expect to find in our galaxy, always provided that our lifetime as a civilization reflects the average, we shall leave this quantity as an unknown in our equation, and simply designate it *L*.

If we do this, and if we combine the numbers in our equation together with the fact that the lifetime of our galaxy is approximately 10 billion years, we obtain the following result:

$$\text{Number of civilizations in our galaxy } (N) = (300 \text{ billion}) \times (0.3) \times (10) \times (0.025) \times (0.5) \times (0.75) \times L/10 \text{ billion}$$

We must bear in mind that we are measuring the lifetime *L* of a civilization with communications ability and desire in years, just as we are measuring the lifetime of our galaxy in years.

We can see that upon inserting these numerical values, the big numbers in the equation tend to cancel each other. The 300 billion stars in our galaxy overwhelm the 10 billion years in the denominator of the last fraction, leaving only a factor of 30 after canceling the first number with the last denominator. Most of the other numbers in the equation are rather close to 1. This fact has great significance, for most of the numbers represent a fraction of planets or stars with some quality that leads toward the establishment of communications possibility. If any of these fractions were tremendously less than 1, for example, 0.0001 instead of 0.5, the total product would fall far below the number we shall find if we assume that the fractions are close to 1. In a similar fashion, our search for the perfect restaurant would be greatly handicapped if, for example, the fraction of restaurants with the right kind of personnel were not 0.5 but 0.0001 instead.

When we multiply all the factors in our equation together, we find that the result for the number of civilizations in our galaxy now with communication ability and desire equals 0.843 times L, where L is the lifetime, measured in years, of a civilization once it has developed the ability and desire to communicate with others. Given the uncertainty in the factors that went into this equation, we can make an accurate statement by simply saying that our estimate puts the number of civilizations in our galaxy now with the ability and desire to communicate, N, equal to the lifetime of such civilizations, L, measured in years: $N = L$ (Color Plate 20).

Table 17.1 recapitulates the numbers that have entered our equation, together with other values for the appropriate fractions suggested by Carl Sagan in 1974. We can see that although Sagan's early estimate of the different fractions differs from our own, a certain balance has kept the best-guessed result approximately the same: *The number of currently active civilizations in our galaxy approximately equals the lifetime of an average civilization with communications ability and desire, measured in years.*

In short, because N approximately equals L, a civilization's lifetime seems to determine just about everything. If we represent an average civilization, and if we destroy ourselves, say, 100 years after we developed radio techniques capable of interstellar communication, then L should equal 100 years, and the number of civilizations in our galaxy now, or at any other time, should be about 100, with our chances for finding one another very slight. If, on the other hand, civilizations discover ways to maintain themselves indefinitely after they have reached our level of technological advance, then L could equal approximately 2 or 3 or 5 billion years, the average lifetime we can expect for the suitable stars in our galaxy. In this case, the number of civilizations now existing in our galaxy that can communicate with one another would reach into the billions. Most likely, an intermediate value is closer to the truth. If L equals, say, 1 million years, then about 1 million civilizations now exist in our galaxy with whom we might communicate.

TABLE 17.1 Estimated Probability Factors Affecting Estimated Number of Milky Way Civilizations

	Sagan (1974)	Our best Estimate	Most Favorable Case	Least Favorable Case
Number of stars	100 billion	300 billion	300 billion	300 billion
Fraction of sunlike stars	1	3/10	1	1/15
Average number of planets per star	10	10	20	5
Fraction of planets suitable for life	1/10	1/40	1/3	1/1000
Fraction of suitable planets where life does acturally arise	1	1/2	1	1/1,000,000
Fraction of planets with life that develop civilizations	1	3/4	1	1/1000
Ratio of civilization's lifetime to galaxy's lifetime	L /10 billion	L /10 billion	L /10 billion	L /10 billion
Product of previous factors—the number of civilizations in the galaxy now, or at any representative moment	L /10	0.8L	300 L	L /100 billion

Let us never lose sight of the fact that this number refers only to our Milky Way galaxy, and that the universe contains at least as many galaxies as there are stars in the Milky Way (Fig. 17.2). We have restricted ourselves to our own galaxy because local communication remains a far easier matter than extragalactic communication. The latter cannot be judged impossible, though we must be prepared to wait much longer times for return messages. The nearest stars in our galaxy would require several years for a round-trip message between ourselves and them, while the farthest stars require tens of thousands of years for a round-trip radio message. In contrast, to exchange a message with a civilization in the Andromeda galaxy demands almost 4 million years, while to exchange a message with one of the galaxies in, for example, the cluster of galaxies shown in Fig. 17.2 would require 1.6 billion years. But we would not require a *dialogue* to be excited by contact with extraterrestrial intelligence. Receiving messages from some civilization that might have vanished long before its wisdom, poetry, or

Figure 17.2 This cluster of galaxies in the constellation Centaurus, about 250 million parsecs away, contains more than a thousand member galaxies, many of them as large as the Milky Way.

music reached us would still be thrilling, just as the works of Plato or Bach can thrill us today. Whatever we may think about the likelihood of receiving such messages, we can agree that our chances for real conversations are much more likely if we limit ourselves to our fellow civilizations in the Milky Way galaxy, who, as we have just estimated, number anywhere from zero to several billion.

How Eager Are Civilizations for Contact?

We have acquired some feeling for the number of civilizations that may exist in our galaxy, and for the importance of estimating their lifetimes if we hope to gain even an approximate idea of the task of detecting them. What about the desire of civilizations for communication? Do human beings, for example, *want* to contact other civilizations? Yes and no; some do and some do not.

If the average lifetime of a civilization capable of interstellar communication far exceeds our 70 years or so, then our civilization cannot now serve as a good representative. Most of a civilization's lifetime would then be spent not in our first flush of technological advance but in the enjoyment of a long period of time during which the question of whether or not to search for others could be debated.

The chief forces that favor interstellar communication are curiosity (the desire to find out what's there), gregariousness (the urge to talk and to listen), and what might be called social avarice (the hope of obtaining "valuable" information). To this we might finally add the more nebulous urge to search simply because we can. Some of the factors that stand in opposition to such attempts at communication are fear (the belief that hostile aliens will enslave, devour, or destroy us in person, or that we will be paralyzed by the shock of simply discovering that other beings are out there); lack of interest (the feeling that we don't want to exert ourselves on this sort of project); and the press of other priorities (scarcities of time and resources, if these commodities begin to be in extremely short supply).

Curiosity and gregariousness seem to be human characteristics that have allowed our society to exist and to dominate our world, while fear and inertia, though always with us, seem to have failed consistently in their confrontations with what we may call "positive" characteristics for social success. With a small leap of faith, we may conclude that "successful" civilizations on other planets should have gained their dominant positions by possessing much the same characteristics. If our curiosity had not tended to overcome our fear, we would not now be able to stare into the night skies and wonder about the wisdom of contact with other civilizations, who may have trod the same basic road of development.

Some civilizations will attempt to contact others, and some will not. Of course, the average civilization's lifetime as a seeker after contact may fall far below the average total lifetime with communications ability. We may find that most civilizations capable of long-term existence will pull in their antennas and concentrate on themselves. Indeed, it may be true that only those civilizations that do not seek contact can last for long; the urge to search could be closely related to various self-destructive impulses. In either case, the key lifetime in the master equation would then turn out to be far less than we might hope.

On the other hand, even a civilization not interested in contact may betray its presence through the "leakage" of radio signals into space, just as we do now. Its communicative lifetime would then be correspondingly lengthened. Finally, there is the interesting possibility suggested by Sebastian von Hoerner that once a civilization has made contact with another more advanced than itself, its chances for long-term survival will increase, because it can learn from the experience of the society it has discovered.

This "positive feedback" will therefore increase *L* and lead to longer average communicative lifetimes.

We have now provided, as best we can, the answers to the first three questions that we posed earlier: How many civilizations exist? How long do they persist? And how eager are they for communication? We have seen that the answer to the first question depends directly on the answer to the second, and we have agreed that the best answer to the third question is simply to forget about civilizations not eager for communication. If a significant fraction of all civilizations that exist do possess the desire to communicate, then our best estimate sets the number of such civilizations in the galaxy now equal to the lifetime of such a civilization, measured in years.

The "least favorable case," which we have also included in Table 17.1, is totally arbitrary. (The reader is invited to experiment with even less favorable cases.) Here we have assumed that only G-type stars nearly identical to the sun are suitable, and we have reduced the figures for the average number of planets and the fraction suitable for life. But the largest changes in our numbers arise from the assumption that the origin of life might be an exceedingly improbable event and that the evolution of intelligent life with a capacity for interstellar communication is even less likely. If these low numbers are correct (and we have no way of proving whether they are or not), then we could expect only 0.1 civilizations to exist in the Milky Way at any given time, and then only if the mean lifetime of civilizations were equal to the lifetime of the galaxy itself! This would make our own civilization not only isolated, but also a statistical fluke! We can use these different estimates of the number of advanced civilizations in the galaxy to confront the last of our four questions.

How Does Communication Proceed?

In order to consider the ways in which interstellar communication might proceed, we must use our estimate of the number of civilizations in the Milky Way to determine how widely separated these civilizations could be. To estimate the average distance among civilizations, we must calculate the fraction of all the stars in our galaxy that now have communication-oriented civilizations nearby. We do know fairly well the average separation between stars, so the fraction of stars with likely conversationalists will provide us with the key distance we are seeking.

How far away, then, is the closest civilization with which we might communicate? Since stars in the sun's region of the Milky Way are spaced, on the average, a bit over 2 parsecs apart, a sphere 20 parsecs in radius, centered at the sun, will contain 33,000 cubic parsecs of volume and about 3000 stars (Fig. 17.3). A sphere around the sun 200 parsecs in radius will have

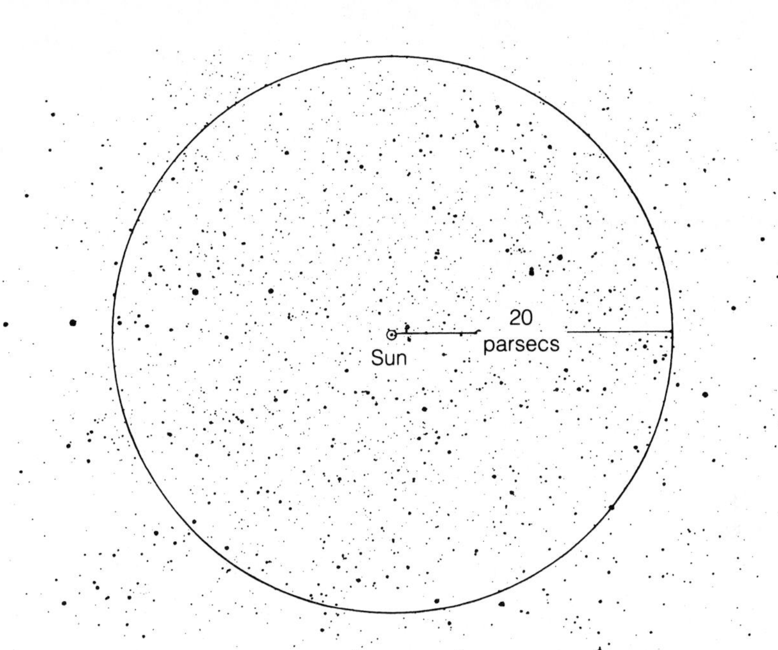

Figure 17.3 A sphere 20 parsecs (65 light years) in radius surrounding the sun contains several thousand stars.

1000 times this volume and will contain some 3 million stars, or about 1/100,000 of all the stars in the galaxy. For greater distances from the sun, we would have to consider the fact that stars in our galaxy are not distributed uniformly in all directions (Fig. 17.4).

Now suppose that the average lifetime of a civilization equals 20,000 years, and that there are 20,000 civilizations in our galaxy. The galaxy's number of stars would then be 15 million times the number of civilizations, so that only one star in 15 million should now have a civilization capable of communication on one of its planets. If these civilizations are located at random in the Milky Way, we should be able to find the nearest such civilization by examining the 15 million or so nearest stars. This would require looking at all the stars within 300 parsecs from the sun—about a hundredth of the distance across the galaxy.

If, on the other hand, most civilizations last for 20 *million* years, then 20 million civilizations ought to exist now in the galaxy. Then one star in every 15,000 should have a civilization on an orbiting planet, so to find the nearest civilization, we would have to search only among the 15,000 nearest

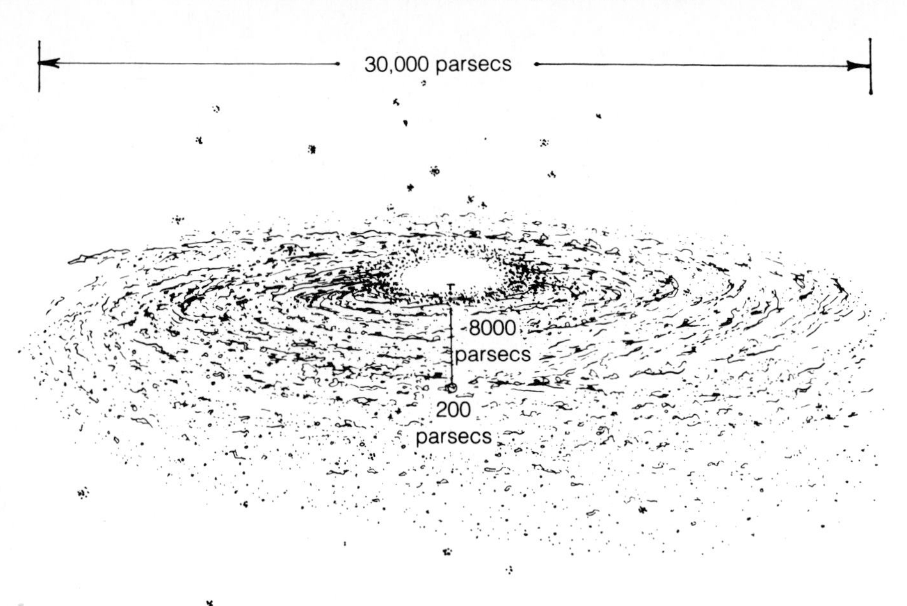

Figure 17.4 In the Milky Way galaxy, which spans about 30,000 parsecs from rim to rim, the sun lies some 8000 parsecs from the center. A sphere 200 parsecs in radius around the sun includes less than 1/1000 of the total volume within the disk of the Milky Way.

stars, or out to a distance of 35 parsecs. (Notice, by the way, that even a lifetime of 20 million years for a civilization barely exceeds a thousandth of the lifetime of most main-sequence stars.) When we consider that the strength of any radio or television signals decreases in proportion to the square of the distance, we can see that to find a civilization at 1/10 the distance means a saving by a factor of 100 in the amount of power needed to produce a detectable signal in either direction. Unfortunately, we simply cannot make an accurate estimate of the number of civilizations—and thus of the distance to the nearest civilization—until we determine the average lifetime of other civilizations, by finding them or their signals.

One more fact ought to draw our attention when we estimate the distance to the nearest civilization. If the average lifetime of a civilization turns out to be relatively short—say, only 1000 years—then the number of civilizations in the galaxy becomes quite small (about a thousand) and the average distance between civilizations increases. If only a thousand stars—one star per 300 million—has a civilization near it, then we must search through 300 million stars to find the nearest such civilization, which would take us out to distances of 1500 parsecs (4900 light years) from the sun. In this case, the time for any messages to travel between neighboring civilizations must, on the average, exceed the average lifetime of civilizations, so that two-way contact would become unlikely.

As Frank Drake first pointed out, in the situation described above, most civilizations would have disappeared by the time that a return signal arrived, even from their nearest neighbors. We must stress that in our terms, the civilization's "lifetime" equals the length of time during which the civilization *can* communicate and *wants to.* If a civilization abandons interest in the outside universe, then although it may grow ever more "advanced," it no longer ranks as a "civilization" as we have defined the term, in order to focus on our interest in communication.

The cutoff point at which civilizations are too widely spaced, on the average, to exchange messages before they disappear, corresponds to an average lifetime of about 3500 years per civilization. Lesser average lifetimes lead to the sad situation of "no time to communicate," while longer average lifetimes allow most civilizations to exchange at least one message with their closest neighbors.

If the average civilization lasts about 3500 years, and if the Milky Way contains about 3500 civilizations at any representative time, neighboring civilizations would on the average be separated by about 630 parsecs (2000 light years). Thus a round-trip message traveling at the speed of light would take 4000 years. If the average lifetime of a civilization falls below 3500 years, then no messages can be exchanged with even the nearest civilization before it (and our civilization, too) will vanish. Conversely, if the average lifetime of civilizations exceeds 3500 years, then the nearest civilization should be less than 630 parsecs away, and several or, if we are lucky, many messages could pass back and forth before one civilization or the other passed into history.

We have based our calculations on our best estimate of the number of civilizations in the galaxy, which we have taken to be equal to a civilization's lifetime (in years). The lower the number of civilizations, the greater the average distance between neighboring civilizations will be. Figure 17.5 shows a graph of the number of civilizations, N, estimated in our galaxy, plotted against the average lifetime of a civilization. The heavy lines show the relationship between the number and lifetime that we expect if $N = L$; that is, if the number of civilizations in the galaxy equals an average lifetime, and also for the two extreme cases of $N = 300\ L$, the most favorable case we care to imagine, and for $N = L/100$ million, an unfavorable case.

The dashed lines in Fig. 17.5 that cut across the heavy lines show the number of two-way message exchanges that can occur for a given combination of separation and lifetime, with a given choice of how N depends on L. Let us suppose, for example, that we believe N equals L. Then if L equals 100,000 years, the average distance between neighboring civilizations will be about 200 parsecs (650 light years), and two neighboring civilizations

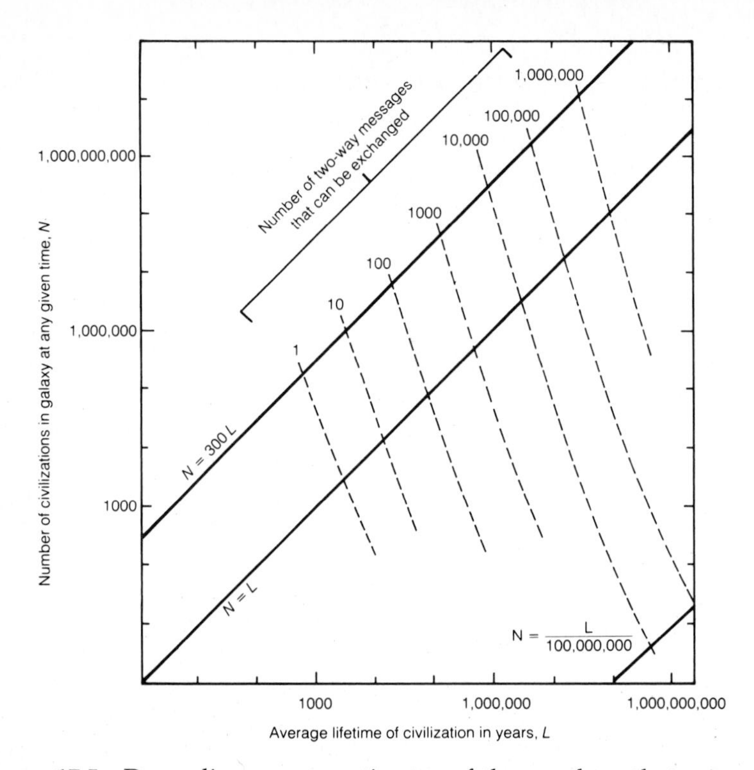

Figure 17.5 Depending on our estimates of the numbers that enter the Drake equation, we will reach different conclusions about the relationship between the average lifetime of a civilization, L, and the number of civilizations that exist in the Milky Way, N. This diagram shows the relationship for the three cases $N = L$, $N = 300\ L$, and $N = L/100{,}000{,}000$. Also shown in the diagram are the number of two-way messages that can be exchanged with another civilization during the lifetime of a given civilization. If we know the relationship between N and L, this number depends only on the value of L.

could exchange about 100 messages back and forth before one of them vanished.

The graph shown in Fig. 17.5 sums up our best answers to the four questions we posed: How many civilizations exist? How long do they last? How eager are they for contact? And how can we communicate with them?

Let us pause to notice that we have considered these questions as if the present moment in time were an entirely representative one. In fact, our galaxy has a finite age (just 10 billion years or so), and the ever-increasing age of the galaxy should see an ever-increasing number of planets where civilizations appear for the first time. The effect of this increase would be to make our use of the Drake equation (page 409) a more complex procedure than we have outlined, as we would have to take into account the fact that many stars have not yet existed for enough time to produce intelligent civilizations.

And how long is the lifetime L? How long does it take for an average communicative civilization to vanish? Although no one on Earth knows the answer to this question, we can imagine some possibilities for the development of a civilization once it has passed the stage that human society now represents.

Further Advances of Earthlike Civilizations

Our speculations about civilizations far more advanced than our own cannot have much accuracy, since we must extrapolate from our own experiences into an unknown future. One possible outcome for human society, global wars that lead to the civilization's destruction, hardly produces what we call a more advanced civilization (unless many cycles of destruction and rebuilding will, after thousands or millions of years, finally produce a stable society). In speculating about our future advances, we can guess what the immediate future may bring, and we can also see certain unavoidable limitations that physical laws impose on any society. Thus, for instance, we can speculate that the first contact with another civilization tends to increase a civilization's lifetime, as its members gain a new respect for their abilities. We can also state that no civilization, no matter how advanced, can build a spacecraft that will travel faster than the speed of light. But it is always possible that realms of physical experience as yet unknown to us may reveal extensions of our physical laws that will show us entirely new concepts, still beyond our imagination.

The prediction most often made concerning human interaction with the rest of the solar system states that our society will attempt to use the spaces that surround us in much the same way that we have used our Earth. That is, we shall try to extract materials of use to us; to build homes, offices, and sports arenas; to manufacture products for other humans to enjoy; and to make pleasure excursions from place to place. In short, we shall colonize space in the solar system, and potentially even beyond.

Gerard O'Neill, a recent proponent of space colonization, has led study groups that predict a population in space colonies of several billion people within 40 years after the time of initial construction. O'Neill says that the time to begin is now, and that, within a decade, matter mined from the moon, launched into space by superconducting slingshots and gathered into construction areas at the moon's distance from the Earth, could yield the first few space habitats (Fig. 17.6). The preference for mining the moon instead of the Earth rests on the fact that to eject a given mass from the moon into space requires only 5 percent as much energy as launching the

Figure 17.6 Gerard O'Neill and other scientists have proposed the construction of space habitats within rotating cylinders, which would use solar power to supply all their inhabitants' needs and the effects of rotation to simulate terrestrial gravity.

same mass from the Earth, because the force of gravity on the moon falls far below that on Earth.

Since the moon's crust contains great amounts of oxygen, silicon, magnesium, aluminum, iron, and copper, O'Neill sees no great difficulty in refining out the useful elements (while retaining the residue for shielding against ultraviolet light and cosmic rays), and then building cylindrical habitats, each a few kilometers long and half a kilometer across. The space cylinders would rotate to simulate the effect of gravity, and each colony would hold a few thousand people (Fig. 17.7). Later versions, much longer

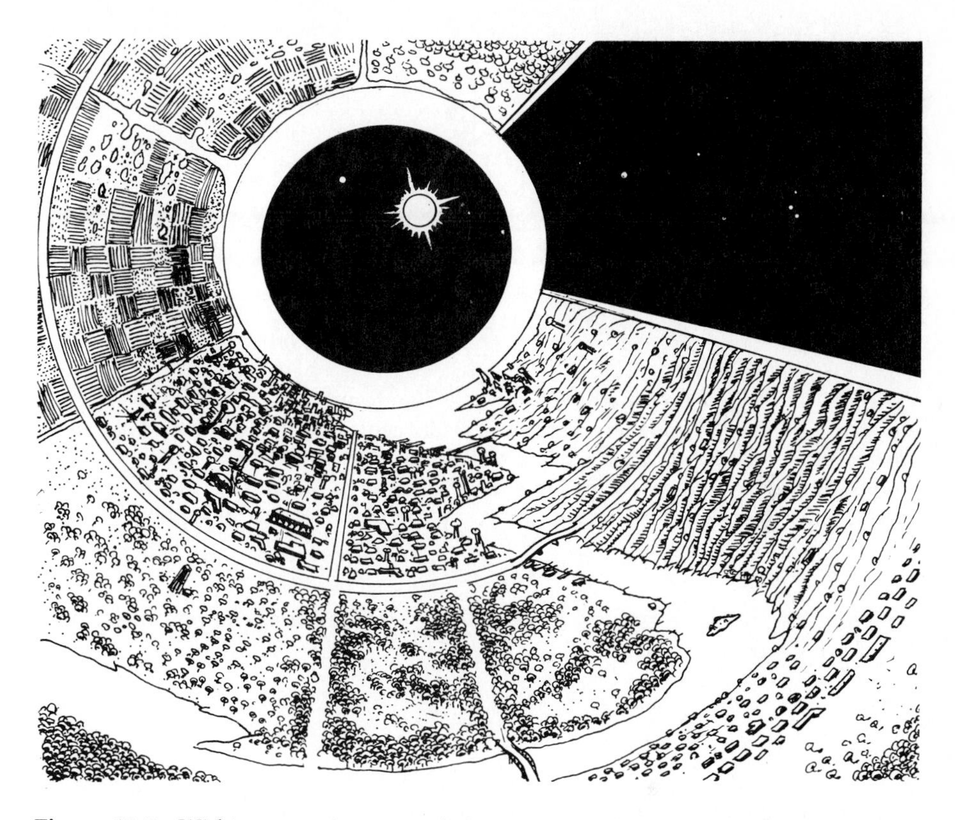

Figure 17.7 Within a rotating space habitat, strips of "land" could alternate with strips of "sky." Each strip of land would feel that it was "down" as the simulated gravity arose from the rotation of the habitat.

and wider, might each support a million inhabitants, using solar power and the material from the moon's surface to supply their needs.

Even the immediate neighborhood of the Earth-moon system contains plenty of room for thousands of space colonies, filled with billions of humans who might pass their entire lives in space. O'Neill suggested that an initial goal for these colonies would be the manufacture, again from lunar materials, of immense power-generating stations that would convert sunlight energy into microwaves, which would be beamed to Earth to supply electrical energy wherever needed. The solar power stations would be placed in synchronous orbits above the Earth's equator at a distance of 36,000 kilometers above the Earth's surface (Fig. 17.8). Objects orbiting at this distance from the Earth take the same time to complete one orbit as the Earth takes to rotate once, so if they move around the Earth in the same

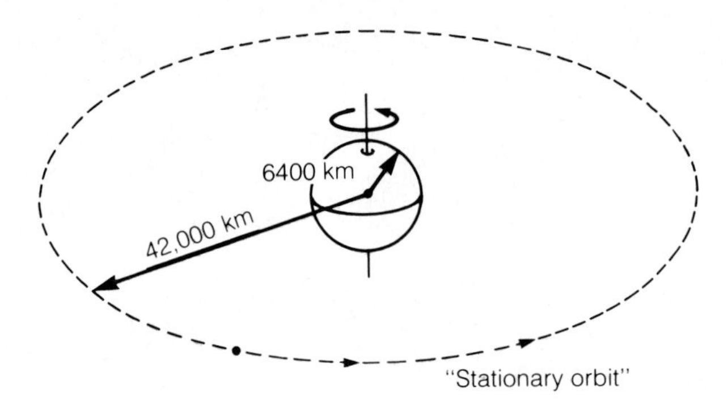

Figure 17.8 As first proposed by Arthur C. Clarke, any object orbiting the Earth at a distance of 42,000 kilometers from the Earth's center (36,000 kilometers above the surface) takes 24 hours to complete each orbit. Since the Earth takes the same amount of time to rotate, satellites moving from west to east in such orbits above the Earth's equator appear "stationary" above a particular point on the equator.

direction as the Earth rotates, they will appear stationary to an observer on the Earth's surface.

Colonizing the Galaxy

For our purposes, the key question raised by O'Neill's proposals refers, as always, to the average civilization, which ours may or may not be. Do most civilizations start to colonize the space around their home planets? And how far does such colonization proceed?

Confronted once again with questions that we really cannot answer, scientists have managed to produce some definitions and some reflections based on our knowledge of physics. Suppose that some civilization does begin to build colony after colony, until a huge population lives in space habitats rather than on the home planet. Why would they do this?

Some scientists speculate that life in space colonies can be more easily manipulated than life on a planet. A space habitat can, at least in theory, produce the climate, the atmosphere, and the political situation most desired by its inhabitants. (Whether some colonies would try to conquer others remains open to speculation.) Space colonies could also provide far more living room than a planet's surface. Most of the Earth lies far within the outermost shell on which we live, and we can mine only the first few kilometers of its 6400-kilometer radius. If we could disassemble and spread out all the matter in a planet like Jupiter at will, we could obtain the raw material not for just a few thousand or a few million space habitats, but for

Figure 17.9 A highly evolved Type II civilization might capture all of its star's energy output for its own use. In this case, we would not see the star at all—only the infrared radiation leaked away from the civilization, which we may assume would operate at a temperature of several hundred K.

trillions upon trillions. These colonies would spread all the way around the sun, thus making the maximum possible use of the photon energy that the sun emits (Fig. 17.9).

The Earth in its orbit now intercepts only about one-billionth of the sun's energy.[1] That is to say, we would need a billion Earths around the sun

[1] Even one-billionth of the sun's energy output amounts to about 10,000 times our present energy requirements on Earth—which is why an easy way to convert sunlight into other useful forms of energy would resolve any "energy crisis" for a long time.

to catch all of the sun's energy. Though we do not have a billion Earths, we could build space colonies that could capture all the light and heat that the sun produces. These space habitats could, at least in theory, support about a billion times the Earth's present population—5 billion billion (5×10^{18}) people, all using solar power to grow crops, to run their machines, and to enjoy life!

This speculation about the eventual growth of space colonies from Earth parallels an earlier idea of the physicist Freeman Dyson. Dyson suggested that an advanced civilization might wish to use the matter from one of the planets in its system to surround its star and "harvest" its energy. In this case, the civilization might remain small in population and confined to the planet of its origin. But now it would have an enormous amount of energy at its disposal. Let's assume that advanced civilizations have developed as either O'Neill or Dyson has speculated that they might. If we then tried to find their central stars, we would not be able to see them. Instead, we would detect a glow of infrared light from the surrounding shells of matter. If space colonies or some type of radiation collectors catch a star's energy output and thus maintain themselves at temperatures above absolute zero—say, at the 300° K temperature that characterizes the Earth's surface—they must radiate infrared emission into space.

We might hope to detect these civilizations by their infrared radiation, just as IRAS detected the disk of material around Beta Pictoris (page 387). We could distinguish a star wrapped in a cocoon of space colonies from one embedded in a cloud of dust by measuring the amount of energy the entire system radiated at several different far-infrared wavelengths. At these long wavelengths (low frequencies) the radiation from a Dyson colony would resemble the heat radiation from a single, large object, whereas a cloud or disk of particles surrounding a star would produce a distribution of energy with wavelengths that would look distinctly different. Nevertheless, confusion could occur, since both small particles and spacecraft could surround the star. It might be difficult to convince the scientific community that an unusual distribution of energy in the spectrum of an unseen object signified the existence of an advanced civilization. Skeptical theorists would surely find alternative explanations for the data that involved strictly natural phenomena.

Thus we cannot count on being able to detect a civilization that has wrapped its parent star in a cocoon of colonies, unless we can share in their radio communications (see Chapter 19). For purposes of easy reference, astronomers call civilizations that use all the energy radiated by their parent star Type II civilizations. This terminology, invented by the Soviet astronomer N. S. Kardashev, designates as Type I those civilizations that, like ourselves, understand the basic laws of physics, can attempt interstellar communication, and have not yet enveloped their stars in a nest of colonies designed to use all of the star's photons. Such a civilization—our own, for example—may use one-trillionth as much energy per second as the parent star liberates. Therefore, if we go by energy usage, Type I civilizations fall at one-trillionth of the level of Type II civilizations.

Having defined Type I civilizations as those much like our own in energy consumption, and Type II civilizations as those that capture all of their star's output, Kardashev went on to define a Type III civilization as one that captures and uses the energy from an entire *galaxy* (Fig. 17.10). Thus a Type III civilization uses about a trillion times more energy per second than a Type II civilization, which in turn uses a trillion times more energy per second than our own Type I civilization. Fantastic? Only from our point of view. Since a giant galaxy may contain almost a trillion stars, a Type III civilization would certainly (by definition) have at its disposal a trillion times the energy output of a single star. As we shall see in Chapter 18, travel between stars presents great difficulties but no impossibilities, provided that we set aside thousands or millions of years for the journeys. Our own Type I civilization has no urge to invest in these great voyages, but a Type II civilization could well take a longer view and decide to take a billion years or so to colonize an entire galaxy.

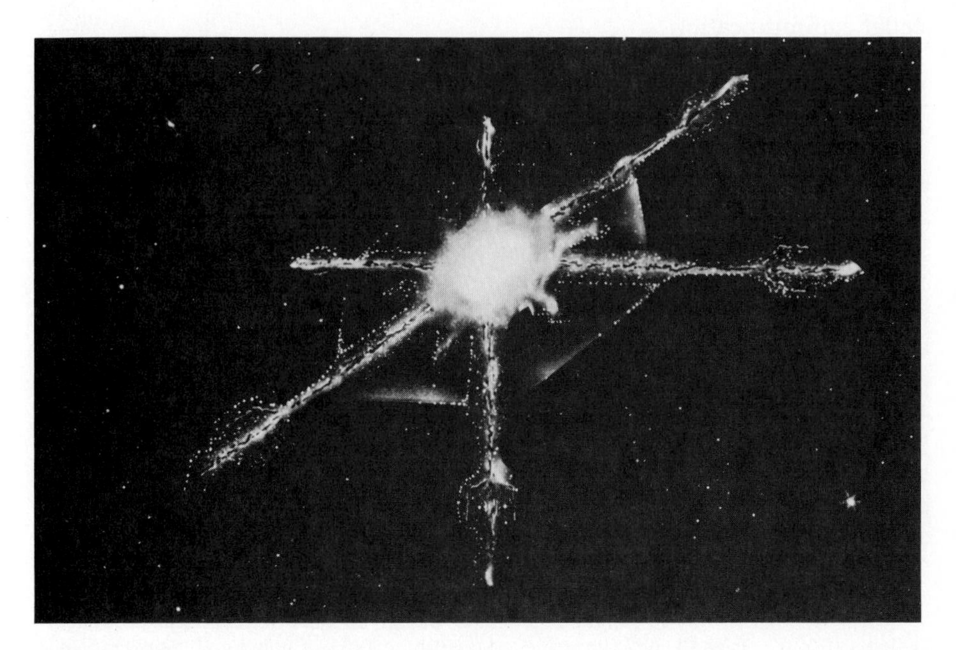

Figure 17.10 A Type III civilization captures the energy output from an entire galaxy of stars. Such a civilization might amuse itself by rearranging the galaxy into patterns it found aesthetically pleasing.

Our galaxy does not contain a Type III civilization. We feel confident in stating this because we can see that our own sun's energy flows freely outward, and we can see outward through the Milky Way in all directions except those along the disk of our galaxy. Of course, it might turn out that a Type III civilization is even now developing in the Milky Way, and we occupy one of the less interesting galactic suburbs, one of the energy sources to be developed later rather than sooner[2]. We cannot reliably estimate the number of Type II civilizations in the Milky Way. In terms of energy use, a Type II civilization would stand as far beyond a Type I as we (a Type I civilization) stand beyond a colony of bees. Since Type II civilizations would need a fairly long time—at least several thousand years—to wrap up their parent stars, we can conclude with some confidence that only those civilizations with a large degree of stability will evolve from Type I to Type II.

Once again we face a crucial lack of information—in this case, our knowledge of the average lifetime of a civilization once it has achieved Type I status. If this average lifetime falls below, say, 500 years, then we would expect few Type II civilizations to exist in our galaxy. If, on the other hand, the average lifetime far exceeds a few thousand years, then conditions appear ripe for many Type I civilizations to evolve toward Type II. But what sorts of plans or devices does a Type II civilization have for interstellar communication?

We won't know until we talk to one. It does appear certain that any Type II civilization will take a longer view of life than we do, and it is possible that all the Type II civilizations in our galaxy are now in communication with one another. Perhaps they have adopted a common policy toward would-be members, much like the state of Oregon, which seeks to discourage overdevelopment of its natural beauty. Perhaps the galactic club of Type II civilizations may have decided that further members cannot be accommodated. If advanced civilizations use highly directed laser beams (or some equivalent) for their mutual communication, it is highly unlikely that a search from Earth would find them.

But we must not allow speculation about modes of development for advanced civilizations to restrict our continued attempts to answer our fourth question: How does interstellar communication proceed? The tentative answers are large enough to fill a chapter, the following one in this book, despite the fact that no interstellar communication has yet been detected.

[2] The fact that we can see hundreds of billions of stars in the Milky Way tells us immediately that a "pure" Type III civilization does not exist here. But we cannot exclude the possibility that millions of stars near the center of our galaxy have been wrapped in an energy-catching cocoon by an expanding civilization that is well on its way to completing its Type II phase.

SUMMARY

When we attempt to estimate the number of civilizations with which we might communicate, we face a formidable problem of making estimates that are not fanciful guesswork. If we limit ourselves to estimating how many civilizations capable of communication exist in our own galaxy at the present time, we can construct a formula that presents this number as the product of seven key terms.

(1)The first number in our formula is the number of stars in the Milky Way galaxy, which we must multiply by (2) the fraction of sunlike stars, those last long enough for life to develop nearby. If we multiply this result (3) by the average number of planets per star, and then by (4) the fraction of these planets with conditions suitable for the origin of life, we obtain the number of planets in our galaxy with conditions suitable to life, in orbit around stars that last long enough for life to develop. Next we must multiply by (5) the fraction of these planets upon which life actually does develop at some point in the planet's history. We must then multiply by (6) the fraction of those planets with life that see the emergence of a civilization with the ability to communicate. Our multiplication of the first six terms then provides us with the number of planets in the Milky Way upon which intelligent civilizations might have appeared at some time during the history of our galaxy. To find the number of civilizations in our galaxy now, we must multiply by one last factor, (7) the ratio of a civilization's lifetime, once it acquires the ability and the desire to communicate, to the total lifetime of the galaxy.

The seven-step multiplication gives us an estimate of the number of civilizations in our galaxy at the present time with whom we might exchange messages. When we insert our best guesses for the relevant numbers, we find that this key number approximately equals L, where L is the lifetime (in years) of a civilization that has achieved the ability and the desire to communicate. If L equals 1 million years, then we expect that about a million civilizations capable of interstellar communication now exist in the Milky Way. If, in contrast, L equals just 100 years, then only about a hundred such civilizations exist at any particular time.

Larger values of L imply more civilizations, which in turn imply a smaller value for the average distance between neighboring civilizations. If L equals, for example, 3500 years, then neighboring civilizations should be about 630 parsecs apart. Values of L smaller than this imply that most civilizations disappear before they can exchange a single round-trip radio message with their closest neighbor! On the other hand, if L equals 20 million years, then the average distance between neighboring civilizations should be a mere 35 parsecs, and civilizations could exchange thousands of messages back and forth during their millions of years of coherent existence. Which of these two possibilities more closely mirrors reality in our galaxy remains a problem that we can answer only by finding other civilizations.

QUESTIONS

1. Why does an error in any of the terms in the Drake equation—for example, overestimating or underestimating the number of planets per star by a factor of two—affect the result of the equation by the same factor as the error in the individual term?

2. Why does our estimate for the average lifetime of a civilization directly affect the estimate of the number of civilizations that *now* exist in our galaxy?

3. Why do scientists believe that some of the terms in the Drake equation can be more accurately estimated than others? For example, why do scientists feel more confident about their estimate of the fraction of planets suitable for life on which life has developed or will develop than they do about their estimate of the average lifetime of an extraterrestrial civilization?

4. How does the average distance between civilizations in the Milky Way affect our chances for communicating with any of these civilizations? In particular, compare the distances we expect for the *closest* civilization to our own in the two cases $N = 10{,}000$ and $N = 1$ million.

5. If the average lifetime of civilizations in the Milky Way turns out to be less than 3500 years, we may be unable to establish two-way communication with other civilizations. Why is this so?

6. Use the approximation $N = L$ (where N is the number of civilizations in our galaxy at any time and L is the average lifetime of a civilization) to estimate how many round-trip radio messages neighboring civilizations could exchange if $L = 1{,}000{,}000$ years (see Fig. 17.4).

7. Why do we think that the parent stars of some advanced civilizations might be totally invisible? What techniques could we employ to search for such civilizations?

8. What is a Type III civilization? What would we expect to see when we look for the galaxy in which such a civilization exists?

9. If we use the approximation $N = L$, how long would an average civilization's lifetime have to be for the civilization to exchange at least 10 round-trip messages with its closest neighbor (see Fig. 17.4)?

FURTHER READING

Kellermann, Kenneth, and George Seielstad, eds. *The Search for Extraterrestrial Intelligence.* Charlottesville, Va.: National Radio Astronomy Observatory, 1986.

Marx, George, ed. *Bioastronomy.* Boston: D. Reidel, 1988.

Ponnamperuma, Cyril, and A.G.W. Cameron, eds. *Interstellar Communication: Scientific Perspectives.* Boston: Houghton Mifflin, 1974.

Sagan, Carl, ed. *Communication with Extraterrestrial Intelligence.* Cambridge, Mass.: MIT Press, 1973.

Science Fiction

Asimov, Isaac. *The Gods Themselves.* New York: Fawcett, 1972.

Benford, Gregory. *In the Ocean of Night.* New York: Dell, 1977.

Brin, David. "The Crystal Spheres," in *The River of Time.* New York: Bantam, 1987.

————. *Startide Rising.* New York: Bantam, 1983.

A Signal from Outer Space

The bright diagonal line running from upper left to lower right is a signal from the Pioneer 10 spacecraft, sent from a distance of 4.5 billion kilometers. Amazingly, the spacecraft transmitted this signal with a power of only one watt, about half the power of a miniature Christmas tree light. Nevertheless, a large radio telescope on Earth easily detected it. The display shown here is a plot of frequency (horizontal axis) against time (vertical axis). The change in the frequency of the signal with time is caused by the Earth's rotation, the random dots are background noise. This display was produced by a multichannel spectrum analyzer of the type NASA will use to look for signals from advanced civilizations over interstellar distances (see Figure 19.11).

18

How Can We Communicate?

Throughout history, human beings who have thought about meeting creatures from other regions of the universe have relied on their human intuition and experience to imagine what such an encounter would be like. This common-sense approach has led them typically to picture a physical interchange with extraterrestrial aliens, a face-to-face confrontation perhaps ending in friendship or in violence. But the insights that astronomy offers suggest that *the best means of interstellar communication consists of radio messages, not spaceflight*. If we can rely on scientific analysis, we should anticipate an exchange of television programs, not a showdown in space, as the first encounter with other civilizations.

The Superiority of Radio and Television

What makes radio messages so superior to space travel in meeting other civilizations? The answer lies in the difficulty of accelerating particles with mass—protons, people, or spaceships—to large velocities. Particles with no mass, and in particular the photons that form light waves and radio waves, always travel at the speed of light, 299,793 kilometers per second. Since we have strong evidence that the speed of light is the greatest velocity that *any* particle can reach, light waves and radio waves travel through space at the cosmic speed limit, that is, at the fastest speed universally attainable.

In addition to their speed, photons have another advantage: They can be produced in large numbers cheaply and easily. To send a radio message of five minutes' duration from the Earth to the moon, or from the moon to the Earth, requires no more than one kilowatt-hour of electrical energy, about the same energy as a household spends each evening. To be sure, the transmitter, antennas, and receivers used in sending and receiving such messages may cost millions of dollars, but they still cost less than 1/1000 the price of space travel. For comparison, the construction of a Saturn rocket capable of sending astronauts to the moon (Fig. 18.1) cost many hundreds of millions of

Figure 18.1 The Saturn rocket, the most powerful ever built by the United States, sent the *Apollo 11* spacecraft on a journey to the moon in 1969.

dollars, plus billions more for developing the systems that supported this effort, and the rocket could be used only once.

When we imagine distances much larger than that from the Earth to the moon, we find that the ratio of costs between a piloted spacecraft and a simple radio message grows steadily larger as we consider greater distances and greater spacecraft speeds. Radio waves keep on traveling at the speed of light, whatever the distance they must travel. Spacecraft, too, will coast at a constant velocity—but only after being accelerated to that velocity, a feat that requires immense amounts of energy. To travel to other planetary systems at the speed of the Apollo journeys to the moon would require hundreds of thousands of years at a minimum. But radio waves can make the trip in only a few years. If we aim to decrease the travel time of a spacecraft journey by increasing the speed, we must expend far more energy, and build far more sophisticated spacecraft, than we have done so far.

Let us therefore give serious consideration to the prospect of exchanging messages by radio waves. One key quantity to consider in this quest is the *intensity* of a radio beam that one seeks to detect. The "intensity" measures the number of photons crossing one square centimeter of area each second at a given location. This intensity decreases in proportion to the *square* of the distance from the photon source (Fig. 18.2). As a result, if we move 1000 times farther from a photon source, the signal decreases in intensity by a factor of 1 million, since 1000 × 1000 (1000 squared) equals 1 million. This means that in order to produce a signal of a given intensity at 1000 times the original distance, we must provide 1 million times more photons each second from the source. Because the intensity of the signal determines the ease with which it can be detected, the fall-off in intensity with distance plays an important role in analyzing the likelihood of detecting a given signal.

Since the production of greater amounts of photons requires greater amounts of energy, communication over ever-increasing distances requires ever-increasing amounts of energy. The energy cost rises as the square of the distance to be covered. Hence communication over interstellar distances, which are millions of times greater than the Earth-moon distance, would require trillions of times more photon energy, if we used the same equipment and the same sorts of messages as were employed by the Apollo

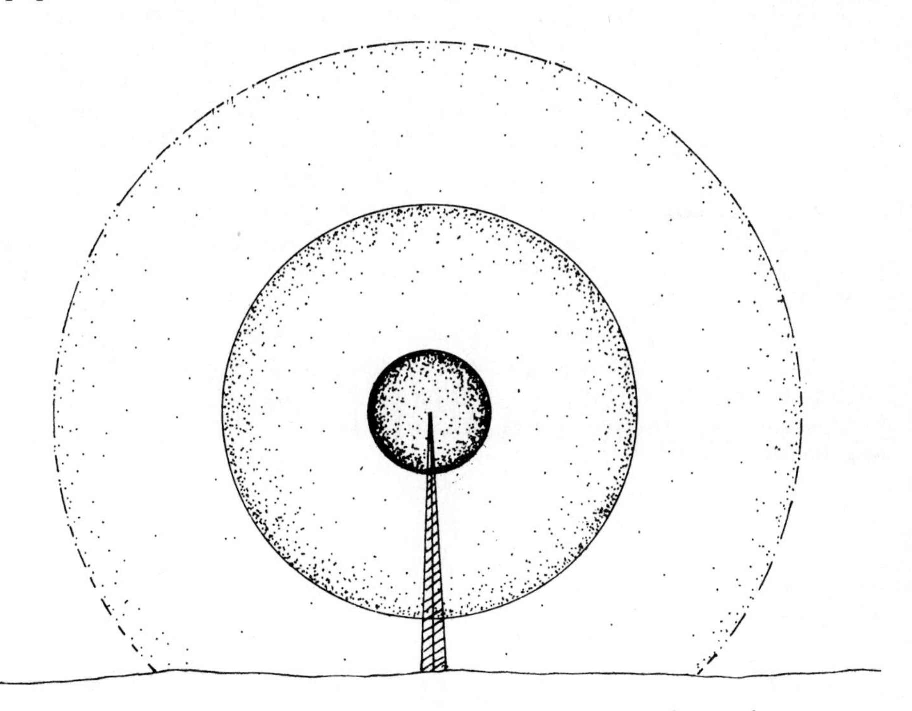

Figure 18.2 The strength of any signal carried by photons, such as radio waves or visible-light flashes, decreases in proportion to the *square* of the distance from the transmitter. At twice the distance, the signal must pass through four times the total area intercepting the signal.

astronauts. Since these costs would rise into the hundreds of billions of dollars just to generate the photons that carry the messages, it is clear that we must instead use shorter messages and significantly more sensitive antennas if we seek to communicate with other planetary systems. In the next chapter we shall discuss the sorts of antennas and receivers that are envisioned for interstellar communication. Because, as we have seen, radio communication costs so much less than travel by spacecraft, we can be sure that if we encounter a situation in which communication by radio becomes expensive, the cost of human space travel would be proportionately larger, thus becoming completely unattainable for the foreseeable future.

Interstellar Spaceships

We have seen that photons can carry information far more efficiently than can any object with mass. In modern society, each of us internalizes this fact by learning to listen to radios and watch television, rather than waiting for the town crier or jester, the governmental courier or traveling troupe, to come by and provide news and entertainment. Nonetheless, our biological heritage and much of our social training continues to insist that personal visits have a special importance. For example, world leaders have special telephone circuits to use for communication by radio, but they make personal visits when they seek to emphasize the importance of their interchanges. These biological and social imperatives may explain the fact that for every report of a strange, possibly extraterrestrial radio or television signal, many thousands of reports are heard of extraterrestrial visitors to Earth (see Chapter 20).

The strength of the human desire for personal visits arises from the curiosity and gregariousness shared by humans everywhere. This desire also reflects an innate human belief that in order *really* to know something, you must experience it *in person.* These traits and beliefs may well characterize other civilizations interested in establishing contact. Therefore we ought to analyze the potential of interstellar spacecraft. This investigation has value not only to convince ourselves that photons will better serve our intentions, but also to try to see how other civilizations, throughout the universe, may have applied the same analysis to reach similar conclusions based on universal principles.

What would a reasonable interstellar spaceship be like? Any such craft must deal with the overwhelming fact of space travel: Distances are enormous. A simple ratio to remember is that *the distance from the sun to the closest stars is 100 million times the distance from the Earth to the moon.* Thus if we imagine the journey from the Earth to the moon (400,000 kilometers) to be equivalent to a trip to the refrigerator, then the distance to Alpha Centauri would itself be like a voyage to the moon!

You may hear someone say that, after all, few people a century ago dreamed that we could build flying machines, let alone rockets to the moon; we must therefore not accept the conclusion that we cannot travel to the stars, by means yet undreamt of, during the next century. However, the two halves of this analysis are founded on different considerations. Long before motor-driven aircraft first flew in 1903, the physical principles of gliding—and of rocket propulsion—had been studied and fairly well understood. What was lacking was the technology—that is, the engine and other components capable of making what had only been a toy into a useful vehicle. In contrast, the difficulties of interstellar spaceflight go beyond technological problems, though those certainly abound. In other words, even if some marvelous new source of energy were discovered, the constraints imposed by human requirements and by the time of flight would remain. Let us examine these limiting effects in greater detail.

Think about the spaceships that you see in science fiction movies and television programs such as "Star Trek." These imaginary spaceships tend to be enormous. Why? Because people need room in which to function, and the ship must carry all the fuel, food, and other required items. The *Skylab* spacecraft (Fig. 18.3), in which astronauts lived with moderate discomfort for several months, measured some 60 meters in length and had a mass of about 10^{10} grams (10,000 tons). The spaceship *Enterprise* in "Star Trek" is at least 10 times longer, and about 1000 times more massive, than Skylab. Impressively large as the *Enterprise* may be in terms of present-day technology, we can easily show from an example developed by Edward Purcell that this ship would in fact be too small for interstellar travel!

To show this, we begin with the assumption that we shall be limited only by the laws of physics and not by technology. We shall therefore use the most powerful source of energy that we can imagine, even though we have no idea at present how we could actually obtain and store the fuel, build the engine, or protect the crew. This ultimate rocket will be powered by the mutual annihilation of matter and antimatter. Antimatter consists of atoms in which the protons have a negative charge (antiprotons) while the electrons are positive (positrons) (see Chapter 2). We have made a few of these antiprotons and positrons in our powerful particle accelerators, but they are destroyed as soon as they contact ordinary protons or electrons. Now we will need to make many tons of this stuff in atomic or molecular form and keep it isolated from ordinary matter until we are ready to bring the two forms together in our engine. The resulting annihilation will convert all of the mass of both forms of matter into energy, following Einstein's famous equation $E = mc^2$. This is 100 times more energy per gram of fuel than the tiny amount liberated by the hydrogen fusion reactions that

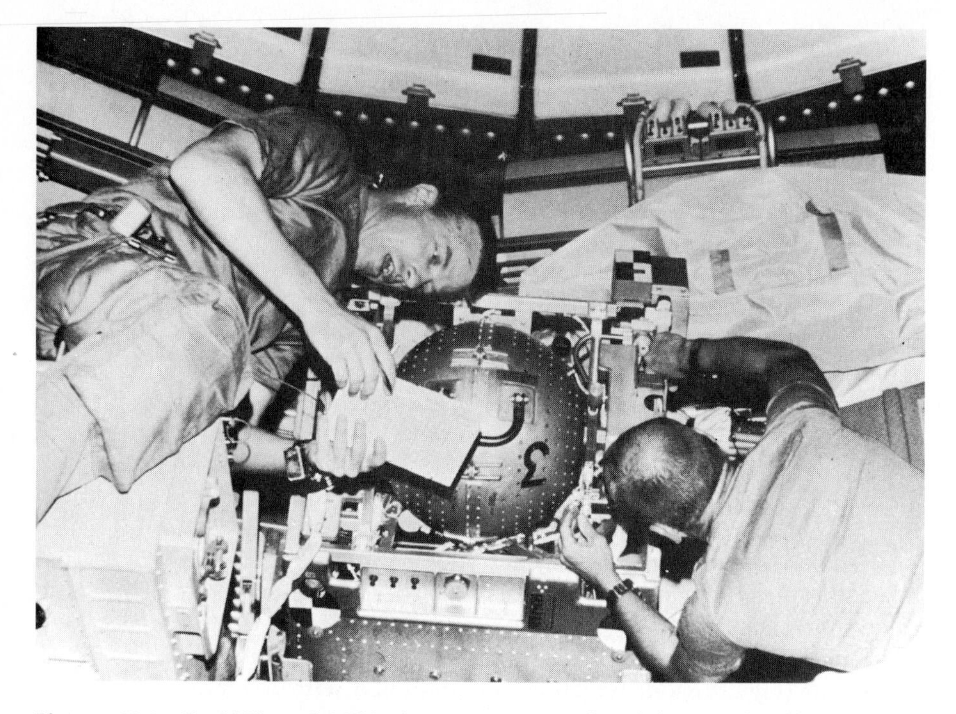

Figure 18.3 In 1973 and 1974, three astronauts lived for nearly three months aboard the *Skylab* spacecraft, confined within a space of about 20 by 8 by 8 meters.

power the sun, the stars and hydrogen bombs. How much matter and antimatter do we need for a journey to another star? We can calculate the answer rather easily, if we assume that the engine works with 100 percent efficiency and the rocket reaches a velocity equal to 99 percent of the speed of light. At this velocity, interstellar journeys between neighboring stars will require several years; at a lower velocity, the travel time would of course be longer.

To reach a velocity of 99 percent of the speed of light by letting matter and antimatter annihilate each other, we must start with 14 times more mass than we shall have when we reach our cruising velocity. If we aim to slow down at the far end of the journey, we need another factor of 14 to provide the deceleration. Thus travel from here to another star at 99 percent the speed of light requires 14×14, or 196 times as much matter-antimatter fuel as the final payload! Finally, of course, we may wish to return home. In a reverse analogue of the journey out, we will require yet another 196 times as much fuel, starting out from the star, as the payload's mass upon our return (Fig. 18.4).

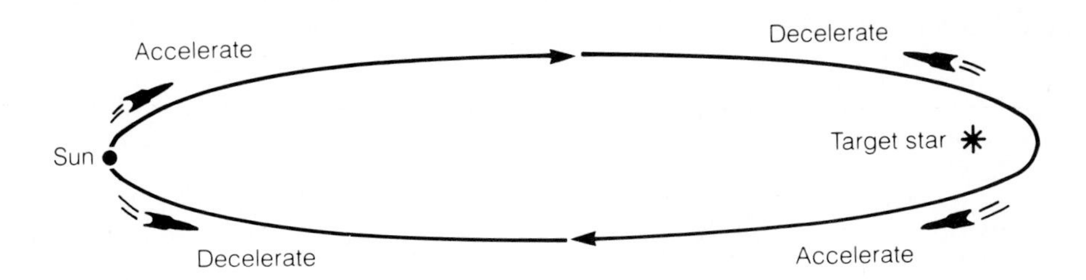

Figure 18.4 A round trip to a distant star requires four separate accelerations or decelerations: up to speed moving away from the Earth, decelerating at the far end, accelerating for the trip home, and decelerating for landing on the Earth.

Therefore, if we hope to make a journey in a self-contained spaceship out to another star, to decelerate there, to start back home, and to stop when we return, we need 196 × 196, or almost 40,000 times, as much mass in the form of matter-antimatter fuel as we use for the payload—crew's quarters, bridge, and so on. To carry a total payload of, say, a modest 10,000 tons, like the Skylab, we must begin with *400 million tons of fuel, half of which is antimatter!* This much fuel would fill a volume of a cubic kilometer, much larger than the entire volume of the starship *Enterprise*.

We therefore conclude that even matter-antimatter annihilation has significant drawbacks, since we need a fantastic amount of fuel to make things work. (Of course, the *Enterprise* uses "di-lithium crystals," about which we know nothing!) For now, we have no way to make significant amounts of antimatter, or to store it for future use in matter-antimatter annihilation. Yet this annihilation process offers us the best source of energy we can imagine, even though at present we have no engine that could use it. Hence every interstellar spacecraft has a tremendous fuel problem. But we must also turn our attention to a second difficulty that science fiction plots typically ignore: the long, not to say tedious, travel times before we arrive anywhere interesting.

This second difficulty about space flight disappears on television and in movies, where travel is made to occur almost instantaneously. In real life, voyages between the stars would require many years, if not many centuries. We must distinguish at least two sorts of imaginary spacecraft: those that could travel to the nearest stars at about the same speed as our best rockets can now, and those that could travel tens of thousands of times faster, at speeds close to the velocity of light. Table 18.1 shows the distances that various speeds will take us in one year, and we can see that the fastest-moving objects that we have made so far, the *Voyager* and *Pioneer* spacecraft, still travel outward from the sun at only one–thirty-thousandth of the speed of light.

TABLE 18.1 Distances Traveled in One Year by Photons and by Various Particles With Mass

Traveler	Speed (km/sec)	Distance Traveled in 1 Year (km)	Number of Years Needed to Cover the Distance that Photons Travel in 1 Year
Human being	0.0003	10,000	1,000,000,000
Automobile	0.03	1,000,000	10,000,000
Jet aircraft	0.3	10,000,000	1,000,000
Pioneer/Voyager	10	333,333,333	30,000
Fusion spacecraft	3,000	100,000,000,000	100
Photons	300,000	10,000,000,000,000	1

With our minds made up, we could probably build the first sort of interstellar spacecraft, capable of travel at Pioneerlike speeds, within the next century or so. Such a spaceship would take over 100 thousand years to reach even the nearest star to the sun. Hence, those who support these missions may face serious problems in convincing others to fund them. Who wants to start a project that *may* produce news of another civilization thousands of generations from now? Yet this kind of mission remains the only technologically feasible one within the foreseeable future. If a choice had to be made between spending money to search for extraterrestrial signals *now* and spending money to send out a space crew to return after, say, 200,000 years, then an immense prejudice in favor of human journeys would be needed to choose the second alternative.[1]

As we mentioned earlier, we cannot plan on building a spacecraft that could travel anywhere close to the speed of light with our present technology. Yet only by traveling at nearly the speed of light can we hope to cover interstellar distances in tens of years. The need to imagine such a spacecraft is in fact so great—at least as far as the consideration of human travel to the stars is concerned—that we shall for the moment imagine that we *could* build a ship capable of traveling almost as fast as photons. Then the crew aboard the spacecraft could profit greatly from a fascinating aspect of travel at this enormous velocity: For the crew, *time would slow down* as the ship moved through space with nearly the speed of light!

[1] To grasp the significance of such an interval of time, try to imagine what your ancestors were interested in 200,000 years ago!

Time Dilation

The theory of relativity predicts that time will slow down or "dilate," an effect that has been repeatedly measured in particle accelerators, where physicists accelerate elementary particles to almost the speed of light every day (Fig. 18.5). But how can time slow down? And how do we know which object has a speed close to the speed of light: Aren't all velocities relative? These questions have answers that are directly relevant to the problem of high-velocity space travel.

First of all, how can time slow down? We measure the flow of time in various ways; for example, by the length of time it takes for a clock to tick, for the Earth to rotate or to orbit around the sun, or for a human being to grow old. All such seconds, days, years, or lifetimes have a standard relationship to one another; thus, for example, each non-leap year contains 31,536,000 seconds. Whatever units of time we employ on Earth, we still think we are measuring the same flow of time, and that two identical clocks will tick at precisely the same rate, even if one clock is in San Francisco and the other in Rome.

What Einstein predicted, however, and what experiments have verified, is that the clocks will tick at the same rate only if neither clock moves with respect to the other. If in fact we put the clock in Rome aboard a jet airplane and fly with it at a speed of 1000 kilometers per hour, then an observer in San Francisco who measures the clock's rate of ticking will find that the clock in motion ticks slightly more slowly than the clock at rest.

For speeds much less than the speed of light, such as 1000 kilometers per hour, the effect can barely be measured, as it amounts to only a tiny fraction of a percent. But for speeds close to the speed of light, 300,000 kilometers per second, the slowing down of time can assume great importance. A clock that passes by an observer at 99 percent the speed of light will tick 7 times more slowly than a clock at rest with respect to that observer.[2] This *time dilation,* the slowing-down of time, will occur whether the clock in question is mechanical, atomic, or biological. Contrary to intuition though it may be, both theory and experiment show that the flow of time within a moving system does indeed slow down, in comparison with the flow of time in a system at rest with respect to the observer.

[2] The algebraic relationship between velocity and time dilation, with v representing the velocity of the moving clock and c the velocity of light, is:

$$\text{Time elapsed in system at rest} = \frac{\text{Time elapsed in moving system}}{\sqrt{1-(v/c)^2}}$$

Figure 18.5 The Stanford Linear Accelerator near Palo Alto, California, which passes beneath Interstate Route 280, routinely accelerates electrons and other elementary particles to 99.9999 percent of the speed of light.

Now comes the truly amazing part: How can we tell which clock is "at rest" and which is "in motion"? Isn't all motion relative? The answer is: Yes, motion is relative, but only if we are talking about unaccelerated motion—motion in a straight line at a constant speed. But any acceleration—a change in speed, in direction, or in both speed and direction—is *not* simply relative, and can be recognized as an acceleration by any observer.

The theory of relativity predicts that particles that are accelerated to almost the speed of light will experience time passing more slowly than they would if they remained at rest. Experiments in particle accelerators have confirmed this prediction to a high degree of accuracy. We may therefore conclude that a per-

TABLE 18.2 Round-Trip Times for Journeys at an Acceleration of 1 g*

Time as Measured by Spacecraft Crew (years)	Time as Measured on Earth (years)	Greatest Distance Reached (light years)	Farthest Object Reached
1	1	0.06	Comets
10	24	9	Sirius
20	270	140	Hyades
30	3100	1,500	Orion Nebula
40	36,000	17,500	Globular cluster
50	420,000	170,000	Large Magellanic Cloud
60	5,000,000	2 million	Andromeda galaxy

* Following an example given by Sebastian von Hoerner, we imagine a spacecraft that accelerates at 1 g; that is, the force of acceleration or deceleration equals the force of gravity at the Earth's surface. After one year, such a spacecraft would be moving at a velocity very close to the speed of light.

son who travels through space at nearly the speed of light and then returns to Earth will age less during the journey than a person who stays at home.

The slowing-down of time in moving systems has a progressively greater effect for speeds closer and closer to the speed of light, 300,000 kilometers per second. If an astronaut travels to Alpha Centauri (4.3 light years away) at a speed equal to 95 percent of the speed of light and returns at the same speed, the astronaut will age by only 3 years, while people on Earth age by 9 years during the astronaut's voyage. If the astronaut travels at 99 percent the speed of light, people on Earth will observe that the journey takes 8.7 years, but the astronaut will age by a year and three months! If we imagine speeds still closer to the speed of light—say 99.999999 percent of the speed of light—we find that a journey that covers 10,000 light years (3000 parsecs) would seem to take 10,000 years to the people left on Earth but would take just over a year from the life of those making the trip. Table 18.2 shows the distances that a spacecraft could cover in various intervals of time, if it could accelerate to speeds closer and closer to the speed of light.

The Difficulties of High-Velocity Spaceflight

It seems evident that spaceflight at speeds near the speed of light is highly advantageous, if we could just devise a way to do it. Suppose we tried another approach. Part of the difficulty we face in interstellar rocket travel is the need to carry the fuel as well as the payload. Robert Bussard has suggested instead that we should consider an interstellar "ramjet" that could scoop up the interstellar gas along its path with some type of magnetic funnel that would require an entrance area hundreds of square kilometers in area. This gas would then serve as the fuel for a fusion reaction engine that would use the same nuclear reactions that make the sun and the stars shine to drive the spaceship. As we discussed in Chapter 4, interstellar gas consists mainly of hydrogen atoms and hydrogen molecules, with some helium (about 10 percent of the hydrogen, by number) and a trace of all the other elements. The average density of matter in interstellar space, an incredibly tiny 10^{-24} grams per cubic centimeter, provides just about 10 atoms in every cubic inch of the interstellar medium. Such a low density of matter can be modeled by placing one tennis ball inside the state of Missouri. Nonetheless, this incredibly rarefied interstellar gas could provide both the potential fuel for an interstellar spaceship and also an almost certain obstacle to travel at speeds close to the speed of light.

Following Bussard, we can easily imagine, if not construct, a rocket ship designed to scoop up hydrogen atoms and molecules over an enormous area as the ship plunges through space (Fig. 18.6). We must certainly worry about the energy required to maintain the magnetic field that collects the particles. This energy would have to be added to the energy requirements for moving the spaceship itself. At present, we must admit that we lack the technology to accomplish either objective, but even if we could, our problems would not be over. As the spaceship's speed increased, the incoming particles of interstellar matter would appear more and more formidable until they became so dangerous as to prohibit further increases in the ship's speed.

Why? The same theory of relativity that predicts the slowing down of time at high speeds also predicts that a particle's kinetic energy will increase dramatically as the particle's speed increases. We all know that in familiar situations, faster-moving particles have more kinetic energy than slower-moving particles do. At velocities far less than the speed of light, a particle's kinetic energy varies in proportion to its mass times the square of its velocity. Thus, for instance, an automobile moving at a speed of 100 kilometers per hour has 16 times the kinetic energy it has at a speed of 25 kilometers per hour, which makes collisions at higher speeds far more destructive. Einstein showed that as particles approach the speed of light, the simple rule of proportionality stated above does not hold, and that particles' kinetic energies increase much more rapidly than they do at lower speeds. A particle moving at 99.9 percent of the speed of light has 3.2 times the kinetic energy it has when it moves at 99 percent of the speed of light. At 99.99 percent of the light velocity, the particle gains another factor of 3.2 in its kinetic energy. In

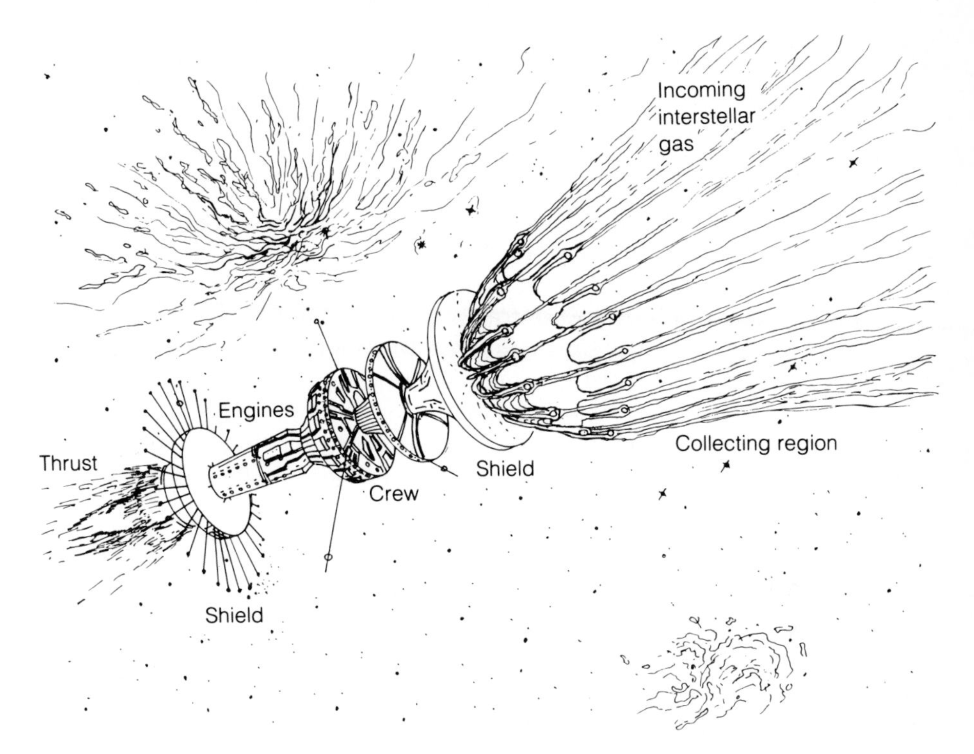

Figure 18.6 Robert Bussard's "ramjet" would use magnetic fields to deflect interstellar charged particles into its fuel supply and could thus travel indefinitely without needing a refueling stop.

order to move at *exactly* the speed of light, any particle with some mass must have an *infinitely* large kinetic energy. This is why physicists conclude that no particle with mass can move as fast as the speed of light.

Consider the effects of this enormous increase in kinetic energy. In order to accelerate a spaceship to almost the speed of light, we must supply the ship with an enormous amount of energy. This creates a problem of supplying such huge energies, which prohibits the use of ordinary fuel carried along with the spaceship. Hence our reliance on interstellar hydrogen for rocket fuel seems ever more necessary.

The highest velocities suggested for future spacecraft may therefore prove to be unattainable in practice, either because of the lack of propulsion systems or because of the danger of traveling through the interstellar medium at speeds within a fraction of a percent of the speed of light. We may be able, someday, to travel through space at perhaps 99 percent of the speed of light, so that a journey of 100 light years would age us only 14 years and thus would return us to the world of our great-grandchildren, but we shall probably never be able to reach the speeds of, say, 99.999999 percent of the speed of light that would enable us to travel through most of our galaxy within a human lifetime.

To keep all these plans in perspective, we would do well to recall that the greatest speed yet attained by a human-made spacecraft equals not 99 percent nor 0.9 percent, but a mere 0.005 percent of the speed of light. Great as our achievements may have been during the last few centuries, we cannot place much confidence in a straightforward extrapolation that suggests to us that within another few centuries we shall be able to travel 20,000 times more rapidly. For the time being, we have no way to do so. Furthermore, it is an ironic fact that the same relativity theory that makes travel at nearly the speed of light important—because of the time dilation effect—also has the effect of making the interstellar medium through which any spacecraft must travel, an ever more dangerous hail of bullets as the spacecraft's velocity nears the universal speed limit.

Automated Message Probes

In the year 1972, the spacecraft *Pioneer 10* set out from Cape Kennedy, accelerated away from the Earth, flashed past the planet Jupiter, and by 1992, coasting at about 10 kilometers per second (36,000 kilometers per hour), was about 5 billion kilometers away from us, farther from the sun than the planet Pluto (see page 432). Along with *Pioneer 11, Voyager 1,* and *Voyager 2,* each moving outward along a separate trajectory, *Pioneer 10* will continue to sail through space at a fairly constant velocity (relative to the sun) until the spacecraft happens to come close to a source of gravitational force, for example, a star (Fig. 18.7). If the star in question were Alpha Centauri, the closest star to the sun, then this encounter could occur in about 100,000 years.

Pioneer 10's trajectory, however, points nowhere near the direction of Alpha Centauri; instead, it is headed for a point in space close to the boundary of Taurus and Orion. Hence, any encounter with another star will take longer than 100,000 years to occur—far longer, in fact, when we consider how empty space is of stars. Because *Pioneer 10* has no guidance system, it can encounter other stellar systems only by chance, and the chance of its coming close to another star within, say, 1 billion years remains infinitesimal. We can calculate that this spacecraft, which takes 100,000 years to cover the average distance between stars, must traverse 100 trillion times this distance if it is to intersect a planetary system like our own. But such travel will require 100 trillion times 100,000 years, or 10 billion billion years, almost a billion times the age of the universe! This enormous waiting period, or, more honestly, the infinitesimally small chance that *Pioneer 10* will ever encounter

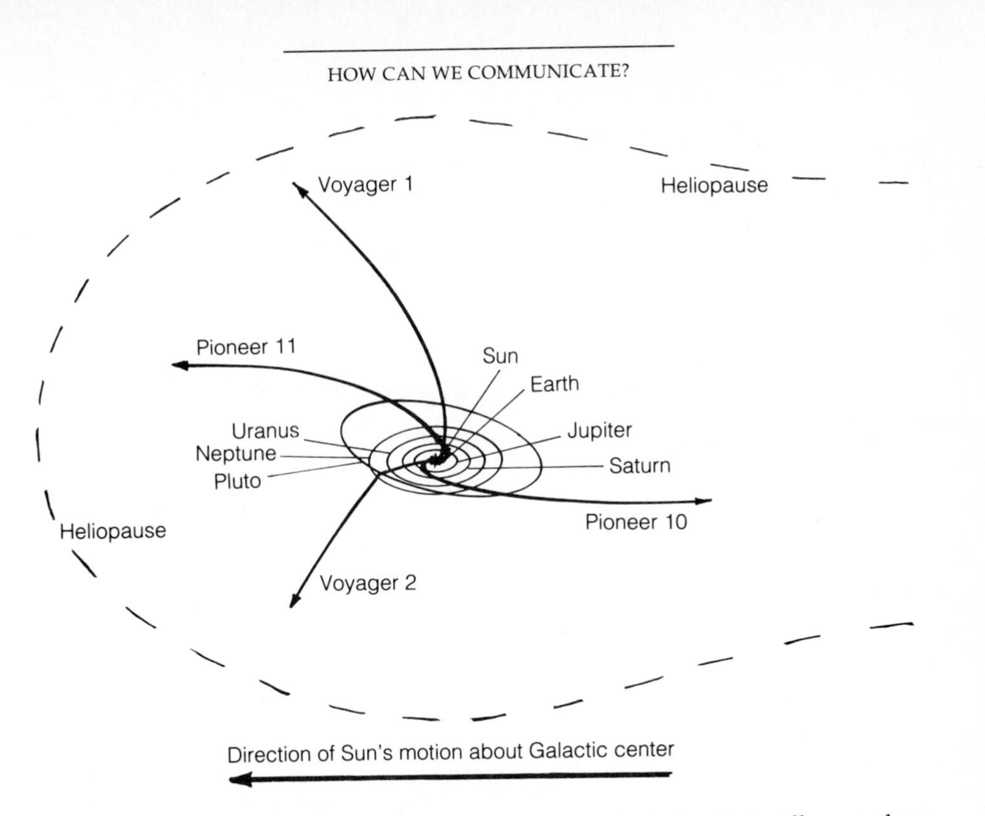

Figure 18.7 Four spacecraft are now moving on trajectories that will carry them into interstellar space: *Pioneers 10* and *11* and *Voyagers 1* and *2*. These are the fastest craft humans have built, but at their speeds of 10 km/sec, they would each require about 100,000 years to reach the nearest stars.

another planetary system (let alone one that supports intelligent life) reminds us, first, that space is mostly empty, and, second, that the message plaque on *Pioneer 10* (see page 143) seems destined to be worn away by interstellar dust before any other civilization sees it. Likewise, the gold-coated records on the *Voyager* spacecraft (see page 378) will sail onward through our galaxy, carrying messages and pictures from Earth, but without much of a chance that any other intelligent species will ever admire them.

Still, we have begun our attempts at communication. The *Pioneer 10* plaque and the *Voyager* records represent humanity's first effort to send interstellar messages by spacecraft, for possible recognition by other civilizations. Now that the idea of communication has arisen, better means can be devised. We might conclude from the unlikely chances of the *Pioneer* and *Voyager* messages being received that we would be foolish to send spacecraft without human control into space, hoping that such messengers could eventually encounter another civilization. A closer look, however, shows that the right kind of spacecraft could perform far better than those we have sent so far.

What we need are spacecraft that send radio messages to possible planets as they pass close to stars. Ronald Bracewell has suggested that sophisticated robot scouts of this type could be in common use by advanced civilizations for routine exploration of our galaxy. There is even a possibility that one or more of these interstellar robots could be in our solar system now, lost among the asteroids, or slowly circling the sun beyond the orbit of Jupiter.

An automated message probe, launched from Earth now, could reach the vicinity of Alpha Centauri after about 100,000 years. Similar probes could be directed, for relatively little cost, to other nearby stars similar to our sun.[3] These spacecraft, in their most sophisticated versions, would each contain a radio receiver and transmitter. The receiver could detect any radio or television messages emitted by a civilization near another star; that is, a small computer aboard the spacecraft could analyze any radio output to see whether the photons arrive in random or in coded patterns. If the spacecraft did detect radio and television signals from another civilization, it would then deliver a prerecorded message at exactly, or nearly, the same frequency as the local transmission. This matching of frequencies would maximize the chance that the spacecraft's signal would be detected and analyzed.

Although the basic purpose of automated probes would be simply to announce that we exist, even a small spacecraft could broadcast a message of several billion "bits" of information, enough to transmit all the information in the *Encyclopaedia Britannica.* With allowances for the difficulty of teaching another civilization how humans communicate their thoughts, our current technological capability would allow us to transmit just about everything we could think up about the Earth with a rather small automated spacecraft in the near future.

Included in the message aboard such a probe would be, of course, a description of its place of origin, if we seek two-way communication. Other automated spacecraft, perhaps deliberately modeled on our own, could be sent back to us by any civilization as advanced as ours, so we might be in danger of exhausting all that we have to say with a single exchange of messages! The chief drawback in setting this scheme in operation remains the long travel times—hundreds of thousands of years to the closest stars, millions or tens of millions of years to those somewhat farther away. Hence the transfer of information would occur, at least to begin with, at an extremely slow rate. In contrast, messages sent directly by radio can travel more than a thousand times more rapidly, and would take proportionately less time for the journey.

[3] If we mass-produced such probes and the rockets to launch them into space, a reasonable cost estimate might be $500 million per probe. Thus for the cost of putting men on the moon we could send several hundred automated messengers toward the stars.

In actuality, only the first message, carried by automated probe, would have to travel so slowly. Afterward, with exact knowledge of where to look, photons could carry the information at the speed of light, and could thus reduce the round-trip travel time to decades or centuries. Automated message probes might make sense as a one-time affair (per star to be investigated). Once the probe nears another star, solar (that is, starlight) energy cells can provide power to the receiver, transmitter, and computer aboard the spacecraft.

The concept of launching automated spacecraft that would travel for thousands or millions of years leads toward an intriguing question: Could we make an "automated" craft come alive? That is, could we arrange to send everything that is alive in ourselves—our creativity, our thought patterns, our biological history, our ability to reproduce and to evolve—on an interstellar journey in the form of, say, a computer program?

The challenge of creating "artificial intelligence" (more precisely, of reproducing what we call human intelligence in computer form) has attracted deep thinkers for decades. Can we ever make a machine whose dialogues are indistinguishable from a human being's? Can we go even further, and arrange for computer programs to evolve into ever more complex and sophisticated ones? In 1991, Thomas Ray, a biologist at the University of Delaware, devised a relatively simple program called "Tierra." Tierra follows a pre-programmed set of rules that create a computer world in which "entities" (basically strings of zeroes and ones) undergo extinction and evolution, with new entities brought into being by the rules governing the evolutionary process. Since Tierra includes a chance element, no two runs of the program will yield an identical result; they do, however, produce a wide variety of entities that have proven successful in reproducing themselves in computer evolution. Tierra is a long way from the complexity of life, but this and similar attempts to model living creatures remind us that eventually nearly all of our human experience could be reduced to electronic form for easier transport.

Automated probes remain a viable concept whether they carry all of our human understanding, or only part of it. If we chose to, we could launch one automated probe every year with little strain on our budget. The reason that we do not do so, and have no plans to do so, comes mainly from the immense travel times that are involved. Few of us support wholeheartedly the sending of messengers to tell their story a million years in the future. Interstellar communication makes a splash in our minds when we think of communication now; communication efforts that may pay off for the ten-thousandth or fifty-thousandth generation tend to be left to our descendants.

But who, then, will begin the process? The answer to this conundrum may come from the possibility of detecting radio waves directly from the advanced civilizations that are producing them, even when interstellar communication is not their goal. Or we may find that other civilizations, less timid, better established, and more secure than ourselves, have performed the same analysis but have decided to proceed with automated probes. Thus, any night now our televisions could light up with an unfamiliar signal—not WABC, KTVU, or KNXT, but EXO, OBOY, or YUHU, signals of far-out programming that will treat us to a real 90-minute spectacular. Then we will know that automated message probes do indeed make sense.

SUMMARY

Although many people assume that interstellar travel must soon become a human reality, analysis using the laws of physics shows that interstellar spaceflight at speeds even a modest fraction of the speed of light must remain incredibly difficult, if not downright impossible, for the foreseeable future. The best (unmanned) spacecraft humans have built until now travel at one thirty-thousandth of the speed of light and would take 100,000 years to reach even the closest star to the sun. If we seek to build much faster spacecraft, we must find new ways to accelerate the craft to greater velocities; such acceleration requires tremendous amounts of fuel, even if we assume that we devise a way to use a 100-percent–efficient annihilation of matter and antimatter.

If we could somehow build a spaceship that traveled at nearly the speed of light, then the time dilation effect, first discussed by Albert Einstein, would allow space travelers to age more slowly than those who stay behind: Time appears to slow down for those who travel close to the velocity of light, in comparison with those who await their return. The same theory that predicts the slowing down of time, however, also states that at speeds close to the speed of light, every tiny particle of interstellar gas or dust would effectively become a bullet of enormous energy, as perceived by the spaceship or by those inside it. Hence, we would also have to devise a way to deflect these bullets, and such deflection adds to the energy problem we encounter in accelerating the spacecraft to near-light velocities.

Knowing the manifold difficulties we meet in trying to move particles with mass at greater and greater speeds, we are finding it more practical to use photons. Because they have no mass and always move at the speed of light, photons achieve speeds that are greater than the speed any particle with mass can ever achieve. Photons do not cost much to produce, and even for the relatively short distances we confront on Earth, photons have come to predominate over particles with mass for sending messages. When we consider the enormously greater distances that must exist between neighboring civilizations, we find that photons have a tremendous advantage over "nuts and bolts" spacecraft. Even though we may take pleasure in thinking about human interstellar space travelers, the physics of the situation—the fact that civilizations are spaced many parsecs apart—argues in favor of photons as the best way to communicate whatever has to be said. If our conclusion has merit, then we ought to try to figure out what sort of interstellar messages are passing through us right now, and how we can join the network of intercommunicating civilizations, should we choose to do so.

QUESTIONS

1. Why do radio waves seem far superior to spacecraft for interstellar communication?

2. Why are radio waves more difficult to detect at greater distances from the radio transmitter?

3. Why must any interstellar spaceship carry an enormous amount of fuel if it is to reach speeds that are a significant fraction of the speed of light?

4. What difficulties stand in the way of a spacecraft that scoops up fuel from interstellar gas and dust as it travels through space at 99 percent of the speed of light?

5. What is the time dilation effect? How could this help with the problem of the immense amounts of time that interstellar journeys require?

6. For velocities, v, that are nearly equal to c, the speed of light, the energy needed to reach a given velocity varies in proportion to the ratio $c/(c\text{-}v)$. How much more energy is needed to accelerate to a velocity $v = .99\ c$, compared with $v = 0.9\ c$?

7. About how long should it take for the *Pioneer* plaque to be carried into another planetary system? What does this imply about the usefulness of sending such messages into space?

8. What advantages could we design into an automated message probe that would render it far superior to the *Pioneer* plaque and the *Voyager* record?

FURTHER READING

Bracewell, Ronald. *The Galactic Club.* San Francisco: W.H. Freeman, 1975.

Mermin, David. *Space and Time in Special Relativity.* New York: McGraw-Hill, 1972.

Morrison, Philip, John Billingham, and John Wolfe, eds. *SETI: The Search for Extraterrestrial Intelligence.* Washington, D.C.: U.S. Government Printing Office, 1978.

Papagiannis, Michael, ed. *Strategies for the Search for Life in the Universe.* Boston: D. Reidel, 1983.

Science Fiction

Gunn, James. *The Listeners.* New York: Scribner, 1972.

Hoyle, Fred, and John Elliott. *A for Andromeda.* New York: Harper & Row, 1962.

19

Interstellar Radio and Television Messages

We have seen that interstellar travel requires large amounts of energy and that the more rapidly we wish to travel, the more energy we must supply to our spacecraft. We do not now have the means to build an effective, piloted interstellar voyager, and even as our technology improves, we—and all other civilizations!—may remain unwilling to spend so much energy since much less expensive means of communication are available. If we can agree that exchanging information rather than face-to-face dialogue represents our primary goal, and if other civilizations have a similar attitude, then we can get down to business and start communicating by using the technology of radio and television.

Let's assume that the question of how interstellar communication begins has almost certainly been answered, at least in its general outlines, by civilizations in our galaxy more advanced than ourselves. These (supposed) advanced civilizations may have acquired interstellar communication abilities not within the last few years, but thousands, hundreds of thousands, millions, or even billions of years ago. Unless such civilizations are extremely rare—either because they are unlikely to develop, or because they destroy themselves rather quickly—most of them that have the desire to contact other civilizations should have done so by now. If this is true, then our focus becomes not how developing civilizations can contact one another, but how a new civilization such as ours can join the network. Is there a test? Or do we simply call "information"?

These questions are based on the assumption that our development as a civilization has followed broadly universal lines. If we are average, then we can hardly be the first, or anywhere near it, in the Milky Way, just as we cannot be among the very last civilizations that will appear in galactic history. Thus we must try to figure out how interstellar radio and television messages can be detected. Courtesy and caution suggest that first we listen, and then we send our own messages.

Of course, if everyone listens and no one transmits, no one will hear anything. We know, however, that our own terrestrial communications leak radio, radar, and television waves into space. These stray photons could be detected at interstellar distances by a suitably sensitive array of antennas and receivers. If the same sort of leakage occurs in other civilizations, we might be able to eavesdrop on their internal radio communications. Furthermore, such advanced civilizations may be beaming information and entertainment to one another across the interstellar distances that separate them. If so, we can hope to intercept these communicaitons if we are in their path. Before we can consider these suggestions in detail, however, we must consider four questions:

1. Where should we point our antennas?
2. Which frequency channels are most likely to be used for interstellar communications?
3. What total range of frequencies might be involved?
4. What sort of messages might be considered "standard" for opening conversations with new members of the galactic network?

Where Should We Look?

On page 392, we listed the nearest stars, some of which seem likely to have planets. Nearby stars basically similar to the sun provide the best chance for finding other civilizations, because the intensity of radio photons emitted by any civilization decreases in proportion to 1 over the square of the distance from it. However, radio telescopes directed toward the apparently "best" of these high-priority stars, Tau Ceti and Epsilon Eridani, have not yet detected any signals that might provide evidence of another civilization.[1] We must therefore settle down for a long effort, prepared to search star after star, before we have a good chance of finding the nearest civilizations. Which stars should we examine first?

All else being equal, the radio waves from nearby civilizations will outshine those from faraway civilizations. Each time we double our distance from a particular radio source, we reduce the intensity of its radio signal—the number of photons that reach any particular antenna each second—by a factor of 4. This fact causes us to think about looking first at the nearest stars, and later at the more distant ones. But it is certainly possible that the Milky Way contains a few civilizations broadcasting far more powerfully

[1] But it must be noted that these searches for signals observed only a few of the millions of possible frequencies (see page 461) with low sensitivity for short periods of time.

than average, and just as detectable at a great distance as a "standard" civilization at a much lesser distance. For this reason, we ought to think about making a *general survey of the Milky Way together with a search directed at the closest stars.*

The closest stars to the sun, listed in Table 16.1, form a logical place to begin our search for other civilizations. But even with our present technology, we can hope to find a civilization far beyond the arbitrary 4-parsec limit of distance in that table. If we choose to invest in the necessary equipment, we could now detect a civilization whose power output from television, radio, and radar transmissions roughly equals our own anywhere among the thousands of stars closest to the sun. Thus, even though closer civilizations require less effort for detection, we need not restrict ourselves to the closest few stars. Instead, we may include the nearest thousands of stars in the Milky Way as candidates for radio signal detection. The question then returns to *which* stars among these thousands merit special attention.

The best answer we can give at present is: stars similar to the sun. Single stars seem better candidates than double- and multiple-star systems, but not by much. Stars whose luminosity resembles the sun's seem superior to low-luminosity M stars, such as Barnard's Star, because of the tiny ecospheres that surround faint stars (see page 395). If we restrict ourselves to stars with a luminosity equal to at least 1 percent of the sun's, we must discard 75 percent of the stars in our galaxy (see Table 16.2). However, the remaining 25 percent will still provide 75 billion stars in the Milky Way, and many hundred thousand within range of the receivers we could build. We certainly must reject the brightest stars, those whose lifetimes fall short of the billion years that we think represents the minimum time in which a civilization can develop. These most luminous stars number less than 1 percent of the stars in our galaxy, so we do not lose much in numbers when we choose to ignore them in our search.

We thus reach the conclusion that *if* all civilizations transmit radio waves at roughly the same power level, the best search strategy to find extraterrestrial signals would be to proceed star by star, searching among both single- and multiple-star systems of spectral types from F5 through K8—that is, among the stars that can last for at least 5 billion years and have sufficient luminosities to produce a reasonably large habitable zone. Our discussion in Chapter 11 suggested that these stars have a high probability of possessing planets, and we can see no fundamental objection to the thought that, on the average, at least one planet in every four planetary systems should have conditions favorable to the development of life, and thus of intelligent civilizations. When we examine the numbers, as we did in Chapter 17, we see that we should prepare to search through at least several thousand stars to have a reasonable chance of finding another civilization that exists now (see page 412). This prospect, although difficult, should not cause despair. If we examine 25 stars each day, we could work through 25,000 stars every three years, so we would soon reach the level that could produce the reward of interstellar contact. The problem thus reduces to one of deciding to commit the resources to construct the computers and receiver systems required to conduct a sufficiently sensitive search.

What Frequencies Should We Search?

How can we hope to determine the frequencies at which we are likely to have the best chance for discovering another civilization? Aren't we facing an insoluble problem in trying to figure out how another civilization would send its messages? The answer, we think, is that other civilizations would send messages much as we would, so we can in fact expect to determine the best frequencies to search.

We must distinguish between messages that a civilization uses for its own purposes, and which we might overhear or "eavesdrop" upon, and messages sent deliberately to other civilizations, or at least into interstellar space, with the hope of detection. Though these two purposes overlap somewhat, we have only to think of our own civilization to see that most radio and television broadcasts have been planned with solely a human audience in mind (unless the broadcasters know more than they let on). Nonetheless, we shall see that radio-television frequencies have such key advantages over other frequencies that we can recommend them for any sort of communication purpose.

Consider the electromagnetic spectrum (Fig. 19.1). The tremendous spread of photon energies, more than a billion billion (10^{18}) times in energy from the highest-energy gamma-ray photons to the lowest-energy radio-wave photons, presents a tremendous range of possibilities in choosing a photon frequency. What are the most striking "natural"—that is, galactic—frequencies of reference? Are there any photon frequencies that would come to mind in any other civilization, whenever they think about how to communicate with other civilizations? The answer seems to be that we can find a most likely range of photon frequencies for interstellar communication, even though we cannot find one single frequency to point to and say: "*That* is the galactic communication channel." In narrowing down the enormous range of possible frequencies, we employ three criteria that we think must be universal: economy, freedom from interference, and cosmic frequency guides. Let us examine each of these in turn.

The first factor in choosing radio photons instead of any other means for interstellar communication deals with the economics of message exchanges, the costs of communication. If we think of a message as made of individual units, then the smallest unit of information, called a "bit" by sci-

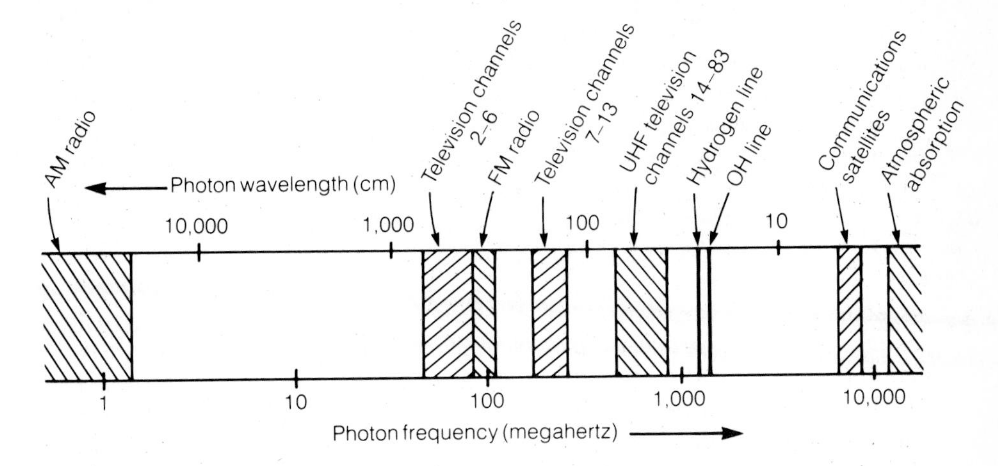

Figure 19.1 Different parts of the electromagnetic spectrum have been given different names. The low-frequency, long-wavelength domain of radio photons includes the frequencies used for AM and FM radio and television broadcasts, as well as radar—radio waves bounced from objects to measure their distances and speeds.

entists who work with the theory of messages, requires at least one photon for transmission. To avoid errors and to ensure detectability, more than one photon per bit may be required, but the number of photons per bit will be about the same for radio messages as for visible-light messages. If we now compare radio waves with visible light, an important truth dawns immediately: Each visible-light photon carries about a million times more energy than a radio photon and therefore, in energy terms, costs a million times more to send. In summary, radio waves are cheap, visible-light waves expensive.

The second factor in choosing a frequency at which to send or receive messages deals with freedom from absorption in interstellar space. Some photons cannot travel as freely through space as others. Here again, radio waves of all but the lowest frequencies have special advantages. To see why this is so, we must consider the absorption of photons as they travel, and the interference of other emission processes with our attempt to detect and to locate a given source of information-carrying radiation.

In the astronomical exploration of our galaxy, the advantage of radio waves over visible light for long-distance communication quickly becomes apparent. We cannot photograph the galactic nucleus, or the spiral arms that lie beyond it, because the intervening gas and dust absorbs the light emitted by stars at these distances in the plane of the galaxy. Yet we can map these features with comparative ease using radio waves, except for a few particular frequencies at which absorption by atoms and molecules occurs. In our own solar system, we have found that radar and microwave radiation penetrate the clouds of Venus easily, while visible light cannot.

The same advantage for radio holds true when we consider interference from other types of emission. Visible-light emission from a planet circling a star (perhaps from a powerful laser paired with a giant telescope) must contend with the immense amount of energy radiated at visible wavelengths by the star itself. We can far more easily send messages at radio frequencies, where the energy level of stellar radiation is much lower. Indeed, as we shall see, *even our relatively primitive civilization already has the capability to send a message at a particular radio frequency with more energy than the sun emits at that frequency.*

However, a cosmic "background" of radio emission exists from two separate sources, the radiation left over from the big bang (see Chapter 2) and the myriad sources of synchrotron radiation caused by electrons spiraling in magnetic fields (see page 72). These two sources define a radio "window" in the electromagnetic spectrum that is relatively free from competing emission, and it is within this window that we may expect interstellar communications among our hypothetical advanced civilizations to be taking place (Fig. 19.2).

At still longer radio wavelengths, we encounter the barrier of our planet's "ionosphere," which reflects most of the radio emission incident upon it (Fig. 19.3). This frequency domain includes the region of commercial AM broadcasts. While the reflecting ionosphere helps to spread our own radio signals around our planet, it greatly hinders extraterrestrial signals from penetrating our atmosphere to reach our receivers. FM radio and television, especially the UHF channels, do not suffer from this barrier. At the short-wavelength end of the radio window, we encounter absorption by Earth's atmosphere. We could avoid both these atmospheric impediments by broadcasting and receiving messages from a station in orbit around the Earth, but as we have seen, radio emission from the galaxy itself would keep this effort from yielding much improvement over ground-based facilities.

We have been guided to the radio band of frequencies by considerations of economics and freedom from absorption and interference. Can we look for cosmic frequency guides within the radio band? Can we tell what particular frequencies are likely to attract any civilization to a universal station on the radio dial?

In the first discussion of the problem in 1959, Philip Morrison and Giuseppe Cocconi pointed out that the single most important radio frequency in the universe appears to be 1420 megahertz (1420 million hertz or MHz), the frequency of the radio waves emitted by the spin-flip process in hydrogen atoms (see page 86). This emission arises from the occasional collisions that result from the random motions of hydrogen atoms in interstellar space. Some of the atoms are continually bumped into the spin-flip state

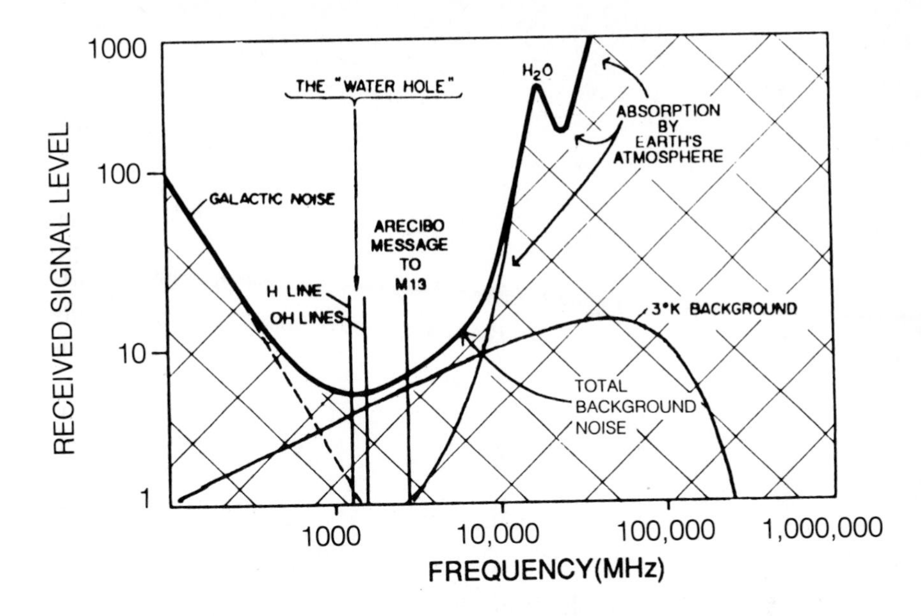

Figure 19.2 If we plot the sources that interfere with our detection of radio signals, we find a trough between the low-frequency, long-wavelength end of the radio spectrum (where noise from our own galaxy increases) and the high-frequency, short-wavelength side (where the Earth's atmosphere tends to absorb radio photons). Within this trough lies the "water hole" that spans the characteristic frequencies at which hydrogen atoms and OH "radicals" radiate radio waves—and hydrogen plus OH makes water.

from which they can emit their characteristic radio signals. Hence, throughout the galaxy in which we live and in other spiral galaxies as well, interstellar gas, 90 percent of which consists of hydrogen, constantly emits radio waves with a frequency of 1420 MHz and a wavelength of 21.1 centimeters. Every intelligent being who studies our galaxy knows about these radio waves, which form the most abundant and most widespread emission of radiation at a definite frequency. Furthermore, the 1420 MHz photons have the ability to travel, and to be detected, over great distances. Finally, the band of frequencies around 1420 MHz remains especially clear from competing radio emission from the cosmos.

Because of the Doppler effect, and the fact that atoms in our galaxy tend to have some motion toward us or away from us, the entire band of frequencies between about 1419 and 1421 MHz contains a great amount of photon emission from the natural processes of hydrogen spin flips. In fact, all the hydrogen atoms in the Milky Way and beyond combine to bathe the Earth in a radio glow at 1420 MHz whose power equals that of a 20-watt

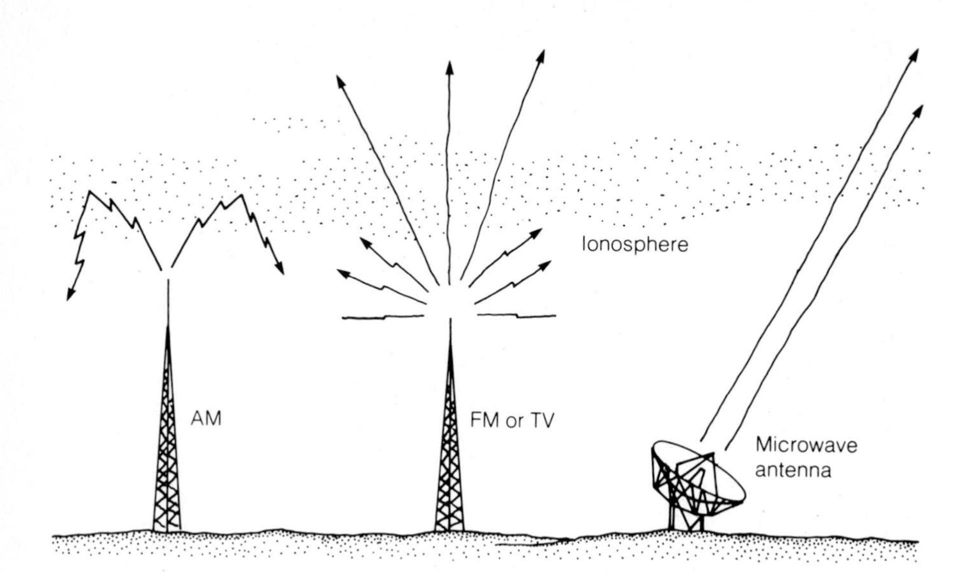

Figure 19.3 The Earth's ionosphere, an atmospheric layer rich in ions, reflects most of the longer-wavelength radio photons that strike it. This includes the wavelengths used for AM broadcasting, but not those employed for FM radio, VHF or UHF television, or microwave transmissions, which have shorter wavelengths.

light bulb! This 1420-MHz emission, affected by subsequent absorption and still later reemission processes, allows us to map the distribution of interstellar hydrogen when we study the frequency range between 1419 and 1421 MHz. But outside this rather narrow band of radio frequencies, much better conditions prevail. In the radio domain somewhat above or somewhat below the 1420-MHz range, relatively few atoms and molecules, as we have seen, provide natural interference to radio signals. Human-made interference is something else again: Only by arranging to protect the few megahertz around 1420 MHz have radio astronomers managed to continue their work on Earth!

Suppose that we, or any other civilization, would "naturally" be led to choose a frequency near 1420 MHz for interstellar messages, and suppose that this sort of frequency is now in use for local message traffic, so that we can hope to "eavesdrop" on civilizations at the same frequencies. We still must face a crucial problem of detail: Which frequency closest to 1420 MHz has the message? How far from 1420 MHz should we tune our dials? And should we look to higher or to lower frequencies?

If we believe that water will be essential for most other forms of life as well as for our own (see Chapter 10), then we may find merit in the suggestion made by the physicist Bernard Oliver. Since each molecule of water

(H_2O) consists of a hydrogen atom (H) plus a molecule of OH, Oliver pointed to the frequency band between 1420 MHz and 1721 MHz as the most likely channel for interstellar communication. OH molecules produce photons at a series of frequencies—1612, 1665, 1667, and 1721 MHz—by an emission process similar to that for hydrogen atoms. The band of frequencies characteristic of photon emission from OH molecules forms a guidepost in the realm of photon frequencies, just as the 1420-MHz frequency of hydrogen-atom emission forms another still more striking marker. If the importance of water occupies a large place in the consciousness of all life forms, then from the fact that H + OH = water, we might indeed find that the range of frequencies between 1420 MHz and 1721 MHz specifies the galactic radio dial—the frequency domain where interstellar communication occurs. Bernard Oliver calls this gap the "water hole": the place (in photon frequency) where galactic civilizations meet (see again Fig. 19.2). (Note that water molecules by themselves emit a complex pattern of radio waves, with no dominant frequency or set of frequencies.)

Frequency Bandpass and Total Frequency Range

If we think that we know the most likely frequency at which interstellar communication should occur, we still must determine the spread in frequencies that the signal covers (Fig. 19.4). No signal occurs at one precise frequency, but instead extends over a range of frequencies, because of the Doppler effect and many other effects that spread the signal slightly. Thus in addition to deciding which frequencies to search, we must also decide on a "bandpass" for our search—the frequency spread that we take as the basic unit. In other words, we must choose a *range* for the search (the total frequency spread to be searched) and also a *bandpass* (the fineness with which we divide the range into frequency intervals). Should we divide the range into tiny intervals, of, say, 1 hertz each? This would give us a better chance to detect a signal that had been deliberately held within an extremely narrow frequency channel. But to search through a range of, perhaps, 1000 megahertz just 1 hertz at a time would require a billion individual searches of each frequency "channel." Fortunately, this problem has a solution, as we shall see.

We face a difficult problem in guessing how large a frequency bandpass another civilization might use for its messages—both local and interstellar—and in estimating the total frequency range over which we should search for interstellar signals. Several factors enter the selection that any civilization will make for the frequency bandpass of its broadcasts, and we must think about them in turn. Two key impulses appear to be the desire to send information rapidly, which suggests a relatively large bandpass, and

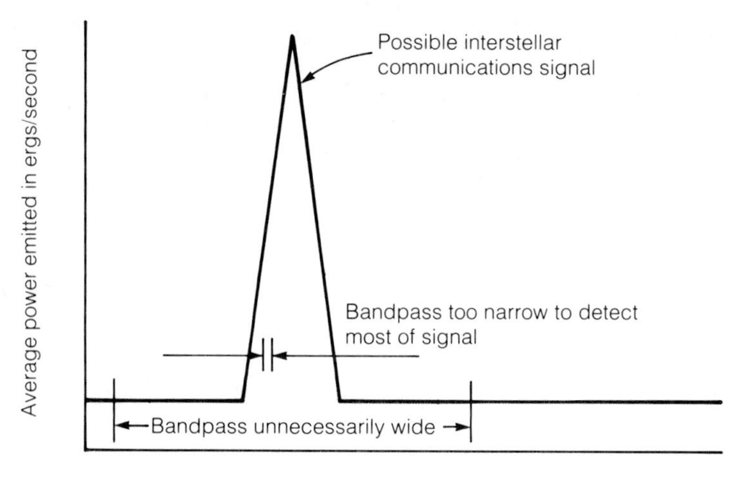

Figure 19.4 A problem arises in searching for a signal that covers an unknown "bandpass" or spread of frequencies. If we search over a frequency region with a much wider bandpass than that of the signal, we will miss faint signals, because we combine that signal with the noise in a much wider bandpass. Conversely, if we use a bandpass much narrower than the signal's, we can detect only a small bit of the signal at any time and will again lose the chance to detect faint signals.

the desire to produce a signal that is recognizable against the background of radio noise, which requires a relatively small bandpass.

When a message arrives at its (unknown) destination, whoever searches for it must be able to separate the signal from the background of "noise"—that is, from the random emission of photons with nearly the same frequency by various cosmic processes (Fig. 19.5). Narrower signals (that is, signals confined to a smaller total frequency range) can be more easily separated from the noise, if we consider signals of the same total radiated power.

A civilization seeking to minimize the costs of sending signals might well attempt to transmit its signal within as narrow a frequency band as possible. Here an important minimum size for the frequency bandwidth arises from "interstellar dispersion." As the radio waves pass through interstellar space, charged particles there (mostly electrons) actually change the frequency of the radio waves as they pass by. Furthermore, the amount of this change varies in a random way as the clouds of electrons move. Therefore, even if we knew the exact frequency at which a radio message had been sent, and even if we knew the precise relative velocity between the source and ourselves so that we could allow for the Doppler shift, we still

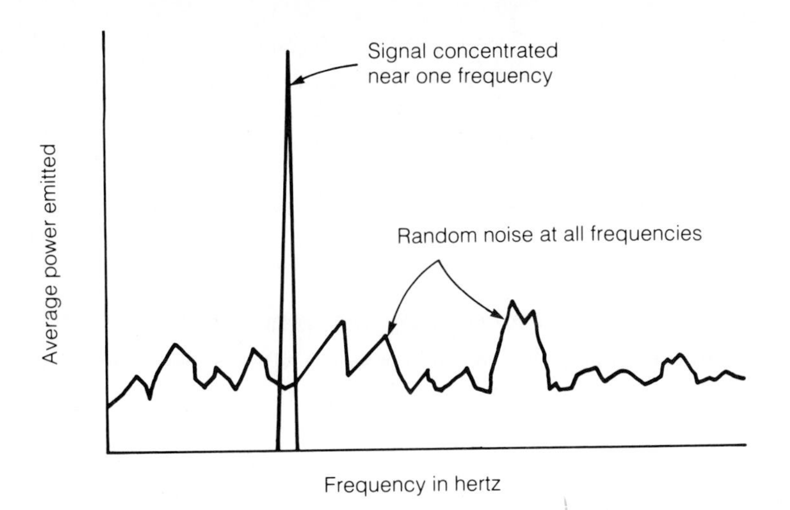

Figure 19.5 Concentrating a signal into a narrower bandpass makes it much easier for the signal to stand out against the background noise that exists at all frequencies.

would find unpredictable changes in the frequency of the arriving radio waves. These changes establish a lower bound on the frequency spread of about 0.1 hertz. Any civilization would in fact be wasting its energy if it tried to confine interstellar messages to a narrower frequency band than this, because the variations in the amount of interstellar dispersion along the path of the message would spread the radio waves in frequency to at least 0.1 hertz. Of course, a civilization with tremendous resources might decide to send as much information as possible, and to use a broad bandwidth. We would need much larger telescopes to detect messages from such a civilization than to find a narrow-bandwidth message.

The electrons in interstellar space create an additional problem: They make any radio wave change its frequency and direction slightly, and in random amounts, as the wave scatters off the electrons. These small changes in direction and frequency are analogous to the "twinkling" of stars seen on Earth, which arises from the scattering of starlight by dust particles in the Earth's atmosphere. Once again, the effect on any search for radio signals is to make it useless to confine the search too narrowly in bandwidth, because even a signal sent into space with practically zero bandwidth will randomly change its frequency by small amounts as it propagates through interstellar space.

Let us, therefore, assume that 0.1 hertz represents about the smallest frequency band that any civilization would use for interstellar communication. Radio broadcasting stations use bandwidths of 10 kilohertz (10,000 hertz) for AM and 200 kilohertz for FM. Television requires much larger bandwidths, about 6 megahertz (6 million hertz) for each channel. However, about half of the total power radiated by a television station resides in the "video carrier signal," which covers less than one hertz of bandwidth. This signal carries no information about the picture, but it would be far easier to detect (because of its narrow bandwidth) than the 6-MHz-wide signal that fills in the picture. A similar sort of signal, highly concentrated in frequency and also carrying about half of the total power, appears in FM radio broadcasting. The trick then becomes how to recognize which frequency bandpass has the message. Must we search through 3 billion frequency channels, each 0.1 hertz wide, to cover the 300 MHz of frequency between 1421 and 1721 MHz? If we use a bandpass of 1 hertz, we must still search through 300 million channels. This may be possible!

Radio engineers who are interested in the possibility of interstellar communication, within the constraints we have described, have designed receiver systems that can analyze from 16 to 32 million frequency channels simultaneously! In other words, instead of having to tune the radio dial to one station at a time to see whether radio messages are arriving from a given direction, we should soon be able to check on many *million* frequencies at once. This still leaves us short of the 300 million frequency channels we have considered, but the gain of 16 to 32 million times in efficiency is not to be sneezed at. If we can indeed analyze 16 to 32 million frequency bands at the same time, we can acquire a reasonable hope of finding a signal fairly quickly—provided that we are looking in the right direction and that our total frequency bandpass does contain the signal frequency. The problem then will become: How can we be sure we have found another civilization's message and not a source of random cosmic noise?

How Can We Recognize Another Civilization?

Our plans to search for other civilizations by radio rests on the assumption that radio photons have such universal usefulness that another civilization would be likely to use them, as we do, for sending messages. Table 19.1, which shows the chief sources of radio-wave photons that now exist on Earth, reveals that most of the radio power arises from television broadcasting, at frequencies between 40 MHz and 850 MHz, and from high-power defense radar systems that sweep the skies with intense radio pulses, at frequencies that are constantly changed for security purposes. If we managed to eavesdrop on another civilization that used photons of similar frequencies for their own communications, what could we hope to detect?

TABLE 19.1 Estimated Power Output of Various Sources of Radio Photons Now Operating at Frequencies Greater than 20 MHz*

Source	Frequency Range (MHz)	Number of Trans-mitters	Fraction of Time Transmitters Emit	*Per Individual Transmitter* Maximum Power Radiated (watts)	*Per Individual Transmitter* Effective Carrier Bandwidth (hertz)*
CB Radios	27	10,000,000	1/100	5	2
Professional mobile radios	20–500	100,000	1/10	20	1
Weather, marine, and air radars	1,000–10,000	100,000	1/100	10,000–1,000,000	1,000,000
Defense radars †	~400	5–10	1/10	200,000,000,000	1,000
FM radio stations	88–108	10,000	1	4,000	1/10
TV carrier	40–850	2,000	1	500,000	1/10

* This table, as well as Figures 19.7, 19.8, and 19.9, follows the results of a study made by W. Sullivan III, S. Brown, and C. Wetherill, published in *Science*, vol. 199, p. 377, 1978.

† We consider only the most powerful defense radars, which dominate the total power output from all such radar systems.

With sufficiently sensitive antennas, we could distinguish one television program from another, and could eventually even determine the content of these programs. This analysis would allow us to decide just how eager we would be to establish two-way contact. At greater distances and with less receiver sensitivity, we could tell that radio photons in great numbers arise from a certain location on the sky, but we could not determine what messages these photons carry. Then we might not feel sure that we had another civilization in our beam, rather than a natural source of radio noise. We could, however, go far toward resolving this issue in the following way: We might notice that the signals are far more concentrated in frequency than natural radio emission. Furthermore, the strength of some of the signals varies in time in a regular and complex manner. We would then turn to a list of the radio frequencies emitted by atoms and molecules. If the frequency under observation is not on this list, we might have found a new molecule—or another civilization.

Planets tend to rotate, as our own does, and this rotation implies that if the sources of radio broadcasting are unevenly distributed around the planet, any observer not directly above the planet's north or south poles will detect a cyclical variation in the arriving photon intensity. Figure 19.6 shows the 2200 strongest television transmitters on Earth, which cluster heavily in the United States, Europe, and Japan. These television stations send far more radio photons roughly along the Earth's surface than straight outward. This means that as the Earth rotates, an observer will see the peak emission from each station as it rises over the edge of the Earth or sets below it (Fig. 19.7). The pattern of change will repeat every day, thus suggesting to the observer the existence of some superslow pulsar (see page 145) or of a nonnatural source of radio signals.

How could an observer distinguish the Earth, which rotates once each day, from a pulsar, which rotates about once each second, with firm conviction? Any observer would obtain a tip-off that more than a slow pulsar must be causing the radio signals from the fact that the signals show Doppler shifts, caused by the Earth's rotation and its motion around the sun, that repeat periodically. An observer located in another planetary system would see the radio waves from each television channel shift back and forth by hundreds of hertz from their average frequency each day and by thousands of hertz each year. An extraterrestrial student of the solar system could thus discover not only that radio waves emerge from the sun's vicinity, but also that these radio waves vary in intensity on a daily cycle, and in

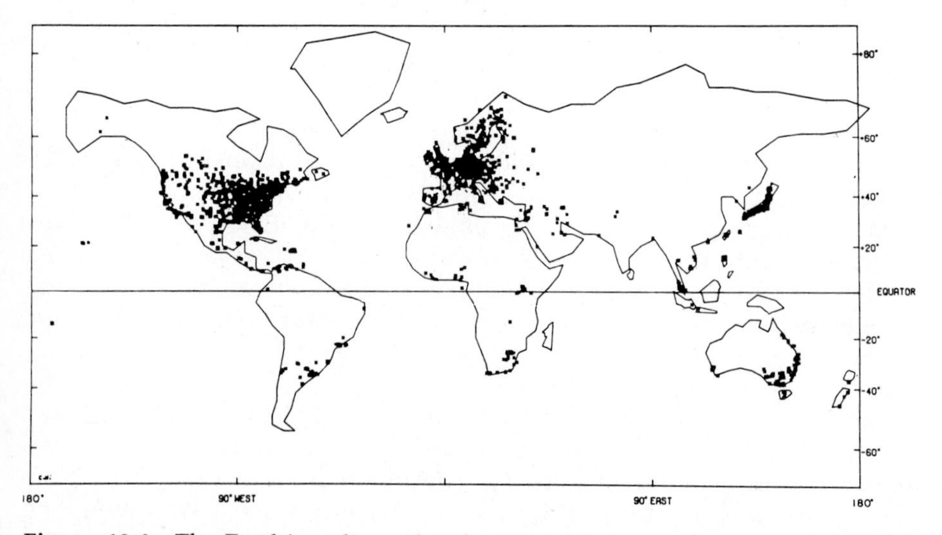

Figure 19.6 The Earth's radio and radar transmitters are concentrated in the United States, Canada, Europe, Japan, and the east coast of Australia.

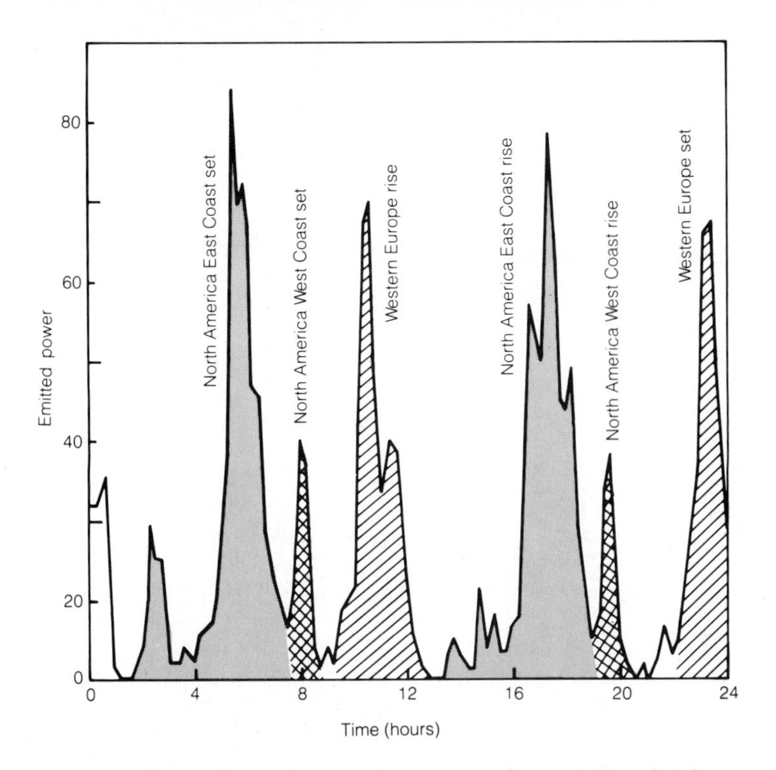

Figure 19.7 As the Earth rotates, an observer not located directly above one of its poles would detect a changing amount of radio emission. As the most powerful transmitters appeared on the horizon or set over it, they would produce especially large amounts of radio power for the observer, because their emission is concentrated in fan-shaped regions parallel to the horizon.

frequency on a yearly cycle. This observer could conclude that the sun has an object in orbit around it that circles the sun once each year, rotating once each day as it does so, broadcasting radio waves that allow these cycles to be determined.

Of course, for us to detect radio and television stations on another planet, these stations must be numerous and powerful. On Earth, the rise of radio has been sudden and recent. Figure 19.8 shows the change in the power of Earth's radio and television broadcasts since the year 1940: A thousandfold increase occurred during the first 30 years, and this increase continues, though at a lower rate than during the 1950s.

The region within 65 light years (20 parsecs) of the sun contains about 3000 stars. Only these stars, and the planets that may exist around them, have had the chance to detect the radio and television photons emitted

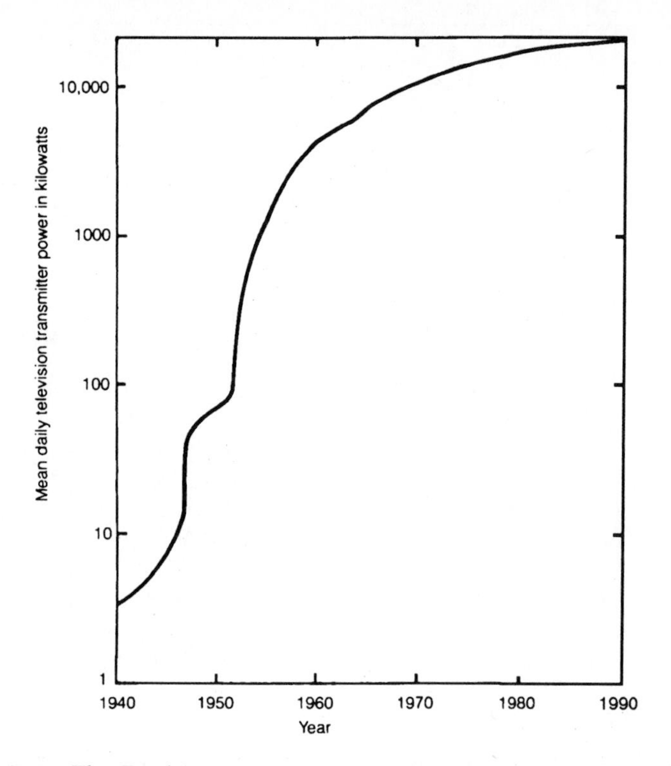

Figure 19.8 The Earth's power output in the radio region of the spectrum has increased many thousandfold since the start of the World War II in 1939.

from the Earth since 1927. These stars lie within the radio bubble that is expanding around the Earth at the speed of light. Only the closer of these stars have had the chance to hear, and faintly at that, "I Love Lucy" shows from the 1950s; years more will pass before they see the Beatles appear on the Ed Sullivan show. Another 400 years must elapse before our radio signals reach the million nearest stars, so these stars still have no news of an intelligent civilization on Earth. When we turn the picture around and think of other "young" civilizations, we see that only the 4000 nearest stars could have signaled their presence to us during the past 70 years. (Note, though, that long-lived civilizations on more distant stars could have sent messages thousands of years ago that are only arriving here now.) As we have seen, 1 million stars may not be a number large enough to include even one other civilization (see page 412). To fulfill our goal of finding other civilizations and exchanging messages with them, we must hope that a fairly large fraction of civilizations last for at least several thousand years after they have developed interstellar communication ability.

The Present State of Radio Searches for Other Civilizations

Until now, Earthbound attempts to locate other civilizations by radio have been rather modest. A few dozen efforts have been made at radio observatories, but these have constituted only a minuscule part of the observatories' activities. Quite understandably, most of the ongoing research at a radio observatory focuses on the natural events that produce radio waves. One radio telescope, a rather antiquated 26-meter antenna near Harvard, Massachusetts, has been "dedicated" to the search for signals close to the "magic" frequency of 1420 MHz or to a frequency just twice 1420 MHz (see Table 19.2). This telescope, funded by a private group called The Planetary Society, has been listening since 1985 without finding anything suggestive of another civilization.

Table 19.2 lists 13 searches that have been made, or are continuing, in the United States, Canada, and the Soviet Union. The calculations that start with the Drake equation reveal that if we hope to use "eavesdropping" techniques to find another civilization whose "leaked" radio power roughly equals our own, and if we then manage to tune to the right frequency band, we still require an antenna collecting area much larger than that of the 26-meter dish in Massachusetts.

Why do we need so large an antenna? Eavesdropping is much more difficult than trying to detect a beamed, highly directed signal arriving from a known direction, at a known frequency, and with a known frequency bandpass. The beamed signal puts much more radio power within its narrow bandpass and its concentrated beam than a signal sent in all directions over a much wider bandpass. The Arecibo radio telescope can now communicate with a similar telescope anywhere in our galaxy, if these three key items—direction, frequency, and bandpass—are known. The Arecibo telescope represents the largest antenna we have built so far, with a diameter equal to the length of three football fields. Its construction was relatively inexpensive because of a convenient topographic configuration (Figs. 19.9 and 19.10 and Color Plate 19).

The antenna problem is a formidable one, but the greatest difficulty in searching for other civilizations lies not in the antenna size but in the search for the right *frequency*. Unless a civilization uses a large fraction of the total radio-frequency range for its communications, and leaks into space about the same amount of power at all radio frequencies, our best hopes for finding a civilization by eavesdropping rest with hitting the frequency band with the most radio power. But as we saw on page 464, literally billions of possible frequencies lie within, say, the "water hole" between 1420 and 1721 MHz.

Again we meet the requirement for a "multi-frequency analyzer." Despite years of minimal or zero funding, in 1988 the scientists at NASA

TABLE 19-2 Some Searches That Have Been Made for Extraterrestrial Radio Signals from Other Civilizations

Year	Scientific Investigator	Antenna Diameter (meters)	Frequency Observed (MHz)	Frequency Resolution (kHz)	Frequency Range (MHz)
1960	Frank Drake	26	1420	0.1	0.4
1968–1982	V.S. Troitskii	14	1000 1800 2500 10,000	0.013	2.2
1970–	V.S. Troitskii		1863 927 600		
1972	G. Verschuur	43 91	1420 1420	7.0 0.5	20.0 0.6
1972–1976	B. Zuckerman & P. Palmer	91	1420	4.0	
1976–1985	S.Bowyer	26	Variable	2.5	20.0
1973–1985	R. Dixon et al.	53	1420	10.0	0.4
1972–1976	A. Bridle & P. Feldman	46	22,235	30.0	
1975–	Frank Drake & Carl Sagan	300	1420 1653 2380	1.0 1.0 1.0	
1977	D. Black, J. Cuzzi T. Clark, & J. Tarter	91	1665 1667	0.005 0.005	1.3 1.3
1977	M. Stull & F. Drake	300	1665	0.0005	4
1978	P. Horowitz	300	1420	0.000015	0.001
1985–present	P. Horowitz	26	1420 1665 2841	0.00005	

involved in SETI (the search for extraterrestrial intelligence) completed a six-year feasibility study of a Microwave Observing Project (MOP) to search for extraterrestrial life. The heart of the MOP will be an advanced receiver

Figure 19.9 The giant radio telescope near Arecibo, Puerto Rico, nestles in a natural limestone bowl where it reflects radio waves to the "feeds" suspended 200 meters above. The antenna is 300 meters in diameter, large enough to hold all the beer drunk on Earth in a given year.

system that can *simultaneously survey from 16 to 32 million different radio frequencies* (Fig. 19.11). Using this receiver system, the NASA scientists hope to begin listening for signals from extraterrestrial civilizations on October 12, 1992—the 500th anniversary of Columbus's arrival in the New World. Their initial efforts will be modest, but by 1996, they hope to have their full system in operation, with a good chance of finding our first neighbor civilization by the dawn of the next millennium.

Armed with this new receiver, the NASA effort to search for extraterrestrial intelligence will not require any new radio antennas. Instead, NASA's SETI project will use existing radio antennas around the world, including the 300-meter Arecibo dish. Using the new type of receiver, NASA will conduct two separate searches, based on two different approaches to answering the question, Where should we look to find intelligent life in the Milky Way?

Figure 19.10 Workers who adjust the panels of the Arecibo reflecting telescope must wear special shoes to avoid damaging the surface, which is maintained in position to a precision better than one centimeter.

The Targeted Search

One effort will concentrate on a "Targeted Search," in which the SETI scientists will point their antennas at 800 nearby stars for 5 to 15 minutes per star. These "most likely 800" among the 300 billion stars in the Milky Way must satisfy the following conditions:

1. They all lie within 25 parsecs (80 light years) of the sun;
2. They are all Population I stars (that is, stars that contain a relatively high fraction of the "heavy elements" thought to be essential to life; see page 140);
3. They are all either single stars or else are members of binary- or multiple-star systems whose member stars are sufficiently widely separated for stable planetary orbits to exist; and
4. They are all main-sequence stars of spectral type F, G, or K; that is, they are likely to have relatively large habitable zones but do not burn themselves out too quickly (less than a billion years) for life to evolve to the stage of "intelligence."

Figure 19.11 A prototype of the "multichannel" receiver that NASA hopes to employ in its SETI search can analyze 16 to 32 million different radio frequencies simultaneously.

Each star will be studied in the frequency range between 1000 and 3000 MHz, thus including the "water hole" (see Fig. 19.2). The Targeted Search of these 800 stars will require approximately six years to complete.

The Sky Survey

Because intelligent civilizations (as we have defined them) might be entirely absent from the 800 "most likely" nearby stars, the SETI scientists also plan to make a survey of the entire sky, hoping to find more distant civilizations that emit far greater amounts of radio power than we do. The "Sky Survey" will also require about seven years of observing time, using radio antennas at Goldstone, California, and Tidbinbilla, Australia, among others. This search will cover the frequency range from 1000 to 10,000 MHz. Because the beam of radio photons reaching one of the giant antennas comes from only a tiny fraction of the sky, each separate direction studied in the Sky Survey will be observed for approximately 1 second. The Sky Survey will test the possibility that our galaxy contains a few civilizations that radiate immense amounts of radio power, so much that they outshine much closer, but relatively far weaker, civilizations.

Between them, the Targeted Search and the Sky Survey represent NASA's best judgment as to how to make a SETI effort that will yield useful results. As a byproduct of this effort, astronomers will learn a good deal more about the 800 stars in the Targeted Search, and about the new radio sources that the Sky Survey is likely to reveal. But the fundamental reason for undertaking the SETI effort lies in the human desire to answer the fundamental questions: *Are we alone? If we are not, how can we find our neighbors and communicate with them?*

What Messages Could We Send or Receive?

Suppose that before this century ends, SETI scientists do find another civilization by eavesdropping on its own radio communications. In that event, astronomers anticipate relatively little difficulty in recognizing it as a civilization. Although various sources of "false alarms" may appear, the most effective way to distinguish a natural from an artificial signal arriving from a particular location is that artificial signals are likely to be concentrated into a narrow frequency band (see page 463). More difficult than recognizing the existence of an artificial radio signal will be the *decoding* of signals identified as "intelligent." This task should provide challenging work for teams of anthropologists (perhaps renamed "xenanthropologists"—those who study "strange persons"), who will seek to reconstruct the language of communication used in a faraway planetary system.

Though decoding the message may take some time, the detection of such a signal will immediately reveal the direction and the preferred radio frequencies of a civilization whose existence thus becomes an established fact. What should we do next? Should we send a message to that civilization? Should we wait cautiously until we decipher their own communications and analyze them as a society? Conversely, should we expect that other civilizations are attempting to signal their existence to the rest of the galaxy, having detected some of these signals themselves?

Until the time when we find another civilization, the answers to these questions must be uncertain. Since we have allowed stray radio photons to leak continuously into space without worrying about the result, why should we hesitate to send a deliberate message to other civilizations? And if this is our view, should we not expect other civilizations to share it? On the other hand, why should we or any other civilization encourage visitors, or even return messages that reach us? Human opinions vary widely on this subject, so we ought to bear in mind that more discussions will be needed to resolve the issue. Meanwhile, you may be surprised to learn that, acting entirely on their own, some astronomers have already sent a few messages toward our unknown cosmic neighbors.

Figure 19.12 shows the most important of these messages, sent on November 16, 1974, from the Arecibo Observatory (Color Plate 19), using the great dish as a radio transmitter. This message, beamed in the direction of the globular cluster M13 in Hercules (Fig. 3.13), contained 1679 bits of information. The bits appear in Fig. 19.12 as 0's or 1's, but in the actual message each bit consisted of either an "on" pulse, broadcast for a tenth of a second at one of two frequencies (separated by 75 MHz) near 2400 MHz or

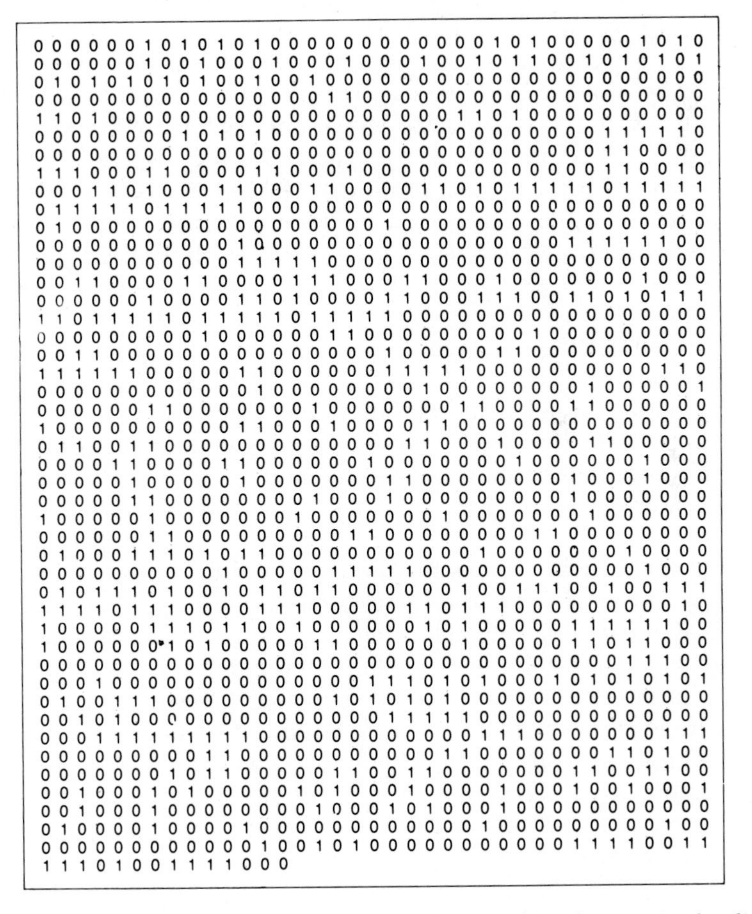

Figure 19.12 The message sent in 1974 from the Arecibo telescope in the direction of the globular cluster M13 consists of 1679 bits of information, either "on" or "off," shown here as 0's and 1's.

the "off" interval separating two of the "on" pulses. The total band of frequencies covered in the message transmission was 10 hertz, 100 times greater than the minimum possible bandpass, but only 1/100 the bandpass of an AM radio station. Since the star cluster is 7700 parsecs from us, the roughly two decades of travel time since transmission have not meant much: The message will reach M13 in about 25,000 years, only then ripe for interpretation and a possible return message that may reach us in 52,000 A.D.

What could we expect another civilization to make of such a message? What would we think if we were the ones receiving it? We would count on another civilization to recognize that it contains just 1679 bits. Then someone might ask, why 1679 bits? Mathematicians would notice that this number equals the product of 23 and 73, and of no other numbers; 23 and 73 are prime numbers, not divisible by any others save themselves and unity, no matter what counting system we use. This fact suggests, at least to us, that we might profit from arranging the bits of information in 73 columns of 23 bits each, or in 23 columns of 73 bits each. The first choice gives no discernible pattern, but the second produces the interesting array pictured in Fig. 19.13, where we have replaced the zeroes by white squares and the ones by black squares to emphasize their contrast. The pattern seems clearly nonrandom, and some study should reveal what the astronomers were trying to say.

The top part of the message gives a lesson about the number system that the astronomers use: It shows the numbers 1 through 10 in binary notation, along with a "number marker" that tells when a symbol represents a number. Since the binary number system forms just about the simplest way to write numbers (although, significantly enough, most of the human race has never heard of it!), this part of the message should be recognizable as a starting point.

Next in the message comes the sequence of numbers 1, 6, 7, 8, and 15. Since this sequence seems odd on purely mathematical grounds, it must be trying to say something—namely, that we pay particular attention to the first, sixth, seventh, eighth, and fifteenth kinds of atoms, listed in order of the atoms' atomic number (number of protons). These atoms are, respectively, hydrogen, carbon, nitrogen, oxygen, and phosphorus. A clever recipient of the message might be able to deduce that these five elements have key importance to us. (Notice that not everyone would assign phosphorus an equal rank with hydrogen, carbon, nitrogen, and oxygen; as we discussed in Chapter 7, the four most important elements seem far more important than the fifth, phosphorus.) Below these numbers we find 12 groups, each of five numbers. Each of these groups gives the chemical formula of molecules important to life, specifying particular elements (hydrogen, carbon, nitrogen, oxygen, and phosphorus) by using the same ordering

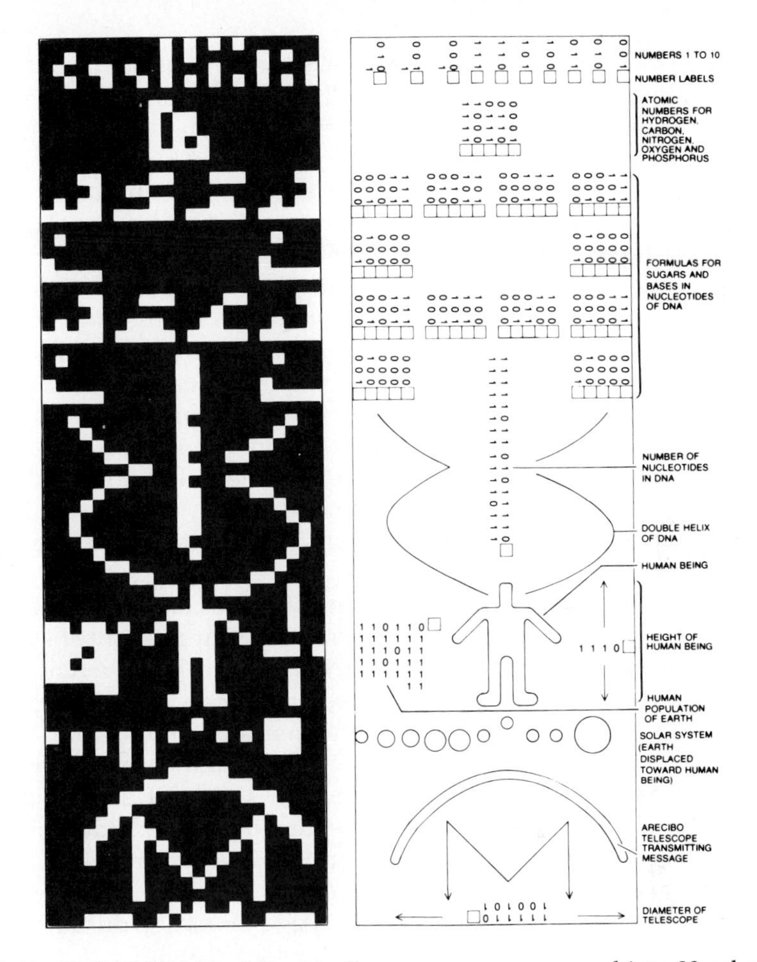

Figure 19.13 If the 1679 bits of the Arecibo message are arranged into 23 columns of 73 rows each, and if the on and off bits are given different colors, a picture emerges that is loaded with information—for those who can decipher it.

of elements as the single group expressed earlier. These molecules include DNA's four key bases—thymine, guanine, adenine, and cytosine—that we discussed on page 163, as well as the phosphate group (PO_4) and the sugar molecule deoxyribose.

Farther down the diagram, we encounter the chemical structure of DNA, the key molecule of life (see page 165). The double helix pattern appears, wound around the number of pairs of bases (about 4 billion) that exist in a single human chromosome, the basic carrier of genetic information. The twin helix ends at a crude picture of a human being, which, at least to the authors of the message, indicates the connection between DNA and the evolution of the intelligent beings who sent the message. To the human's right we see a line extending from head to foot, together with the number 14 (in binary notation), so the human must be 14 units tall. The only unit we share in common with the recipient of our message will be the wavelength of the photons that carry the news, so humans must be 14 wavelengths tall, or 14×12.6 cm = 176 cm, or 5 feet 10 inches. To the left of the human figure, the number 4 billion (actually a bit different, because binary notation does not round off in the same way that the decimal system does) gives the human population at the time that the message was sent.

Below the human, we find a sketch of the solar system, with the sun at the right and nine planets to its left. The third planet (Earth) stands out of line, which shows that the Earth is something special; its displacement toward the human figure establishes the connection that humans live on Earth. Finally, below the solar system, we see a drawing of a telescope, whose function can be clearly seen from the fact that it "focuses" photons to a central point. The last line of information gives the size of the telescope, 2430 wavelengths or 306 meters.

Not bad for 1679 bits of information! We have conveyed (at least to ourselves) the idea that those who sent the message have a number system, consider certain elements to be of primary importance, judge certain molecules made from these elements to be also of key importance, attach great meaning to a certain spiral form that springs from the central figure, whose size and numbers appear on either side, live on the third of nine planets around a star, and build telescopes 300 meters across. Even if the receiving party could not understand all of this message, certain facts should emerge clearly: We are here, primed with information and eager to talk about it. The intensity of photons upon arrival would make their point of origin appear as the brightest "star" in our galaxy at the message frequency during the time of reception. This would indicate a high technological ability.

This message, traveling at the speed of light, took only an hour to travel farther from the sun than *Pioneer 10,* which had been on its journey out of the solar system for more than two years. Furthermore, the cost of sending this message was about 1/10,000 the cost of *Pioneer 10.* The message has now traveled a distance greater than the distance to dozens of nearby stars. Recall that the message is not directed toward any of these nearby stars—perhaps a bit of caution on the part of the astronomers who sent it—though some relatively nearby stars happen to lie between ourselves and M13. If a civilization that orbits some star between ourselves and M13 or in the globular star cluster itself should receive it, we may eventually have a reply. Meanwhile, we may pause to wonder how many of these "Hello—are you there?" sorts of messages may be sailing through the galaxy now, and the impact upon human civilization if another world's 1679 bits appeared in the daily newspaper. Would we consider it just another crossword puzzle, or would we stop to look at that strange figure in the middle of the diagram?

SUMMARY

Once we judge that photon messages are the best way for civilizations to find and to communicate with one another, we must ask certain important questions, and determine their correct answers, before we can hope to enter into communication with our closest neighbors. First of all, we must consider the most likely photon frequencies to be used for messages. The generally favored frequencies lie in the radio domain of the spectrum, because most other spectral regions suffer from propagation difficulties, and radio photons are cheaper to produce than visible-light photons. Just which radio frequency will be used for deliberate searches remains a mystery, but it may be resolved by the fact that the 1420-MHz frequency of the photons emitted by hydrogen atoms represents the single most important frequency in the universe, and should therefore be known to any other civilization that studies the cosmos. Hence, many astronomers incline to the view that interstellar messages travel at frequencies close to 1420 MHz (or perhaps at an even multiple, or a simple fraction, of this basic frequency). If water is as important to another civilization as it is to ours, then the "water hole" of frequencies between 1420 MHz (H) and 1721 MHz (OH) may be the frequency range to study.

We must also answer the question of the bandwidth of frequencies over which the message signal spreads, if we hope to conduct our search for other civilizations' messages with high efficiency. Furthermore, we must decide on a total range in frequency over which we are willing to search—and hope that this includes the frequencies at which messages are actually traveling.

If we resemble other civilizations in our judgment and knowledge, then we stand a good chance of finding messages sent by the nearest civilizations if we search the most likely nearby stars, at the most likely frequencies, using the most likely frequency bandpass. However, it might be an equally good (or better) strategy to search throughout the Milky Way for a rare but extremely bright civilization. Human beings have now sent messages into space with the deliberate, though hardly unanimous, attempt to announce our presence to any civilization that intercepts them. Such messages can be reassembled into a crude picture by any civilization as intelligent as ours. A few thousand bits of information suffice to tell another civilization the basic chemistry, location, population, and physical size and shape of the senders.

Even if other civilizations tend not to broadcast messages deliberately designed to reveal their presence, they may still, like ourselves, produce photons for their own local communications, some of which inevitably leak into space. Although the detection of such leaked photons presents a far more difficult task than the discovery and analysis of a signal designed for interplanetary communication, we could now construct antennas and receivers capable of eavesdropping on a civilization like our own anywhere among the thousands of nearest stars. Though we might not determine the actual content of the radio programs or whatever messages are being sent, the simple discovery of another civilization by eavesdropping techniques would have profound consequences for our own attitude toward life on Earth.

QUESTIONS

1. What factors—in particular, those that enter the Drake equation—make it more likely that the Milky Way might contain a network of intercommunicating civilizations? What factors make it less likely? What difficulties exist in attempting to answer this question on Earth today?

2. Which stars seem to be the best candidates to have planets on which civilizations have developed? Why?

3. What leads astronomers to judge radio waves superior to other types of photons for interstellar communication?

4. What do radio astronomers mean by the "water hole"? Why do some astronomers think that the water hole may include interstellar radio transmissions?

5. Why does finding other civilizations by eavesdropping require more sensitive detection devices than communication with a civilization whose existence is known?

6. How could we hope to distinguish a planet that uses radio waves in much the same way that we do from a natural source of radio emission, such as a pulsar? Does such a search strategy make sense to you? Why or why not?

7. In its search for extraterrestrial signals, NASA plans to make both a Sky Survey and a Targeted Search, the latter directed at the nearest sunlike stars. Why does NASA's strategy include both of these programs?

8. Light waves and radio waves travel at 300,000 kilometers per second. How long did it take the radio message sent from Arecibo in 1974 to overtake the *Pioneer* spacecraft, which had then been traveling for two years at a speed of 10 kilometers per second? What does this tell us about the relative merits of sending messages by spacecraft and by radio?

9. Suppose that we received an apparent message from another civilization that consisted of "on" and "off" pulses that repeated after a total of 2117 "ons" and "offs." How could we begin to interpret this message? Is there more than one way to make such an interpretation?

FURTHER READING

Billingham, John, ed. *Life in the Universe.* Cambridge, MA: MIT Press, 1982.

Morrison, Philip, John Billingham, and John Wolfe, eds. *SETI: The Search for Extraterrestrial Intelligence.* Washington: U.S. Government Printing Office, 1978.

Regis, Edward, ed. *Extraterrestrials: Science and Alien Intelligence.* London and New York: Cambridge University Press, 1985.

Rood, Robert, and James Trefil. *Are We Alone?* New York: Scribner, 1981.

Wetherill, Chris, and Woodruff Sullivan III. "Eavesdropping on the Earth," *Mercury* (March/April 1979).

Science Fiction

Gunn, James. *The Listeners.* New York: Scribner, 1972.

Hoyle, Fred, and John Elliott. *A for Andromeda.* New York: Harper & Row, 1962.

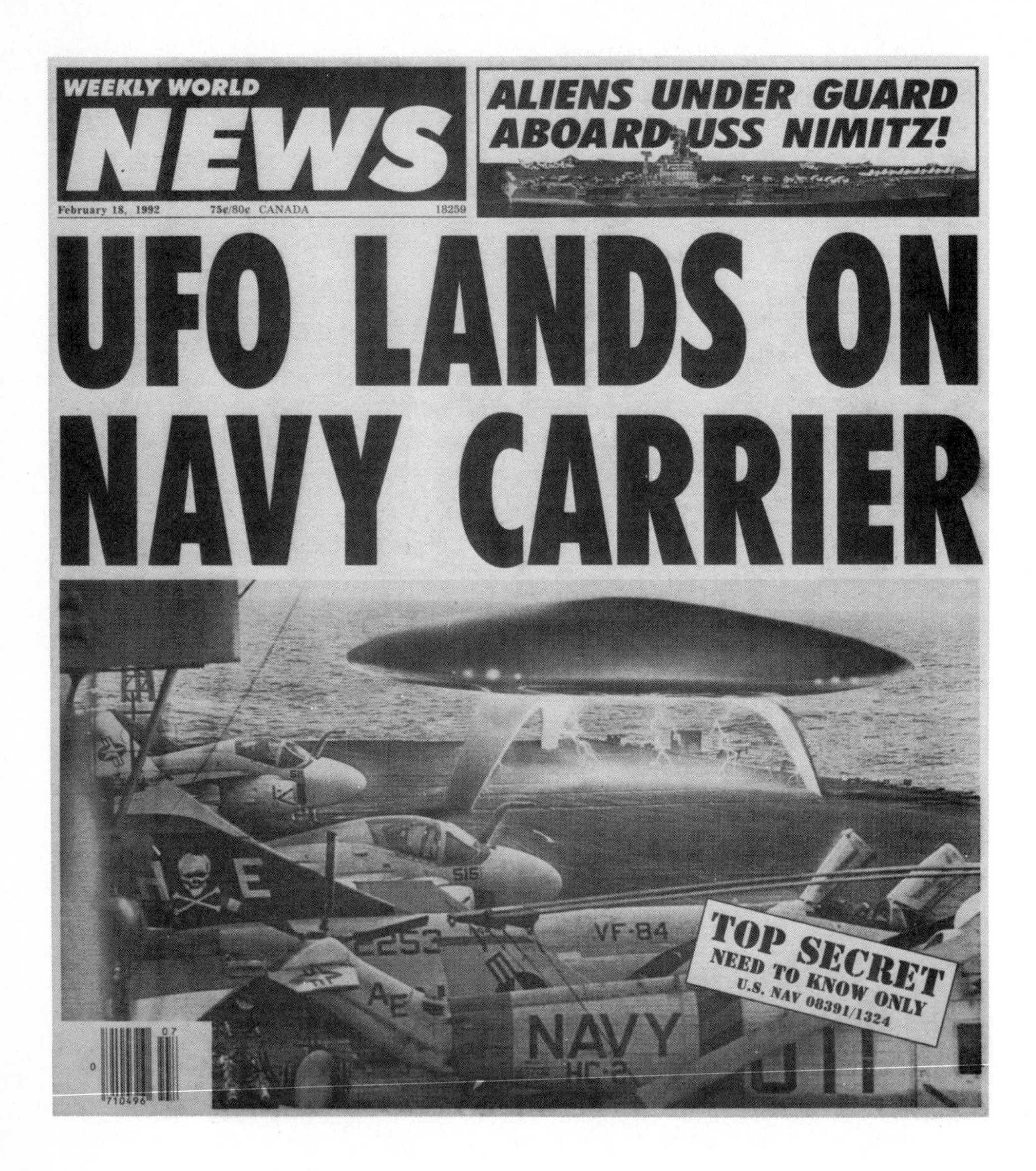

WEEKLY WORLD

NEWS

February 18, 1992 75¢/80¢ CANADA 18259

ALIENS UNDER GUARD ABOARD USS NIMITZ!

UFO LANDS ON NAVY CARRIER

TOP SECRET
NEED TO KNOW ONLY
U.S. NAV 08391/1324

20

Extraterrestrial Visitors to Earth?

"SPACE ALIEN BABY FOUND ON MOUNT EVEREST!" screams the headline in the tabloid newspaper at the checkout stand. Could it be so? Could extraterrestrial visitors in fact be in touch with our leaders? Would this explain much that is apparently unfathomable about the world situation? And most important of all questions in science: *How can we hope to decide whether or not it is true?*

The previous three chapters have provided a scientific discussion of how to find other civilizations and how to communicate with them. Would it not be marvelous to discover that such a plodding and rather dry analysis is unnecessary? In considering the possibilities of communicating with life elsewhere, human nature tends to insist not on science but on the simplest of all responses: Why not let *them* find *us*? Why spend so much time and energy to send messages, or to listen for stray signals, if we could merely open our eyes and discover extraterrestrial visitors on Earth?

UFOs (Unidentified Flying Objects) and the beings that might travel in them rank among the most provocative and enduring symbols of our urge to end our cosmic loneliness. To many, UFOs offer the possibility of establishing direct contact with creatures far more capable and intelligent than ourselves. A more exciting prospect can scarcely be imagined—if only UFOs do indeed carry extraterrestrial visitors! The reports of visitors from other worlds, no matter how erroneous they may be, continue to signal an abiding belief that we are special, definitely worthy of an inspection tour, if not an extended holiday. No matter what we conclude about the evidence for extraterrestrial visitors, we can admire the inbred human tendency that leads most of us to assume not only that "we are not alone," but also that someone else ought to do the hard work of traveling.

What Evidence Do We Seek?

When we examine the evidence for extraterrestrial visits to our planet, we would like to begin with what would surely be the best evidence of all:

an actual visitor, or group of visitors, visible to crowds of people and ready for shaking hands, polite conversation, and dancing. No such visitors have appeared on Earth recently. How do we know? Because such a visit would receive immediate, worldwide attention. (We shall later meet the paranoid approach to this issue.) The next best evidence for extraterrestrial visitation would come from some device of clearly nonterrestrial origin, such as an antigravity machine or an interstellar spacecraft. Again, no such artifact exists, despite some charlatans' claims to the contrary. (You might think it would be difficult to recognize such an artifact. But consider that thousands of meteorites have been spotted as unusual by farmers and other people with little experience at recognizing them.) Third best would be indisputable photographic or other "hard" evidence demonstrating that spacecraft capable of interstellar flight have passed close by the Earth's surface. Since astronomers and geologists by now have taken millions upon millions of photographs to survey the sky and the Earth, the absence of images of spacecraft from any of these pictures argues against extraterrestrial visitors, at least against those who might travel in a vehicle that can be photographed.

Fourth best in the list of possible evidence comes the set of UFO reports from humans who claim to have seen extraterrestrial visitors or their spacecraft. These reports are inevitably subject to human error, especially since they are often made under emotionally stressful conditions. The potential for error increases with the strangeness and suddenness of the observation, as is known to any police officer, judge, or jury dealing with eyewitness accounts of traffic accidents, assaults, and murders. Furthermore, errors in recollection increase as time passes.

When we consider the evidence now available from sightings of UFOs, we find that we lack the first, second, and third best types of evidence—appearance on television or before large groups, physical artifacts from another civilization, and reliable photographs or similar permanent records. Instead, we must deal with eyewitness testimony, with all its contradictions and personal biases.

To gain an appreciation of the difficulties here, consider the confusion that followed the assassination of President John F. Kennedy. There were dozens of reliable witnesses, sound recordings, a movie, and an intensive and thorough investigation involving many experts. Nevertheless, there are still many people who doubt that Lee Harvey Oswald was indeed the lone assassin. Now imagine yourself trying to interpret an interview with an excited rural policeman who saw something in the sky he didn't understand: Was it a spacecraft of unknown form and dimension? The planet Venus at its brightest? A wayward bat?

To sift through all the varied kinds of UFO reports; to interview known and potential witnesses; to attempt to evaluate all possible natural explanations (such as planets, clouds, and birds) and all possible human artifacts (sunlight reflected from weather balloons and airplanes, artificial satellites, unusual aircraft lights) represents an immense task, often one that may reveal the dark side of human nature, as in those cases where UFO reports turn out to be the result of pranksters. Hence it is not surprising that few scientists have devoted much effort to a study that they see mainly as the realm of sociologists and psychologists. Indeed, most scientists are poorly trained to deal with problems in which the observer plays a key role in the phenomenon. One of the best-known, most thorough investigators of UFO sightings is Philip Klass, a senior editor of the journal *Aviation Week & Space Technology*. In our discussion of UFOs, we have drawn heavily on the evidence that Klass has assembled to interpret some famous UFO reports.

Recurrent Themes in UFO Sightings

Although people have seen puzzling things in the sky for millennia, the UFOs commonly discussed are a relatively modern phenomenon. They were first reported soon after World War II ended in 1945. This may not be a mere coincidence in time. The rapid growth in technological capabilities during the six years of that conflict, coupled with the constant scanning of the skies for enemy aircraft, gave many people a sense of awe and danger in the skies, along with a feeling that anything might be possible. If radar, television, jet aircraft, rockets, and the atomic bomb could all appear in less than a decade, why not interstellar spacecraft? Furthermore, the troubled state of the world may have inclined many people, then as now, to hope that superior extraterrestrial beings might intervene to help solve the world's problems.

The first UFO report came from a private pilot named Kenneth Arnold, who said that he had seen mysterious "flying disks" near Mount Rainier in Washington. Arnold's 1947 sighting, which may have been a mirage, led to the coinage of the term "flying saucers," which has hung over UFO reports ever since. Arnold's story was followed by a burst of "flying saucer" sightings. Six months after Arnold's report, Air Force Captain Thomas Mantell crashed his jet airplane while chasing a giant UFO that was probably a new type of military balloon. Mantell's fatal accident was almost surely the result of his failure to use an oxygen mask while climbing toward the unidentified object; he "blacked out" from oxygen deprivation during the chase. But Mantell's death produced the assertion that flying saucers had the ability to destroy those who pursued them.

A year after the "flying disks" near Mount Ranier, on July 24, 1948, an Eastern Airlines DC-3 airliner was flying near Montgomery, Alabama, in the early hours before dawn. At an altitude of 5000 feet, the captain and copilot had a nearly full moon to light the night sky, which was clear except for a few broken clouds. Without warning, the crew spied what they thought was a giant jet aircraft coming from the east, which passed by at a distance estimated to be 700 feet, at a speed of 500 to 700 miles per hour. Both the captain and copilot said later that the object had two rows of windows that appeared to be lit from within; the captain said that "You could see right through the windows and out the other side." Both men guessed that the object was about 100 feet long, and 25 to 30 feet across (Fig. 20.1). The object passed by for only 10 seconds, long enough for a good look but not long enough to arouse other potential witnesses. One passenger aboard the plane reported seeing a similar streak of light, while one other aircraft in the same vicinity saw a kind of jet or rocket trail at the same time.

Both the pilot and copilot saw rows of windows on the UFO, and both disagreed with the suggestion that they might have seen a meteor. Furthermore, the pilot said that the UFO had "pulled up with a tremendous burst of flame out

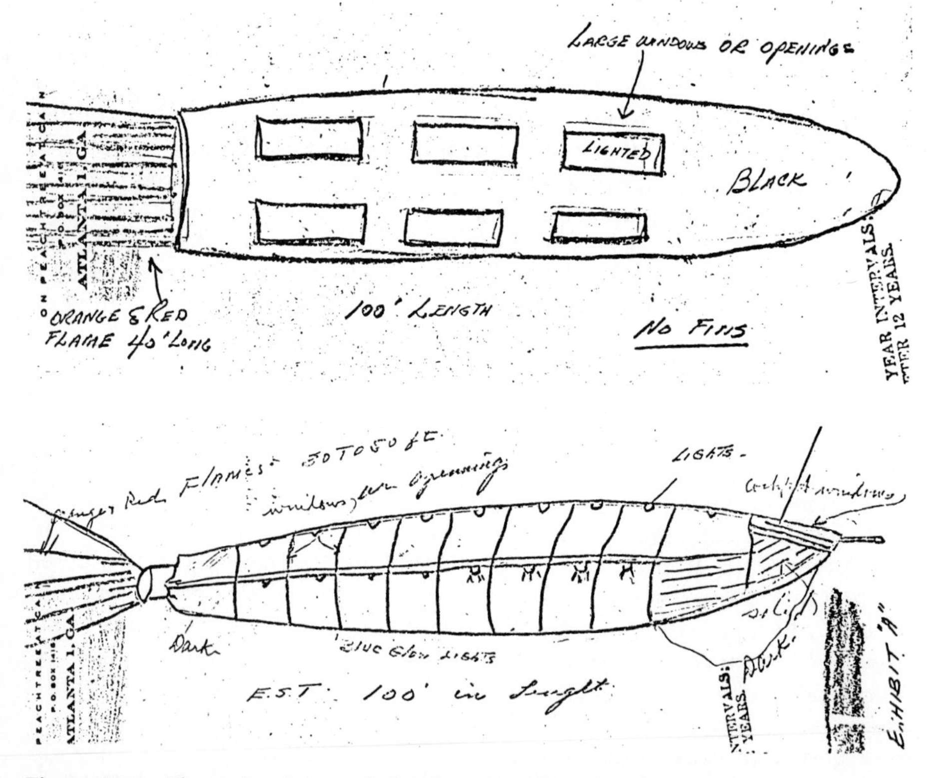

Figure 20.1 These sketches made by the pilot (above) and copilot (below) of a DC-3 aircraft on July 24, 1948, show a vehicle with a line of windows or similar regular openings. Note how differently the two observers saw this object.

of its rear and zoomed up into the clouds." The Air Force spent a considerable effort in investigating possible explanations of this incident. One of their consultants, Dr. J. Allen Hynek, wondered whether "the immediate trail of a bright meteor could produce the subjective impression of a ship with lighted windows." But the investigators felt that the pilot and copilot could not have been so grossly wrong as to see a meteor as a craft with double rows of windows.

Nevertheless, exactly this seems to have happened 20 years later, in an incident that was reported by observers in several states and provides a rare "experiment" that we can use to test a UFO report. On the night of March 3, 1968, three reliable witnesses in Tennessee saw the same kind of phenomenon from the ground: A bright light neared the trio, spouted an orange-colored flame from behind, and passed overhead at an altitude that the three said was 1000 feet or less. In complete silence, the UFO floated by, "like a fat cigar," said one of the witnesses, with at least 10 large, square windows that seemed lit from inside (Fig. 20.2). The three observers talked

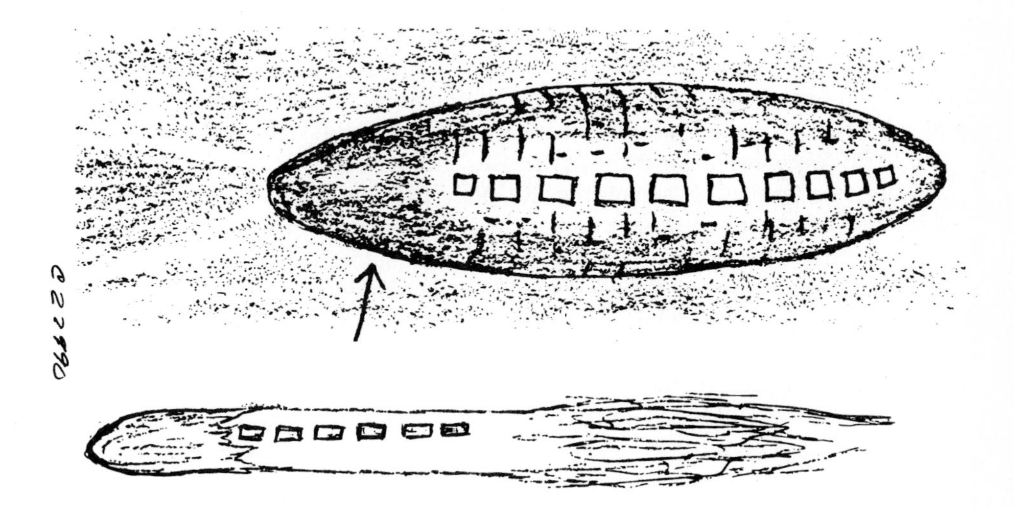

Figure 20.2 Twenty years after the 1948 sighting, a Soviet rocket consumed by friction in the Earth's atmosphere over the Midwest induced two observers, one in Tennessee (top) and one in Indiana (bottom), to sketch a craft with a line of lighted windows.

about what they had seen and agreed that the object could only be either a secret military aircraft or an extraterrestrial spacecraft.

Now it happens that the same UFO was seen by six people in Indiana, 200 miles to the north, who also saw a cigar-shaped object with many brightly lit windows (see again Fig. 20.2). This object, the observers said, was 150 to 200 feet long, and passed at treetop level without a sound. Other entirely trustworthy witnesses, far away in Ohio, saw three similar craft, also passing in silence.

It turned out that on the night before these sightings, the Soviet Union had launched a set of rocket boosters to help send a spacecraft into orbit. On the following night, one of these rockets fell back to Earth, burning up in our atmosphere as an artificial meteor. American radars had kept track of this object from the time of its launching, and no doubt exists that the rocket booster was passing several dozen miles above the observers at the time that they reported seeing one or more craft with flaming rear exhaust and illuminated windows soaring silently by at low altitudes above their heads.

We may conclude from the 1968 incident that ordinary, reliable people may not record what they see with good accuracy, because human minds supply details that "ought" to belong, in order that we may view the world in a way that seems logical. Our minds rely on a background of previous experience to interpret what we meet: If we see strange glowing objects moving together in the sky, we may tend to see "windows"; if we hear no noise from what looks like an airplane fuselage, it must be a spacecraft.

What about the 1948 sighting by the Eastern Airlines pilots? No Soviet booster rockets existed then! Yet the drawings by the two sets of witnesses are definitely similar, and the witnesses' interpretations of what they saw are nearly identical. It seems highly likely that the pilots were reacting to the sight of a meteor, rather than a rocket booster, breaking apart in the Earth's atmosphere during its descent to the surface. This explanation of the 1948 UFO report grows even more likely from the fact that July 24, the date of the sighting, marks the peak of the Delta Aquarid meteor shower. In both of these instances, separated by 20 years, human eyes saw something startling and unusual, and human brains, seeking to fit the information into known patterns, added details that were actually missing from the "raw" observations. The aircraft pilots assigned the meteor a distance of several hundred feet rather than several dozen miles, and substituted illuminated windows for a chain of glowing pieces as the meteor broke up in the atmosphere (Fig. 20.3).

Without photographs, we could never be sure of the assertion that observers saw a meteor and deduced an extraterrestrial spacecraft. Even with photographs, arguments will surely continue (Fig. 20.4), but we at least have a piece of physical evidence rather than dim human recollections.

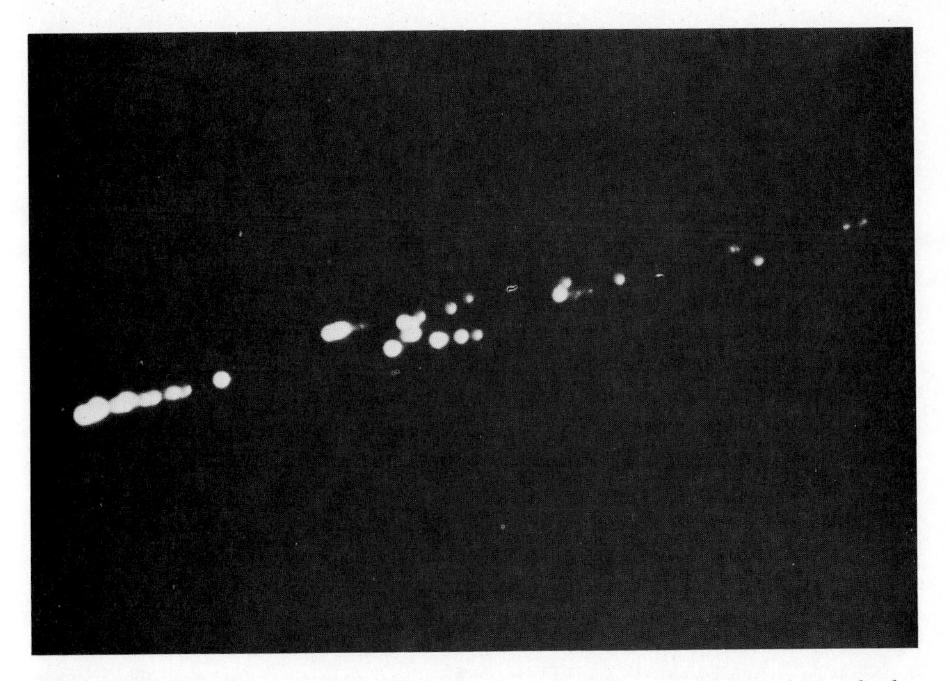

Figure 20.3 A meteoroid breaking apart during its passage through the atmosphere produces a string of glowing pieces that the human eye and brain may tend to align and to space at regular intervals.

Figure 20.4 These photographs have been cited as the most convincing evidence for UFOs as extraterrestrial spacecraft. They were taken near McMinnville, Oregon, in May 1950 by the husband of a woman who had seen UFOs on several previous occasions, "but [she said] no one would believe me."

The Lubbock Lights

One hot night in August 1951, three members of the faculty of Texas Tech. University were looking at the sky from one of their backyards for the well-known Perseid meteor shower. All three, professors of science or engineering, were trying to count meteors when suddenly they saw 15 to 20 faint, yellowish white lights passing from north to south. An hour later, they saw another group of lights, moving in a semicircular formation. Near midnight, more than two hours after the first sighting, a third group passed overhead.

The three scientists estimated the altitude of the lights to be 50,000 feet, and their speed to be 5 miles per second. They telephoned the Lubbock *Evening Avalanche,* and their sighting received nationwide publicity as the "Lubbock Lights." But further investigation by the professors themselves solved the Lubbock mystery.

What were these mysterious lights that seemed to travel more rapidly than any airplane? They were birds! Armed with binoculars, the men discovered that they had been looking at migrating plover, whose undersides reflected the light from the newly installed mercury vapor street lamps of Lubbock. Knowing the size of a plover, the professors realized that the birds were much closer than the observers had thought, and therefore were moving much less rapidly. Similar sightings have been (mistakenly) reported by astronomers, who certainly should know better. We should therefore hardly be surprised if untrained observers with a UFO interpretation already "programmed" into their thinking can be equally confused. We simply cannot tell the altitude, and thus the speed, of an unknown object. Whenever you hear someone say, "It was a mile away," or, "It was a hundred yards away," of something seen in the sky, you are entitled to ask,"How do you know?" You will usually find that human intuition, rather than any scientific means of measurement, has provided the answer.

Venus in Georgia

During the fall of 1967, police officers in more than 10 locations in Georgia reported sighting UFOs. The first such case, typical of the remainder, began in the early hours of October 20, when an officer saw a "bright-red, football-shaped light" near the horizon, chased it with his partner in their patrol car for eight miles, and lost sight of it only to find, as they returned to Milledgeville, that the UFO had again caught up with them! The UFO was so bright, the two officers said, that they could read the hands on their watches by its light. A third officer joined them at headquarters, where the three noted that the UFO had gained altitude. They watched for about half an hour as the UFO changed in color from bright red to orange to white, climbing in the sky until it resembled "a star."

The police report sparked a local interest in UFOs, and on the following dawn not one but two UFOs were seen in the east. On the fourth night, an aircraft aided the police officers in their chase and also saw the UFO, plus a second, fainter object above the eastern horizon; both objects seemed to "back off" and keep "moving higher and away from us" as the airplane chased them toward the east. More police officers in other Georgia towns reported observations similar to those made in Milledgeville. For a few weeks, central Georgia percolated with UFO sightings.

These observations recorded the early-morning rising of the planet Venus, which was then particularly bright, accompanied by a fainter planet, Jupiter. Venus has often been misidentified as a UFO, and in fact in 1967, the number of reports from Georgia of Venus as a UFO set a record for a single state in a single year (Fig. 20.5).

Figure 20.5 This photograph shows Venus and the crescent moon in the eastern sky just before sunrise, viewed from the summit of Mauna Kea, Hawaii. Seen by itself, especially at times when it is bright enough to cast a shadow, Venus has frequently been reported as a UFO.

This fact testifies to the ability of one report of a strange phenomenon to trigger another, thanks to the influence that each of us exerts on his fellow humans. Once you have spent some time looking at the night sky, you will recognize that any celestial object appears to "back off" as you approach it, and it will also seem to follow you wherever you go. In the heat of a police chase, however, this simple fact may be forgotten. The police report of a UFO so bright that the officers could read the hands of a watch by its light refers to another astronomical fact: Venus shines remarkably brightly in our skies, sometimes able to cast a shadow.

Landing in Socorro

On April 24, 1964, Lonnie Zamora, a patrolman in Socorro, New Mexico, was following a speeding motorist when he heard a loud roar and saw a "flame in the sky" over a mesa less than a mile away. Zamora drove up a steep road to the mesa top, where he saw a "shiny-type object" with "two people in white overalls" nearby. Zamora parked his police car about a hundred feet from the object, got out, and heard a "very loud roar" as the object slowly rose from the mesa with a blast of flame underneath it. As Zamora ran back to his car, he knocked off his glasses, but he heard a brief, sharp whine, and saw the UFO fly off to the southeast.

Zamora drew sketches of the UFO as he recalled it, making it look something like a large egg with legs, and he found four indentations in the ground, plus some evidence of burned vegetation. The chief Air Force consultant on UFOs, J. Allen Hynek, interviewed Zamora and found him "basically sincere, honest, and reliable."

What can we make of Zamora's report? On the one hand, Zamora said that he saw something highly unusual. On the other hand, the evidence to support Zamora's sightings—small indentations and a small amount of burned vegetation—could never be called convincing, if it had to stand alone. Even more damaging, a man living with his wife 1000 feet south of the "landing site" was home and heard nothing, though his doors and windows were open.

The Socorro incident is important because it is often included among the "most reliable" cases that offer evidence of extraterrestrial visitors to Earth. Although Zamora's report is intriguing, no court of law would consider the case proven, especially since there is conflicting testimony from another witness. As is the case with most UFO sightings, the evidence consists almost entirely of an eyewitness sighting, in this case by a single individual whose account is not supported by others.

Difficulties in Verifying the Spacecraft Hypothesis

With a fair degree of justice, we can summarize the best-known UFO reports by saying that they are no more convincing than the Socorro incident. More spectacular reports exist, but the trustworthiness of those making the report remains open to question. A subject as loaded with emotion as the possibility of extraterrestrial visitors must inevitably draw its share of liars, charlatans, and pranksters, and all of these have appeared in abundance in UFO reports. A good investigator can hope to uncover such frauds, and many hoaxes have indeed been exposed (see again Fig. 1.7). A few cases remain where the report seems trustworthy as an accurate description of what observers think they saw. The difficulty here, as we have said, is that the human mind *interprets* its surroundings; that is, our minds insist upon imposing human intuition—drawn from experience—upon the scenes that they take in.

Even a group of people who observe the same event can influence one another, or can all be subject to much the same sort of interpretive interference. We saw this in the case of the reentering spacecraft, and it appears in many UFO reports submitted by two or more people, often close relatives. An example of this effect appears in the sighting of UFOs (probably the planets Venus and Jupiter) in New Guinea in 1959, in the presence of an Anglican missionary and 37 other people. Twenty-five of the 37 later signed the missionary's statement attesting to what they saw. However, their relationship to the missionary (pupils, assistants, parishioners) made them far from independent in their interpretation and memory—and it is far from clear that they knew just what they were asserting as true. The fact remains, simply stated, that *eyewitness accounts can never by themselves prove whether a given event did or did not occur.* The problem of mental interpretation grows especially acute with strange objects in the sky, because most people nowadays are unfamiliar with natural celestial objects—hence the numerous reports of Venus as a UFO.

Classification of UFO Reports

In the face of unreliability of eyewitness reports, one thing that we can do is to classify the reports of unidentified flying objects. This entirely scientific procedure draws support from all sides of the extraterrestrial visitor argument. Moreover, in case after case of scientific inquiry, the classification of evidence by itself has helped to resolve a problem. During the late 1950s, J. Allen Hynek devised a sixfold classification for UFO sightings:

1. Nocturnal lights: bright lights seen at night
2. Daylight disks: usually oval or disklike
3. Radar-visual: visual sightings also detected by radar
4. Close encounters of the first kind: visual sightings of an unidentified object
5. Close encounters of the second kind: visual sightings plus physical effects on animate and inanimate objects
6. Close encounters of the third kind: sightings of "occupants" in or around the UFO.

Notice that Hynek's list omits "close encounters of the fourth kind," actual physical contact with occupants of UFOs. Such encounters would fall in our category of the best evidence of a visit to Earth by extraterrestrials. In other words, aside from the claims of a few people who are either untrustworthy or mentally unbalanced, close encounters of the third kind are as near as the reports come to actual contact. This suggests either that normal human imagination stops short of physical contact, or that extraterrestrial visitors shun such close encounters, or that we have simply never had such visitors.

Hynek derived an interesting general fact about the UFO reports that he classified in his scheme: Less unusual events are correlated with more reliable reports, and vice versa. In other words, some reports have a great degree of strangeness (for example, that a flying saucer landed in a cornfield), but a low degree of credibility (for example, that only one person with limited eyesight saw this happen). Other reports have a low degree of strangeness (for example, that a strange light appeared to hover in the sky, then accelerate to great speed as it disappeared), but a great degree of credibility (for example, that five reliable witnesses all saw the same sequence of events). The elusive case remains the one with great strangeness and great credibility.

Consider the fact that the first three of Hynek's six categories—nocturnal lights, daylight disks, and radar-visual—are commonplace occurrences, distinguished chiefly by the degree of surprise that the observer reports. Airplanes, meteors, and planets (Venus most of all) account for the majority of nocturnal lights. Daylight disks usually turn out to be blimps, weather balloons, or clouds. Occasional hoaxes, especially in the days immediately following a well-publicized UFO report, often involve balloons lit by candles or covered with aluminum foil. The later explanation of these hoaxes rarely receives as much media coverage as the original report. Sometimes UFO enthusiasts even reject a hoaxer's confession, as was true for the mysterious crop circles in England (see again Fig. 1.7). Finally, false radar echoes remain a common problem for every radar operator. By themselves, the radar spottings of objects in locations with no known aircraft do not provide much evidence for extraterrestrial spacecraft: Flocks of birds, swarms of insects, and what radar operators call "angels," or anomalous propagation of radar waves (caused by fluctuations in air density), may be at work.

None of this discussion proves that *all* UFO reports have a completely natural explanation as misidentified objects, or that *all* UFO observers have made human errors. But the numerous UFO reports that involve natural or humanmade objects show that the possibility of human error will remain high if the UFO report is based only on eyewitness testimony.

Arguments for the Spacecraft Hypothesis

Without clear-cut physical evidence, the proponents of the extraterrestrial hypothesis for UFOs face a difficult task when they seek to prove that we are being visited by other civilizations. Basically, they must use theoretical analysis rather than concrete evidence to reach this conclusion. Theoretical reasoning is not to be dismissed—consider how much of this book relies on it!—but it lacks credibility if it does not flow from evidence and from physical laws on which nearly everyone can agree, *and* if it cannot be tested by experiment.

The proponents of the extraterrestrial-spacecraft hypothesis usually rely on the following line of argument:

1. The number of civilizations in our own galaxy, let alone those in other galaxies, may reach an enormous figure.
2. Even if most of these presumed civilizations have no interest in or aptitude for interstellar spaceflight, some of them must have both the inclination and the ability.
3. Since a more advanced civilization might well possess incredibly sophisticated technology, the civilizations that do decide to visit can send as many spacecraft as they like, and can perform whatever feats they choose with them.
4. Hence the most likely explanation of mysterious objects with no natural cause that we can determine must be that extraterrestrial visitors are repeatedly visiting us.

Many scientists would agree with points 1 and 2 of this argument. On point 3, as we have seen, a scientific analysis suggests that spaceflight always requires a huge amount of energy, though we must certainly allow for the possibility that another civilization may know some enormous secret we don't know. Therefore, we might be led to conclusion 4, that the sky is not the limit when we consider spaceflight by extraterrestrial civilizations. But in fact the scientific argument is that the sky *is* the limit, and it is an important limit. We have presented this argument in Chapter 18.

Like all scientists, the authors of this book are intrigued and excited by the fact that we don't know everything. We might be completely wrong about the difficulties of space flight, and extraterrestrials might be able to visit Earth far more easily than scientists imagine. But—again like other scientists—we insist on real evidence before we accept that such visitors are among us. After all, most people, for solid reasons, are equally skeptical about far less earthshaking claims. They rightly want to be sure that a new automobile or television will really perform as advertised, and do not simply accept as valid the claims of others—especially those with a financial or emotional interest in making them.

A Cover-up?

Before leaving the subject of UFO reports, we ought to address the possibility of a massive "cover-up." UFO enthusiasts often assert that the government (typically the United States government) knows the "truth" about UFOs—indeed, that it even has "little green men" preserved in a morgue following the crash of a flying saucer—but refuses to let the public know, for fear that we may prove unable to stand such news or might catch on to the fact that extraterrestrials are in effect supervising the planet. Such assertions are notoriously difficult to disprove, for the more the government issues denials, the more these denials sound like a cover-up.

We may note, however, that whatever the truth about UFOs may be, the fact that the United States (and other governments) may or may not have engaged in a cover-up does not prove that they have something to cover up. Recent history shows, instead, that even trivial facts may seem so dangerous to government officials that they deal with them in a time-honored way, by concealing them. Quite possibly, if extraterrestrial spacecraft did land in some out-of-the-way location, our military and political leaders would try to cover this up. But these leaders would also try to cover up evidence of incompetence, misjudgment, or just plain laziness on their part. All of these possibilities (and more) enter the explanation of why the government's opinion on UFOs does not carry much weight with most scientists.

We cannot rule out the possibility that our military officers know all about UFOs—that, for example, they have already entered into communication with other civilizations—and find it useful to conceal this fact from us. This supposed conspiracy could include scientists too, so that this book might, for instance, have been written by extraterrestrial command (and its sometimes skeptical tone could be a particularly clever way to trap the reader). Indeed, it is worth wondering sometimes whether or not everyone you meet might not be from another planet, all collaborating to maintain a "normal" environment in which they can study you better. Or, to complete the picture, you too could be from another solar system, but have not yet grown up sufficiently to realize it.

Are UFOs a Modern Myth?

Since this sort of approach leads us nowhere, let us return to the arena where UFO reports have their greatest impact and importance: the human mind. Whatever their provoking cause, UFO reports arise in minds of those who make them, and must therefore be considered a purely human and terrestrial phenomenon until we conclude that extraterrestrial events, or extraterrestrial visitors, have produced them. Some investigators of UFO reports consider the UFO *experience*—the human involvement with UFO sightings and UFO reports—to have more importance than the attempt to discover the "reality" of the UFOs. If we try to generalize what various people say they saw in close encounters of the third kind, we find that most saw a spacecraft with inhabitants who looked a good deal like humans: two arms, two legs, a head, and so forth. If the inhabitants had a sex, that sex is almost always male. These facts have a ready explanation if people see what they might like to see, or expect to see. If we accept a psychological explanation of UFOs, this apparent "projection" of an internal idea actually supports, rather than contradicts, the reality of the UFO experience. In this analysis, the reality is emotional rather than physical.

Human beings long for some kind of special connection with the vast dark space beyond the sky. The greatest gods in most religions usually dwell somewhere in the heavens, and they or their emissaries often make visits to their chosen subjects. The Judeo-Christian tradition speaks of angels, for example, but there are many similar heavenly messengers in other cultures. The psychologist Carl Jung has pointed out that there is a strong, nearly universal human tendency, in times of trouble, to look heavenward as people search for signs and symbols that will give them guidance. Jung suggested that the UFO phenomenon is simply a modern myth, a mechanized, high-tech manifestation of this deep-seated human tradition.

As we look at past records of miracle-working "little people" and celestial visitors, we can see a continuity in human attitudes. This does not resolve the question of what UFO reports may imply, but we may reach the conclusion that the same response has occurred, in slightly different forms, for as long as humans have contemplated the vast and mysterious cosmos in which we live. The strand that unites the continuing flow of stories, tales, and reports consists of human attempts to connect our own experience with the universe at large, in a basically human-centered, trusting, and straightforward manner. We assume that the inhabitants of unknown parts of the universe care about us, play with us, abduct us, or dance and sing for our benefit and entertainment.

Von Däniken: Charlatan of the Gods?

To take the theory of extraterrestrial visitors one step farther, we may consider the possibility that the record of what we consider to be human progress may consist of little more than outside intervention into a relatively dull and unimaginative human population. (If we take this theory one step too far, we all become visitors from other planets, as we discussed above). Approximately 50 million copies of Erich von Däniken's books—*Chariots of the Gods?*, *Gold of the Gods*, and several other titles—have been sold to readers who apparently like to learn that ancient astronauts visited the Earth (so von Däniken says), leaving behind many relics of their interactions with human beings.

The relics are most obvious to von Däniken when he examines "primitive" civilizations, those which he judges clearly incapable of large engineering feats. Thus, for example, von Däniken says of ancient Egypt: "Great cities and numerous temples . . . pyramids of overwhelming size—these and many other wonderful things shot out of the ground, so to speak. Genuine miracles in a country that is suddenly capable of such achievements without any recognizable prehistory."

To distort history as von Däniken does requires a sweeping ignorance, not to mention a bold charlatanism. In fact, Egypt has a long and quite recognizable prehistory, going back well before the time in which the "miracles" of which von Däniken writes were built. The same holds true for the ancient Sumerians, whose legend of Oannes, the bringer of writing and many other inventions, has been cited as an example of possible extraterrestrial intervention.

Although von Däniken on occasion rises to wholesale invention, as when he describes a trip that he never made into South American caves filled with ancient gold treasures, his most impressive feat consists of citing "amazing" coincidences which any reader could demonstrate to be false. To impress us with the impossibility that the "heathen" Egyptians built the great pyramid of Khufu at Giza (Fig. 20.6), von Däniken asks: "Is it really a coincidence that the height of the pyramid of Cheops [Khufu] multiplied by a thousand million—98,000,000 miles—corresponds approximately to the distance between the Earth and the sun?"

Well, let's see. The pyramid's height equals 481 feet. If we multiply this by 10^9 (1000 million), and then divide by the 5280 feet in a mile, we obtain

Figure 20.6 The three great pyramids at Giza were built more than four millennia ago in Egypt. The ancient Egyptians built many pyramids on the west bank of the Nile, of which these three are the largest and best preserved.

91,000,000 miles. In fact, this gives an even better result than von Däniken found because the true distance from the Earth to the sun is 93,000,000 miles. Is this a coincidence, or does this give an example of ancient beings with special knowledge?

One test is to look at the heights of other buildings that were constructed before the distance from the Earth to the sun was well known. A nice example appears in the tower of the cathedral at Rouen, France, completed in the thirteenth century. If we multiply this tower's height of 485 feet by 1000 million, we obtain an answer of 92,000,000 miles, in better agreement with the Earth-sun distance than the great pyramid! Reasoning along von Däniken's lines, one might suggest that the builders of Rouen cathedral also had special knowledge, but we think not.

Once we start multiplying things by some really large numbers, we can find many interesting coincidences. For example, is it merely chance that the length of a common felt-tip pen, 5 7/8 inches, when multiplied by a million million (10^{12}), gives the *exact* distance to the sun, 93,000,000 miles?[1] Are

[1] If you fail to perform this multiplication for yourself, you should ask yourself why you believe this book instead of von Däniken's.

pen manufacturers eager to show their extraterrestrial knowledge? And why do they produce a better product than the great pyramid? After all, if ancient astronauts really had built the pyramid, surely they could have added another 10 feet to its height, and thus furnished the exact value for the Earth-sun distance! Strange, too, that they didn't even try with the other two pyramids at Giza!

Another favorite ploy of von Däniken is the presentation of an ancient drawing or sculpture as portraying spacemen from another world, as in his claim that a Mayan illustration shows an astronaut with a backpack (Fig. 20.7). In fact, these drawings show well-known religious motifs, which become far more apparent when more of the drawing than von Däniken presents are included. Expressing wonder at the patterns of lines scraped from the surface soil on the Nazca plains of Peru, von Däniken suggests that these must have been landing strips for extraterrestrial spacecraft. These geometrical markings are clearly humanmade, but it is hard to see why an advanced civilization would use such complicated figures to guide their landings. Far more likely is the explanation that these carvings expressed respect for the totemic figures of ancient cultures, figures that we also find on their pottery. Some of the straight lines may have an astronomical significance, others may have been made for ceremonial reasons or simply for amusement.

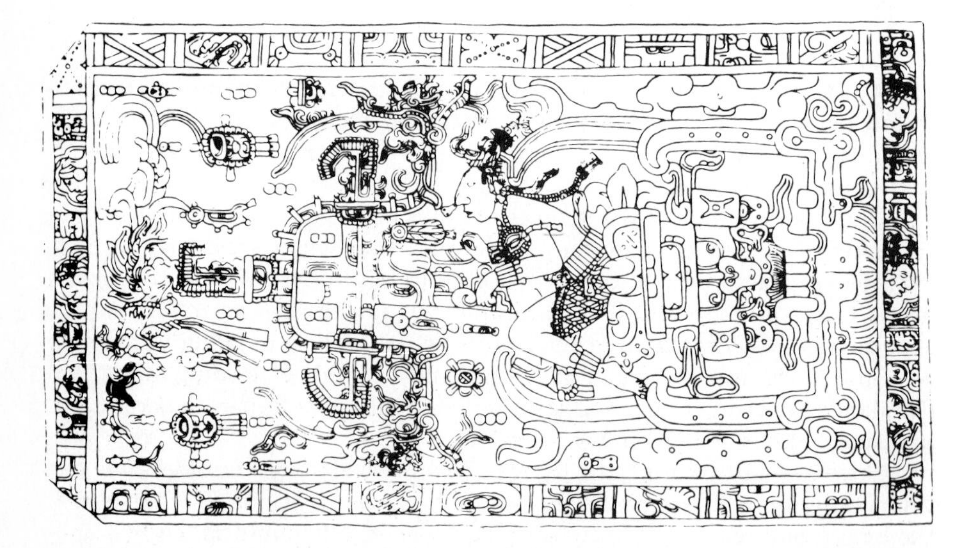

Figure 20.7 This tomb lid from Palenque, Mexico, shows a Mayan king, Pacal, together with important symbols of Mayan culture. Von Däniken sees Pacal as an astronaut riding a rocket, with his foot on the pedals, his hands on the controls, and an oxygen mask supplying his respiratory needs.

Why do so many people take von Däniken seriously? Two basic factors provide the explanation. First, von Däniken gives a human dimension to the cosmos, connecting with our desires to find friends in the vast heavens around us and to believe that a benevolent providence looks after us. His assertions connect so well with the longing of many of us to see ourselves as a part of the cosmos, not adrift in a huge and uncaring universe, that he knows that most readers will not want to check up on him. On the contrary, their desire will help them believe he is correct. Second, many people have a lingering belief that ideas such as von Däniken's could not be printed unless they were true: The magic power of the printed word persists. In actual fact, whole forests have been sacrificed to various propaganda causes, and although this may be laudable in a free society, we should not fall prey to the intuitive belief that what appears in books automatically has more validity than what you hear on the street, or what you can figure out for yourself.

Many authors bank on readers' intuitive beliefs that the world is a magic place, that if we can only look with different eyes, we shall see delights beyond mere amazement. The liberating joy of seeing the world in a new way, combined with the respect that we give to books, has made millionaires out of many a charlatan. But before we can feel entitled to judge these men (and notice that these supersalesmen have nearly all been male) to be wrong, we must consider the grounds on which we assess the truth or falsity of an assertion about the world. If we believe because it "sounds right," we deserve to be taken for a few dollars' worth of gullibility (see page 482). If we believe because we swallow the "proof" that such authors provide, we receive low marks for critical assessment of what "proof" may be, as in the felt-tip pen example we mentioned earlier. Contrast such belief with what you get if you believe something to be true only because you have considered arguments for and against it, and have actually done some calculations and analysis for yourself.

We should not, however, leave the subject of "ancient astronauts" without noticing its relevance to the problem of interstellar communication. When we puzzle over the pyramids, or try to read and to understand ancient Egyptian inscriptions, we are "receiving messages" from a civilization that flourished over 4000 years ago (Fig. 20.8). The fact that we cannot send messages in the other direction may be discouraging, but it does not dampen our interest in this culture. We would have the same experience in receiving messages from an advanced civilization on a star 4000 light years away. Once again, we might not have the chance for a pleasurable dialogue, but these messages from another civilization would surely fascinate, delight, and instruct us.

Figure 20.8 The enigmatic Sphinx at Giza, Egypt, more than 4000 years old, embodies attitudes from a long-vanished civilization. According to von Däniken, the figure represents an extraterrestrial visitor to Earth.

SUMMARY

The evidence concerning unidentified flying objects (UFOs) consists mainly of eyewitness accounts of strange objects in the sky whose presence, motion, color, or appearance seems to rule out natural or humanmade explanations to the observer who reports them. Even the best eyewitness testimony remains subject to the attempts of human minds to impose a familiar reality upon strange objects, and to the degradation of memory with time. Such evidence therefore remains unconvincing in providing proof of an extraterrestrial origin of UFOs.

Photographs would be much better, but the photographs of UFOs that exist are either unreliable or unpersuasive. They do not demonstrate that anything extraterrestrial or even particularly strange has been photographed. Although photographs can be faked, they do at least provide evidence that can be studied with care. In contrast to human observations, photographic evidence does not change with time nor suffer from the filtering effects of human consciousness. Thus if scientists ever become convinced that an extraterrestrial explanation of UFO reports has some merit, it will probably occur through photographs rather than through eyewitness accounts.

The weight of the evidence we now have suggests that almost all UFO reports represent natural events, often misinterpreted but with no fakery on the part of the observer. The most impressive UFO sightings, from the point of view of proving that extraterrestrial visitors have reached the Earth, rely on far less verifiable data than the "typical" UFO sightings of meteors, weather balloons, or the planet Venus, and may in fact include hoaxes and the unfortunate projections of unstable personalities.

Public interest in UFOs as possible evidence of extraterrestrial visitors connects with a long human history of interest in the heavens and belief that the cosmos contains friendly and protective beings who watch over us. The success of pseudoscientific efforts such as those of Erich von Däniken shows the power of this feeling, since the "evidence" that von Däniken cites in favor of his notion that extraterrestrial visitors built the pyramids and similar ancient monuments falls apart upon examination, or proves to be outright falsehood. The ancient monuments of Earth have much to tell us about communication across the centuries, but they deal with communication among humans, not with the interstellar communication we seek.

QUESTIONS

1. What are the chief factors that make eyewitness accounts of strange events unreliable?

2. Why are photographs superior to human observation for scientific analysis of a past event?

3. How can we explain the fact that observers who see a meteor many miles away may believe that it is moving silently at an altitude of only a few hundred feet?

4. Why is Venus the single object most commonly reported as a UFO?

5. What would you consider to be the minimum reliable evidence that the Earth is being (or has been) visited by extraterrestrials? Why?

6. Suppose that the average lifetime of a civilization in our galaxy equals 10,000 years, and that therefore about 10,000 civilizations exist in the Milky

Way at any time. If one civilization in 10 goes exploring, how many stars must each one explore to cover all 300 billion stars in our galaxy?

7. The distance from the Earth to the moon is 240,000 miles. See if you can find the height of some ancient (or modern!) monument that is a simple fraction of this distance.

8. Suppose that the Milky Way contains 1 million civilizations. What fraction of all stars then have civilizations? Suppose that one civilization in 100 explores the galaxy, and that—like the Star Trek spacecraft *Enterprise*—it visits one planetary system per hour. How long would it take this spacecraft to visit the 300 billion stars in the Milky Way? Does a visit to one planetary system per hour strike you as likely to occur? How many *Enterprises* would it take to reduce the exploration of the Milky Way to a total time of 100 years?

FURTHER READING

Bracewell, Ronald. *The Galactic Club.* San Francisco: W.H. Freeman, 1975.

Christian, James, ed. *Extraterrestrial Intelligence: The First Encounter.* Buffalo: Prometheus Books, 1976.

Finney, Ben, and E. Jones, eds. *Interstellar Migration and the Human Experience.* Berkeley: University of California Press, 1985.

Jung, Carl. *Flying Saucers.* Princeton, NJ: Princeton University Press, 1978.

Klass, Philip. *UFOs Explained.* New York: Vintage, 1974.

———*UFOs: The Public Deceived.* Buffalo: Prometheus Books, 1983.

———*UFO Abductions: A Dangerous Game.* Buffalo: Prometheus Books, 1988.

Papagiannis, Michael, ed. *The Search for Extraterrestrial Life: Recent Developments.* Boston: D. Reidel, 1985.

Sagan, Carl, and Thornton Page, eds. *UFOs—A Scientific Debate.* Ithaca, N.Y.: Cornell University Press, 1972.

Shaeffer, Robert. *The UFO Verdict: Examining the Evidence.* Buffalo: Prometheus Books, 1986.

21

Where Is Everybody?

The previous chapters of this book have presented scientific arguments against the likelihood of visitors coming to Earth from another civilization. In true scientific fashion, we may now reassess this conclusion by asking a famous question, the title of this chapter, that was first posed by Enrico Fermi at a lunch with some of his fellow physicists in the summer of 1950 (Fig. 21.1). Fermi was an extraordinarily brilliant scientist, one of the early pioneers of the "new physics" of the 1920s, and later responsible for producing the first controlled nuclear chain reaction, at the University of Chicago in 1942. He was also one of the key figures in the development of the U.S. atomic bomb during World War II. Another atomic-bomb scientist, Edward Teller, soon to become famous as the "father" of the U.S. hydrogen (fusion) bomb, was a participant in this lunchtime conversation. The expertise on nuclear energy gathered around that table was formidable!

Not surprisingly, Teller recalled many years later that the scientists were speculating about the probability of discovering a way to travel faster than light. These physicists were well aware that the energy they had learned to liberate from atomic nuclei would not by itself solve the enormous problems posed by interstellar travel. On the other hand, they also realized that no fundamental physical laws forbid interstellar journeys. Nothing discovered during the 40-plus years since that lunch contradicts this conclusion. Furthermore, from our present perspective, the chances for the widespread origin of life and its evolution into intelligent life seem good. Why then, do we have no evidence for the existence of any other advanced civilizations in our galaxy? *Where is everybody?*

We can suggest three possible answers: (1) Maybe no one is out there! (2) Other civilizations exist, but are not interested in communication. (3) Civilizations exist and communicate—but they use signals we can't yet detect. Let us examine each of these three alternatives.

Figure 21.1 Enrico Fermi won the Nobel Prize for physics in 1936. He is shown here in 1950 examining some secret documents at his desk at Los Alamos, New Mexico, where he played a key role in developing the first atomic bomb.

We May Be Alone, or Nearly So

In our discussion of the Drake Equation in Chapter 17, we emphasized that the equation yields widely different values of *N*, the number of civilizations in the Milky Way. Civilizations may be far less abundant than we have concluded. Our equation for estimating the number of civilizations, *N*, may be correct, but the last term in the equation, *L*, the average lifetime of a civilization with communications ability and interest, may fall far below our estimates, thus reducing the number of civilizations in the Milky Way at any given time to a mere handful. If *N* is extremely small, this implies that our own civilization's lifetime is likely to be short. But we on Earth might not be an average civilization. We may have missed some important point about the way in which life has evolved on Earth, or in which civilizations can develop on a planet with life, that stamps our own civilization as unique, or nearly so. In other words, we may have unconsciously adopted a sort of religious belief in our averageness because we harbor some deepseated hope of finding contact or even wisdom in the form of advanced civilizations throughout our galaxy and beyond.

Until we intercept interstellar messages, we can do well to use our new perspective to consider life on Earth in a cosmic context. When we contemplate the sweep of biological history on our planet, we cannot fail to be impressed by the speed with which great changes in life have occurred in recent eras. Most of the history of life on Earth consists of a two-billion-year (or longer!) process in which some prokaryotes evolved into eukaryotes, and in which an oxygen-poor atmosphere changed into an oxygen-rich atmosphere. Once these events had taken place, the abundance of oxygen provided a tremendous source of free chemical energy, and eukaryotes were ready to develop ever more complex structures based on the use of this gas. Given this history, we can well imagine that the transition from prokaryotic to eukaryotic life may be more difficult than the evolution of intelligent life from eukaryotes, or the evolution of long-lived civilizations from intelligent forms of life. We might therefore expect that life takes the longest for the first steps toward complex life, on planet after planet.

Now suppose that life on Earth is not an average example of life in the universe because our planet has had one of the most rapid developments of intelligence from primordial soup. Perhaps the *average* time needed to pass from self-replicating molecules to eukaryotes (or their equivalent) is actually double the three billion years that characterized the Earth. Then we would have little hope of finding intelligent life on most planets. Recent studies of stellar evolution suggest that the oldest Population I stars like our sun may be only 6 or 7 billion years old. If such stars are essential to produce life-bearing planets (see Chapter 16), and if most planets need 6 billion years to produce eukaryotes from self-replicating molecules, then we may occupy a point in our galaxy's lifetime too early to find many advanced civilizations. In this event, only those few civilizations with unusually rapid development will be ready for communication now. Of course, with our present state of ignorance, we can assign an equal likelihood to the opposite scenario: If we are one of the *slowest* civilizations in development, and the average planet needs only 1.5 billion years to produce what took Earth 3 billion years, then the galaxy could be teeming with intelligence far beyond our own.

Humans have yet to find a single example of life outside the Earth. Because of this, efforts to apply the tools and knowledge of science to estimating the probability that life exists elsewhere—a field of endeavor often called exobiology—could be considered a scientific discipline with no subject matter. We can reply that many fields of science began with just this handicap. For example, before astronomers could send equipment above our atmosphere, they had no way to detect gamma rays from celestial objects, since these high-energy photons could never penetrate to the ground. Gamma-ray astronomy was then a discipline with no subject matter, no observations, no way of checking theory against fact. But the theoretical framework that astronomers constructed before the era of satellites proved immensely useful when gamma-ray observations could at last be made. Perhaps the same will turn out to be true for the "science" of exobiology.

We cannot resolve these issues until we find other forms of life. Our experience does suggest in the strongest terms that since we do not consider the Earth unique among planets, or the sun unique among stars, whenever we have "world enough and time," life should develop and intelligence should evolve. But contact with another civilization is the only way we can make this conclusion certain. We must perform the experiment to be sure!

Meanwhile, consider some of the social problems that any civilization must solve in order to achieve a long lifetime. Chief among these is the issue of population growth. On Earth, humans have spread as widely as possible rather than attempt to regulate the population. We may call this approach "colonization instead of regulation." We now confront the limitations of such a policy on a planet with a finite area (and all planets have finite areas): massive overpopulation.

The number of humans on Earth has been doubling about every 40 years for the past two centuries. In 1810, there were only a quarter of a billion humans; in 1850, half a billion; in 1890, 1 billion; in 1930, 2 billion; in 1970, 4 billion; in 1990, 5 billion. We now increase in numbers every four years by the total population of the world in 1810! In fact, the *rate* of population growth has actually increased during the past 50 years over the rate of the previous century, so the present doubling time is only 35 years. Calculations made by some scientists show that if the rate of increase in population continues on its present course (that is, if the doubling time continues to decrease), then by the third decade of the next century, the number of people on Earth will become infinite (Fig. 21.2)! Of course, this is impossible, yet the calculation was made in 1960, and 30 years later, in 1990, world census figures indicated that we were ahead of schedule! This faster-and-faster doubling of the population represents the most fundamental problem faced by contemporary human society, and any developing civilization will have to solve it in order to survive.

We can appreciate this problem with a biological analogy. Let us adopt a poetic point of view and think of our planet as a single organism. Then all the various forms of life on Earth are simply cells or organs in this small cosmic creature, which endlessly circles the stellar fire that keeps it alive and warm. Looking at the Earth today, we see signs of sickness in this organism: The antibodies are failing to protect it against the accumulations

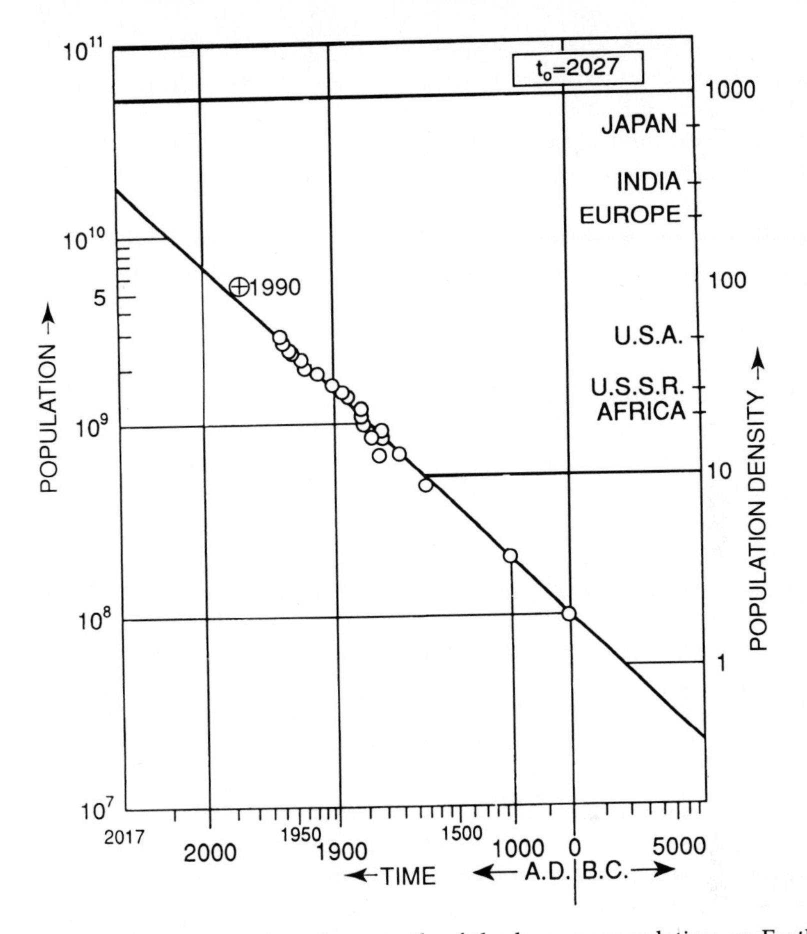

Figure 21.2 If we extrapolate the growth of the human population on Earth, we can predict that the population will become infinite in the year 2027. Note that time runs from right to left, and that both the time axis and the population scale are logarithmic. The diagonal line represents a prediction derived from the population data available in 1960. The point for the 1990 census indicates that we are doing even better at achieving an infinite population than this extreme prediction! This figure is based on a study made by H. von Foerster, P. M. Mora, and L. W. Amiot, published in *Science,* vol. 132, p. 1291, 1960.

of various toxins. Chlorofluorocarbons and acid rain are upsetting the pulmonary system; petrochemicals ranging from crude oil to polychorinated biphenyls, plus a host of rotting debris, are poisoning the bloodstream; the natural cleansing action of organs evolved for this purpose can't cope with this onslaught. The temperature of the entire organism is rising; it is developing a fever as the concentrations of carbon dioxide, methane and other trace gases increase. When we look for the causes of these problems, the diagnosis comes quickly: The patient is suffering from a kind of cancer, in which one type of cell is endlessly multiplying, spreading throughout the organism, destroying surrounding tissues and bone. As with any cancer, a real danger exists that the entire organism may die. The only hope is that we have made an early diagnosis and we know the identity of the malignant cells: *We* are the cancer that is destroying our planet.

We have been our own enemy in other ways, most notably in building an arsenal of thermonuclear weapons capable of killing most life on our planet. Yet now we seem to be retreating from the edge of that abyss; perhaps we shall soon come to terms with the population problem. We must simply recognize that a finite world demands a stable population. In this case, our descendants may indeed acquire a long-term perspective. If sufficient resources are available, they might even choose interstellar spaceflight as a reasonable, though high-risk, venture, despite the length of the journeys involved. But the same willingness to act upon the principle of finite resources might remove the expansionist urge that leads to the colonization of other worlds. This might also be true of other civilizations: Once they grow "mellow," they could be happy without exploring their galactic environment. Would they then still send signals into the cosmos?

Civilizations May Have Little Interest in Communication

A developing civilization that cannot solve its social problems may be compelled to deal with them forever. This could leave few resources for activities such as space exploration that seem to have no immediate social benefit. We humans certainly do not take advantage of every technological opportunity that appears. We remain ambivalent about the use of supersonic aircraft, and despite great needs for energy, nuclear power has not captured the popular imagination. In exploring our cosmic backyard, we have seen that human success in landing on the moon has not produced a continuing program of space exploration by humans (Fig. 21.3). Have we entered an era comparable to the pause that occurred after the discovery of the New World, before humans began to colonize what they regarded as open territory? Or are we adopting an attitude that other, more pressing priorities must engage our attention and our resources, long into the future? China offers us an historical example of this latter approach. The Ming Dynasty (1368–1644) decided to abandon an active program of exploration and trade by sea, retreating within the country's borders to pursue a path that still remains largely inner-directed, some four centuries later. Our society may already be caught in this mood. Are we therefore justified in abandoning the idea that extraterrestrial civilizations make direct visits? The answer appears to be

Figure 21.3 This unused Saturn V rocket was produced as one of the series of rockets that took astronauts to the moon between 1969 and 1972 (see again Fig. 18.1). It now rusts in the grass outside the Johnson Space Center in Houston, Texas, while the moon moves through our skies, currently nearly as inaccessible to humans as it was when the ancient Egyptians built the first pyramids 4500 years ago.

that we can expect interstellar spaceflight only among those civilizations that are immortal, or nearly so. Low-thrust spacecraft that take millions of years to travel from star to star have little appeal for us: We won't take a trip that long if the message we carry can cover the same distance in a few decades. We probably wouldn't go even if we knew how to freeze ourselves, or could otherwise induce a near-zero rate of living, because the cost in energy would be too large. Nor is our society likely even to send automated probes that might generate a response in several million years (and where would we send them?) when we have such pressing problems on our own planet. Only if we developed a much longer view of our existence would we be ready for million-year journeys, either by ourselves or our spacecraft. So we may speculate that only civilizations that have solved their immediate problems and have achieved the certainty of great lifetimes are likely to make interstellar voyages. There may be *no* such civilizations anywhere in our galaxy, or there may be *many*. Can we decide which possibility lies closer to reality?

To a limited extent, we can. *If* we believe that the simple possibility of spaceflight inevitably leads to its use in exploration and colonization, then we must, on the evidence, be alone in the Milky Way. In other words, if—and this is a big "if"—civilizations as they develop continue to follow the expansionist tendencies that our own civilization has shown on its own planet, and spread through their neighboring planetary systems to the best of their ability, then the 10 billion-year age of the Milky Way may have provided plenty of time for the Earth to have been colonized. Yet as of today, our planet remains free from extraterrestrial settlers (unless we believe that humans are such immigrants—but the molecular resemblances among all life on Earth rule this possibility completely out). Does the absence of extraterrestrial colonization really show that no other civilizations exist in our galaxy?

If we think about the steps that would actually accompany a civilization's spread through the galaxy by colonization, we can see that repeated shifts in attitude would have to occur. First, such a civilization would have to develop expansionist tendencies; this much seems natural, if we use the Earth as a guide. Then the civilization must learn to deal with an ever-growing population and must achieve enough stability and surplus wealth to build and to use interstellar spacecraft. Colonization can never relieve population pressure by itself, since the room made available by colonization tends to be quickly filled by the still-growing population. Thus the construction and use of interstellar spacecraft requires dedication and direction to that end on the part of the civilization that considers interstellar spaceflight. This activity will not solve the population problem. Once the descendants of the colonists from such a "mature" civilization have reached the planets of other solar systems, they must again revert to an expansionist attitude, in order to gather the resources from the asteroids or planets and build new ships for the next set of expeditions. And thus a galaxywide program of colonization requires alternating expansionist and "mature" periods. This may prove to be natural, but it may also turn out that no ci- vilization will embrace enough cycles to spread through an entire galaxy. If this is so, then the absence of colonists from other civilizations here on Earth does not prove that no long-lived civilizations exist in the Milky Way.

Using ourselves as an example, we can also recognize that humans do not occupy all the space available on Earth. We could live in colonies in Antarctica, on the ocean bottoms, or in our deserts; all of these are environments orders of magnitude more hospitable, accessible, and familiar than space colonies would be. But we don't. Our experience shows that our civilization flourishes best when it is least constrained.

This argument does not rule out the possibility that a "mature" civilization could send automated message probes throughout the galaxy, carrying messages or reconnaissance devices, perhaps even programmed to "seed" new life wherever possible. Alien probes could be watching us even now, enjoying our antics while cleverly guarding themselves against detection, so that Earth represents one habitat in the "cosmic zoo." But we must put this "zoo" hypothesis with the idea that there are new realms of physics so far outside our knowledge that advanced civilizations, who know these realms, can do simply anything. Both concepts, though intriguing, are empty of practical impact, since they cannot be tested in any effective way. We are led back to the search for other civilizations as the only means to discover the general rules by which civilizations develop.

Because we still lack our first contact, we have had to discuss the motivations of hypothetical civilizations, including those far more sophisticated than ourselves, in a complete vacuum. Consider, for example, the difficulty of guessing how our own society will develop in the future. Famous novelists, such as George Orwell (in *1984*), Aldous Huxley (in *Brave New World*), and Herman Hesse (in *Magister Ludi*), have offered plausible visions of "advanced" societies in which the pursuit of scientific knowledge—let alone the exploration of space!—forms no part of human activity.

We Are Still a Primitive Civilization

If we leave aside the question of motivation, our scientific knowledge does tell us that moving matter costs enormously more energy than sending radio waves, and this reality will be clear to every civilization. A preference for interstellar radio over interstellar spacecraft seems hard to avoid, except under the specific circumstances (immortality) that we have described. In this case, we can hardly be surprised that we have not yet found other civilizations: The dawn of the radio era occurred so recently, both for detecting and "leaking" radio waves, that Earth has barely begun to participate. We can speculate endlessly on the possibilities for the existence of other civilizations and for their activities. But science rests on a framework of experiments: The only way to discover whether or not we are alone is to pursue a systematic search for evidence of extraterrestrial intelligence at work.

The Drake equation of Chapter 17 provides our summary and our guide. If we are an average advanced civilization, except for being extremely young, then many other civilizations exist and our chances of making contact across the depths of space seem good. If the current communicative lifetime of our civilization, some 70 years, is the average value for all civilizations, then so few civilizations exist at any one time that we will probably never find any. This short lifetime could be ours if we rapidly move from open, satellite-relayed television broadcasts to strictly cable-carried television, if we disassemble our powerful military radars, and if we generally begin behaving like a mature, energy-conscious civilization, which chooses not to send messages across interstellar distances. We would also have a rather short communicative lifetime if we proceed to blow ourselves up in the next few years, if we fail to solve our population and pollution problems, or if we succumb to an AIDS-like epidemic.

Perhaps we could find a shortcut in solving these problems if we could make contact with, and then learn from, an advanced civilization that has long since overcome them. But it is easy to imagine contact without such benefits. After all, we know most of what we must do to improve our lot on Earth. The problem is in persuading ourselves to do it. Quite possibly, a message from the Andromeda galaxy telling us, "Recycle your aluminum" would be ignored as thoroughly as the ones that we read everyday. On the other hand, a message that says, "Recycle your aluminum or we'll blow up your planet" would probably receive greater attention. The Old Testament in the Bible is replete with such warnings, a reminder of what it takes to capture the human imagination.

If, as we optimistically suspect, we eventually solve all these problems, and if we represent an average example of planetary life, then we can expect civilizations in our galaxy to number at least many hundred thousand, because their average lifetime will span at least a few hundred thousand years. In this case, our 70-year-old civilization ranks among the youngest in the Milky Way at the present time. The many truly advanced civilizations may have entered into a second communicative phase, in which communications are now deliberate, consisting of beacons and messages, elaborate cosmic games, music, shared feelings and physics, and many other possibilities. Our abilities to guess what such a typical civilization may be doing could be compared to the efforts of a six-year-old child of the Stone Age trying to speculate about civilization in the world today.

The Pacific Islands offer a useful example. The people who inhabit the uplands of New Guinea were completely unaware of the existence of the rest of the world until some German traders reached them in the 1930s. This is certainly strange, since all those New Guineans needed to do was to turn on a simple radio!

We, like those highlanders, have lacked the necessary equipment to make contact across the great void that separates us from our nearest neighbors. Even as you are reading these words, messages from some distant civilization may be passing through this page in the form of radio waves, along with the broadcasts from terrestrial radio and television stations and emissions from interstellar molecules whose identities we already know. We have simply lacked the appropriate instrumentation to tune into these cosmic messages and marvel over their mysterious contents. Still, amazingly enough, we are now building this equipment, and can thus move from our awesomely small experience to a better understanding of what is happening in our galaxy. Eavesdropping on our closest neighbors is relatively straightforward. This is directly analogous to the New Guineans' use of a simple AM radio to detect other civilizations on Earth. But a complication arises in trying to intercept messages beamed directly between civilizations: The solar system must lie "in the beam" (Fig. 21.4). This could occur as the sun and the other civilizations orbit the center of the Milky Way, but because the distances between stars are so enormous, we cannot expect to intercept many of these deliberately beamed transmissions at any given time, unless the number of communicating civilizations is extremely large. This fact could provide another answer to the question posed in the title of this chapter. We shall have much better chances of finding them if some of these civilizations sweep the galaxy with radio beacons, inviting newcomers to join their club, much as used car dealers use searchlight beams to attract customers.

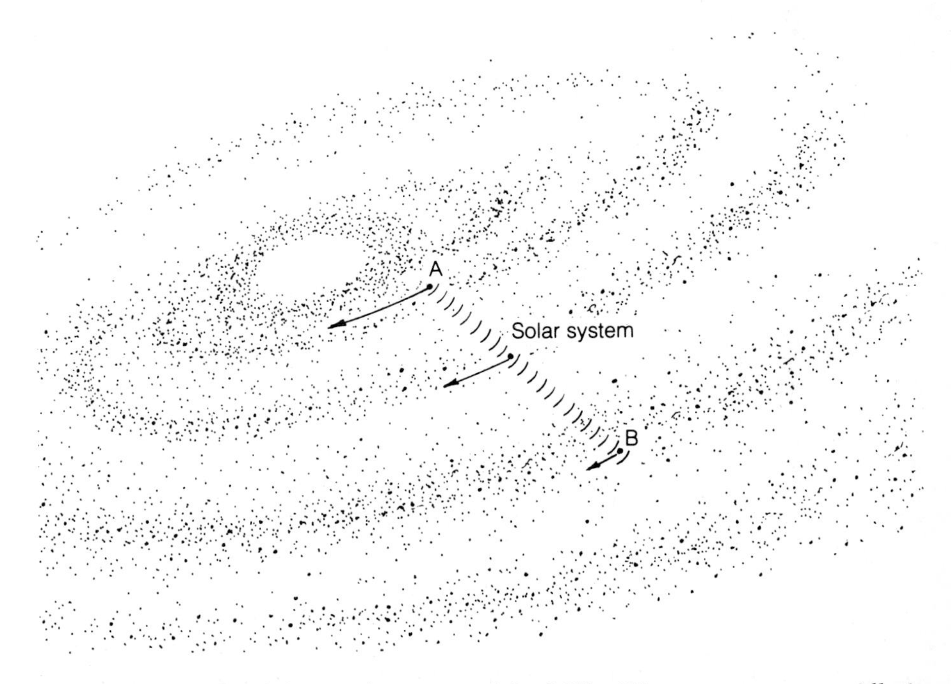

Figure 21.4 A star (A) near the center of the Milky Way moves more rapidly in orbit than another star (B) farther from the center, just as Mercury orbits the sun more rapidly than Jupiter. As the sun moves at its own orbital velocity, the solar system may sometimes pass directly between A and B. Only then could we intercept messages beamed between civilizations (if any exist!) on planets orbiting close to those stars. This diagram is highly schematic. While each photon follows a straight line, a snapshot of the message link would show it as a curved line, since both A and B are moving.

Despite these uncertainties, the basic difference between rockets and radio waves remains significant. While direct exploration lies far beyond our present capabilities, our visible-light and radio telescopes have already brought us into contact with the distant reaches of the universe. The fact that radio astronomers keep discovering new molecules in the interstellar medium is indicative of the rapid increase in our knowledge. The next critical step for exobiology is to convert it into a true experimental science, to move from centuries of conjecture into a mode of thoughtful exploration. If we mean to be serious about finding other civilizations, our avenues are clear: Keep looking, keep listening, keep thinking, and our chances for eventually joining the galactic club will increase. We may even find new friends for humanity.

To succeed at last, we must embrace many failures. To many scientists, the most heart-wrenching failure would be the refusal to search in earnest, simply because we cannot be sure that the search will yield useful results. Consider the following excerpts from an interchange in the House of Representatives between the late Congressman Silvio Conte of Massachusetts and Congressman William Hefner of North Carolina in June 1990, when NASA was seeking budget approval for its SETI project:

> *Conte:* At a time when good people of America can't find affordable housing, we shouldn't be spending precious dollars to look for little green men with misshapen heads . . .
>
> *Hefner:* Mr. Chairman, it has just been on the AP wire, they have located some extraterrestrial beams [probably a southern way of saying "beings"] and they are wearing striped coats.
>
> *Conte:* Of course there are flying saucers and advanced civilizations in outer space. But we don't need to spend $6 million this year to find evidence of these rascally creatures. We only need 75 cents to find a tabloid at the local supermarket.

Congressman Conte went on to insert in the *Congressional Record* articles from the *Weekly World News* about UFOs poised to land during a Chicago Bears–Philadelphia Eagles football game, about the construction of Noah's Ark by visitors from outer space, about aliens destroying Earth's tadpoles for their research, about a UFO emerging from the Indian Ocean, and about a magic ray from a UFO that produces miraculous cures. He concluded with the basic human reaction: "[Let us] prove that there is still intelligent life on Earth. Let us save our hard-earned money and let the space aliens spend their currency to find us."

Of course, Congressman Conte did not spend the taxpayers' money on reprinting these articles without a motive: He knew that most of his constituents would regard funding for SETI as completely frivolous, either because they believe that aliens are already visiting Earth, or because (as Congressman Conte implied) no earthly good could arise from contact with another civilization. Fortunately we live in a democratic society, in which many voices can be heard, and experience shows that a good idea can eventually convince the majority that it deserves support.[1] We hope that each reader of this book can decide for himself or herself whether SETI is worth funding, and if so, to what extent. This grand adventure is simply waiting for support from our society.

Epilogue: The Search Continues

To provide an idea of how the search for other civilizations has proceeded in the absence of direct governmental funding, we can do no better than to read Frank Drake's account of one of his sessions at the giant radio telescope of Arecibo, searching for extraterrestrial intelligence (Fig. 21.5):[2]

> It is 1976. The night surf shimmers and breaks rhythmically on a dark Puerto Rican beach. In the neat, 30-room hotel above the beach, a sleepy night attendant (is he as old as Methuselah or does he only seem so?) for the third night in a row reluctantly awakens two Americans at the unthinkable hour of 4 A.M. He asks no questions, but looks on with hostility and suspicion. Any visitors at this resort who arise at four in the morning must be embarking on some evil deed. While the beautiful people sleep soundly, their expensive sunburns permitting, the two Americans drive off into the night. Chickens and toads scramble to escape the road in time. Carl Sagan is propped up in the front seat, eyes closed. He munches laboriously on scraps of dried-out garlic bread rescued from last night's dinner. It is all the breakfast there will be until the morning's observations are over.
>
> An hour later the sky turns pink, and a Puerto Rican toad abandons its bell-like musical aria and crawls beneath an orchid plant for the day. The sun changes the wet misty quiet into a humid torpor. The brightness of the sunlight which warms the toad, however, will be equal only to that on Mars, for the toad is shielded by a canopy of 18 acres of perforated aluminum sheets—the reflector of the Earth's largest radio telescope. Five hundred feet over the canopy, murmuring machinery moves the 300-ton superstructure to follow a distant galaxy, the Great Nebula in Triangulum, across the sky. A nearby building houses radio receivers, far more sensitive than those of 1960 and capable of covering 3024 frequency channels at once rather than the single channel covered in 1960. Scientists and telescope operators have meticulously arranged this vast ensemble of electronic equipment to fulfill the demanding requirements of a SETI program. It has taken years for the talented people to construct the devices and computer programs that will deal with all the subtleties of the search such as the con-

[1] Not without some difficulty, however. Earlier in this debate, Congressman Ronald Machtley from Rhode Island produced the following authoritative statement: "We have no, and I repeat no, scientific evidence that there is anything beyond our galaxy except we do have some curiosity."

[2] From *Technology Review* 78:22 (June 1976).

Figure 21.5 Prof. Frank Drake of the University of California, Santa Cruz, is an expert in radio astronomy. He has also been a pioneer in the search for extraterrestrial intelligence for more than three decades. This photograph shows him with his wife Amahl, one of the organizers of the conference whose participants are shown in Color Plate 20.

tinuously changing effects of the Earth's spin and its orbital sweep around the sun. Every 30 seconds, 100,000 transistors invisibly transmit the recorded information to the observatory's memory.

The information captured in each of the 3024 channels spiashes a swath of twinkling green points across the face of an oscilloscope. It takes a hundredth of a second for the telescope to duplicate the two months of work done in 1960.

A split second of excitement. The astronomers, garlic bread and hunger now long forgotten, search the swath of points for a pattern which could not have been made by nature alone. When first their minds and later their hearts convince them that no such pattern is present, the command is given to move on to look at another part of that distant galaxy—another billion stars—and to start the electronics searching and recording again. The hours fly by. One hundred billion stars will be searched before the task is completed and the hotel night attendant will no longer have to awaken at 4 A.M. No signals will be found this time. But there will come a time when that most thrilling of all cosmic treasures, news of life elsewhere, will be plucked from the vast vault of space.

The SETI observing program that has just begun to become organized in 1992 will be millions of times more sensitive than the observations described by Frank Drake in 1976. One of the scientists involved in this effort has suggested that the new, highly sensitive multichannel receivers and the computers that manage them will allow the SETI project to accomplish more in its first minute of operation than the sum of the work done by all of the previous experimenters. It will take a decade to complete this program, during which a host of scientists will spend many hours examining their monitor screens and poring through computer output. We can be sure that there will be further flashes of adrenaline, followed by disappointment, but the search will continue. Then one day, we may well have contact, and we shall no longer wonder whether we are alone in the universe.

FURTHER READING

Hart, Michael, and Ben Zuckerman, eds. *Extraterrestrials—Where Are They?* London: Pergamon Press, 1982.

Regis, Edward, ed. *Extraterrestrials: Science and Alien Intelligence.* Cambridge: Cambridge University Press, 1985.

Smith, David, ed. *SETI Pioneers.* Tucson: University of Arizona Press, 1990.

Calvin and Hobbes copyright 1989, Watterson. Dist. by Universal Press Syndicate. Reprinted with permission. All rights reserved.

CREDITS

Cover Illustration: Courtesy of Jon Lomberg.
Part I: Facing page 1, Lick Observatory. Page 2, NASA.
Chapter 1: 1.1, Hale Observatories, California Institute of Technology. 1.2, Courtesy of Nafi Toksoz, MIT. 1.3, Lick Observatory. 1.4, Lowell Observatory. 1.5, courtesy of *The New York Times.* © 1938 by The New York Times Company. 1.6, Courtesy of Paramount Pictures, from the movie *"War of the Worlds"*. Copyright © 1991 by Paramount Pictures. 1.7, Courtesy of South West News Service (SWNS), Bristol, England. 1.8(3 images), © Yerkes Observatory.
Part II: Page 19, National Optical Astronomy Observatories. Page 20, National Optical Astronomy Observatories. Page 22, Courtesy of the Bancroft Library, University of California, Berkeley.
Chapter 2: 2.9, National Optical Astronomy Observatories. 2.13, Courtesy of Professor Margaret Geller and the Center for Astrophysics. Page 53, Courtesy of Sidney Harris. © 1991 by Sidney Harris—*Physics Today.*
Chapter 3: Page 54, National Optical Astronomy Observatories. 3.1, National Optical Astronomy Observatories. 3.2, Yerkes Observatory. 3.3, NASA. 3.10, National Optical Astronomy Observatories. 3.1, Lick Observatory. 3.12, (left) Hale Observatories; (right) National Optical Astronomy Observatories. 3.13, Hale Observatories. 3.15, Lick Observatory. 3.16, National Radio Astronomy Observatory. 3.17, Hale Observatories. 3.18, National Optical Astronomy Observatories.
Chapter 4: 4.2, National Optical Astronomy Observatories. 4.3, Lick Observatory. 4.4, NASA. 4.5, Lick Observatory. 4.7, (left) National Radio Astronomy Observatory; (right) National Optical Astronomy Observatories.
Chapter 5: 5.1, Hale Observatories. 5.2, Lick Observatory. 5.6, Hale Observatories. 5.7, Hale Observatories.
Chapter 6: 6.4, Hale Observatories. 6.5, Lick Observatory. 6.10, National Optical Astronomy Observatories. 6.11, Hale Observatories. 6.12, NASA. 6.15, Lick Observatory.
Part III: Page 151, Photograph: Lick Observatory. Page 151, Poem by William Butler Yeats, "He Wishes for the Cloths of Heaven," by courtesy of Macmillan Company. Page 152, NASA. Page 153, From Dylan Thomas's *Poems of Dylan Thomas.* Copyright © 1945 by the Trustees for the Copyrights of Dylan Thomas, reprinted by permission of New Directions Publishers. Page 154, By permission of the Rijksmuseum, Amsterdam, Netherlands.
Chapter 7: 7.6, Courtesy of Professors John Wolfson and David Dressler, Harvard University Medical School.
Chapter 8: Page 178, By permission of the Uffizi Galleries, Florence, Italy. 8.4, Courtesy of Professor J. William Schopf, University of California, Los Angeles. 8.5, Donald Goldsmith. 8.11, Courtesy of Professor Sidney Fox.
Chapter 9: 9.1, Photograph by Professor Elso S. Barghoorn, Harvard University. © President and Fellows of Harvard College. 9.4, From Levi-Setti, *Trilobites: A Photographic Atlas.* © University of Chicago Press. 9.5, Courtesy of Professor Allison Palmer. 9.6, From Foster, *Earth Science.* © 1982 Courtesy of The Benjamin/Cummings Publishing Co. Reprinted with permission of Addison-Wesley, Reading, MA.
Part IV: Page 249, NASA. Page 250, NASA. Page 252, Courtesy of Jon Lomberg.
Chapter 11: 11.1, National Optical Astronomy Observatories. 11.5, Hale Observatories. 11.7, Courtesy of H. Reitsema and A. Delamere, Ball Aerospace Corporation. 11.8, NASA. 11.9, NASA. 11.10, Meteor Crater Enterprises, Inc. 11.11, Courtesy of Mr. Allen Parker. 11.12, Courtesy of Dr. Carl Orth. 11.13, Courtesy of Professor Cyril Ponnamperuma. 11.14, NASA. 11.15, NASA. 11.16, NASA. 11.17, From D. Goldsmith, The Evolving Universe, 2d ed. © 1985 The Benjamin/Cummings Publishing Co. Reprinted with permission of Addison-Wesley, Reading, MA.
Chapter 12: 12.1, International Planetary Patrol Photographs, furnished courtesy of Lowell Observatory. 12.2, NASA. 12.4, NASA. 12.7, Based on *Moons and Planets,* Second Edition, by William K. Hartmann. © 1983 by Wadsworth, Inc. Adapted by permission. 12.8, NASA. 12.9, NASA. 12.10, NASA.

Chapter 13: 13.1, NASA. 13.2, International Planetary Patrol Photographs, furnished courtesy of Lowell Observatory. 13.3, Courtesy of Editions Hermann, Paris. 13.4, NASA. 13.5, NASA. 13.8, NASA. 13.9, NASA. 13.10, NASA. 13.1, NASA. 13.12, NASA. 13.13, NASA. 13.14, NASA. 13.15, NASA. 13.16, NASA.
Chapter 14: 14.1, NASA. 14.4, NASA. 14.5, NASA. 14.6, NASA. 14.8, NASA. 14.9, Courtesy of the State of New Hampshire.
Chapter 15: 15.1, NASA. 15.2, NASA. 15.3, NASA. 15.4, NASA.
15.5, NASA. 15.8, NASA. 15.9, NASA. 15.10, NASA. 15.11, NASA. 15.12, NASA. 15.13, NASA. 15.14, From D. Morrison and T. Owen, *The Planetary System.* © 1987 The Benjamin/Cummings Publishing Co. Reprinted with permission of Addison-Wesley, Reading, MA. 15.16, NASA. 15.17, NASA.
Part V: Page 378, (top) Breitkopf & Haertel, Wiesbaden, excerpt from J. S. Bach, Brandenburg Concerto No. 2, PB 4302. Page 378, (bottom) NASA. Page 379, From *Poems* by George Seferis, translated by Rex Warner. Copyright © by Little, Brown and Company in association with the Atlantic Monthly Press, 1960. Page 380, NASA.
Chapter 16: 16.1, National Optical Astronomy Observatories. 16.3, NASA.
Chapter 17: Page 402, Main Photograph: National Radio Astronomy Observatory. Page 402, Inset Photograph: NASA. 17.2, National Optical Astronomy Observatories.
Chapter 18: Page 432, NASA. 18.1, NASA. 18.3, NASA. 18.5, Stanford Linear Accelerator Center.
Chapter 19: 19.6 and 19.7, Courtesy of Professor Woodruff T. Sullivan, III. 19.9 and 19.10, National Astronomy and Ionosphere Center. The Arecibo Observatory is part of the National Astronomy and Ionosphere Center operated by Cornell University under contract with the National Science Foundation. 19.11, Courtesy of NASA and Dr. Seth Shostak. 19.13, Courtesy of *Scientific American*, from "The Search for Extraterrestrial Intelligence," by Carl Sagan and Frank Drake. Copyright © 1975 by *Scientific American.* All rights reserved.
Chapter 20: Page 482, *Weekly World News.* 20.1 and 20.2, Courtesy of Mr. Philip Klass and United States Air Force. 20.3, Courtesy of Mrs. Veronica Schwartz. 20.4, Mr. Paul Trent. 20.5, Courtesy of Dr. David Morrison, NASA Ames Research Center. 20.6, Courtesy of Dr. E.C. Krupp. 20.7, Courtesy of Instituto Nacional de Antropologia e Historia de Mexico. 20.8, Tobias Owen.
Chapter 21: 21.1, Los Alamos National Laboratories. 21.3, NASA. 21.5, Donald Goldsmith. Page 519, From *Calvin & Hobbes.* © 1989 Watterson. Dist. by Universal Press Syndicate, reprinted with permission. All rights reserved.
Color Plates: Plate 1, Hale Observatories. Plate 2, Photography by David Malin. © Anglo-Australian Telescope Board. Plate 3, NASA. Plate 4, © Dr. Richard Wainscoat, Institute for Astronomy, University of Hawaii. Plate 5, same as Plate 2. Plate 6, Photography by David Malin. © ROE/AATB. Plate 7, Courtesy of Professor John Stolz, Duquesne University. Plate 8, Donald Goldsmith. Plates 9-13, NASA. Plate 14, Courtesy of Dr. Christopher McKay, NASA Ames Research Center. Plates 15-18, NASA. Plate 19, Photograph by Robert W. Madden © National Geographic Society. Plate 20, Donald Goldsmith.

Index

Absorption lines, 34, 116, 117
Abundances of elements, 41, 42, 95, 121, 136, 137, 141, 157, 173, 174
Acceleration, 72
Accretion disk, 75-77
Acetylene, 231, 359, 368
Adenine, 163-168, 190, 193, 194, 272, 477
Alanine, 93, 160
Aldebaran, 118
Alpha Centauri, 28, 29, 35, 36, 74, 207, 386, 391-396, 443, 446, 448
Alpher, Ralph, 43
Aluminum, 157, 315
Alvarez, Walter, 269
Amino acids, 93, 160-168, 174, 175, 192-194, 197, 201, 271, 272, 330, 381
Ammonia, 88, 191, 192, 232, 235-238, 246, 327, 355-361
Anders, Edward, 276
Andromeda galaxy, 22, 35, 57, 58, 413, 443, Color Plate 1
Angular size, 23, 24
Antares, 118
Antarctic soil, 330, 334, 437, Color Plate 14
Antimatter, 437
Antoniadi, Eugenio, 305
Apollo, 434, Color Plate 10
Apparent brightness, 35
Arecibo message, 475-478
Arecibo radio telescope, 389, 469, 471, 472, 475, 517, Color Plate 19
Argon, 186, 316, 367
Arnold, Kenneth, 485
Arrhenius, Svante, 88
Arsine, 359
Asteroids, 253, 266-270, 276, 280, 293, 321, 375
Atoms, 32
Automated message probes, 446-450

Bacteria, 205, 206, 210, 218, 229
Bandpass, 461-463
Barnard's Star, 392-396, 455
Beta Pictoris, 388, 389, 426
Beryllium, 126
Betelgeuse, 118
Big bang, 40, 41
Bigfoot, 343
Big Joe, 337
Binary stars, 385, 392
Bipolar outflows, 98, 99
Black clouds, 240-242, 247
Black holes, 75-77
Blue-green bacteria, 185
Bracewell, Ronald, 448
Brightness, *see* Apparent brightness, Luminosity
Bussard, Robert, 445

Calcium, 157, 158, 223, 315
Calcium carbonate, 185, 223, 289
Callisto, 363-367
Cambrian era, 211, 212, 226
Camel Rock, 344
Carbohydrates, 169, 175, 193
Carbon, 34, 93-95
  in comets, 261-263
  in life, 158, 162, 230-234, 246
  in stars, 126, 133
Carbon cycle, 109n
Carbonaceous chondrites, 271
Carbon dioxide, 158, 171, 183-185, 223, 233, 263, 286, 288-292, 298, 303, 304, 316, 318, 326, 332-334, 367, 368, 373, 382
  in greenhouse effect, 286-290, 298
Carbon monoxide, 88, 183, 286, 333, 355, 356, 359, 368, 373
Carbonyl sulfide, 286
Cassini-Huygens spacecraft, 353, 369

Catalyst, 161, 199
Cell, 163, 164, 205, 208, 226, 236
Cellulose, 96
Challenger, 5, 336
Charon, 349, 363
Chiron, 363
Chladni, Ernst, 12
Chlorine, 315
Chlorofluorocarbons, 326, 509
Chlorophyll, 207
Chloroplasts, 172
Chondrites, 271
Christy, James, 363
Chromosomes, 208, 209
Chryse basin, 312
Civilizations, 244, 403-429, 448, 510-516
Clark, Benton, 196
Clarke, Arthur, 424
Clay, 197, 198
Clouds, interstellar, 61, 83, 88-91, 96, 97
Clusters
  of galaxies, 35-37, 67, 75, 414
  of stars, 69, 70, 78
Coacervates, 200
Cocconi, Giuseppe, 458
Coma, 261
Comet Rendezvous Asteroid Flyby (CRAF), 265
Comets, 97, 182, 253, 256, 259-265, 273, 276, 280, 375, 384, 443
Communication with other civilizations, 9, 14, 217, 244-247, 374, 378, 411-430, 453-480
Conte, Silvio, 516
Copernicus, Nicolaus, 25, 27, 28, 39
Copper, 159
Cosmic Background Explorer (COBE), Color Plate 3
Cosmic background radiation, 42, 43, 50
Cosmic rays, 141, 142, 148, 169
Crab Nebula, 140, 145, 147
Crab Nebula pulsar, 145, 147
Craters, 267, 273, 274, 304, 311
Crop circles, 13
Cyanoacetylene, 260
Cyanogen, 93, 194
Cytosine, 163-168, 193, 272, 477

DNA (deoxyribonucleic acid), 163-169, 170, 175, 193, 197-200, 205, 208-210, 218, 226, 230, 477
Dark matter, 43, 44, 51, 63-65
Darwin, Charles, 15, 16, 170, 190, 221
Deimos, 319-322
Density of matter
  in interstellar clouds, 90-92
  in planets, 258, 259
  in universe, 48-51
Density-wave pattern, 59-61
Deuterium, 290, 291, 316
Deuteron, 106, 107
Dinosaurs, 220
Distance measurements, 24-30, 35
Doppler effect (Doppler shift), 35-38, 50, 73, 99, 387, 459, 462
Double stars, 385, 392
Drake Equation, 408-416, 429, 506, 513
Drake, Frank, 142, 222, 242, 517-519, Color Plate 20
Dust grains, 87-89
Dyson, Freeman, 426

Earth
  age of, 179, 181, 188, 199
  atmosphere of, 113, 181-185, 222, 224, 289, 290, 383
  composition of, 157
  crust of, 5, 157, 181, 273, 297
  description of, 477, 478
  greenhouse effect on, 289
  history of, 16, 179-188
  human population of, 508, 509
  life on, 5, 15, 16, 155-227, 327, 506-510
  orbit of, 28, 227, 256, 382, 383
  primitive conditions on, 179-191, 202, 273, 276, 381
  surface temperature of, 224, 289, 382
Ecosphere. *See* Habitable zone.
Einstein, Albert, 14, 30, 106, 437, 441, 444
Eiseley, Loren, 221
Electromagnetic forces, 110, 111, 130, 131, 134, 239
Electromagnetic spectrum, 30, 456, 457
Electron, 32, 33, 87, 108, 109, 134, 437
Element, 33
  abundances of, 41, 42, 95, 121, 136, 137, 141, 173, 174
Elliptical galaxies, 65-67
Energy, in living organisms, 189-193, 205-208, 240-242, 381
Energy of mass, 106-109

Energy of motion (kinetic energy), 106-109, 444, 445
Enterprise, 437, 439
Enzymes, 161, 233
Epsilon Eridani, 392, 394, 396, 397, 454
Ethane, 359, 368
Ethyl alcohol, 93
Eukaryotes, 208-211, 226
Europa, 363-365, Color Plate 17
Evolution, 15, 141, 169-175, 188-203, 211-227, 507
  chemical, 96
Exclusion principle, 130, 131
Expansion of the universe, 39, 40, 50

Face on Mars, 343-345
Fatty acids, 160, 169, 272
Fermi, Enrico, 505, 506
Fluorine, 159
Formaldehyde, 88, 260, 333
Formic acid, 334
Fossils, 5, 206, 327
Fraunhofer lines, 116
Frequency, 31-33, 36, 87, 456-464
Fusion, nuclear, 106-111, 125, 126, 135, 136, 358

Gaia, 224, 225, 227, 245
Galaxies
  clusters of, 35-37, 67, 75, 414
  distances of, 35-38
  elliptical, 65-67, 69, 78
  formation of, 58
  irregular, 67, 68, 78
  radio, 71, 72
  recession velocities of, 38, 39
  spiral, 56-65, 78
Galileo spacecraft, 266, 353, 355
Galileo, 14, 83, 274, 363
Gas chromatograph mass spectrometer (GCMS), 329-337
Gas exchange experiment (GEX), 331-334, 336, 346
Gamma rays, 31, 32, 169, 244, 508
Gamow, George, 43
Ganymede, 363-367
Gaspra, 266, 267, 355
Geller, Margaret, 45
Gene, 166, 205, 209, 218
Genetic code, 166, 168
Germane, 359
Giant planets, 257-259, 349-375
  chemistry on, 359-361
  composition of, 354-359
  possible life on, 361, 362
  satellites of, 362-375
Giotto, 264
Globular star clusters, 69, 70
Glycine, 93, 192, 194
Glycogen, 161
Gravitation, 48, 61, 109-112, 125, 130, 131, 134, 239, 244-247, 279
Greenhouse effect, 286-290, 298, 393
  runaway, 290-293, 298, 392
Guanine, 163-168, 193, 272, 477

HI and H II regions, 83n, 85
Habitable zone, 392-395, 399
Haldane, J.B.S., 190
Hall, Asaph, 319
Halley's comet, 260-265
Heat capacity, 235, 236, 246
Heat of vaporization, 235, 236, 246
Heavy elements, 42, 117, 121, 157
Hefner, William, 516
Helium, 41, 42, 106-109, 125, 126, 157, 257, 258, 355, 356
Helium flash, 127, 128, 130, 148
Herman, Robert, 43
Hertzsprung-Russell (H-R) diagram. *See* Temperature-luminosity diagram
Hesse, Herman, 513
Horsehead Nebula, 254
Hoyle, Fred, 96, 97, 240
Hubble, Edwin, 38
Hubble's Law, 38, 39, 48-50, 73
Hubble Space Telescope, 387
Huchra, John, 45
Huxley, Aldous, 513
Huygens, Christian, 3, 7
Hydrazine, 232
Hydrochloric acid, 286
Hydrofluoric acid, 286
Hydrogen, 42, 86-88, 106-109, 156-158, 257, 258, 355, 356
Hydrogen chloride, 237
Hydrogen cyanide, 158, 194, 263, 359, 368
Hydrogen peroxide, 232, 334, 336
Hydrogen sulfide, 207, 237
Hynek, J. Allen, 487, 492-494

Iapetus, 370, 371

Impacts, 195, 196, 199, 267-270, 276
Inert gases, 182
Inflationary theory, 43, 44, 51
Infrared radiation, 31, 58, 287, 288, 298
Infrared Astronomy Satellite (IRAS), 83, 85, 387, 426
Inner planets, 257-259
Intelligence, 218-220, 227
  definition of, 218, 219
Interstellar clouds, 61, 83, 88-91, 96, 97
Interstellar gas, 60, 61, 65, 66, 81-101, 240-242, 254-256, 276, 444-456
Interstellar dispersion, 462
Io, 363-366, Color Plate 16
Ion, 116
Ionization, 85, 116, 117
Ionosphere, 460
Iridium, 269, 270
Irregular galaxies, 67, 68
Iron, 133-135, 157, 315
Ishtar, 245

Jupiter, 99, 349, 351-362, 374, 375
  atmosphere of, 233, 355-361
  colors on, 360, 361, Color Plate 15
  convection currents in, 356, 357
  Great Red Spot on, 352, 357-360, 375
  possible life on, 246, 361, 362
  satellites of, 361-368
  temperature of, 356-358

Kardashev, N. S., 426, 427
Kepler, Johann, 319, 320, 387
Kelvin temperature scale, 110n
Kinetic energy, 106-109, 444, 445
Kowal, Charles, 363
Klass, Philip, 485

Labeled Release Experiment, 331-334, 346
Lagoon Nebula, 82
Large Magellanic Cloud, 67, 443
Leeuwenhoek, Anton, 327
Life
  definition of, 155, 156
  detection of, 325, 326
  on Earth, 155-226
  effect of supernova on, 137, 141, 142
  elemental composition of, 41, 95, 136, 156-158
  energy sources for, 189-193, 205-208, 240-242, 381
  evolution of, 15, 141, 169-175, 188-203, 211-227, 507
  in interstellar clouds, 96, 97, 240-242
  lifetimes of stars, effect upon, 215, 84, 390-396, 455, 472
  nonchemical, 239-245
  number of sites for, 48, 55, 97, 103, 115, 155, 245, 246, 253, 292, 298, 342, 381-399, 403-429, 455, 456
  on neutron stars, 242-244, 247
  origin of, 4-6. 179-202, 253, 264
  on planets,
Light, 31
  speed of, 30, 433
Light year, 28, 29
Lightning, 189, 190, 360
Lin, Robert, 360
Local Group, 35, 36
Locoweed, 159
Lovelock, James, 224, 227
Lowell, Percival, 8, 9, 11, 305
Luminosity, 29, 30, 35

Magellan, 29, 294-297
Magellanic clouds, 67
Magnesium, 133, 157, 159, 315
Magnetic fields, 72, 144-148, 445
Main sequence, 119, 122, 127, 130
Mammals, 219, 220, 226
Mantell, Thomas, 485
Margulis, Lynn, 224, 225, 227, Color Plate 7
Mariner spacecraft, 304-307, 310, 322, 329
Mars, 301-347, 381, 382, Color Plates 11-13
  atmosphere of, 303, 304, 310, 313, 316-318, 322, 325, 326, 329-336, 341, 342
  canals on, 8-11, 301, 305, 326
  polar caps of, 302, 303, 307, 317-319, 322, 337-339
  rotation of, 301, 305-308, 383
  search for life on, 6-10, 305, 325-347
  seasons on, 301-303
  surface of, 310-317, 322
  surface pressure of, 303, 304, 318, 339-342
  temperature on, 303, 304, 316, 335, 338
  volcanoes on, 305, 308, 318, 322
  water on, 302, 303, 306, 307, 316-319, 322, 338, 339
  wave of darkening on, 301-303
Mars Observer, 340

Mass extinctions, 269
Maxwell, 295
Mercury, 273-276, 280, 366, 367
Meteor Crater, 268
Meteors, 486-490
Meteorities, 193-196, 270-273, 381
Meteoroids, 253, 266-269, 280, 384
Methane, 88, 191, 192, 231, 233, 246, 326, 329, 332, 355, 356, 359, 361, 367, 368
Methyl alcohol, 235-237, 246, 263, 327
Microbes, 187, 276, 327-336
Microwave Observing Project (MOP), 470
Milky Way galaxy, 35, 36, 55, 83-101, Color Plates 2 & 3
  age of, 410, 411
  disk of, 59, 85
  motions of stars in, 59-63
  number of civilizations in, 404-429, 505-516
  number of stars in, 55, 398, 410, 416-419, 429
  structure of, 56-60, 63, 64
Miller, Stanley, 191
Miller-Urey experiment, 191-194, 264
Molecular clouds, 90-96
Molecules, 32
Momentum, 61
Monomers, 159-161, 164, 166, 173, 197
Montmorillonite, 197
Morrison, Philip, 458
Moon, 273-281
  formation of, 274
  importance to life on Earth, 278-281, 383, 384
Multifrequency analysis, 469-473
Murchison meteorite, 271, 272, 276, 321
Murray meteorite, 272, 276
Mutations, 141, 142, 148, 169, 170, 218

Natural selection, 170, 221
Nealson, Kenneth, Color Plate 7
Neon, 42, 157, 355, 356
Neptune, 349, 352-354, 361-363, 374, 375
Neutrino, 41, 106-108, 136, 144
Neutron, 32-34, 134, 136, 144
Neutron star, 136, 138, 144, 148, 242-244, 247
Newton, Isaac, 14, 219
Nitric oxides, 185
Nitrogen, 42, 158, 183-186, 316, 329, 356, 367, 368, 372, 373
Nontronite, 197, 315
Nuclear fusion, 106-111, 125, 126, 135, 136, 358
Nucleic acids, 163, 168, 193
  see also DNA, RNA
Nucleotides, 163, 164, 192, 381
Nucleus,
  of atoms, 32-34, 243, 244
  of cells, 208, 226
  of comets, 260-262

Oannes, 498
OH radical, 459-461, 471
Oceans, 184
Oliver, Bernard, 460, 461
Olympus Mons, 305, 308
O'Neill, Gerard, 42. 422-426
Oort Cloud, 256
Oparin, Alexander, 190
Open star clusters, 69, 70
Orbits
  of comets, 260-263
  of planets, 256, 266, 274, 350, 386, 391-395
  of stars and star clusters, 59-63
Organic molecules, 189
Organelles, 209, 226
Orgel, Leslie, 198
Orion Nebula, 92, 93, 443, Color Plate 5
Oró, John, 193
Orwell, Geoge, 513
Oxidation, 183, 186, 331
Oxygen, 42, 158, 173, 184-186, 195, 207, 208, 316, 368
Ozone, 184, 238, 293, 304

Pacal, 500
Pangaea, 181
Panspermia, 188
Parallax effect, 27-29
Parsec, 28
Penzias, Arno, 43
Peptides, 198
Peroxides, 334, 336, 346
Phobos mission, 321
Phobos, 319-322
Phosphate, 163, 193, 477
Phosphine, 359, 360
Phosphorus, 158, 476, 477
Photodissociation, 183
Photon, 30, 106-108, 433
  in cosmic background radiation, 42, 43, 50

Doppler effect on. *See* Doppler effect.
gravitational force on, 75
interactions with atoms, 32-34
Photosynthesis, 171-175, 206, 207, 210
Pioneer spacecraft, 142-147, 350, 351, 374, 439, 440, 446, 447, 478
Planck, Max 30
Planetary nebulae, 129
Planetesimals, 257
Planets
around other stars, 98, 99 257, 259, 384-399, 408, 409
formation of, 98, 253-259, 384-387
orbits of, 256, 266, 274, 350, 386, 391-395
See also names of individual planets.
Plate tectonics, 5, 179, 181, 273, 297
Pleiades, 70, 71, 78, Color Plate 6
Pluto, 246, 260, 349, 353, 363, 372, 375
Polymers, 160, 161, 166, 169, 173, 196-202, 205
Ponnamperuma, Cyril, 272
Population I stars, 140, 472, 507
Population II stars, 140
Positron (antielectron), 106-109, 437
Primitive atmosphere of Earth, 183
Procyon, 118, 391-394
Prokaryotes, 205-211, 226
Proteins, 161, 164, 166, 169, 174, 197, 209, 330
Protogalaxies, 68
Protons, 32, 33, 87, 106-110, 128, 134
Proton-proton cycle, 106-110
Protoplanetary disks, 100, 388, 389
Protostars, 98, 110, 111, 120
Proxima Centauri, 392, 394, 396
PSR 1829-10, 390
Pulsars, 142-148, 389-391
Pyramids, 219, 498-500
Pyrolitic Release (PR) experiment, 331-336, 436

Quasars (quasistellar radio sources), 72-74, 77, 78

RNA (ribonucleic acid), 168, 169, 93, 199, 200, 205, 226, 230
Radar, 310
Radioactivity, 179, 190, 273, 332-335
Radiation. *See* Electromagnetic radiation.
Radio galaxies, 71, 72
Radio telescopes, 325, 389, 469, 471, 472, 475, 517
Radio waves, 31, 32, 58, 72, 283, 284
from civilizations, 415, 433-450, 454-480
from Earth, 465-468
from hydrogen atoms, 86-88
Ray, Thomas, 449
Red giant stars, 120-122, 129, 148
Redshift, 37
Reducing atmosphere, 194, 331
Relativity, theory of, 441
Replication, 165-169
Reproduction, 163, 209, 210
Rings, 362, 363
Roche, Edwin, 362
Roche limit, 362
Rosette Nebula, 85
Rubin, Vera, 64
Runaway greenhouse effect, 286-290, 298, 393

Salpeter, Edwin, 362
Sagan, Carl, 142, 213, 276, 362, 518
Satellites. *See* names of planets and their satellites.
Saturn, 3, 351, 359, 362, 363, 367, 374, 375
satellites of, 366-371, 375
Saturn rocket, 433, 434, 511
Schiaparelli, Giovanni, 7, 8, 305
Secondary atmospheres, 258, 375
SETI (Search for Extraterrestrial Intelligence), 9, 14, 217, 244-247, 374, 378, 411-430, 453-480, Color Plate 20
Sexual reproduction, 209, 210
Shklovsky, Josef, 213
Silane, 232-234, 329
Silicates, 234
Silicon, 232-234
Silicon dioxide, 233
Sirius, 29, 118, 132, 391-394
Skylab, 437, 438
Sky Survey, 473
Slipher, Vesto, 38
Small Magellanic Cloud, 67
Smith, Bradford, 388
Sodium, 42, 157, 364
Solar nebula, 256
Solar system, 25-27
age of, 271, 280
formation of, 181-183, 254-259, 354, 355
size of, 27, 255-257
Solvents, 230, 234-238, 246, 327
Space, curvature of, 46, 47

Space Telescope. See Hubble Space Telescope
Spectrum, 30, 34, 37, 38, 456-461
  of starlight, 115-117, 129
  radio and television, 456-465
Spiral arms, 59, 60
Spiral galaxies, 56-65
Stanford Linear Acclerator Center, 442
Star Trek, 437
Stars
  brightness of. *See* Apparent brightness, Luminosity.
  central temperatures of, 110-115, 125, 126
  clusters of, 68, 69
  collapse of central regions of, 134-142, 148, 189
  composition of, 125, 155, 156
  contraction of core in, 125-127, 130
  densities within, 129-131
  distances of, 27, 29, 436, 467, 468
  double and multiple, 385
  energy production in, 103-115
  evolution of, 126-141
  explosion of, 132-141, 148
  formation of, 61, 66, 81-83, 98, 254, 385
  gravitational forces in, 109-112, 122, 125
  lifetimes of, 103-105, 115, 129, 130
  luminosities of, 103, 113-120, 127, 128, 131, 148
  masses of, 105, 112-114, 119, 133
  motions of, 59-63
  nuclear fusion in, 106-114, 122, 125, 126, 133
  planetary systems of, 98, 99 257, 259, 384-399, 408, 409
  populations of, 140
  red giant, 120-122, 129, 148
  spectra of, 115-118
  spectral types of, 116-118, 215, 384, 390-396, 455, 472
  surface temperatures of, 116-120
  temperatures within, 108-114, 133, 134
  types of, 115-122
  white dwarf, 120-122, 130, 131, 135, 148
Stationary orbit, 424
Stromatolites, 185, 186, 206, 210
Strong forces, 109, 110, 239, 243, 244
Submillimeter radiation, 31, 32
Sulfur, 360, 364
Sulfur dioxide, 95, 364, 366
Sulfuric acid, 122, 286, 297, 298
Sun
  abundance of elements in, 157
  age of, 103
  energy source for life, 171-173
  evolution of, 254-257
  lifetime of, 104
  luminosity of, 118
  orbit of, 62
  representative nature of, 132
  spectrum, 115, 116
  temperature in, 114
  ultraviolet radiation from, 183, 184, 189-191, 199, 261, 274, 316, 326, 339, 355, 359, 360
Sunspots, 104
Supernova explosions, 105, 132-142, 148, 189
  Type I and Type II, 135
Supernova 1987A, 138
Surface tension, 236
Swift, Jonathan, 320
Synchrotron radiation, 72, 145

Targeted Search, 472-474
Tau Ceti, 392, 396, 397, 454
Television, 433, 466
Teller, Edward, 505
Temperature
  on Earth, 224, 289, 382
  Kelvin scale for, 110n
  of liquid solvents, 235
  on neutron stars, 242
  of planets, 116-120, 283-294, 298, 303, 304, 316, 335, 338, 356-358, 367, 372
  in stars, 108-114, 133, 134
  on stellar surfaces, 116-119
Temperature-luminosity diagram, 118-120
Terrile, Richard, 388
Thiococcus, 229
Thomas, Lewis, 218
Thymine, 163-168, 193, 272, 477
Tides, 279, 383
Time dilation, 441-453, 450
Titan, 349, 353, 366-371, Color Plate 18
Trace elements, 159
Trilobites, 212, 213, 218
Triple-alpha process, 125, 126
Triton, 260, 349, 353, 371-375
Tsiolkovsky, Constantin, 15
Twenty-one centimeter emission, 86-88

UFOs (unidentified flying objects), 11, 12, 483-503
Ultraviolet radiation, 31, 32, 129, 183, 184
  interaction with living organisms, 184, 189-191, 238
Universe
  center of, 40
  density within, 40, 41, 49, 50
  early moments of, 40-45
  expansion of, 39, 40, 50
  future of, 48-51
  inflation of, 43-44
  models of, 39, 40, 46, 47
  particle types in, 41, 42
  size of, 46-48
  temperature of, 41
Uracil, 193, 272
Uranium, 133n
Uranus, 352, 362, 363, 374, 375
Urey, Harold, 191

Vega, 29
Venera 10, 287, 295
Venus, 283-298, 381, 382
  atmosphere of, 285-287, 316
  greenhouse effect on, 286-289
  life on, 292, 298
  rotation of, 284, 285, 293, 294, 383
  temperature of, 283-294, 298, 316
  as UFO, 484, 490, 491
  water on, 289-293
Viking spacecraft, 301, 307-317, 321, 322, 327-347, 381
Virgo Cluster, 35
Viruses, 205
Visible light, 31, 287, 288
Volatiles, 182, 183, 278, 329, 342
Volcanoes, 185, 187, 189, 273, 275, 305, 319, 322, 364
Voltaire, 3, 319, 320
von Däniken, Erich, 498-508
Voyager spacecraft, 25, 142, 350, 351, 355, 362, 366, 372, 374, 378, 439, 440, 446, 447

Water, 183, 184, 232, 238, 290-293, 302, 303, 306, 307, 316-319, 321, 326, 327, 355, 357, 364, 384
  as essential to life, 235-237, 292
Water hole, 459-461
Wavelength, 31-33, 36, 37
Weak forces, 109, 134
Welles, Orson, 9, 10
Wells, H. G., 9, 10, 343
Whipple, Fred, 260
White dwarfs, 120-122, 130, 131, 135, 148
Wickramasinghe, Chandra, 96, 97
Wilson, Robert, 43
Woosley, Stan, 390

X rays, 31, 32, 58

Yeti, 343

Zamora, Lonnie, 492
Zinc, 159
Zodiac, 26, 27
Zygote, 210